Bio-Inspired Smart Machines: Structure, Mechanisms and Applications

Bio-Inspired Smart Machines: Structure, Mechanisms and Applications

Editors

Yanjie Wang
Xiaofeng Liu
Aihong Ji
Shichao Niu
Bo Li

MDPI

Basel • Beijing • Wuhan • Barcelona • Belgrade • Novi Sad • Cluj • Manchester

Editors

Yanjie Wang
Hohai University
Changzhou
China

Xiaofeng Liu
Hohai University
Changzhou
China

Aihong Ji
Nanjing University of
Aeronautics and Astronautics
Nanjing
China

Shichao Niu
Jilin University
Changchun
China

Bo Li
Xi'an Jiaotong University
Xi'an
China

Editorial Office
MDPI
St. Alban-Anlage 66
4052 Basel, Switzerland

This is a reprint of articles from the Special Issue published online in the open access journal *Machines* (ISSN 2075-1702) (available at: https://www.mdpi.com/journal/machines/special_issues/BioInspired_SmartMachines).

For citation purposes, cite each article independently as indicated on the article page online and as indicated below:

Lastname, A.A.; Lastname, B.B. Article Title. *Journal Name* **Year**, *Volume Number*, Page Range.

ISBN 978-3-0365-7264-2 (Hbk)
ISBN 978-3-0365-7265-9 (PDF)
doi.org/10.3390/books978-3-0365-7265-9

Cover image courtesy of Dong Mei and Yanjie Wang

Contents

About the Editors

Yanjie Wang

Yanjie Wang is currently working as a Professor at the College of Mechanical and electrical Engineering, Hohai University, Jiangsu Province, China. He received his Ph.D. degree in Mechanical Engineering from Xi'an Jiaotong University (XJTU), Xi'an, China, in 2015 and then joined Hohai University. His research interests include polymer-based sensors and actuators, smart materials and structures, and advanced bionic systems and robotics.

Xiaofeng Liu

Xiaofeng Liu received the B.S. degree in electronics engineering and M.S. degree in computer application from the Taiyuan University of Technology, Taiyuan, China, in 1996 and 1999, respectively, and the Ph.D. degree in biomedical engineering from Xi'an Jiaotong University, Xi'an, China, in 2006. He joined as an Associate Professor with the College of Information and Electrical Engineering, Shandong University of Science and Technology, in 2006. From 2008 to 2011, he held a post-doctoral position with the Institute of Artificial Intelligence and Robotics, Xi'an Jiaotong University. From 2011, he has been with the College of IoT Engineering, Hohai University, Changzhou, where he is currently a Full Professor and the Vice Director of the Changzhou Key Laboratory of Robotics and Intelligent Technology. His current research interests focus on the study of nature-inspired navigation, human–robot interaction, and neural information processing.

Aihong Ji

Aihong Ji is currently working as a Professor at the College of Mechanical and Electrical Engineering, Nanjing University of Aeronautics and Astronautics, Nanjing, China. He received his Ph.D. degree in Nanjing University of Aeronautics and Astronautics in 2007. His research interests include biomechanics, climbing robots, Micro Aerial Vehicle, Unstatic Aerodynamics, Insects flying, animals locomotion.

Shichao Niu

Shichao Niu is currently a professor at Key Laboratory of Bionic Engineering (Ministry of Education), Jilin University. He received his PhD degree (2014) in Bionic Science and Engineering from Jilin University. His current research focuses on bio-inspired materials with micro/nanoscale hierarchical structures, the diversified optical functional surfaces in nature, bio-inspired functional materials for energy and optical applications and the bionic machinery design and manufacture.

Bo Li

Bo Li is a currently an associate professor in School of Mechanical Engineering, Xi'an Jiaotong University, Xi'an, China. He received his PhD degree in Xi'an Jiaotong University in 2012, and served as a postdoctral research fellow in National University of Singapore from 2013–2014. His research interests include design and fabrication of soft actuator and robot with potential applications in medical operation and inspection.

machines

Editorial

Bio-Inspired Smart Machines: Structure, Mechanisms and Applications

Yanjie Wang [1,*], Dong Mei [1], Xiaofeng Liu [2], Aihong Ji [3], Shichao Niu [4] and Bo Li [5]

[1] College of Mechanical and Electrical Engineering, Hohai University, Changzhou Campus, Changzhou 213022, China; dongmeicz@hotmail.com
[2] College of Internet of Things (IoT) Engineering, Hohai University, Changzhou Campus, Changzhou 213022, China; xfliu@hhu.edu.cn
[3] College of Mechanical & Electrical Engineering, Nanjing University of Aeronautics and Astronautics, Nanjing 210016, China; meeahji@nuaa.edu.cn
[4] Key Laboratory of Bionic Engineering, Ministry of Education, Jilin University, Changchun 130022, China; niushichao@jlu.edu.cn
[5] School of Mechanical Engineering, Xi'an Jiaotong University, Xi'an 710049, China; liboxjtu@xjtu.edu.cn
* Correspondence: yj.wang1985@gmail.com

With the long-term evolution of nature, each creature has its unique structure and function, which can adjust to unstructured environments with diversity. These structures, mechanisms, and potential principles from biology will definitely excite new ideas for improving and optimizing conventional machine designs and control. By imitating certain characteristics of these creatures, such as grasp, twist, locomotion, or flying, etc., bio-inspired smart machines can engage in certain difficult tasks instead of human beings, such as medical surgery, rapid manufacture and assembly, disaster search and rescue, and scientific investigation. Therefore, bio-inspired smart machines have an important research significance and broad developmental prospects.

In light of this, this Special Issue provides an international forum for professionals, academics, and researchers to address some of the latest theoretical and technological advances in bio-inspired smart machines and their structure, mechanisms, and applications. After a stringent peer review process, eighteen papers were finally included in this Special Issue, which cover the following aspects: (1) bio-inspired machines and robotics, (2) bio-inspired modeling and control, and (3) bio-inspired sensors and active materials. A summary of the accepted papers is outlined below.

In the context of bio-inspired machines and robotics, a weevil-inspired jumping mechanism was designed in [1]. In this paper, a miniature prototype was designed to reproduce a weevil's jumping mechanism with its working principle and anatomical structure to verify how weevils' jumping mechanisms work; it performed well in terms of its jumping height. This paper presented the anatomical structure and working principle of the weevil jumping mechanism, followed by an explanation and analysis of its kinematics and dynamics, then performing virtual prototype simulations to compare different design schemes, with results guiding the parameter optimization and subjecting a prototype machine into a height test. In comparisons among the existing jumping mechanisms whose jumping method is bio-inspired, the present design weighs 44.7 g and can jump to a maximum height of 2 m.

To meet the requirements for the flexible end-effectors of industrial grippers and climbing robots, inspired by the animal attachment mechanism, a bio-inspired adhesive unit (Bio-AU) was designed in the second study [2]. Based on the lamination mold casting process, the "simultaneous molding and assembly" method was established, which can be applied to form and assemble complex cavity parts simultaneously. Moreover, the dovetail tenon-and-mortise parting structures were analyzed and designed. Furthermore, the adhesion between the parting surfaces can be improved using plasma surface treatment technology. By applying the above methods, the assembly accuracy and pressure-bearing

Citation: Wang, Y.; Mei, D.; Liu, X.; Ji, A.; Niu, S.; Li, B. Bio-Inspired Smart Machines: Structure, Mechanisms and Applications. *Machines* **2023**, *11*, 405. https://doi.org/10.3390/machines11030405

Received: 15 March 2023
Accepted: 17 March 2023
Published: 20 March 2023

capability of the complex flexible cavities are improved, which reduces the individual differences between the finished products.

In [3], the authors proposed two new flexible perching grasping mechanisms inspired by a bird's claw, one of which is articulated and the other is resilient. The difference being that the former has a pin-articulated claw structure and uses a double fishing line to perform the grasping and resetting action, while the latter uses a resilient linking piece, a single fishing line, and a resilient linking piece to perform the grasping and resetting action. The results of the experiments show that the two types of perch gripping mechanism have their own advantages and disadvantages. Another work, [4], was also inspired by birds, which proposed and fabricated a single-joint worm-like robot with a centimeter scale, the motion of which can be easily guided by a magnet. The robot consists of a pneumatic deformable bellow and a permanent magnet fixed in the bellow's head that will generate magnetic force and friction. The experiment exhibits its excellent environmental adaptability. Moreover, the robot's motion was successfully guided under the presence of the magnetic field, which shows a great potential for pipeline detection applications.

In [5], a four-track twin-rocker bionic rescue robot with an inner and outer concentric shaft was designed to achieve the best obstacle surmounting performance of a mobile robot in the rescue environment. From the viewpoint of dynamics, the motion process of the mass center of the robot when climbing steps forward and backward was studied. The maximum obstacle height of the robot was calculated. The relationship between the elevation angle of the car body, the swing angle of the rocker arm, and the height of the steps was analyzed using a simulation. The simulation results show that the maximum forward and reverse obstacle crossing heights were 92.99 mm and 155.82 mm, respectively.

A tubular stiffening segment based on layer jamming was proposed in [6], which can temporarily increase the stiffness of the soft robot in the desired configuration. First, the authors provided the details of the TSCR design, including the mechanical design and bio-inspired compliant spine mechanisms. Then, an analytical model of a two-layer jamming structure was proposed, as well as extending predictions to many-layer jamming structures. Experimental tests show that the bending stiffness of the initial TSCR increased by more than 15× at 0°, 30× at 90°, and 60× in compressive stiffness.

In [7], the effect of exoskeleton assistance on preferred speeds was tested. The U-shaped oxygen consumption and lower limb muscle activity curve with the minimum at preferred frequency were obtained by inviting participants to participate in the test, which indicated that the resonant condition existed under the preferred condition. Average metabolic reductions of 4.53% and 7.65% were found in the preferred condition compared to the general and comparison condition, respectively.

In [8], with the aim to improve the mobility and adaptability of WCR to complex urban operating environments and expand the application scope of the robot according to the clinging characteristics of different creatures, a double propeller wall-climbing robot (DP-Climb WCR) with a hybrid adhesion system was designed that can be internally transitioned based on the principle of a biomimetic design from the perspective of robot dynamics. Through mechanical and aerodynamic experiments, it was verified that the robot's actual output pulling force can meet the transition motion demand, and the robustness and adaptability of the WCR to complex application environments are improved.

In [9], a six-wheeled wire-rope-climbing robot was proposed. Under the condition of its own weight of 3.8 kg, the robot can carry a maximum of 8 kg of working tools for online laser cleaning and the maintenance of steel wire ropes and a visual safety inspection. The developed climbing robot is composed of separable driving and driven trolleys. It adopts the spring clamping mechanism and the wheeled movement method and can also easily adapt to the narrow working environment and the different diameter ranges of the sluice wire rope.

Within the context of bio-inspired modeling and control, a novel process parameter optimization approach for laser cladding is proposed in [10], based on a multi-objective slime mold algorithm (MOSMA) and support vector regression (SVR). In particular, SVR is used as a bridge between target and process parameters for solving the problem of lacking accurate information regarding the function relationship. The performance of the proposed approach was evaluated using the TOPSIS method, based on actual laser cladding data, and was compared with several well-known approaches. The results indicate that the optimal process parameters obtained by the proposed approach have a better process performance.

In [11], a bio-inspired compound continuum robot (CCR) combining the concentric tube continuum robot (CTR) and the notched continuum robot was proposed to design a high-dexterity minimally invasive surgical instrument. A kinematic model, considering the stability of the CTR part, was established. The unstable operation of the CCR is avoided. The simulation of the workspace shows that the introduction of the notched continuum robot expands the workspace of the CTR. The dexterity indexes of the robots are also proposed, and the simulation shows that the dexterity of the CCR is 1.472 times that of the CTR.

In [12], an improved ANCF lower-order plate element was used to increase the accuracy of the Yeoh model and characterize the geometrical structure of silicone rubber fingers, taking into particular consideration the effect of volume locks and multi-body system constraints. First, the improved Yeoh model based on the ANCF plate element is introduced; then, on the basis of the above-mentioned theory, the computation of a silicone rubber finger is presented. The simulation results showed that the motion performance of the flexible finger can be characterized by the proposed model effectively, including the expansion, deformation, bending status, and so on, under different air pressures.

In order to analyze the effect of higher-order harmonics on the hub load, a rotor that can realize individual blade pitch control was designed in [13]. The Glauert inflow model was introduced to calculate the induced velocity of rotor blades in a rotor disk plane, and the Leishman Beddoes (L-B) unsteady dynamic model was employed to calculate the aerodynamic forces of each section of a rotor blade. The results showed that the influence of each high-order harmonic control on the reduction in the individual blade vibration load is similar in different advanced ratios.

In [14], an impact load identification method based on impulse response theory (IRT) and a BP (back propagation) neural network is proposed. By extracting the peak value in the rising oscillation period of response, it transformed the excitation and response signals into the same length. First, it is deduced that there is an approximate linear relationship between the discrete time integral of the impact load and the amplitude of the oscillation period of the response. Second, a BP neural network was used to establish a linear relationship between the discrete time integral of the impact load and the peak value in the rising oscillation period of the response. Third, the network was trained and verified. The results show that this method has a high accuracy and application potential.

In [15], a control system for desktop experimental manipulators based on an audio-visual information fusion algorithm was designed. The robot could replace the operator to complete some tedious and dangerous experimental work by teaching it the arm movement skills. The system is divided into two parts: skill acquisition and movement control. For the former, the visual signal was obtained through two algorithms of motion detection, which were realized by an improved two-stream convolutional network. The latter employed motor control and grasping pose recognition, which achieved precise controlling and grasping.

In the process of trajectory tracking using the linear quadratic regulator (LQR) for driverless wheeled tractors, a weighting matrix optimization method based on an improved quantum genetic algorithm (IQGA) was proposed in [16] to solve the problem of weight selection. A kinematic model of the wheel tractor was built based on the Ackermann steering model, the state weighting matrix in the LQR controller was optimized using IQGA, and, finally, a joint simulation was performed using Carsim and MATLAB. The simulation results after comparing the other four optimization algorithms showed that the proposed IQGA speeds up the algorithm's convergence, increases the population's diversity, improves the global search ability, preserves the excellent information of the population, and has substantial advantages over other algorithms in terms of performance.

As to the context of bio-inspired sensors and active materials, a new strategy was proposed for developing electronic skin with tactile sensing and pain warning in [17]. A bionic artificial receptor with innocuous sensing and damage warning functions based on the coordination of the ion–electric response principle and mechanical signal attenuation was fabricated. The ion-sensing film provides the carrier of touch or pain perception, while the PDMS layer as a soft substrate is used to regulate the perception ability of the receptor. The sensing voltage of the artificial receptor primarily derives from the bending deformation of two IPMC sensory layers. Experiments show that the sensitivity of the touch response and pain response is 0.136 mV/Pa and 0.026 mV/Pa, respectively. Additionally, the distinction ability of touch and pain becomes more pronounced under a higher elastic modulus and larger thickness.

To obtain the lattice structure with an excellent energy absorption performance, the structure of a loofah inner fiber was studied by [18] to develop the bionic design of the lattice structure by an experiment and simulation analysis method. From the compression experiment of the four bionic multi-cell lattice structures (bio-45, bio-60, bio-75, and bio-90) and VC lattice structures, its shown that all are made of PLA and fabricated by the fused deposition modeling (FDM) 3D printer. The comprehensive performance of the bio-90 lattice structure is the best in the performance of the specific volume energy absorption (SEAv), the effective energy absorption (EA), and the specific energy absorption (SEA). Based on the experimental result, the energy absorption performance of the bio-90 lattice structure was then studied using a simulation analysis of influence on multiple parameters, such as the number of cells, the relative density, the impact velocity, and the material.

As Guest Editors of this Special Issue, we wish to express thanks to all of the authors of the published articles, as well as the authors of articles who were interested in publishing in the Special Issue. We especially thank the reviewers, who examined all the submitted articles in detail and, utilizing their scientific experience, expressed their opinion on the solved issues and provided invaluable advice for improving the presented results of the authors' studies.

Author Contributions: Conceptualization, Y.W. and D.M.; methodology, Y.W.; software, Y.W.; validation, D.M. and X.L.; investigation, A.J., S.N. and D.M.; data curation, D.M., B.L. and Y.W.; writing—original draft preparation, Y.W. and D.M.; writing—review and editing, X.L., B.L. and Y.W.; visualization, A.J. and S.N.; supervision, Y.W.; project administration, Y.W. All authors have read and agreed to the published version of the manuscript.

Data Availability Statement: Not applicable.

Acknowledgments: We appreciate all the authors and anonymous reviewers who contributed to this Special Issue. Meanwhile, we would like to thank the support from the Editorial Staff for our Special Issue.

Conflicts of Interest: The authors declare no conflict of interest.

References

1. Yu, Z.; Zeng, Y.; Guo, C. Mechanical Design and Performance Analysis of a Weevil-Inspired Jumping Mechanism. *Machines* **2022**, *10*, 161. [CrossRef]
2. Zhang, L.; Wang, L.; Weng, Z.; Yuan, Q.; Ji, K.; Wang, Z. Fabrication of Flexible Multi-Cavity Bio-Inspired Adhesive Unit Using Laminated Mold Pouring. *Machines* **2022**, *10*, 184. [CrossRef]
3. Zhu, Y.; He, X.; Zhang, P.; Guo, G.; Zhang, X. Perching and Grasping Mechanism Inspired by a Bird's Claw. *Machines* **2022**, *10*, 656. [CrossRef]
4. Mei, D.; Zhao, X.; Tang, G.; Wang, J.; Zhao, C.; Li, C.; Wang, Y. A Single-Joint Worm-like Robot Inspired by Geomagnetic Navigation. *Machines* **2022**, *10*, 1040. [CrossRef]
5. Xu, X.; Wang, W.; Su, G.; Liu, C.; Cai, W.; Zhang, H.; Ran, Y.; Tan, Z.; Luo, M. Obstacle Modeling and Structural Optimization of Four-Track Twin-Rocker Rescue Robot. *Machines* **2022**, *10*, 365. [CrossRef]
6. Liu, Z.; Xu, L.; Liang, X.; Liu, J. Stiffness-Tuneable Segment for Continuum Soft Robots with Vertebrae. *Machines* **2022**, *10*, 581. [CrossRef]
7. Zhang, Z.; Wang, H.; Guo, S.; Wang, J.; Zhao, Y.; Tian, Q. The Effects of Unpowered Soft Exoskeletons on Preferred Gait Features and Resonant Walking. *Machines* **2022**, *10*, 585. [CrossRef]

8. Zhang, Q.; Gao, X.; Li, M.; Wei, Y.; Liang, P. DP-Climb: A Hybrid Adhesion Climbing Robot Design and Analysis for Internal Transition. *Machines* **2022**, *10*, 678. [CrossRef]
9. Fang, G.; Cheng, J. Design and Implementation of a Wire Rope Climbing Robot for Sluices. *Machines* **2022**, *10*, 1000. [CrossRef]
10. Zhang, Y.; Gong, B.; Tang, Z.; Cao, W. Application of a Bio-Inspired Algorithm in the Process Parameter Optimization of Laser Cladding. *Machines* **2022**, *10*, 263. [CrossRef]
11. Zhang, G.; Du, F.; Xue, S.; Cheng, H.; Zhang, X.; Song, R.; Li, Y. Design and Modeling of a Bio-Inspired Compound Continuum Robot for Minimally Invasive Surgery. *Machines* **2022**, *10*, 468. [CrossRef]
12. Xing, Y.; Liu, L.; Liu, C.; Li, B.; Wang, Z.; Li, P.; Zhang, E. Mechanical Deformation Analysis of a Flexible Finger in Terms of an Improved ANCF Plate Element. *Machines* **2022**, *10*, 518. [CrossRef]
13. Yang, R.; Gao, Y.; Wang, H.; Ni, X. Reducing Helicopter Vibration Loads by Individual Blade Control with Genetic Algorithm. *Machines* **2022**, *10*, 479. [CrossRef]
14. Huang, D.; Gao, Y.; Yu, X.; Chen, L. The Feature Extraction of Impact Response and Load Reconstruction Based on Impulse Response Theory. *Machines* **2022**, *10*, 524. [CrossRef]
15. Li, C.; Chen, X.; Ma, X.; Sun, H.; Wang, B. Skill Acquisition and Controller Design of Desktop Robot Manipulator Based on Audio–Visual Information Fusion. *Machines* **2022**, *10*, 772. [CrossRef]
16. Fan, X.; Wang, J.; Wang, H.; Yang, L.; Xia, C. LQR Trajectory Tracking Control of Unmanned Wheeled Tractor Based on Improved Quantum Genetic Algorithm. *Machines* **2023**, *11*, 62. [CrossRef]
17. Zhao, X.; Tang, G.; Zhao, C.; Mei, D.; Ji, Y.; Xiang, C.; Li, L.; Li, B.; Wang, Y. Bio-Inspired Artificial Receptor with Integrated Tactile Sensing and Pain Warning Perceptual Abilities. *Machines* **2022**, *10*, 968. [CrossRef]
18. He, G.; Yang, H.; Chen, T.; Ning, Y.; Zou, H.; Zhu, F. Lattice Structure Design Method Aimed at Energy Absorption Performance Based on Bionic Design. *Machines* **2022**, *10*, 965. [CrossRef]

machines

MDPI

Article

Mechanical Design and Performance Analysis of a Weevil-Inspired Jumping Mechanism

Zhiwei Yu *, Yifan Zeng and Ce Guo

College of Mechanical and Electrical Engineering, Nanjing University of Aeronautics and Astronautics, Nanjing 210016, China; congzz@nuaa.edu.cn (Y.Z.); guozc@nuaa.edu.cn (C.G.)
* Correspondence: yuzhiwei@nuaa.edu.cn

Abstract: Jumping mechanisms constitute an important means of resolution in applications such as crossing uneven terrain and space exploration. However, the traditional design mainly uses engineering design thinking, but seldom studies the structural characteristics of organisms themselves and lacks biomimetic research basis, which leads to the difference between jumping mechanism and biological structure and its jumping ability. On the other hand, it lacks in-depth study on biological jumping mechanism from the view of engineering. Weevil has excellent jumping performance, and its key jumper structure is specially designed by biologist. To investigate the motion mechanism and working mechanism of the jumping mechanisms, this paper takes the weevil as the bionic object, and designs a weevil-inspired jumping mechanism. A miniature prototype is designed to reproduce weevil's jumping mechanism with its working principle and anatomical structure to verify how weevil's jumping mechanisms work, and turns out to perform well at jumping height. This paper is presented the anatomical structure and working principle of the weevil jumping mechanism, followed by explanation and analysis of its kinematics and dynamics, then performing virtual prototype simulations to compare different design schemes, with results guiding the parameter optimization and subjecting a prototype machine into a height test. In comparisons among existing jumping mechanisms whose jumping method is bio-inspired, the present design, which weighs 44.7 g and can jump to a maximum height of 2 m. The present research establishes a biologically inspired working principle and provides a new practical archetype in biologically inspired studies.

Keywords: bio-inspired robot; jumping mechanism; mechanical design; weevil

Citation: Yu, Z.; Zeng, Y.; Guo, C. Mechanical Design and Performance Analysis of a Weevil-Inspired Jumping Mechanism. *Machines* **2022**, *10*, 161. https://doi.org/10.3390/machines10030161

Academic Editor: Giuseppe Cabone

Received: 13 January 2022
Accepted: 9 February 2022
Published: 22 February 2022

1. Introduction

Overcoming obstacles and rough terrain is a difficult but common challenge faced by miniature jumping robots. In general, biological creatures do so by flying, climbing, or jumping, of which jumping is superior in that it (i) uses less energy and is quieter than flying and (ii) is simpler and more reliable than climbing. The decisive factor for a mechanism's jumping performance is its mechanical design, and there has been much research into methods for storing and releasing energy. In those works, some structures such as planar six-bar mechanism and cams have emerged as being particularly useful and have been used widely.

Among them, the planar six-bar mechanism is the structure that is used most widely. A typical planar six-bar mechanism comprises bars in a hexagonal configuration that has mirror symmetry both vertically and horizontally. However, the upper and lower edges of the hexagon are usually realized in the form of joints or planes, and linear tension elements are usually fixed on their left and right endpoints. Anderson learned through research that elastic energy storage and utilization can enhance jumping efficiency much more than overall jumping performance [1,2]. So it is useful to increase the jumping range with energy storing strategies by adding springs. To maximize the stored energy, springs usually appear in parallel; they store energy by widening the structure while compressing its height and lowering its center of mass (COM). This design is the most universal and the one most likely to achieve high energy destiny. Jung used it in their JumpROACH and its previous type [3,4], and Woodward used it in their Multimo-Bat [5]. Because of its good performance, this classical design has been altered variously to adapt it to more requirements. In their wheeled mechanism, Ye converted the linear rods into arcs to allow the wheels to rotate [6]. In their Jump-Flapper, Truong broke the linear springs into two disconnected ones and changed their hanging locations [7]. In their MSU Jumpers, Zhao removed the linear springs and instead used torsion ones on the joints to gain space inside the hexagon [8–10].

Cams are used widely as triggering solutions. A cam stores energy as its radius increases and then releases that energy freely when its radius jumps from maximum to minimum. As such, the two degrees of freedom (DOFs) of storing energy and triggering are combined into one using only one component, and therefore cams are very suitable for robots for which the main priority is to minimize the weight. In their work, Shen, Zhang and Kovac used cams for triggering [11–13].

Another way to design a mechanism that has superior jumping ability is to use biologically inspired design. Because many insects can jump to heights of tens or even hundreds of times their body length, potential designs can come from studying their jumping locomotion or anatomical structure. However, most bio-inspired jumping mechanisms to date have been designed by focusing on limb locomotion, the aim being to maximize the jumping height by enlarging the gain in displacement during launch. Feature points extracted using high-speed photography are recombined into the skeleton of the mechanism. Typical examples are (i) the galago-inspired SALTO series by Haldane and (ii) numerous flea- and locust-inspired robots, such as that proposed by Zhang [14–19]. Rarer in the literature is bio-inspired design based on an anatomical launching principle, even though such an approach is essential for improving how energy is stored and released [17]. Of this type exists the flea-inspired robot developed by Noh [20–22].

According to Nadein, weevils (Coleoptera: Curculionidae: Rhamphini) can reach a maximum speed of 2.0 m/s when jumping, which is quite large among jumping insects. With their hind legs being relatively short, their jumping ability is attributed instead mainly to their special jumping mechanism [22].

However, this special structure is yet to be realized into an engineering design. To fill in this research gap, we have designed a jumping mechanism according to the anatomical structure of weevils and involving no classical structures. With few but useful improvements to achieve better engineering performance, it well explains jumping principle of a weevil's jumping mechanism. It represents a new structure type in the research into bio-inspired jumping mechanisms, and shows feasibility and superiority of this structure by operating smoothly and performing well in jumping height.

In this study, with weevil's jumping mechanism as the research object, designs a weevil-inspired jumping mechanism, its kinematic and dynamics were explanted and analyzed, virtual prototype simulation and experiment were then conducted. This study provides a new practical archetype in biologically inspired studies.

2. Design of Jumping Mechanism

We present a new jumping mechanism with the bionic idea from weevil, and it is not like the existing jumping mechanism.

2.1. Biological Working Principle

With limbs that are relatively short, the weevil's jumping ability comes mainly from its jumping mechanism. From anatomical observations, Nadein proposed a functional model of the weevil's jumping mechanism [22].

In the initial phase, the extensor muscle is relaxed, and contraction of the flexor muscle (fm) minimizes the angle between the femur and tibia until the flexion state, to restore the jumping mechanism. During this period, the tibial flexor sclerite (TFS) is dragged over an internal protrusion (ip) (Figure 1a,b).

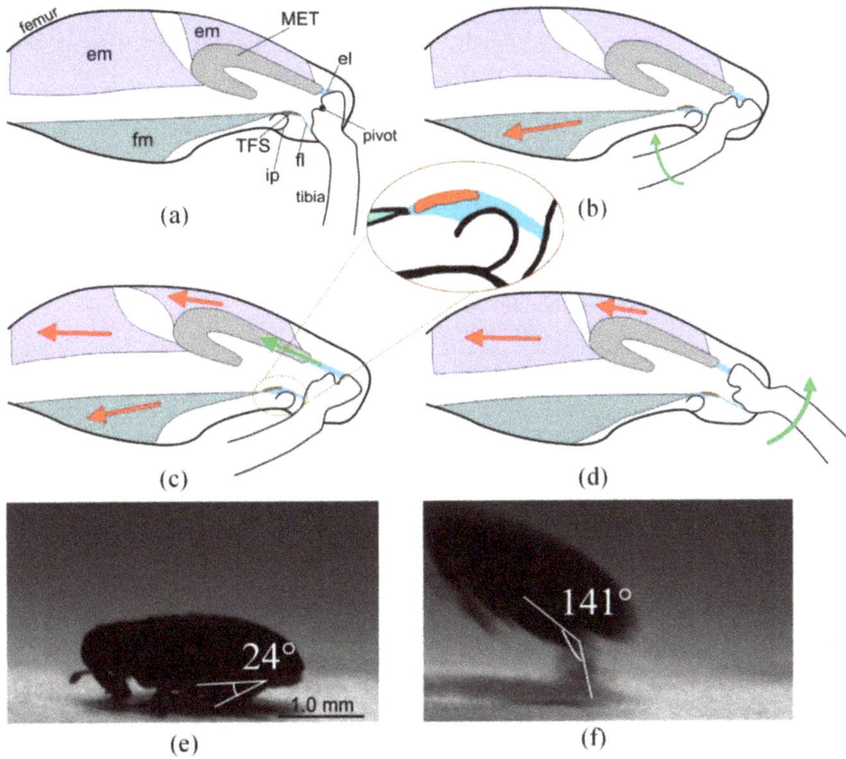

Figure 1. Working period (**a–d**) with initial and terminal phases (**e–f**) of weevil's jumping mechanism. Red arrays stand for forces in muscles, green arrays stand for motions. Abbreviations: em, extensor muscle; el, extensor ligament; fm, flexor muscle; fl, flexor ligament; ip, internal protrusion; MET, metafemoral extensor tendon; TFS, tibial flexor sclerite [22].

Next, the extensor muscle begins to contract, pulling the metafemoral extensor tendon (MET) to store energy by stretching the extensor ligament (el). The fm is now kept contracted so as not to displace the TFS (Figure 1b,c).

When the energy stored in the el reaches a certain amount, the fm is relaxed. This causes the TFS to be dragged over the internal protrusion again to the lower side. Losing the pull provided by the fl, the tibia rotates freely around the pivot, driven by the huge extension force provided by the el (Figure 1c,d). As the tibia rotates relative to the femur, the body of the weevil is sent into the air.

Because of the support from the ip to the tibial flexor sclerite in the configuration shown in Figure 1c, a relatively small force is required to stop the tibia from rotating. Compared to structures with no sclerite or internal protrusion, the weevil's jumping mechanism requires less power in the fm.

2.2. Mechanical Redesign

Based on the working principle of the weevil's jumping mechanism and aimed at better engineering implementation, the jumping mechanism is redesigned into a mechanical structure. In our design, the pull from the flexor muscle is removed through a mechanical position-limiting design. Representing the tibial flexor sclerite, a triangular slider (TS) is driven by rectangular slider (RS) through the upper string (US), and the RS is driven by the motor, thereby saving a degree of freedom.

This design can be approximated as a two-dimensional model as shown in Figure 2, where the parts are named. The shell of the femur becomes the Support board (SB), which contains structures including the guide rail of the RS, the sliding bolt-nut pair (NBP), the string hole, and the protrusion structure (PS). The latter corresponds to the internal protrusion in the biological jumping mechanism. The biotical TFS becomes the TS mechanical structure. The slider is designed to move downward freely in its guide rail beneath the PS, and go through the gap between the RS and the PS. While above the PS, it is also being credible to touch the PS at its cup while strings are tensioned.

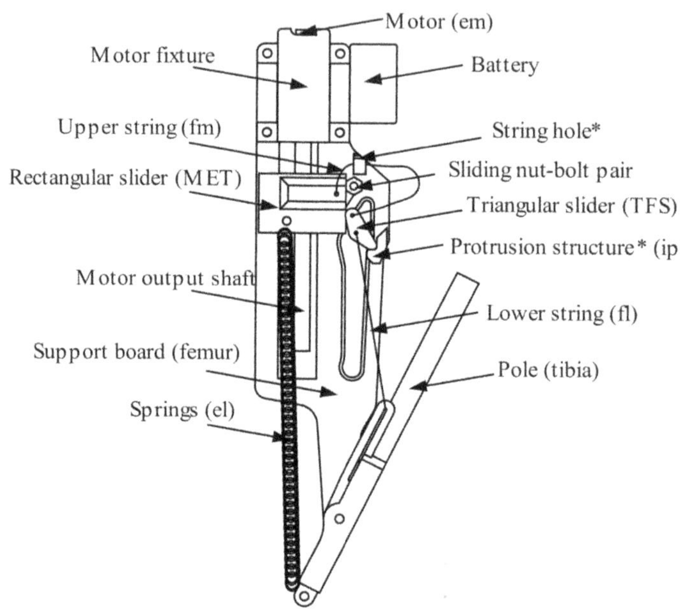

Figure 2. Naming of parts in a side view of the device. In brackets are biological structures in corresponds. *: structures on support board (SB).

The metafemoral extensor tendon becomes the RS. Inside this is a nut that is driven by the motor so that the RS moves in a straight line along the output shaft, driving the RS to move together with it on its guide rail.

The tibia becomes the pole, which was not designed precisely. It is connected to the SB with a revolute joint, enabling it to rotate freely around the joint. Hollow structures are designed so that the pole does not interfere with the SB. The guide rail of the lower string (LS) is to the right of the joint, and springs hang on the other side.

The scalar design is arranged to match the length of the motor output shaft (MOS) in scale, and detailed shape design to fit the angular range of output movement with the one of biological mechanism, also to enlighten the mechanism as much as possible. Besides, the length of the pole is decided according to a rule described in Section 3.2 to maximize the jumping height.

2.3. Working Cycle

Figure 3 shows the working cycle of the mechanism. The springs are loaded for a relatively long time, during which a large amount of elastic energy is stored. Soon after, it is released in a moment so that the pole is driven freely by the springs, transforming the stored energy into kinetic energy. Throughout the whole cycle, the springs always have a preload.

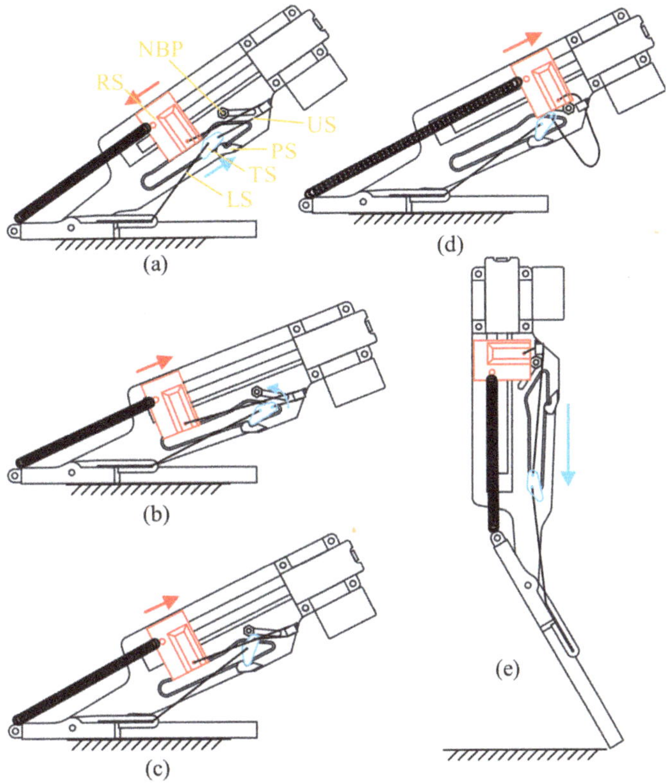

Figure 3. Five typical stages of the mechanism with some components marked: (**a**) the RS moves downwards and pulls the TS moving upwards through the US; (**b**) the RS moves upward and the TS reaches the L-shaped PS, whereupon it rotates because of the unbalanced torque; (**c**) the RS continues to move upward and the pole is unable to rotate because of the pull provided by the LS; (**d**) the RS continues to move upward, pushing the NBP upward in its guide rail; (**e**) the TS falls freely down through its guide rail and the mechanism executes the launching process.

Between the stages shown in Figure 3a,b, the RS moves downwards and pulls the TS moving upwards through the US. Size of the PS, TS and RS are adjusted carefully to let the TS get through the gap between the RS and PS smoothly.

Between the stages shown in Figure 3b,c, the RS moves upward and the two strings are both tensioned. The TS moves downward from its highest position until it reaches the L-shaped PS, whereupon it rotates because of the unbalanced torque until it comes into contact with the sliding nut–bolt pair. Now, the tension in the lower string and the support from the sliding nut–bolt pair and the protrusion structure form a balance in the TS, stopping it from displacing. Also, the pole is unable to rotate because of the pull provided by the LS.

Between the stages shown in Figure 3c,d, the RS continues to move upward, pushing the NBP upward in its guide rail and taking the place of the NBP in the position limit of the TS.

These two processes, in which the RS moves upward, stretching the springs while the pole stays locked in the pre-triggering stage, reflect the processes in the biological jumping mechanism whereby the flexor muscle holds the tibia and the tensed extensor muscle stretches the extensor ligament, storing elastic energy therein.

Between the stages shown in Figure 3d,e, the RS continues to move upward. At a critical position, it loses contact with the TS, whereupon the position limit for the TS is lost. The TS then falls freely down through its guide rail. During this sliding action, the lower string slides to the other end of its guide rail on the pole. This is the launching process, whereby the pole is driven violently by the springs releasing their elastic energy in a moment.

The corresponding process in the biological structure is when the fm relaxes and the TFS loses the forces that were stopping it from moving toward the tibia, whereupon the tibia is driven by the el to rotate with great acceleration.

After the stage shown in Figure 3d, the RS continues to move upward until its upper limit, whereupon the motor rotates in the opposite direction, sending the sliders to their opposite ends and restoring the stage shown in Figure 3a. During this period, the lower string slides to the end farthest from the revolute joint, and the sliding nut–bolt pair slides to the lower end. This is the resetting process that marks the end of the cycle.

To describe how the movements of the RS, TS, and pole are related, the whole mechanism is simplified into a geometrical model as shown in Figure 4. The origin of the coordinate system is dot O, the revolute joint connecting the SB and the pole. The y axis is parallel to the motor output shaft, and the x axis is vertical. The string hole and the TS become a dot to simplify the whole model. The variables appear in the form d_{MN}; for example, d_{OP} is the distance between dots O and P. The value of such a variable is calculated using the Pythagorean method and the coordinates of each dot.

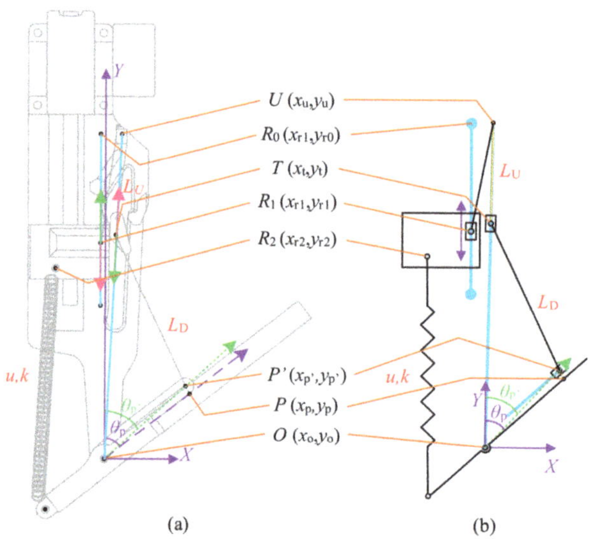

Figure 4. Geometric model describing transmission relationship: (**a**) practical realization; (**b**) abstracted model. The thick blue lines indicate the orbit of the associated sliding dot. Meanings of dots: R_1, hanging point of upper string on rectangle slider (RS), whose orbit is expressed in blue; R_2, hanging point of spring on RS; U, string hole; T, triangle slider (TS); P′, position where lower string passes through the pole; P, intersection point of the extension cord of lower string and the direction of the pole (from dot O). Meanings of constants and variables: k, stiffness coefficient of serried springs; u, length of springs, L_{D}, length of lower string, equals d_{OP} in value; L_{D}, length of upper string, approximately equals d_{UR1} +d_{TR1} in value; θ_{P}, angle between y axis and OP; $d_{\mathrm{OP'}}$, angle between y axis and OP′.

The variable η is defined to express the displacement of the TS on its orbit, the line UO, through the constraint in the upper string. The value of η is calculated as

$$\eta = \frac{d_{\mathrm{UT}}}{d_{\mathrm{UO}}} = \frac{L_{\mathrm{u}} - \sqrt{(x_{\mathrm{r1}} - x_{\mathrm{u}})^2 + (y_{\mathrm{r1}} + y_{\mathrm{u}})^2}}{\sqrt{x_{\mathrm{u}}^2 + y_{\mathrm{u}}^2}} \tag{1}$$

In the triangle formed by points T, O and P, we have

$$\cos(\theta_{\,\mathrm{p'}}) = \frac{d_{\mathrm{OP'}}^2 + [(1-\eta)d_{\mathrm{UO}}]^2 - L_{\mathrm{D}}^2}{2d_{\mathrm{OP'}}(1-\eta)d_{\mathrm{UO}}} \tag{2}$$

As θ_{p} reaches a critical value at which the lower string is almost perpendicular to the $\overrightarrow{OP}$ direction, the lower string slides from its present end to the other. We call this critical angle θ_{c}. Therefore, jumps occur in d_{OP} and $(\theta_{\mathrm{p}} - \theta_{\,\mathrm{p'}})$, leading to the jump in the curve of $(y_{\mathrm{r0}} - y_{\mathrm{r1}})$ versus θ_{p} shown in Figure 5. The maximum and minimum values of θ_{p} are 146° and 26°, respectively, very close to those in the biological structure, namely 141° and 24°, respectively [22].

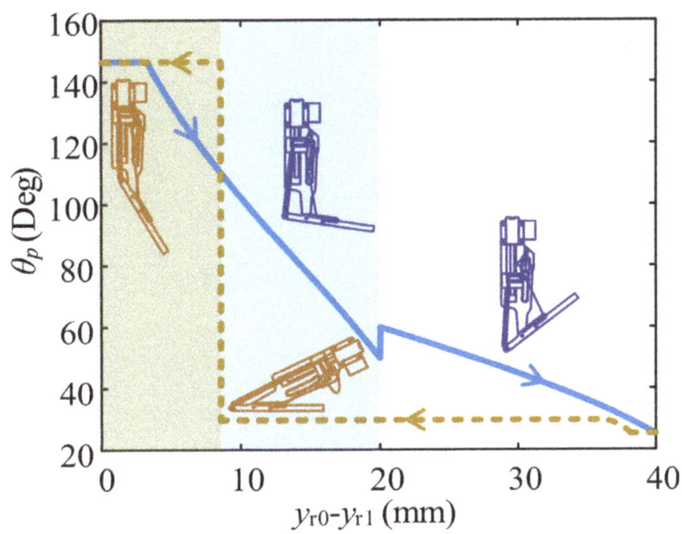

Figure 5. Curve of displacement of RS $(y_{\mathrm{r0}} - y_{\mathrm{r1}})$ vs. angle between pole and motor output shaft (θ_{p}). The thick blue curve is the resetting process in which the RS moves downward. The area with a colored background is for $\theta_{\mathrm{p}} > \theta_{\mathrm{c}}$, and that with the white background is for $\theta_{\mathrm{p}} < \theta_{\mathrm{c}}$. The yellow dotted curve is the launching process in which the RS moves upward. The area with the yellow background is for the post-launch period, and the rest is for the pre-launch period.

2.4. Theoretical Dynamical Analysis

Komarsofla has proposed a novel one-legged hopping robot mechanism with aflat foot [23]. By researched the transitions between different modes and the conditions for these transitions, including stance, taking-off under-actuated, flight, landing under-actuated, and recovery, they derived the potential and kinetic energies of the system and other terms of the Euler–Lagrange equations. On this basis, we established the dynamic equation. Adopting the same coordinates as those in Figure 4 (O as the origin; the motor output shaft as the y axis), the mechanism is abstracted into a simpler model for dynamical analysis of the ejection process as shown in Figure 6.

The mass of the whole mechanism excluding the pole is m_2. The COM of this part is marked with the bigger black-and-white icon, and distance from it to point O is L_{M}. The distances from point O to the two ends of the pole are L_1 and L_2, and that to the COM of the pole is L_{P}. The mass of the pole is m_1, with its COM on the smaller black-and-white icon. The angle between the horizontal plane and the pole is θ_1, and that between the pole and line OM is θ_2.

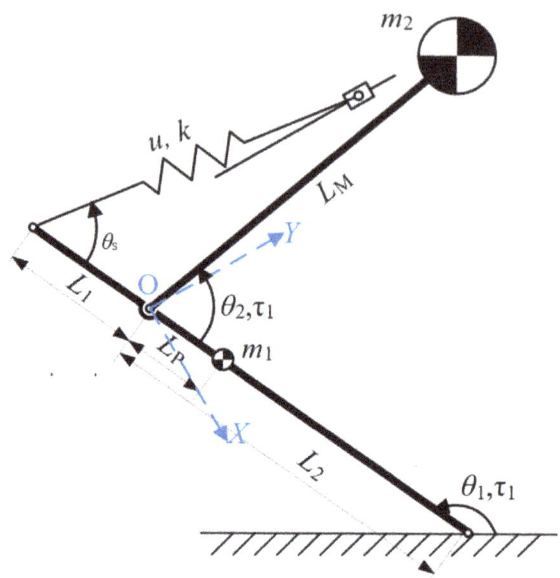

Figure 6. Dynamic model of proposed mechanism for ejection process.

Lagrange dynamical equations are set up to carry out dynamical analyses. Here τ_1 refers to values of torques on the point where the pole contacts with the ground, and τ_2 refers to torque on the point O.

The dynamic equation of the mechanism can be written in the form:

$$\begin{bmatrix} \tau_1 \\ \tau_2 \end{bmatrix} = \begin{bmatrix} D_{11} & D_{12} \\ D_{21} & D_{22} \end{bmatrix}\begin{bmatrix} \ddot{\theta}_1 \\ \ddot{\theta}_2 \end{bmatrix} + \begin{bmatrix} D_{111} & D_{122} \\ D_{211} & D_{222} \end{bmatrix}\begin{bmatrix} \dot{\theta}_1^2 \\ \dot{\theta}_2^2 \end{bmatrix} + \begin{bmatrix} D_{112} & D_{121} \\ D_{212} & D_{221} \end{bmatrix}\begin{bmatrix} \dot{\theta}_1\dot{\theta}_2 \\ \dot{\theta}_2\dot{\theta}_1 \end{bmatrix} + \begin{bmatrix} D_1 \\ D_2 \end{bmatrix} \quad (3)$$

where $\begin{bmatrix} D_{11} & D_{12} \\ D_{21} & D_{22} \end{bmatrix}$ is the 2 × 2 mass matrix of the mechanism, $\begin{bmatrix} D_{111} & D_{122} \\ D_{211} & D_{222} \end{bmatrix}$ is the Coriolis coefficients, $\begin{bmatrix} D_{112} & D_{121} \\ D_{212} & D_{221} \end{bmatrix}$ is an 2 × 2 matrix of centrifugal coefficients. $\begin{bmatrix} D_1 \\ D_2 \end{bmatrix}$ is an 2 × 1 vector of gravity terms. Therefore, we have

$$D_{11} = m_1(L_2 - L_P)^2 + m_2\left(L_2^2 + L_m^2\right) - 2L_2L_m\cos\theta_2 \quad (4)$$

$$D_{12} = m_2(L_m^2 - L_2L_m\cos\theta_2) \quad (5)$$

$$D_{122} = m_2L_2L_m\sin\theta_2 \quad (6)$$

$$D_{112} = 2m_2L_2L_m\sin\theta_2 \quad (7)$$

$$D_1 = m_1g(L_2 - L_P)\cos\theta_1 - m_2g_2(L_m\cos(\theta_1+\theta_2) + L_2\cos\theta_1) \quad (8)$$

$$D_{21} = m_2(L_m^2 - L_2L_m\cos\theta_2) \quad (9)$$

$$D_{22} = m_2L_m^2 \quad (10)$$

$$D_{211} = -m_2L_2L_m\sin\theta_2 \quad (11)$$

$$D_2 = -m_2\mathrm{g}\mathrm{L}_m\cos(\theta_1 + \theta_2) \quad (12)$$

$$D_{111} = D_{121} = D_{222} = D_{212} = D_{221} = 0 \quad (13)$$

Other dynamical indicators that describe the motion during the ejection process are given in Equations (14)–(16) and are plotted in Figure 7. The angle θ_s between the springs and the pole is

$$\theta_s = \arctan\left[\frac{x_{r2} - L_1\sin(\theta_p)}{y_{r2} - L_1\cos(\theta_p)}\right] - (90^\circ - \theta_p) \tag{14}$$

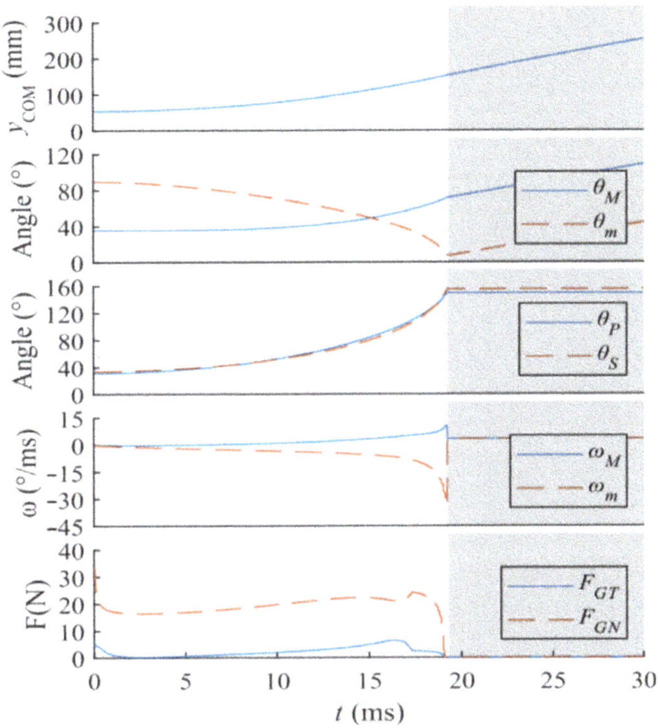

Figure 7. Curves of dynamical indicators vs. time. Area with white background is when mechanism is in contact with the ground. Area with gray background is when mechanism is in the air.

The torque T_{s2p} on point O applied by the springs is

$$T_{s2p} = k(u - u_0)L_1\sin(\theta_s) \tag{15}$$

The tension F_{ls} in the lower string is

$$F_{ls} = \frac{k(u - u_0)L_1\sin(\theta_s)}{(1 - t)d_{UO}\sin(\theta_p)} \tag{16}$$

Based on the model roughly designed through kinematics analyses, a virtual prototype is built in detail to study how some dynamical paraments of this mechanism vary via time. The result shows that angular kinetic energy is ignorable relative to translational kinetic energy, so that little energy loss is caused by spinning.

3. Virtual Prototype Simulation

3.1. Introduction to Simulation

With kinematics analysis above and to fit the size of motor output shaft and range of the angle between tibia and femur during a weevil's jump, we have arrived at an abstract structural design. To have precise adjustments on some parameters having considerable impact on jumping ability of the mechanisms, we carried out virtual prototype simulations to learn about how these parameters impact with jumping orbit of the mechanism.

Some virtual prototypes with differences in certain variables are built and tested in the MSC Adams software package (version 2018). The simulation results were imported and processed in the MATLAB software package (version R2019a) and then plotted to determine how to improve the performance of the mechanism by controlling these variables.

A typical virtual prototype is built according to the parameters defined in Figure 8. The scale is defined to match the length of the MOS. L_{c1} is set to 73 mm, θ_{c1} to 5.5°, L_M to 84 mm, θ_{c2} to 5.2°, L_{c2} to 2.3 mm, L_{c3} to 30 mm, L_D to 45 mm, L_2 to 73 mm, and k to 0.64 N/mm. Here, k is the stiffness coefficient of the springs, and θ_p is the angle between the motor output shaft and the horizontal plane. The value of the latter just before ejection is defined to be θ_{pinit} and is 26.2° here.

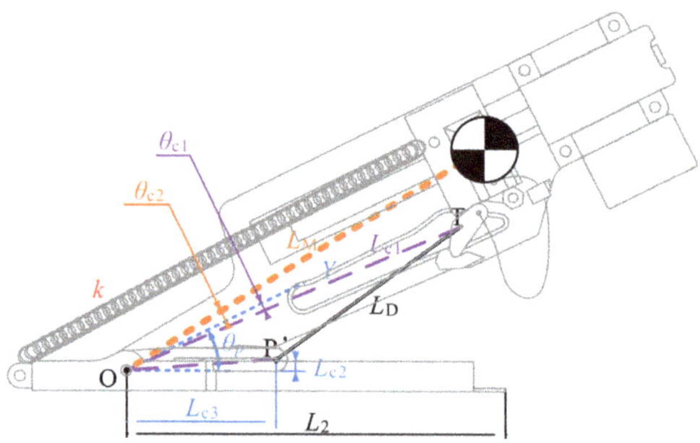

Figure 8. Five typical stages of the mechanism with some components marked.

3.2. Spepated Variable Simulation

In this section, the parameters k, L_D, and L_2 are altered in turn to see how they influence the performance of this mechanism. Only one or two variables are adjusted simultaneously. After altering each variable separately, L_D and L_2 are adjusted together while keeping the COM right above the front end of the pole. Ding have mentioned that horizontal off-set between the COM and the contacting point during the launch leads to horizontal displacement of the flight [24]. This cross-variable-comparison is proposed in supplementary to test how can we maximize its jumping height while setting the similar limit to the horizontal displacement. Unless stated otherwise, a variable takes its value from the original prototype. Table 1 shows the featured dynamical parameters that can reflect the jumping performance of the jumping mechanism. Figure 9 shows how these variables impact on the jumping performance.

Table 1. Featured Dynamical Parameters.

Variable Name	Definition	Practical Significance
x_m	Displacement on x axis at landing	Idealized horizontal displacement
y_m	Displacement on y axis at apex	Idealized jumping height
F_{kmax}	Maximum spring tension force	Stored elastic energy and deformation of the spring
F_{kmin}	Minimum spring tension force	Unreleased elastic energy
$\overline{F_{gt}}$	Average of tangential component of force between ground and mechanism with time	Initial horizontal velocity
$\overline{F_{b2t}}$	Average of tangential force between SB and TS with time	Interaction forces between components
$\overline{\omega^2}$	Average angular velocity of SB	Measure of spinning

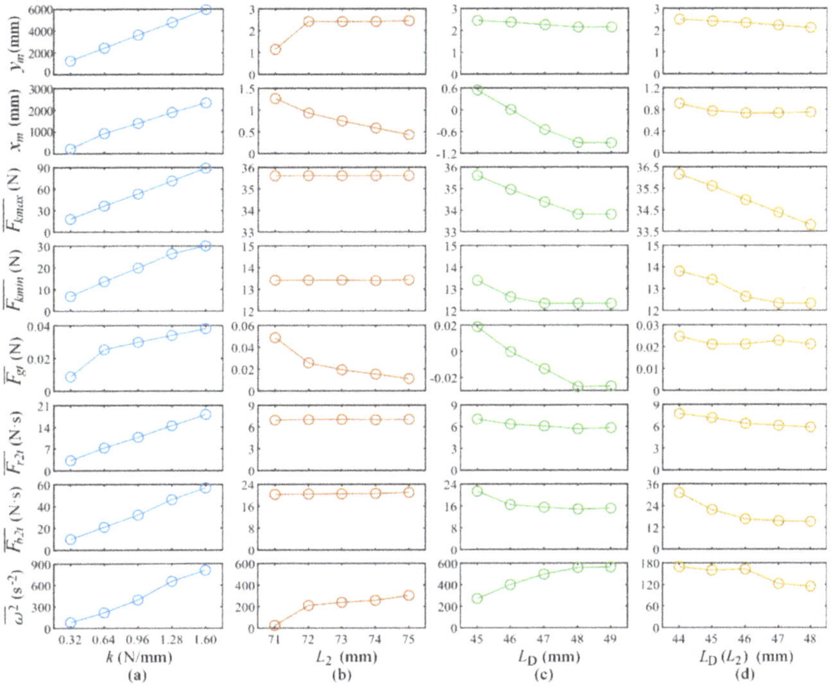

Figure 9. Jumping performance influenced by changing the design parameters (abscissa: altered parameter; ordinate: influence indicator): (**a**) the indicators of ejection performance influenced by the change of the parameter k; (**b**) the indicators of ejection performance influenced by the change of the parameter L_2; (**c**) the indicators of ejection performance influenced by the change of the parameter L_D; (**d**) the indicators of ejection performance influenced by the change of the parameters L_D and L_1.

In Figure 9a, k is changed to analyze how it influences the chosen indicators of ejection performance. In the physical prototype, this variable can be changed by changing the number of springs or using springs with different stiffness.

In Figure 9b, L_2 is changed by modifying the length of the thin slice attached to the end of the pole in the direction in which the pole is pointing. This operation hardly changes the mass distribution. When L_2 is set as 71 mm, the mechanism starts turning over before ejection because the projection of the COM on the ground is outside the supported area; in this case, the device slips upon ejection. By setting L_2 as 72 mm or larger, the device stands in a stable manner.

In Figure 9c, L_D is changed such that θ_{pinit} is 26.5°, 30.3°, 33.5°, 36.4°, and 39.1°. With L_D set to 44 mm, the mechanism falls so quickly that the ejection is considered to fail.

In Figure 9d, because 43 mm is the minimum L_D value for the whole running cycle to finish (otherwise the TS would be unable to form the position limit), L_D is assigned starting with 44 mm, so θ_{pinit} is correspondingly 21.8°, 26.5°, 30.3°, 33.5°, and 36.4°. Meanwhile, L_2 is set as a function of L_D, with the constraint condition provided by kinetic analyses, aiming at keeping the projection of the COM on the ground near the front edge of the pole. Here, the best L_2 value is to be determined.

Each comparison shown in Figure 9 is analyzed below in order. With increasing k, the jumping performance of the mechanism grows almost linearly. However, the internal reaction forces also grow, thereby drawing more motor power.

The value of L_2 has little effect on how the components interact with each other, but it has a large effect on the ejection orbit. A critical value Lc appears when the projection of the COM on the ground coincides with the front edge of the box on the pole. With L_2 greater than L_c, the height performance worsens slowly, the landing position moves in the −x direction, and the angular velocity increases slowly. With L_2 less than L_c, slips occur that lead to ejection failure, and consequently the jumping height decreases sharply. Therefore, the value of L_2 should be set so that the projection of the COM on the ground is close to the front edge and securely inside the supported area.

Keeping L_2 steady but decreasing L_D in the range that enables the operation cycle to finish, the maximum elastic energy is stored, thereby leading to greater displacement and forces being produced, as well as faster spinning. L_D is expected to be smaller, but a sufficient margin is required to avoid the risk of turning over caused by uneven terrain and vibrations while the mechanism is running. For the case shown in Figure 8, suitable values are 44 mm for L_D and 30° for θ_{pinit}.

For cases in which the COM is right above the front edge of the pole, because torque applied to the COM is reduced by minimizing arms of forces causing spinning, the rotations are reduced adequately.

In conclusion, a larger k value should be adopted to maximize the force used for launching, L_2 should be controlled as a function of L_D to keep the subpoint of the COM near the front edge of the pole. In the meantime, L_D should be kept as short as possible to lower the COM before the launch, releasing more energy during the ejection.

3.3. Dynamical Quantities vs. Time

To observe the instantaneous status of the mechanism, some parameters were sampled during the ejections of two chosen virtual prototypes. These cases have the same geometrical design but with k values of 0.64 and 1.28 N/mm, the latter being used for the physical prototype. To plot smooth curves of these parameters versus time, the data were over-sampled and filtered by a moving-average filter at a length of nine. Figure 10 shows the time-varying trends during ejection and while in the air.

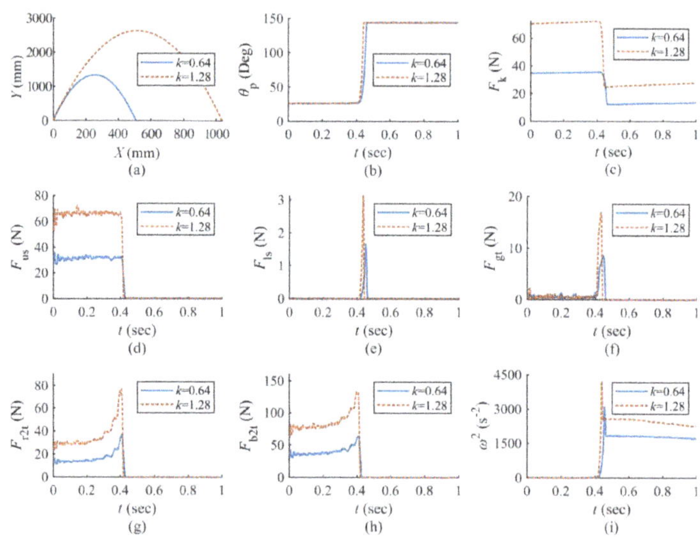

Figure 10. Curves vs. time t [s] of some dynamical parameters, with k = 0.64 and 1.28 N/mm: (**a**) orbit of COM of the mechanism [displacement on y axis (Y) vs. displacement on x axis (X)]; (**b**) time-varying trend of θ_p; (**c**) time-varying trend of elastic force in serried springs (F_k); (**d**) time-varying trend of tension force in upper string (F_{upstr}); (**e**) time-varying trend of tension force in lower string (F_{lowstr}); (**f**) time-varying trend of tangential component of contact force between pole and ground (F_{gt}), equal to friction force between mechanism and ground (normal component of this force has the same trend after subtracting the weight of the mechanism); (**g**) time-varying trend of interaction force between RS and TS (F_{r2t}); (**h**) time-varying trend of interaction force between support board (SB) and TS (F_{b2t}); (**i**) squared angular velocity of motor (ω^2).

The curves in Figure 10 show that the ejection starts at 0.4 s and lasts for a very short while. In a short period before and after the ejection, for the two different k values adopted, the chosen parameters have the same trend. With bigger k value adopted, the curve grows linearly, and time spent in the ejection declines.

The maximum horizontal and vertical displacements grow in the same ratio as k is doubled. The tension in the springs is established as the RS moves upward and is then released rapidly at the ejection. θ_p increases rapidly during the ejection but is constant before and after this process.

The tension in the upper string is small before the ejection, and a pulse for accelerating the TS appears during the ejection. This pulse can be weakened by lengthening the upper string, to transfer this force into collision between the TS and SB. The lower string is tensioned before the ejection and released at the ejection. If L_D is set too short, then a pulse can appear at the ejection to stop the pole.

The spinning accelerates swiftly then decelerates markedly because of the separate actions for each part during ejection of the pole. This spinning then decelerates slowly during the in-air process.

To have a jump that is gentle in damage but strong in displacement, a larger value of k should be set, the surfaces of the sliders should be polished to reduce friction, and the strings must be long enough to avoid inducing intense interior pulse forces.

4. Experimental Studies

4.1. Introduction

Based on the chosen scheme of virtual prototype, a physical prototype (Figure 11) was built using three-dimensional printing. Fishing wire was used for the upper and lower strings, and the total stiffness of the serried springs was measured to be 1.28 N/mm. The motor adopted is in the GA12-N20 type, with an output shaft in the length of 55 mm. The Li-ion battery is with an output voltage of 12 V.

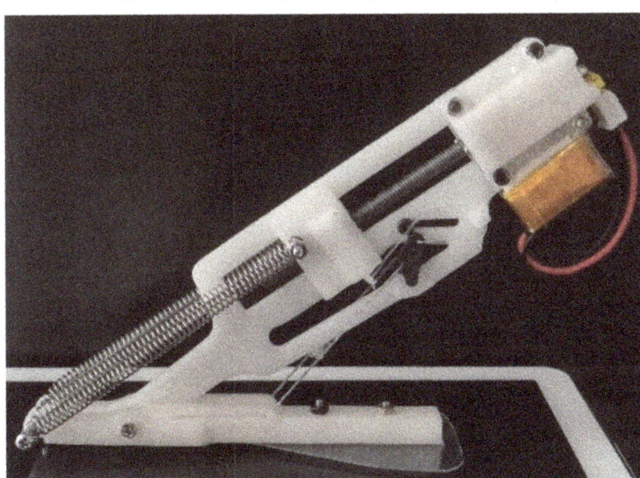

Figure 11. Physical prototype.

To deal with uneven terrain and vibration of the motor while running, a PVC slice was attached to the pole through nut–bolt pairs, thereby enlarging the supported area and preventing the mechanism from falling over. Being convenient to install and uninstall, we used it as our method for adjusting the variable L_2.

In the posture shown in Figure 4b, the total height of the mechanism is 9 cm, and the height of the COM is 5 cm. This is a long-lasting state just before the ejection. With four springs in parallel (k = 1.28 N/mm in total), this mechanism weighs 44.7 g in total (Table 2).

Table 2. Weight budget of proposed jumping mechanism.

Part	Mass[g]	Portion
Motor	13.7	30.6%
Battery	5.0	11.2%
Motor fixtures	3.0	6.7%
Pole	6.8	15.2%
SB	6.3	14.1%
RS	1.9	4.3%
TS	0.2	0.4%
Springs	4.0	8.9%
Nuts and bolts	3.8	8.5%
Total mechanism	44.7	100%

4.2. Orbit Test

We subjected the mechanism to an ejection test to verify its jumping performance. In the recording, a scale in centimeters was set as the background. Because of installation error, according to the first frame in Figure 12 for t = 0 ms, the height of the COM seemed to be approximately −7 cm. For normalization, we added 12 cm to all the sampled height data. The recording was done using a camera with a frame rate of 480 frames per second; frames during the in-air process were sampled uniformly and are arrayed in Figure 12 The time cost of the catapult action is estimated as being 12.5 ms, which is six times that in the biological structure, namely 2.1 ms [22].

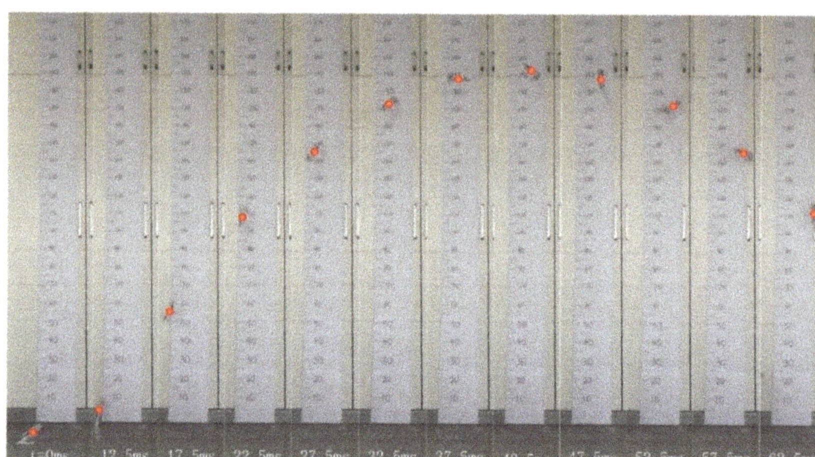

Figure 12. Arrayed frames during in-air process.

The vertical displacement of the COM during a typical jump was sampled uniformly and is plotted in Figure 13 as the blue dots fitted with the red parabola in MATLAB (version R2019a) through a polynomial model with a degree of two. And height-to-time curves of other jumps distributes in the area surrounded by imaginary curves and colored with pink.

The height of the COM was 5 cm before the ejection and reached approximately 205 cm at apex. Thus, the maximum vertical displacement that was achieved is 200 cm, which corresponds to an increase in potential energy of 0.876 J.

Compared with the simulation results, the experimental jumping height is smaller than the simulation results, because the actual test needs to consider friction or other objective conditions, this can be seen from the comparison between the red line peak value of $k = 1.28$ in Figure 10a and the experimental jump height in Figures 12 and 13. The overall results are consistent, and the superior performance of the jumping mechanism is verified by simulation and experiment.

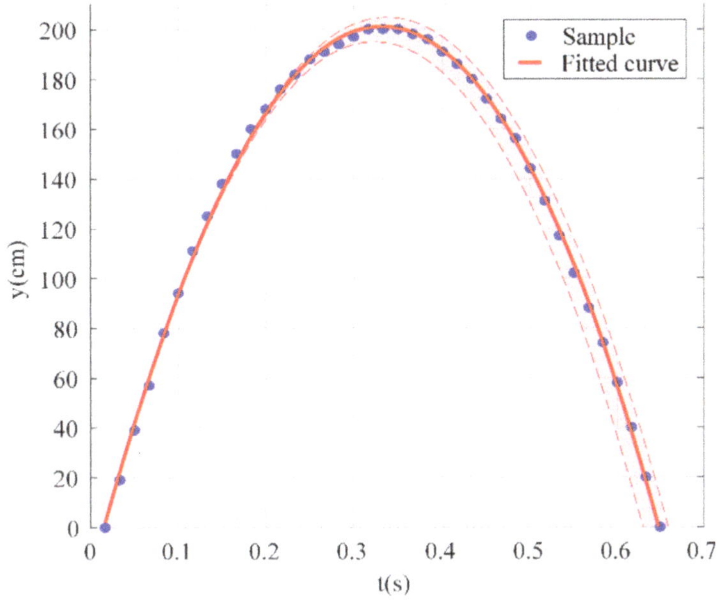

Figure 13. Sampled vertical displacement with curve-fitting result.

5. Discussion

Two topics are to be discussed in this section. The first one is comparisons in similarity of the corresponding biological structure, followed by comparisons in novelty of adopted elements. The second one is statements on measurements of proposed mechanism, followed by comparisons in jumping height performances. Each comparisons are carried out between this work and other miniature jumping mechanisms.

This work is introducing few structures that does not exist in biological structures, and all of them plays an auxiliary role in the working principle (namely the NBP), or being unavoidable in recent researches (namely motor and its relevant elements). Except for this work, few can do the same. The flea-inspired mechanism TAUB [25,26], as a miniature mechanism that is bio-inspired in jumping method which performs the best in jumping altitude according to our survey, uses a hook as its triggering method, BounceBot operates in a similar way [27]. However, no corresponding structure exists in a flea's jumping mechanism. Even the mechanism well explains working principle of a flea's jumping mechanism propose by Koh is adopting planar six-bar mechanism to provide the leaping force [21].

Most of resent bio-inspired works focus on limb locomotion during an insect's jump, and the biological archetype adopted are usually fleas or locusts. Ones using traditional mechanical design methods can hardly avoid using planar six-bar mechanism or other more complex linkage structure [28]. Therefore, new mechanical structure and archetypes introduced at present can promote researches on jumping mechanisms. Inspired by weevils, a kind of insect nobody has built a jumping mechanism according to, a new strategy for storing and releasing energy was enabled herein, providing a new type of structure for miniature jumping robots. Unlike most bio-inspired miniature jumping mechanisms focused on increasing the gain of displacement output by imitating the external appearance of the creature concerned, our work has proposed a new strategy for storing and releasing energy, one that is rarely seen but that has the potential to be combined with other inspirational ideas.

The design was improved to (i) reduce the weight and (ii) simplify the control. The number of DOFs was reduced from two to one by redesigning the structure into one that is dissimilar in appearance compared to the biological structure. The COM of the physical prototype was lifted 200 cm in a test. The present device weighed 44.7 g in total, and is with a height of 9 cm in total when compressed.

Well-known miniature jumping mechanisms proposed previously and that have good jumping performance are listed in Figure 14, and labeled with jumping height. Only bio-mimic mechanisms that are driven mechanically while launching are mentioned; other mechanisms (e.g., ones launched using chemical explosion or electro-deformation) and robots in large size, especially multilegged ones are not mentioned. Also, only the robot with best jumping-height performance is shown here among ones proposed by the same team. It could be seen that this work performs relatively well among all jumping mechanisms in the similar size.

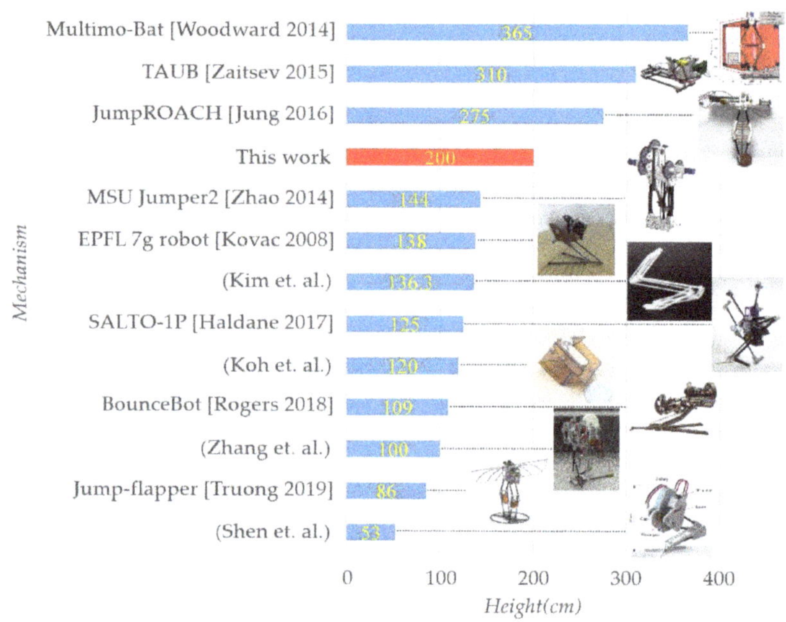

Figure 14. Listing of jumping-height performance for well-known miniature jumping mechanisms proposed previously.

There are still some problems to be improved in our proposed mechanism. For example, it can't make rapid continuous jumps at present, and it needs a little time to restore energy after each jump. What's more, the stability control in the process of mechanism jumping also needs to be solved. At present, there are some novel solutions, such as adding tail and other measures to achieve reliable jumping direction [29], which is also the direction of our next stage improvement.

6. Conclusions

Inspired by the working method of the weevil's jumping mechanism, where two sliders and a lever serve as the core, a new type of weevil-inspired jumping mechanism was proposed herein, thereby providing a new structure, together with its working principle and optimizing suggestions for future studies of bio-inspired miniature jumping mechanisms.

According to the kinematic analysis and simulations of the virtual prototypes and within the limits of the structural strength and motor power, the performance of the mechanism can be improved by doing the following. (1) A spring combination with greater stiffness should be adopted. (2) The projection of the COM on the ground should be kept near the front edge of the supported area but securely inside. (3) The lower string should be shortened as much as possible to lower the COM before launch. (4) The upper string should be long enough to allow the pole to be stopped by colliding with the SB.

The mechanism proposed herein works in a similar pattern with the biological structure, and in a similar appearance. Only few but effective improvements are made to the mechanism to fit engineering demands, to preserve the original working principle and apparent structure as well as possible. As a result, only one DOF is required for this mechanism, and basic requirements such as coping with uneven terrain as the launching surface or adjusting the jumping height and distance are met in this work. This 44.7 g mechanism can lift its COM by 200 cm and remain safe upon a hard landing. Its jumping performance ranks fifth among existing miniature jumping mechanisms.

Author Contributions: Made substantial contributions to the research and investigation process, reviewed and summarized the literature, wrote and edited the original draft, Y.Z.; Performed oversight and leadership responsibility for the research activity planning and execution, as well as developed ideas and evolution of overarching research aims, Z.Y.; Performed critical review, commentary and revision, as well as provided technical, and material support, Z.Y., C.G. All authors have read and agreed to the published version of the manuscript.

Funding: This work was funded by the National Key Research and Development Program of China (Grant No. 2019YFB1309600), the National Natural Science Foundation of China (Grant Nos. 52075248, 51475230, 51875282).

Institutional Review Board Statement: Not applicable.

Informed Consent Statement: Not applicable.

Data Availability Statement: Not applicable.

Conflicts of Interest: The authors declare no conflict of interest.

References

1. Anderson, F.C.; Pandy, M. Storage and utilization of elastic strain energy during jumping. *J. Biomech.* **1993**, *26*, 1413–1427. [CrossRef]
2. Anderson, F.C.; Pandy, M.G. A Dynamic Optimization Solution for Vertical Jumping in Three Dimensions. *Comput. Methods Biomech. Biomed. Eng.* **1999**, *2*, 201–231. [CrossRef] [PubMed]
3. Jung, G.-P.; Casarez, C.S.; Lee, J.-E.; Baek, S.-M.; Yim, S.-J.; Chae, S.-H.; Fearing, R.S.; Cho, K.-J.; Fearing, R.S.; Chae, S.-H. JumpRoACH: A Trajectory-Adjustable Integrated Jumping–Crawling Robot. *IEEE/ASME Trans. Mechatron.* **2019**, *24*, 947–958. [CrossRef]
4. Jung, G.-P.; Casarez, C.S.; Jung, S.-P.; Fearing, R.S.; Cho, K.-J. An integrated jumping-crawling robot using height-adjustable jumping module. In Proceedings of the 2016 IEEE International Conference on Robotics and Automation (ICRA), Stockholm, Sweden, 16–21 May 2016; pp. 4680–4685. [CrossRef]
5. Woodward, M.; Sitti, M. Multimo-bat: A biologically inspired integrated jumping-gliding robot. *Int. J. Robot. Res.* **2014**, *33*, 1511–1529. [CrossRef]
6. Ye, C.; Wang, B.; Wei, B.; Tang, B. Modeling and Analysis of a Jumping Robot with Deforming Wheeled Mechanism. In Proceedings of the 2018 IEEE International Conference on Mechatronics and Automation (ICMA), Changchun, China, 5–8 August 2018; pp. 980–985. [CrossRef]
7. Truong, N.T.; Phan, H.V.; Park, H.C. Design and demonstration of a bio-inspired flapping-wing-assisted jumping robot. *Bioinspiration Biomim.* **2019**, *14*, 036010. [CrossRef] [PubMed]
8. Zhao, J.; Xu, J.; Gao, B.; Xi, N.; Cintron, F.J.; Mutka, M.W.; Xiao, L. MSU Jumper: A Single-Motor-Actuated Miniature Steerable Jumping Robot. *IEEE Trans. Robot.* **2013**, *29*, 602–614. [CrossRef]
9. Zhao, J.; Zhao, T.; Xi, N.; Mutka, M.W.; Xiao, L. MSU Tailbot: Controlling Aerial Maneuver of a Miniature-Tailed Jumping Robot. *IEEE/ASME Trans. Mechatron.* **2015**, *20*, 2903–2914. [CrossRef]
10. Zhao, J.; Yan, W.; Xi, N.; Mutka, M.W.; Xiao, L. A miniature 25 grams running and jumping robot. In Proceedings of the 2014 IEEE International Conference on Robotics and Automation (ICRA), Hong Kong, China, 31 May–7 June 2014; pp. 5115–5120. [CrossRef]
11. Shen, Y.; Ge, W.; Mo, X.; Hou, Z. Design of a locust-inspired miniature jumping robot. In Proceedings of the 2018 IEEE International Conference on Robotics and Biomimetics (ROBIO), Kuala Lumpur, Malaysia, 12–15 December 2018; pp. 2322–2327. [CrossRef]
12. Zhang, J.; Song, G.; Li, Y.; Qiao, G.; Song, A.; Wang, A. A bio-inspired jumping robot: Modeling, simulation, design, and experimental results. *Mechatronics* **2013**, *23*, 1123–1140. [CrossRef]

13. Kovac, M.; Fuchs, M.; Guignard, A.; Zufferey, J.-C.; Floreano, D. A miniature 7g jumping robot. In Proceedings of the 2008 IEEE International Conference on Robotics and Automation, Pasadena, CA, USA, 19–23 May 2008; pp. 373–378.
14. Plecnik, M.M.; Haldane, D.W.; Yim, J.K.; Fearing, R.S. Design Exploration and Kinematic Tuning of a Power Modulating Jumping Monopod. *J. Mech. Robot.* **2016**, *9*, 011009. [CrossRef]
15. Haldane, D.W.; Plecnik, M.M.; Yim, J.K.; Fearing, R.S. Robotic vertical jumping agility via series-elastic power modulation. *Sci. Robot.* **2016**, *1*, eaag2048. [CrossRef] [PubMed]
16. Haldane, D.W.; Yim, J.K.; Fearing, R.S. Repetitive extreme-acceleration (14-g) spatial jumping with Salto-1P. In Proceedings of the 2017 IEEE/RSJ International Conference on Intelligent Robots and Systems (IROS), Vancouver, BC, Canada, 24–28 September 2017; pp. 3345–3351.
17. Zhang, J.; Song, G.; Qiao, G.; Li, Z.; Wang, W.; Song, A. A novel one-motor driven robot that jumps and walks. In Proceedings of the 2013 IEEE International Conference on Robotics and Automation, Karlsruhe, Germany, 6–10 May 2013; pp. 13–19.
18. Chen, G.; Tu, J.; Ti, X.; Hu, H. A Single-legged Robot Inspired by the Jumping Mechanism of Click Beetles and Its Hopping Dynamics Analysis. *J. Bionic Eng.* **2020**, *17*, 1109–1125. [CrossRef]
19. Zhang, Z.; Chang, B.; Zhao, J.; Yang, Q.; Liu, X. Design, Optimization, and Experiment on a Bioinspired Jumping Robot with a Six-Bar Leg Mechanism Based on Jumping Stability. *Math. Probl. Eng.* **2020**, *2020*, 1–23. [CrossRef]
20. Noh, M.; Kim, S.-W.; An, S.; Koh, J.-S.; Cho, K.-J. Flea-Inspired Catapult Mechanism for Miniature Jumping Robots. *IEEE Trans. Robot.* **2012**, *28*, 1007–1018. [CrossRef]
21. Koh, J.-S.; Jung, S.-P.; Noh, M.; Kim, S.-W.; Cho, K.-J. Flea inspired catapult mechanism with active energy storage and release for small scale jumping robot. In Proceedings of the 2013 IEEE International Conference on Robotics and Automation, Karlsruhe, Germany, 6–10 May 2013; pp. 26–31. [CrossRef]
22. Nadein, K.; Betz, O. Jumping mechanisms and performance in beetles II. Weevils (Coleoptera: Curculionidae: Rhamphini). *Arthropod Struct. Dev.* **2018**, *47*, 131–143. [CrossRef] [PubMed]
23. Komarsofla, A.K.; Yazdi, E.A.; Eghtesad, M. Dynamic Modeling and Control of a Novel One-Legged Hopping Robot. *Robotica* **2021**, *39*, 1692–1710. [CrossRef]
24. Ding, Y.; Park, H.-W. Design and experimental implementation of a quasi-direct-drive leg for optimized jumping. In Proceedings of the 2017 IEEE/RSJ International Conference on Intelligent Robots and Systems (IROS), Vancouver, BC, Canada, 24–28 September 2017; pp. 300–305. [CrossRef]
25. Zaitsev, V.; Gvirsman, O.; Ben Hanan, U.; Weiss, A.; Ayali, A.; Kosa, G. A locust-inspired miniature jumping robot. *Bioinspiration Biomim.* **2015**, *10*, 066012. [CrossRef] [PubMed]
26. Beck, A.; Zaitsev, V.; Ben Hanan, U.; Kosa, G.; Ayali, A.; Weiss, A. Jump stabilization and landing control by wing-spreading of a locust-inspired jumper. *Bioinspiration Biomim.* **2017**, *12*, 066006. [CrossRef] [PubMed]
27. Rogers, J.; Page-Bailey, K.; Smith, R. BounceBot: A One-Legged Jumping Robot. In Proceedings of the 19th Annual Conference, TAROS 2018, Bristol, UK, 25–27 July 2018; pp. 52–63. [CrossRef]
28. Kim, M.-J.; Yun, D. Kinematic design and system implementation of jumping robot legs. In Proceedings of the 2017 IEEE International Conference on Multisensor Fusion and Integration for Intelligent Systems (MFI), Daegu, Korea, 16–18 November 2017; pp. 592–596. [CrossRef]
29. Saab, W.; Rone, W.S.; Ben-Tzvi, P. Robotic tails: A state-of-the-art review. *Robotica* **2018**, *36*, 1263–1277. [CrossRef]

machines

MDPI

Article

Fabrication of Flexible Multi-Cavity Bio-Inspired Adhesive Unit Using Laminated Mold Pouring

Linghao Zhang [1,†], Liuwei Wang [1,2,†], Zhiyuan Weng [1], Qingsong Yuan [1], Keju Ji [1] and Zhouyi Wang [1,2,*]

[1] Institute of Bio-inspired Structure and Surface Engineering, Nanjing University of Aeronautics and Astronautics, Nanjing 210016, China; zhanglh@nuaa.edu.cn (L.Z.); wangliuwei@nuaa.edu.cn (L.W.); wengzy@nuaa.edu.cn (Z.W.); yuan-qs@nuaa.edu.cn (Q.Y.); jikeju@nuaa.edu.cn (K.J.)
[2] Nanjing University of Aeronautics and Astronautics Shenzhen Research Institute, Shenzhen 518063, China
* Correspondence: wzyxml@nuaa.edu.cn
† These authors contributed equally to this work.

Abstract: To meet the requirements for the flexible end-effectors of industrial grippers and climbing robots, inspired by the animal attachment mechanism, a bio-inspired adhesive unit (Bio-AU) was designed. Due to its fluid-driven operating characteristics and multi-level adhesive structure, its fabrication and molding is challenging, including the assembly and molding of complex cavities with good pressure-bearing capability, mechanical properties of multi-level materials with variable stiffness, etc. In this study, based on the lamination mold casting process, the "simultaneous molding and assembly" method was established, which can be applied to form and assemble complex cavity parts simultaneously. Moreover, the dovetail tenon-and-mortise parting structures were analyzed and designed. Furthermore, the adhesion between the parting surfaces can be improved using plasma surface treatment technology. By applying the above methods, the assembly accuracy and pressure-bearing capability of the complex flexible cavities are improved, which reduces the individual differences between finished products. Additionally, the maximum pressure-bearing value of the sample was 83 kPa, which is 1.75 times that before optimization. the adhesive structure with different stiffness components was fabricated at low cost using silicon rubber substrates with different properties, which met the requirements of multi-level material with variable stiffness of the Bio-AU. The bending angle of the optimized molding product was about 50.9° at 80 kPa, which is significantly larger than the 24.6° of the lighting-cured product. This indicates that the optimized lamination mold casting process has a strong inclusion of materials, which improves the deformation capacity and self-adaptability of Bio-AUs and overcomes the defects of 3D printing technology in the formation of large, flexible, and controllable-stiffness structures. In this study, the effective fabrication of flexible multilayer adhesive structures was accomplished, and technical support for the development of Bio-AUs was provided, which met the requirements of bionic climbing robots and industrial adhesive grippers for end-effectors.

Keywords: fabrication of flexible multi-cavity; lamination mold pouring process; bio-inspired adhesive unit; variable stiffness of materials

Citation: Zhang, L.; Wang, L.; Weng, Z.; Yuan, Q.; Ji, K.; Wang, Z. Fabrication of Flexible Multi-Cavity Bio-Inspired Adhesive Unit Using Laminated Mold Pouring. *Machines* **2022**, *10*, 184. https://doi.org/10.3390/machines10030184

Academic Editors: Yanjie Wang, Xiaofeng Liu, Aihong Ji, Shichao Niu, Bo Li and Dan Zhang

Received: 27 January 2022
Accepted: 2 March 2022
Published: 3 March 2022

1. Introduction

A flexible end-effector is an irregular flexible structure with a large deformation ability; it is typically composed of flexible materials (such as silicone and rubber). When driven by fluids and electromagnetics, among others, they can be attached to the target with high adaptivity. Therefore, the flexible end effector has unique application prospects in many fields [1–3], such as robot feet and flexible grippers for industrial sorting. In recent years, with the in-depth research on the bio-adhesion mechanism and the development of micro-nano manufacturing technology [4–7], biomimetic dry adhesion technology has been introduced in the field of the attachment of end-effectors. As a result, a fluid-driven flexible end-effector with an adhesive function has been developed [8,9], which can adhere to objects that are extremely difficult to grasp by friction in conventional environments, such as objects with small surface curvature, smoothness, and easy damage, thereby expanding the workable boundary of the flexible end-effector. The performance of flexible end-effectors is affected by a combination of the material properties, cavity structure, and actuation modes [10]. Based on current research on the bio-adhesion mechanism, the animal's flexible adhesion system [11,12] is a multi-layer and cross-scale variable stiffness structure, which achieves good adhesion through the cooperation of various functional components [13]. Inspired by the multi-layer and cross-scale variable stiffness structure, the fluid-driven multi-layer adhesive cavity structure was creatively designed, of which the fabrication and molding are challenging, such as the molding of irregular cavity structures of flexible materials, the assembly and sealing of complex cavities, and the stiffness and toughness of the forming structure must meet the requirements of adhesive deformation.

Existing fabrication technologies for flexible multicavity structures are mainly divided into 3D integrated printing technologies and traditional mold casting technologies. The most used 3D integrated printing technologies, such as stereolithography (SLA) [14], fused deposition modeling (FDM) [15], and poly-jetting [16,17], are limited by the materials used in the process. A structure prepared using the SLA process (light-cured material) can withstand a maximum strain of 40%, whereas that prepared by traditional mold casting (silicone rubber material, PDMS) can withstand more than 600% strain [10]. Similarly, the shore hardness of materials used in FDM, such as Ninja-flex or TPU, is usually 80~90 A, and a large external load is required to achieve the target deformation. The material used by the poly-jetting technology is closer to PDMS than the above two technologies, which has better flexibility and elasticity. Although the poly-jetting technology can realize a mixed and integrated printing of materials with different stiffnesses, the expensive equipment has a low performance-price ratio. Traditional mold casting technologies, which include lamination molding [10,18], lost-wax casting [19], and rotational molding [20], can be used to fabricate simple structural models. Meanwhile, this technology is inexpensive, and the material used is flexible and elastic. However, in view of the fabrication requirements of the Bio-AU, its complex multicavity structure cannot be fabricated integrally by lamination mold casting, and it needs to be poured in layers and bonded using adhesives, which challenge the assembly accuracy and pressure-bearing capability of the complex multi-cavity structure. Lost wax casting technology, which makes the paraffin inner mold first and dissolves the inner mold through a high temperature or a water bath after the pouring is completed, can be used to fabricate the complex multi-cavity structure integrally. However, this method involves complicated processing steps, and it is difficult to completely remove the residual paraffin, which ultimately affects the adhesion performance of the Bio-AU. Compared with other fabrication processes, the lamination mold casting process is highly sensitive to the complexity of the structure, making it difficult to fabricate the cavity integrally using this process, although it is more inclusive of materials and can be applied to a wide range of materials. Therefore, cavity assembly technology can be improved based these considerations, which can improve the fatigue resistance and pressure-bearing capability of the cavity to meet the fabrication requirements of the Bio-AU.

In this paper, owing to the preparation and molding requirements of complex multi-cavity Bio-AUs, the existing processing technologies were compared. It is understood that the existing traditional lamination mold casting cannot guarantee the assembly accuracy and pressure bearing capacity of the cavity, and the 3D printing technology is limited by the base material. Therefore, based on the lamination mold pouring (divided cavity preparation, combined assembly) technology, the cavity assembly process was optimized. The parting surface structure is directly related to the assembly accuracy and structural bearing performance of the multi-layer cavity, thereby affecting the pressure bearing capability and fatigue resistance of the toe, and also affecting the design of the mold. First, the parting surface and mold design of the flexible multi-cavity Bio-AU were carried out, and the structural design of the parting surface of the dovetail tenon and tenon was carried out to improve the fatigue resistance and compression resistance of the cavity. Then, the preparation and molding of the flexible Bio-AU were introduced from the selection and formulation of the substrate and the specific production process. Three process methods were introduced: the lamination mold casting ("divided cavity molding-integrated assembly," DCMA) process; the optimized lamination mold casting ("simultaneous molding and assembly," SMA) process; and the stereolithography (SLA) 3D printing technology. Finally, in order to evaluate the performance of the optimized lamination mold casting process, the bending and pressure-bearing properties of the three types of samples prepared by the three processes were compared and tested, and the application of the Bio-AU was demonstrated.

2. Parting Surface and Mold Design of the Bio-AU with Flexible Multi-Cavity

2.1. Forming Requirements

The design of the soft multi-cavity Bio-AU was inspired by the geckos' cross-layer variable stiffness adhesion system, as well as its reversible adhesion-detachment behavior. It is a flexible multi-cavity structure with a sheet-like bionic flexible skin flap attached to the abdomen. It expands to bend towards the abdomen when driven by positive pressure, enveloping, and adhering to the target. Additionally, it contracts to bend toward the back side when driven by negative pressure, detaching from the target. These behaviors are similar to the adaptive stable adhesion and valgus detachment of geckos. As a type of adhesive unit, the Bio-AU is required to have good intrinsic flexibility to adaptively envelope and adhere to the object, together with a certain rigidity to drive the underlying flap structure and lift the Bio-AU adhesion (Figure 1). Therefore, the preparation and molding of flexible multi-cavities of Bio-AUs with variable stiffness characteristics (complex soft multi-cavity with excellent flexibility, high pressure-bearing capability, and significant differences in stiffness and deformation at each layer) is the basis of their function, which is a significant challenge for future research on bionic climbing robots. To ensure that the multi-cavity structure has the required compression and deformation capabilities, dovetail tenon and tenon parting surface structures were designed to enhance the bonding performance and the pressure-bearing capability of the Bio-AU. The mold design and specific fabrication process of the Bio-AU were established to realize the fabrication and molding of the AUs' complex soft multi-cavities.

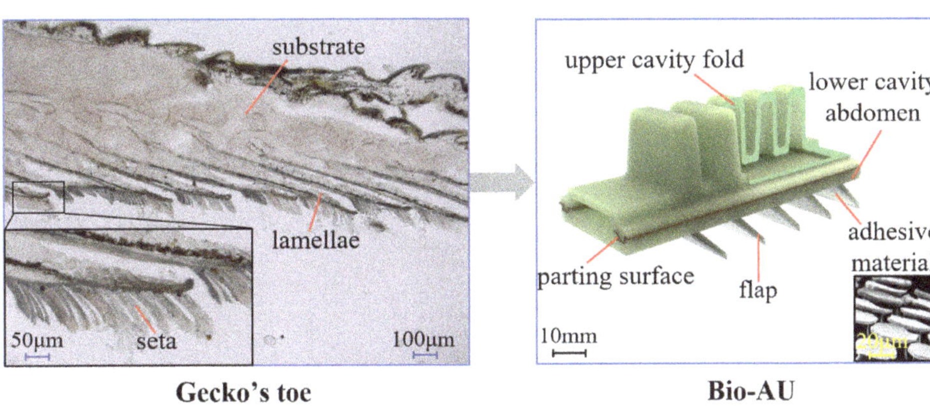

Figure 1. Design inspiration and structure of bio-inspired adhesive unit (Bio-AU).

2.2. Parting Surface Design

Based on lamination mold casting technology, the Bio-AU was fabricated in separate cavities and assembled together. The structure of the parting surface is directly related to the assembly accuracy of the multi-cavity, thereby affecting the compression and fatigue resistance of the cavity. Inspired by the traditional stable tenon-and-mortise structure, a dovetail tenon-and-mortise parting surface is designed.

First, the influence of different parting surfaces on the mechanical properties of interfacial bonding was studied. The failure mode of the cavity expansion to the parting surface is usually vertical interfacial separation. When the inner cavity of the Bio-AU was loaded, the effect on the parting surface could be simplified by applying a vertical interfacial separation force on the parting surface. To simulate the above process, a simplified two-dimensional finite element model of the parting surface was established in Abaqus (SIMULIA, Dassault System, Providence, RI, USA), as shown in Figure 2b. The lower parting surface was fixed, while the upper parting surface was displaced along the vertical direction until it was completely separated. The interfacial separation force and avulsion energy (See the Supplementary Materials for measurement methods of both), which were generated throughout the separation process, were extracted and used as evaluation criteria for the stable performance of the parting surface structure. The Mooney-Rivlin model was used to describe the silicone rubber material used for the fabrication of the Bio-AUs. The cohesive force model (bilinear constitutive model) was introduced to describe the adhesive connection and failure behavior between the two interfaces. The maximum nominal stress criterion was selected as the evaluation criterion for damage initiation, in which the maximum stress value was set to 0.15 MPa, and the maximum avulsion displacement was set to 5×10^{-4} mm. Finite element models of three different parting surface structures (plane, right-angle tenon-and-mortise (RTM), and dovetail tenon-and-mortise (DTM)) are shown in Figure 2a,b. Under the same displacement, the failure states of the interfaces were analyzed and compared. Figure 2c shows that the maximum interfacial separation force required for the three parting surface structures is approximately the same, whereas the DTM structure with the mechanical locking feature results in the rebound of the interfacial separation force during the damage failure stage. Meanwhile, Figure 2d shows that the average energy required for the complete separation of the DTM parting surface structure is the largest.

To optimize the pressure-bearing capability of the DTM parting surface, four structural parameters were studied: the length of the short side of the tenon W, the height of the vertical tenon H_t, the length of the long side of the tenon W_t and the fixed overall length H (Figure 2b) (three levels of W, H_t and W_t were set as shown in Table S1 of the Supplementary Materials). The orthogonal experimental method was used to extract representative combinations for comparison (Table 1). As shown in Figure 3a,b, the maximum interfacial separation forces of the nine parting surfaces are approximately the same. However, the best performance was from the test No. III parting surface, where its interfacial separation force has an obvious rebound during the damage stage, and its avulsion energy reached 0.43 mJ, which is nearly 1.6 times that of the test No. VIII parting surface (Figure 3c). Therefore, the test no. III parting surface was selected for fabricating the Bio-AU cavity.

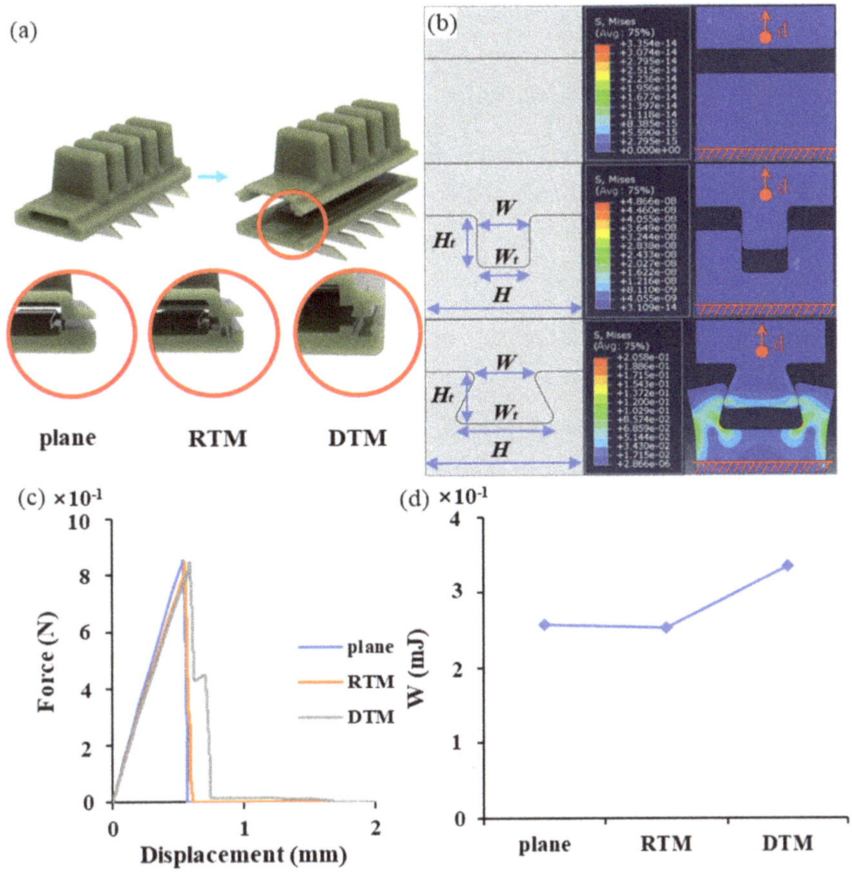

Figure 2. Finite element analysis and corresponding results of plane models for three different parting structures (plane, right-angle tenon-and-mortise (RTM), and dovetail tenon-and-mortise (DTM)): (**a**) diagram of three parting structures; (**b**) finite element analysis of plane models for three parting structures under the same displacement (0.6 mm); (**c**) interfacial separation force; and (**d**) avulsion energy of three parting structures generated during interface separation.

Table 1. Orthogonal experiment method design.

Test Number	Factors			Level Combination
	W	H_t	W_t	
I	1	1	1	$W_1H_{t1}W_{t1}$
II	1	2	2	$W_1H_{t2}W_{t2}$
III	1	3	3	$W_1H_{t3}W_{t3}$
IV	2	1	2	$W_2H_{t1}W_{t2}$
V	2	2	3	$W_2H_{t2}W_{t3}$
VI	2	3	1	$W_2H_{t3}W_{t1}$
VII	3	1	3	$W_3H_{t1}W_{t3}$
VIII	3	2	1	$W_3H_{t2}W_{t1}$
IX	3	3	2	$W_3H_{t3}W_{t2}$

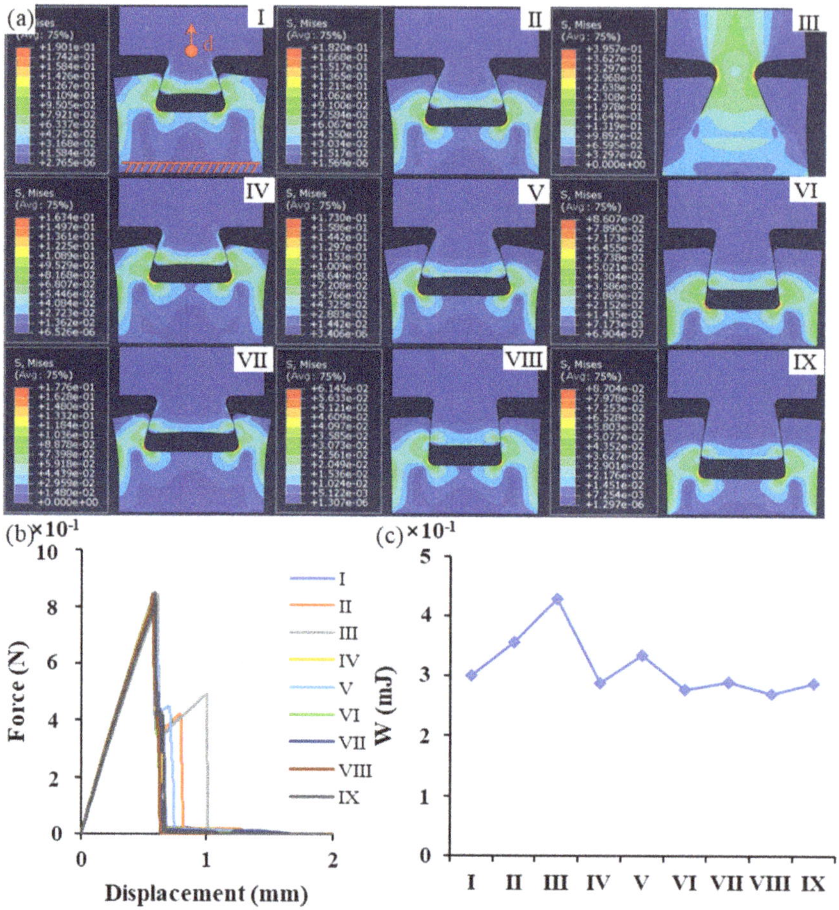

Figure 3. Finite element analysis and corresponding results for plane models of nine DTM parting structures: (**a**) finite element analysis of plane models for different DTM parting structures under the same displacement (0.6 mm); (**b**) interfacial separation force of nine DTM parting structures generated during interface separation; and (**c**) avulsion energy of nine DTM parting structures generated during interface separation.

2.3. Mold Design

Based on the lamination mold casting technology, the Bio-AU is divided into three parts: upper cavity fold, lower cavity abdomen, and skin flap. These were independently fabricated and assembled. The upper cavity fold was combined with the lower cavity abdomen through the DTM parting surface. Due to the Bio-AU with complex structure, the following challenges will be encountered during the fabrication process: (a) efficient filling of silicone during the pouring of the complex structures: such as thorough release of air bubbles, rapid replenishment, etc. (b) effective fabrication and rapid demolding of large slenderness ratio structures (such as thin flap layers).

Regarding the above challenges, the following solutions were proposed:

1. Arranging the position and number of vents and feeding grooves reasonably: To avoid the complex multi-cavity structure, the vents were mostly located at the joint surface. In addition, the vents could be used as injection ports and exhaust ports simultaneously. Garbage chutes were set around the joint surface, which could

accelerate the release of air bubbles and the replenishment of the mold from all around (Figure 4a).

2. Using tapered pins and screws to locate and lock between male and corresponding female molds. Taper pins are precisely located and easy to separate, while screws can not only ensure the stability of mold clamping, but also facilitate disassembly.
3. To take out the finished product from the mold smoothly, a certain draft angle was set at the position of complex structures such as cavity folds. And notches were designed on each mold to separate the male and female molds using tools.
4. Large slenderness ratio structures (such as thin flap layers) are difficult to demold due to their different axial strengths but uniform radial stiffness. In order to ensure the quality of finished products, the molds with a radial demolding direction were designed, which were symmetrically divided along the axis. In this way, the axial tearing is avoided during demolding (see Figure 4c).

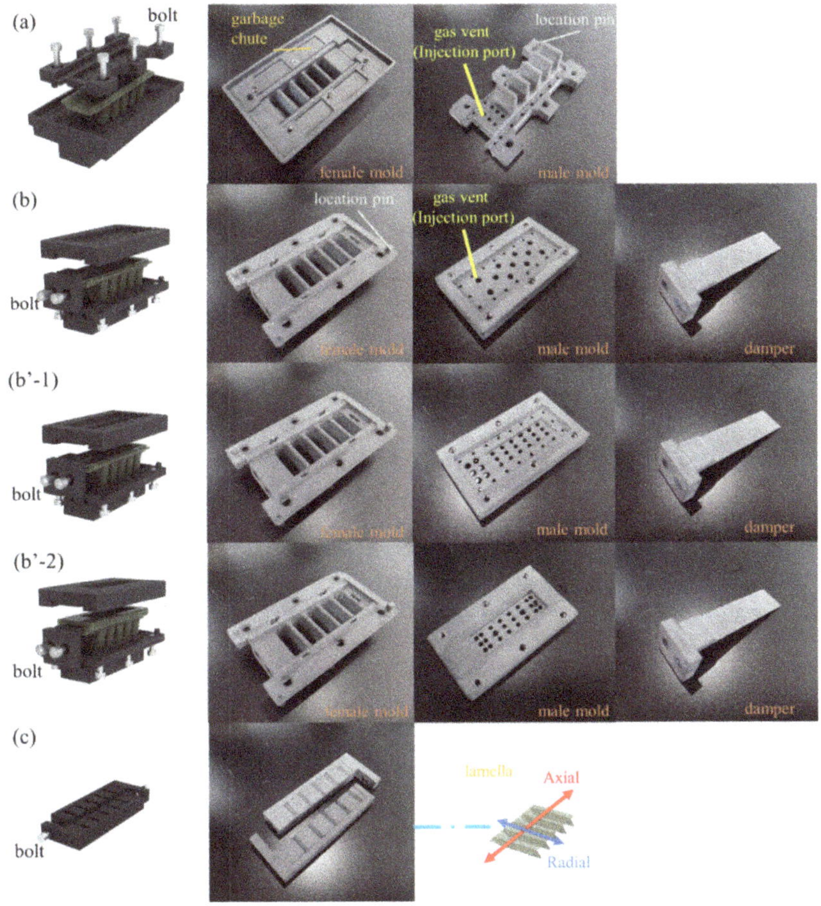

Figure 4. Exploded and physical diagrams of mold: (**a**) upper cavity fold molds; (**b**) lower cavity abdomen molds (unembedded objects); (**b'-1**) first set of lower cavity abdomen molds (embedded objects); (**b'-2**) second set of lower cavity abdomen molds; and (**c**) flap molds.

The above molds are all laser sintered by Shenzhen Future Factory using 7500 high-performance nylon material, which has the characteristics of high temperature resistance, high strength, high surface quality, and easy demolding (See the Supplementary Materials for calculations related to mold design).

3. Fabrication and Molding of Bio-AUs

3.1. Selection and Formulation of Variable-Stiffness Substrates for Bio-AUs

The Bio-AU has requirements of high deformation and outstanding adaptability, which presents challenges to the fabrication process. When driven by pressure, the upper cavity fold should be easily deformed, while the lower cavity abdomen should maintain a certain stiffness and toughness to drive the underlying flap structure. Furthermore, the flap should also maintain a certain stiffness for rapid detachment. Therefore, the Bio-AU should be a multi-layer structure with variable stiffness, which means that the selection and formulation of materials are determined by the requirements in stiffness and deformation of Bio-AUs' each layer.

Polydimethylsiloxane (PDMS) is a commonly used flexible silicone rubber material. It is known that its elastic modulus ranges from 0.83 to 1.23 MPa under different curing conditions [21], which is difficult to meet the Bio-AU's requirements of structural variable stiffness. Studies [22] have shown that the elastic moduli of materials can be significantly improved by adding second phase particles (such as SiO_2, Fe_3O_4, etc.) to PDMS. Meanwhile, the addition of Ecoflex material, which belongs to the same type of silicone but has a lower elastic modulus and higher toughness than PDMS, can make the Bio-AU maintain high flexibility as well as a certain toughness [23]. Overall, PDMS (Dow Corning 184) was selected as the base material, while the Ecoflex 0030 series (Smooth-On Company) and the nano-silica fume SiO_2 (Maclean Company) were used as the auxiliary materials. Finally, component A and B materials were formulated by using the above materials (see Table 2).

Table 2. Flexible gripper material composition.

	PDMS	Hardener	SiO_2	Ecoflex-A	Ecoflex-B
Component A	50%	5%	5%	20%	20%
Component B	66.7%	6.7%	6.7%	10%	10%
Component C	carbon fiber board T700.				

According to the standard for dumbbell tensile samples specified in D412-98a, the elastic modulus of component A and B materials were studied. Four dumbbell tensile test samples were respectively cast for each component materials using a mold (Figure 5a). At the experimental temperature of 24 °C, the tensile testing machine (Figure 5b) was controlled to stretch at a constant rate of 500 $\pm$ 50 mm/min. The results show that the elastic moduli of component A and B materials are respectively 0.9 MPa and 1.74 MPa (Figure 5c), which illustrates that flexible materials with different stiffness can be formulated by using auxiliary materials. Therefore, in order to meet the mechanical performance requirements of the Bio-AU, the upper cavity fold was endowed with component A material, while the lower cavity abdomen and skin flap structure were endowed with component B material.

Additionally, the stiffness adjustment technology of the flexible cavity (embedded objects) was also studied, which can meet the variable stiffness requirements of the Bio-AU with multi-layer structure under different working conditions. The component C material in Table 2 is T700 Carbon Fiber Board (elastic modulus $\geq$ 200 GPa; tensile strength $\geq$ 3500 MPa), whose stiffness is much higher than that of both component A and B materials. The local stiffness can be significantly increased by cutting carbon fiber plates into sheets and evenly embedding them in the flexible cavity. And the stiffness can be adjusted within a certain range by changing the distribution density of the carbon fiber sheets. Therefore, when the technology is applied to the lower cavity abdomen, its strength can be significantly improved, thereby ensuring the adaptability of the Bio-AU and improving its adhesion.

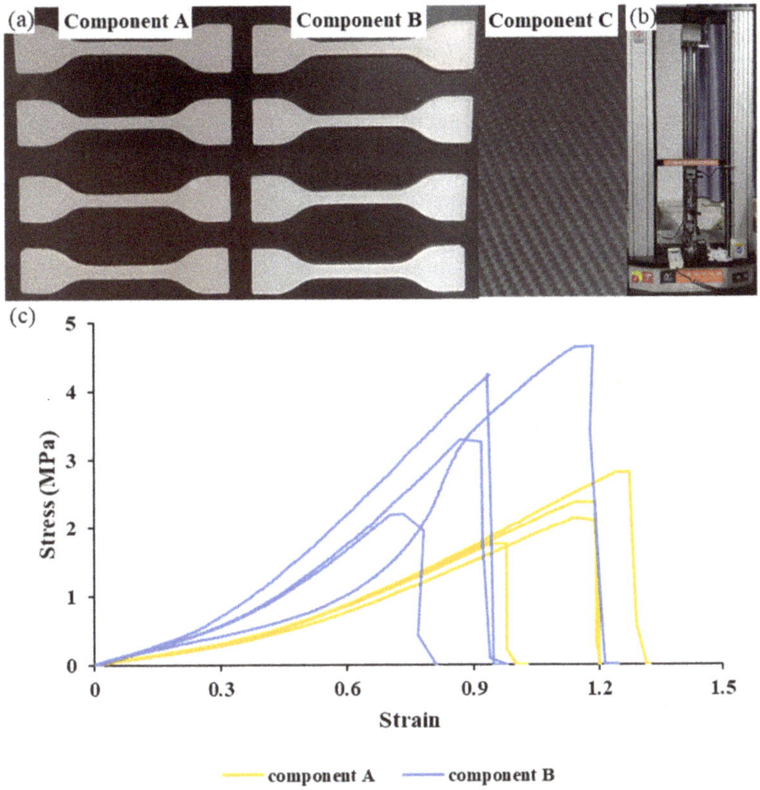

Figure 5. Tensile test for elastic moduli of materials: (**a**) three component materials; (**b**) tensile testing machine; and (**c**) results of tensile test.

3.2. Fabrication Process

The lamination mold casting usually adopts the process of "divided cavity molding-integrated assembly," i.e., after the flexible cavities are poured independently in layers, the cavity is assembled using the adhesive (silicon with the same properties). This type of process has low assembly accuracy and uneven bonding, resulting in cavities with large differences in individual performance and low pressure-bearing capability, which cannot meet the requirements of bonding and assembly. For this reason, the lamination mold casting process was optimized by adopting the process of "simultaneous molding and assembly" (Figure 6b). In this process, the upper cavity is fabricated in advance, and the assembly of the upper and lower cavities is accomplished simultaneously during the curing process of the lower cavity. This process can ensure the assembly accuracy of cavities, and the adhesion between the parting surfaces is enhanced using plasma surface oxidation treatment technology, which improves the pressure-bearing capability of the Bio-AU.

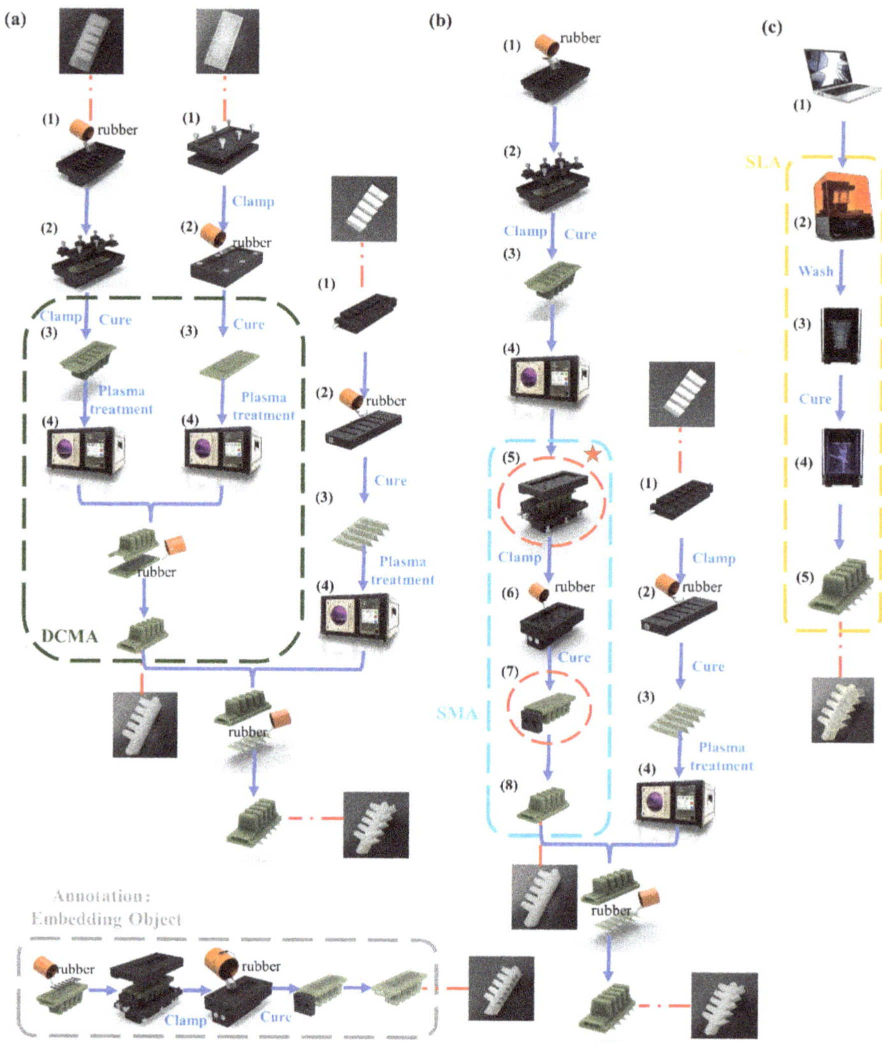

Figure 6. Three fabrication processes: (**a**) lamination mold casting ("divided cavity molding-integrated assembly," DCMA); (**b**) optimized lamination mold casting ("simultaneous molding and assembly," SMA); and (**c**) stereolithography (SLA) 3D printing technology.

For the purpose of reflecting the characteristics of the proposed "simultaneous molding and assembly" process, three processing methods (the DCMA process (Figure 6a), SMA process (Figure 6b), and SLA technology (Figure 6c) were introduced and compared.

1. DCMA: The Bio-AU is divided into three parts: upper cavity fold, lower cavity abdomen, and skin flap. These were independently fabricated and assembled. Subsequently, the cavity and the skin flap were assembled sequentially using the adhesive (the same flexible material). The process is as follows:

Divided cavity molding: (1) Spraying release agent on the surface of upper cavity fold/lower cavity abdomen/skin flap mold respectively, and pouring silicone rubber material into the corresponding female mold; (2) Clamping the mold and removing the bubbles from the mold (4~6 h in the environment of vacuum degree $\geq$ 100 kPa); (3) Oven heating and curing (100 °C for 50 min), followed by demolding and taking out the finished product; (4) Putting the finished product into the plasma surface treatment machine for surface oxidation treatment;

Integrated assembly: Cavity and flap assembly are performed sequentially. During the assembly process, the adhesive (component B in Figure 5) should be evenly applied to the cavity parting surface, while the flap is bonded to the lower surface of the cavity. Before the heat curing step (30 min at 100 °C), each assembled part needs to be pressed tightly. The release agent material used in the process is Release 200 spray release agent (Smooth-On Company). In addition, studies have confirmed that the bonding effect between PDMS treated with oxygen plasma will be improved [24].

2. SMA: Steps (1) to (4) are consistent with the DCMA, which involves making the upper cavity fold finished product; (5) Spraying the release agent on the surface of the lower cavity mold and the baffle, and installing the upper cavity in the female mold of the lower cavity, followed by the baffle and the lower cavity mold, and finally locking the molds; (6) pouring silicone rubber into the lower cavity mold and releasing air bubbles from the mold (4–6 h in an environment of vacuum degree $\geq$ 100 kPa); (7) oven heating and curing (100 °C for 50 min), followed by demolding and removing the finished cavity with a baffle plate; (8) pulling out the baffle to obtain the final cavity product. In addition, the fabrication process of the flap was the same as that for the DCMA. The flap was bonded to the lower surface of the cavity using an adhesive (component B in Figure 5).

The core steps of this process are Steps (5)–(8). Specifically, the upper cavity folded product is embedded in the lower cavity abdominal mold, and the baffle plate (The molds are shown in Figure 4b) prevents the flexible material from blocking the upper cavity fold. This process ensured that the assembly of the upper and lower cavities was accomplished simultaneously during the curing process of the lower cavity. However, because of the difficulty in demolding the flap with a complex and fragile structure, it is difficult to apply the SMA process to fabricate the flap. Therefore, a flexible adhesive was used to assemble the cavity and skin flap. Furthermore, the variable stiffness requirements of the Bio-AU with the multi-layer structure under different working conditions can be satisfied by combining the stiffness adjustment technology (embedded objects) and the SMA process, as shown in the lower left corner of Figure 6 (The molds are shown in Figure 4(b'-1,b'-2)).

3. SLA: By applying this technology, the Bio-AU can be integral fabricated without dividing it separately. The process is as follows: (1) importing the 3D model and printing; (2) taking out the finished product to be processed; (3) cleaning the finished product to be processed; (4) processing the finished product support and performing secondary curing; (5) taking out the final product.

In conclusion, when the SMA process is applied, the cavity can be fabricated without using the adhesive. Additionally, based on the plasma surface treatment technology, the assembly accuracy of the cavity is ensured, and the pressure-bearing capability of the cavity structure is improved. Although the steps of light-curing integrated molding technology are simple, the materials used are not as good as traditional silicone rubber materials in terms of strain capability and tensile strength. According to the deformation requirements of the Bio-AU at different layers, the lamination mold pouring process can formulate materials with different elastic moduli at a low cost, while the mixed-material 3D printer equipment is expensive and cost-effective. Therefore, when facing the formation of irregular cavity structures based on flexible materials, the fabrication and assembly of complex cavities, and the demand for flexible adhesive cavities with variable stiffness and high toughness, the optimized lamination mold casting process proposed in this study can provide a feasible solution.

4. Performance Tests and Results

In this section, the performance of the finished samples fabricated using the above three processes were studied and compared respectively, which can test the feasibility of the DCMA process. It includes the bending performance (bending angle α) and pressure-bearing capability (pressure-bearing range) of the Bio-AU under different air pressure values. The definition of the pressure-bearing range is as follows: when the positive pressure is broken, the air pressure value of the Bio-AU is the maximum positive pressure value; when the Bio-AU is upturned under negative pressure and the deformation is stable, the initial air pressure value of the Bio-AU is the maximum negative pressure value. In this study, the machine with SLA technology is Form3 (Formlabs, Boston, USA). The material used is flexible resin elastic 50 A, the elastic modulus of which is about 2.89 Mpa, and its hardness after secondary curing is 50 A. The weights of the finished Bio-AU samples fabricated by the three processes above were measured to be 26.9 (±0.53) g, 26.7 (±0.31) g and 27.9 (±0.16) g, respectively. The performance test platform includes the pneumatic drive system (Figure 7), the Bio-AU fixture and the image acquisition device (See the Supplementary Materials for instructions for use of the pneumatic drive system).

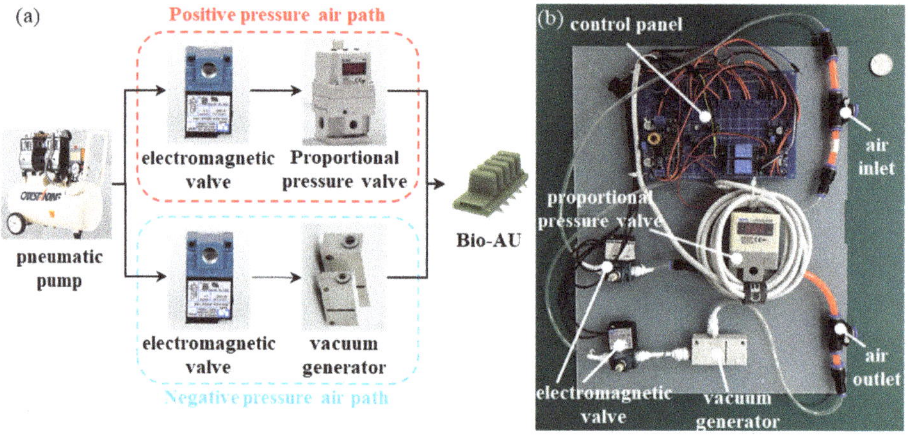

Figure 7. Pneumatic drive system: (**a**) functional block diagram; and (**b**) physical connection diagram.

4.1. Comparison of the Bending Performance and Rupture Failure

Figure 8 shows the bending and pressure-bearing properties of the Bio-AU samples fabricated using the above three processes. The angle between the line connecting the first and last points of the Bio-AU and the horizontal line was defined as the bending angle α of the Bio-AU (Figure 8a). As shown in Figure 8a, because both the DCMA and SMA processes belong to the lamination mold casting process, the bending angles of the respective samples are similar under the same pressure, and the rupture failure of the respective samples occurs at the parting surface. The size of the rupture opening of the DCMA sample is significantly larger than that of the SMA sample, which illustrates that the cavity fabricated by the SMA process has a higher assembly accuracy and pressure-bearing capability. When the driven pressure was 80 kPa, the bending angle of the SMA sample was approximately 50.9°, which was significantly larger than that of the SLA sample (24.6°), indicating that the bending performance of the SMA sample was better than that of the SLA sample under the same pressure.

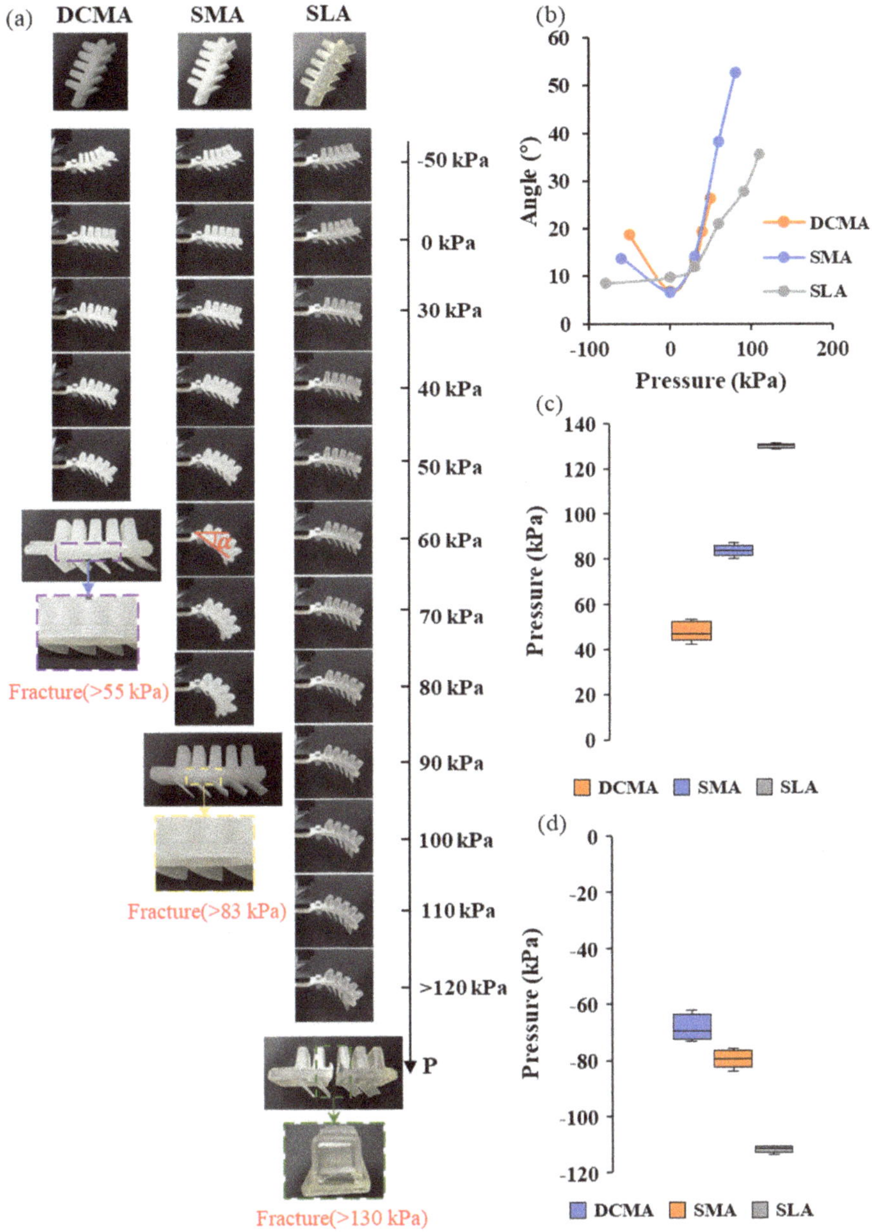

Figure 8. Respective performances of Bio-AU samples fabricated via three processes: (**a**) comparison of bending performance and rupture failure of Bio-AU samples; (**b**) bending performance of Bio-AU samples; (**c**) box diagram of positive pressure-bearing ranges of Bio-AU samples; and (**d**) box diagram of negative pressure-bearing ranges of Bio-AU samples.

4.2. Comparison of the Pressure-Bearing Capability

The pressure-bearing capability is another important performance index of the Bio-AU. Therefore, the pressure-bearing capabilities of the samples fabricated using these three processes were studied. In this way, five samples were made by using each process, leading to a total of 15 samples, whose pressure-bearing ranges were tested. Specifically, the average pressure-bearing ranges of the DCMA, SMA, and SLA samples were −68.24 to 47.98 kPa, −79.44 to 83.72 kPa, and −111.73 to 130.11 kPa, respectively. Because of the inherent advantage of the one-piece molding process, the pressure-bearing range of the SLA sample was better than that of the DCMA and SMA samples. Meanwhile, the SMA process effectively increased the pressure-bearing capability of the sample by simultaneously forming and assembling the cavity. The results show that the maximum pressure bearing of the SMA sample is nearly 1.74 times that of the DCMA sample, which narrows the gap between the SMA and SLA samples. The results in Figure 8c,d show that the individual differences between the SMA and SLA samples are small. Because the cavity of the DCMA sample is assembled by adhesive bonding, the DCMA samples have low assembly accuracy, uneven bonding, and large individual performance differences.

4.3. Application

To verify the rationality of both the structure design and the adhesion performance of the Bio-AU, the adhesion-locking mechanism with the Bio-AU was used to carry out the adhesive grasping experiment on targets with different structures (Figure 9). As shown in Figure 9, for objects with irregular shapes and rigid or flexible objects, the adhesion-locking mechanism not only can achieve stable adhesion by self-retracting and self-locking, but also accomplish rapid detachment by extension. The results show that the Bio-AU has good surface adaptability. Therefore, the new adhesion operation mode with the good surface adaptability and excellent adhesion capability can effectively reduce the damage to the target while working stably.

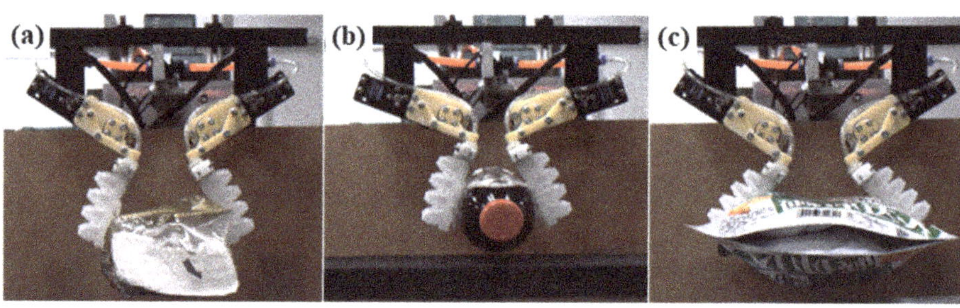

Figure 9. Application of the Bio-AU: (**a**) flexible hand-pulled paper bag weighing 54 g; (**b**) coke weighing 482 g; and (**c**) snacks weighing 200 g.

5. Conclusions

Owing to difficulties in the fabrication and molding of fluid-driven multi-layer variable-stiffness adhesive structures, a lamination mold casting process using "simultaneous molding and assembly" method was proposed, which can be applied to form and assemble complex cavity parts simultaneously. Compared with traditional lamination mold casting process, this process can solve the dependence of cavity structure assembly on adhesive performance, resulting in the improvement of the assembly accuracy of the complex flexible cavities. Meanwhile, the fatigue resistance and pressure-bearing capability of the complex flexible cavities can be improved by the application of the dovetail tenon-and-mortise parting structures combined with the plasma surface treatment technology. The results of pressure-bearing tests show that the individual differences between finished products prepared by optimized process are small, and the maximum pressure-bearing value of the sample was 83 kPa, which is 1.75 times that before optimization. Additionally, the Bio-AU was fabricated using silicon rubber substrates with different properties, whose elastic modulus of the upper cavity fold was 0.9 MPa, and that of the lower cavity abdomen and flaps was 1.74 MPa, which met the requirements of multi-level material with variable stiffness of the Bio-AU. Compared with 3D printing, the optimized lamination mold casting process has a strong inclusion of materials, which improves the deformation capacity and self-adaptability of Bio-AUs and overcomes the defects of 3D printing technology in the formation of large, flexible, and controllable-stiffness structures. The results of bending performance tests show that the bending angle of the optimized molding product was about 50.9° at 80 kPa, which is significantly larger than the 24.6° of the light cured product.

In this study, the effective fabrication of flexible multilayer adhesive structures was accomplished, and technical support for the development of the Bio-AU was provided, which meets the requirements of bionic climbing robots and industrial adhesive grippers for end-effectors. However, owing to the challenge of miniaturization of the cavity structure with variable stiffness of complex multi-level materials, it will involve the processing and forming of the mold, the assembly of the cavity and the efficient demolding of the finished product, etc., and it is necessary to optimize or explore a new type of lamination mold casting process.

Supplementary Materials: The following supporting information can be downloaded at: https://www.mdpi.com/article/10.3390/machines10030184/s1, Figure S1: Bilinear Constitutive Model, Table S1 Three factors and their corresponding levels.

Author Contributions: L.Z. and L.W. made equal contributions to this study; Z.W. (Zhouyi Wang) and L.Z. conceived the ideas and designed the methodology; Z.W. (Zhouyi Wang) and K.J. provided guidance; L.Z. and L.W. carried out the lab experiment and completed simulation analysis; L.Z. and L.W. wrote manuscript; Z.W. (Zhiyuan Weng) and Q.Y. collected and analyzed the data. All authors have read and agreed to the published version of the manuscript.

Funding: This work was supported by the National Key R&D program of China (2019YFB1309600), National Natural Science Foundation of China (Grant No. 51975283 to Zhouyi Wang and 52075249 to Keju Ji) and Basic Research Program of ShenZhen (JCYJ20210324122810033).

Data Availability Statement: The data presented in this study are available on request from the corresponding author. The data are not publicly available due to [the unpubulished work about the design of the Bio-AUs].

Acknowledgments: We thank Xiao Yin for participating in the experiments. We also appreciate Yi Song and Jiwei Yuan for modifying the manuscript.

Conflicts of Interest: The authors declare that they have no competing interests.

Abbreviations

Bio-AU: bio-inspired adhesive unit; SLA: stereolithography, FDM: fused deposition modeling; DCMA: divided cavity molding-integrated assembly process; SMA: simultaneous molding and assembly process; RTM: right-angle tenon-and-mortise; DTM: dovetail tenon-and-mortise; PDMS: Polydimethylsiloxane.

References

1. Jing, Z.; Qiao, L.; Pan, H.; Yang, Y.; Chen, W. An Overview of the Configuration and Manipulation of Soft Robotics for On-Orbit Servicing. *Sci. China Inf. Sci.* **2017**, *5*, 1–19. [CrossRef]
2. Cianchetti, M.; Laschi, C.; Menciassi, A.; Dario, P. Biomedical Applications of Soft Robotics. *Nat. Rev. Mater.* **2018**, *3*, 143–153. [CrossRef]
3. Zhang, B.; Xie, Y.; Zhou, J.; Wang, K.; Zhang, Z. State-Of-The-Art Robotic Grippers, Grasping and Control Strategies, as Well as their Applications in Agricultural Robots: A Review. *Comput. Electron. Agric.* **2020**, *177*, 105694. [CrossRef]
4. Wang, Z.Z.; Xu, Y.; Gu, P. Adhesive Behaviour of Gecko-Inspired Nanofibrillar Arrays: Combination of Experiments and Finite Element Modelling. *J. Phys. D Appl. Phys.* **2012**, *45*, 14. [CrossRef]
5. Gorb, S.; Varenberg, M.; Peressadko, A.; Tuma, J. Biomimetic Mushroom-Shaped Fibrillar Adhesive Microstructure. *J. R. Soc. Interface* **2007**, *4*, 271. [CrossRef] [PubMed]
6. Gorb, S.N.; Varenberg, M. Mushroom-Shaped Geometry of Contact Elements in Biological Adhesive Systems. *J. Adhes. Sci. Technol.* **2007**, *21*, 1175–1183. [CrossRef]
7. Sitti, M.; Fearing, R.S. Nanomolding Based Fabrication of Synthetic Gecko Foot-Hairs. In Proceedings of the 2002 2nd IEEE Conference on Nanotechnology, Washington, DC, USA, 28–28 August 2002.

8. Glick, P.; Suresh, S.A.; Ruffatto, D.; Cutkosky, M.; Tolley, M.T.; Parness, A. A Soft Robotic Gripper with Gecko-Inspired Adhesive. *IEEE Robot. Autom. Lett.* **2018**, *3*, 903910. [CrossRef]
9. Hao, Y.; Biswas, S.; Hawkes, E.W.; Wang, T.; Zhu, M.; Wen, L.; Visell, Y. A Multimodal, Enveloping Soft Gripper: Shape Conformation, Bioinspired Adhesion, and Expansion-Driven Suction. *IEEE Trans. Robot.* **2020**, *37*, 350–362. [CrossRef]
10. Mosadegh, B.; Polygerinos, P.; Keplinger, C.; Wennstedt, S.; Shepherd, R.F.; Gupta, U.; Shim, J.; Bertoldi, K.; Walsh, C.J.; Whitesides, G.M. Soft Robotics: Pneumatic Networks for Soft Robotics that Actuate Rapidly (Adv. Funct. Mater. 15/2014). *Adv. Funct. Mater.* **2014**, *24*, 2109. [CrossRef]
11. Autumn, K.; Sitti, M.; Liang, Y.A. Evidence for Van Der Waals Adhesion in Gecko Setae. *Proc. Natl. Acad. Sci. USA* **2002**, *99*, 12252–12256. [CrossRef] [PubMed]
12. Autumn, K.; Gravish, N. Gecko Adhesion: Evolutionary Nanotechnology. *Philos. Trans. R. Soc. A Math. Phys. Eng. Sci.* **2008**, *366*, 1575–1590. [CrossRef] [PubMed]
13. Tian, Y.; Wan, J.; Pesika, N.; Zhou, M. Bridging Nanocontacts to Macroscale Gecko Adhesion by Sliding Soft Lamellar Skin Supported Setal Array. *Sci. Rep.* **2013**, *3*, 1382. [CrossRef] [PubMed]
14. Peele, B.N.; Wallin, T.J.; Zhao, H.; Shepherd, R.F. 3D Printing Antagonistic Systems of Artificial Muscle Using Projection Stereolithography. *Bioinspiration Biomim.* **2015**, *10*, 55003. [CrossRef] [PubMed]
15. Yap, H.K.; Ng, H.Y.; Yeow, C.H. High-Force Soft Printable Pneumatics for Soft Robotic Applications. *Soft Robot.* **2016**, *3*, 144–158. [CrossRef]
16. Drotman, D.; Jadhav, S.; Karimi, M.; de Zonia, P.; Tolley, M.T. 3D Printed Soft Actuators for a Legged Robot Capable of Navigating Unstructured Terrain. In Proceedings of the 2017 IEEE International Conference on Robotics and Automation (ICRA), Singapore, 29 May–3 June 2017.
17. Yirmibesoglu, O.D.; Morrow, J.; Walker, S.; Gosrich, W.; Cañizares, R.; Kim, H.; Daalkhaijav, U.; Fleming, C.; Branyan, C.; Menguc, Y. Direct 3D Printing of Silicone Elastomer Soft Robots and their Performance Comparison with Molded Counterparts. In Proceedings of the 2018 IEEE International Conference on Soft Robotics (RoboSoft), Livorno, Italy, 24–28 April 2018.
18. Wang, Z.; Or, K.; Hirai, S. A Dual-Mode Soft Gripper for Food Packaging. *Robot. Auton. Syst.* **2020**, *125*, 103427. [CrossRef]
19. Marchese, A.D.; Katzschmann, R.K.; Rus, D. A Recipe for Soft Fluidic Elastomer Robots. *Soft Robot.* **2015**, *2*, 7–25. [CrossRef] [PubMed]
20. Zhao, H.; Li, Y.; Elsamadisi, A.; Shepherd, R. Scalable Manufacturing of High Force Wearable Soft Actuators. *Extrem. Mech. Lett.* **2015**, *3*, 89–104. [CrossRef]
21. Boesel, L.F.; Greiner, C.; Arzt, E.; Del Campo, A. Gecko-Inspired Surfaces: A Path to Strong and Reversible Dry Adhesives. *Adv. Mater.* **2010**, *22*, 2125–2137. [CrossRef] [PubMed]
22. Tian, Y. *Design and Friction Adhesion Properties of PDMS Gecko Foot Biomimetic Composites*; Harbin Institute of Technology: Harbin, China, 2016.
23. Martinez, R.V.; Branch, J.L.; Fish, C.R.; Jin, L.; Shepherd, R.F.; Nunes, R.M.; Suo, Z.; Whitesides, G.M. Whitesides. Robotic Tentacles with Three-Dimensional Mobility Based on Flexible Elastomers. *Adv. Mater.* **2013**, *25*, 205–212. [CrossRef] [PubMed]
24. Xu, Q.; Dai, B.; Jiao, Z.; Hong, R.; Yang, Z.; Zhang, D.; Zhuang, S. Fabrication of Large Micro-Structured High-Numerical-Aperture Optofluidic Compound Eyes with Tunable Angle of View. *Opt. Express* **2018**, *26*, 33356–33365. [CrossRef] [PubMed]

machines

MDPI

Communication

Application of a Bio-Inspired Algorithm in the Process Parameter Optimization of Laser Cladding

Yingtao Zhang [1,†], Benxiang Gong [1], Zirong Tang [1] and Weidong Cao [2,*,†]

1 College of Mechanical and Electrical Engineering, Hohai University, Changzhou 213022, China; zhangyt@hhu.edu.cn (Y.Z.); gongbx@hhu.edu.cn (B.G.); tang@hhu.edu.cn (Z.T.)
2 College of Internet of Things Engineering, Hohai University, Changzhou 213022, China
* Correspondence: cwd2018@hhu.edu.cn
† These authors contributed equally to this work.

Abstract: The process parameter optimization of laser cladding using a bio-inspired algorithm is a hot issue and attracts the attention of many scholars. The biggest difficulty, at present, is the lack of accurate information regarding the function relationship between objectives and process parameters. In this study, a novel process parameter optimization approach for laser cladding is proposed based on a multiobjective slime mould algorithm (MOSMA) and support vector regression (SVR). In particular, SVR is used as a bridge between target and process parameters for solving the problem of lacking accurate information regarding the function relationship. As a new metaheuristic algorithm, MOSMA is to obtain the Pareto solution sets and fronts. The Pareto solution sets are optimized process parameters, and the Pareto fronts are optimized objectives. Users can select the corresponding optimized process parameters according to their needs for the target. The performance of the proposed approach was evaluated by the TOPSIS method, based on actual laser cladding data and compared with several well known approaches. The results indicate that the optimal process parameters obtained by the proposed approach have better process performance.

Keywords: bio-inspired algorithm; laser cladding; process parameter optimization; multiobjective slime mould algorithm; support vector regression

Citation: Zhang, Y.; Gong, B.; Tang, Z.; Cao, W. Application of a Bio-Inspired Algorithm in the Process Parameter Optimization of Laser Cladding. *Machines* **2022**, *10*, 263. https://doi.org/10.3390/machines10040263

Academic Editor: Antonios Gasteratos

Received: 15 March 2022
Accepted: 6 April 2022
Published: 7 April 2022

1. Introduction

With the growing requirements for the mechanical properties of transmission parts, traditional hardening technology is not remarkable in terms of production cost, benefit and performance improvement. Laser cladding is a new surface hardening technology with great potential application value. Under the irradiation of a high energy laser, the cladding alloy powder is melted and added to the surface of the substrate. The cladding layer has good mechanical properties after cooling [1]. In addition, it also has the advantages of high precision, high quality and the low thermal effect of the substrate [2,3].

Laser power P, laser scanning speed V and powder feeding rate F (voltage Fv) are the three key process parameters, which are easy to control and have a strong impact on the quality of the cladding layer [4]. Optimal process parameters vary with different substrate materials and laser cladding powders. In order to obtain a high quality cladding layer, it is of great significance to perform parameter optimization for each laser cladding process.

An empirical statistics method has been used to optimize the process parameters of laser cladding. The relationship between cladding geometry (width, height, depth of molten and dilution rate) and process parameters (laser power, powder feeding rate and scanning speed) has been studied to obtain the linear relationship by the regression method [5]. According to simulation results of molten pool temperature distribution, the laser power and scanning speed had a great influence on the temperature field and geometry of the molten pool [6]. There are specific functions that have been proposed to describe the geometry of a laser cladding layer based on the recursive model and experimental results. Thereby, the complete geometry of the laser cladding layer can be predicted from the basic process parameters [7].

In this study, T15 high speed steel powder was deposited on a 42CrMo steel substrate with the laser cladding process. It greatly improved the surface hardness and wear resistance of the substrate and further improved the service life of mechanical parts [8]. On this basis, heuristic algorithms can be used to solve and optimize process parameters. However, there is no accurate formula between the machining objectives (dilution ratio, powder utilization rate, etc.) and process parameters. This causes significant difficulty for the multiobjective optimization of process parameters.

A slime mould algorithm (SMA) is an up to date bio-inspired method, which was presented by Chen et al. [9] in 2020. It simulated the behavior of slime mould and established the distinct mathematic model for an outstanding exploratory capacity and exploitation propensity. SMA was evaluated using various benchmark data and achieved superior results. A multiobjective slime mould algorithm (MOSMA) employed the same underlying SMA mechanisms for convergence, combined with an elitist nondominated sorting approach to estimate Pareto optimal solutions [10].

In this paper, a novel process parameter optimization approach for laser cladding is proposed based on MOSMA and SVR (MOSMA-SVR-POLC). MOSMA is used to optimize the process parameters of laser cladding (Pareto solution set). Support vector regression (SVR) is used to build the relationship between process parameters and machining objectives. The actual laser cladding data are utilized to assess the performance of MOSMA-SVR-POLC. Simultaneously, the data fitting method (DFM) [5], the response surface method (RSM) [11], MODA (multiobjective dragonfly algorithm)-SVR-POLC [12], MOEA/D (multiobjective evolutionary algorithm based on decomposition)-SVR-POLC [13] and other variant methods are compared with the proposed approach.

There are two innovations. First, the ingenious combination of MOSMA and SVR is modeled in the process parameter optimization of laser cladding. SVR is used to predict the optimization objectives based on the past data of laser cladding, and the predicted results are involved in the iteration process of MOSMA. Secondly, the proposed approach is tested using actual machining data, which can well verify the feasibility and effectiveness of the method.

The remainder of this paper is arranged as follows. Section 2 shows the research methodology and analysis methods. The MOSMA-SVR-POLC is described in detail. The case study is shown in Section 3. The discussion and conclusion are provided in Sections 4 and 5, respectively.

2. MOSMA-SVR-POLC Approach

In this section, the new MOSMA-SVR-POLC hybrid approach is proposed to deal with the problem of the process parameter optimization of laser cladding. The main idea is as follows: (a) carry out the process parameter population representation of laser cladding, (b) obtain the current Pareto solution set and front based on the actual machining data as the training set and SVR as the fitness function, (c) update the population via the Pareto solution set, Pareto front and MOSMA. Steps (b) and (c) are repeated until the cut-off condition is met. The cut-off condition is generally set to reach the maximum iteration. During theorisation, there are two issues that need special attention. The first is how the process parameter population of laser cladding is represented. The second is the membership of the Pareto front. In MOSMA-SVR-POLC, the discussions of these issues are shown in the following.

1. The process parameter population representation of laser cladding: The population X is made up of individuals X_i. $X = \{X_1, X_2, \ldots, X_n\}$, where n indicates the population size. The attributes of X_i are determined by the process parameters of laser cladding (P, Fv, V).
2. The membership of Pareto front: The dilution ratio $D\%$, powder utilization rate PU and machining efficiency ME are usually used to evaluate the effect of laser cladding.

Therefore, these indicators can constitute the membership of Pareto front. Thus far, there is no accurate employing these objectives and process parameters; hence, the ϵ-SVR is applied to predict the objectives. In addition, PU and ME are special (in essence, bigger is better). It is necessary to obtain the reciprocal of PU and ME for unifying the characteristics of all the indicators (smaller is better).

In this work, ϵ-SVR [14] is used to predict the dilution ratio $D\%$, powder utilization rate PU and machining efficiency ME, because there is no accurate calculation formula employing the objectives ($D\%$, PU and ME) and the process parameters (P, Fv and V) in the field of laser cladding. The development kit LIBSVM offers the source codes of ϵ-SVR [15]. The radial basis function (RBF) kernel is very efficient, especially in the prediction of high dimensional samples [16]. The function is listed below.

$$K(\cdot) = \exp(-\gamma|\mathbf{u} - \mathbf{v}|^2) \quad (1)$$

where γ is a coefficient to be set manually, $\mathbf{u}$ is the prediction result vector, and $\mathbf{v}$ indicates the actual value vector.

The main process of the MOSMA-SVR-POLC approach is depicted in Figure 1. The specific steps are shown in the following.

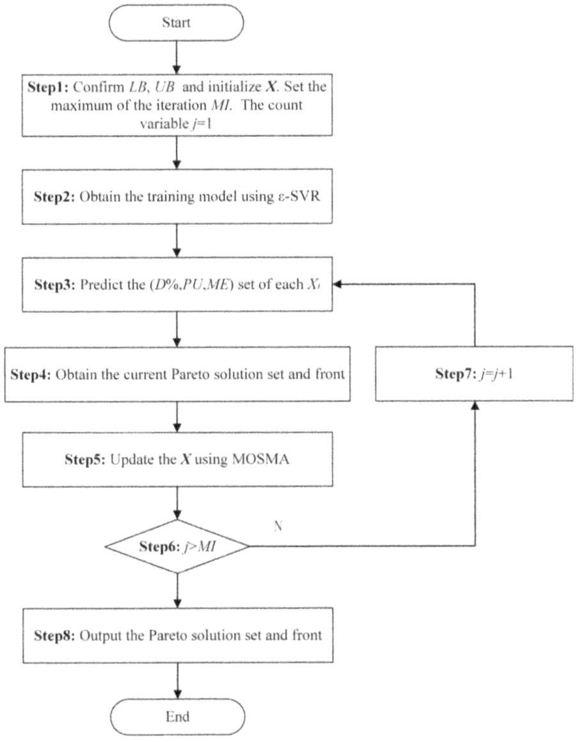

Figure 1. Flowchart of MOSMA-SVR-POLC.

Step 1 Confirm the lower and upper limits of P, Fv and V to form the lower limit sets LB and the upper limit sets UB. Thus, X is initialized randomly within LB and UB. The maximum of the iteration is marked as MI. The iteration count variable j is set to 1.

Step 2 Based on training data, obtain the training model using ϵ-SVR.

Step 3 Based on the training model, predict the $(D\%, PU, ME)$ set of each X_i.

Step 4 Obtain the current Pareto solution set PS_X and front PS_F with the $(D\%, PU, ME)$ sets.

Step 5 Update X with the operations of MOSMA.

Step 6 If $j > MI$, go to Step 8; otherwise, go to Step 7.
Step 7 $j = j + 1$. Go to Step 3.
Step 8 Output the Pareto solution set and front.

Multiobjective Slime Mould Algorithm

SMA was presented by Chen et al. [9] in 2020. It was motivated by the oscillating patterns of slime mould. For forming the optimum path to connect food with outstanding exploratory capacity and exploitation propensity, a distinct mathematic model was presented using accommodative weights to imitate the process of generating positive and negative feedback of the propagation wave supported by a bio-oscillator. MOSMA employed the same underlying SMA mechanisms for convergence, combined with an elitist nondominated sorting approach to estimate Pareto optimal solutions [10]. The operations of MOSMA required for this work are shown in Figure 2.

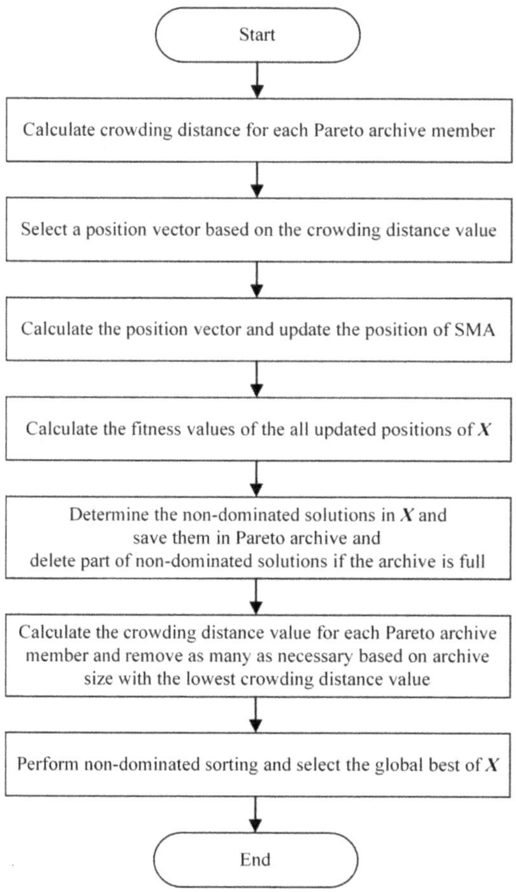

Figure 2. Flowchart of the required operations of MOSMA.

3. Case Study

3.1. Feasibility Experiment

The experimental conditions involve (i) a PC with Matlab R2018a, (ii) the LIBSVM software package, and (iii) the data sets, which were obtained from the practical laser cladding processing experiment. In this study, 42CrMo steel with dimensions of (100 × 100 × 15) mm was used as the substrate. The steel was treated by quenching and tempering. The cladding powder was T15 high speed steel powder with a particle size of (30–100) μm. The chemical compositiosn of the T15 powder and 42CrMo steel are represented in Table 1.

The laser model was a TruDisk4002, as shown in Figure 3. A single layer cladding layer was prepared by the coaxial powder feeding method with T15 high speed steel as the cladding powder, argon as the powder carrier gas and shielding gas in the cladding process. Experimental process parameters are listed below: laser power (P) selection of 1400 W, 1700 W, 2000 W and 2300 W; the scanning speeds (V) were 6 mm/s, 7 mm/s, 8 mm/s and 9 mm/s; similarly, there were four rates of powder feeding voltage (Fv): 40 V, 50 V, 60 V, 70 V. Other parameters include: defocus: 16 mm, lap rate: 30%, and protective gas flow: 20 L/min. The complete experimental method was adopted. A total of 64 machining experiments were carried out. They were cut with a sample size of (15 × 15 × 15) mm. The inlaying, grinding, polishing and other processes were carried out to make the sample surface show a mirror effect and no obvious scratches. Then, the surface of the samples were etched with a solution with HNO_3. Lastly, the 64 samples were placed in sequence under an electron microscope for observation and photography. The geometric shape of the cladding layer of each sample was recorded (as shown in Figures 4 and 5, where w is width, h is height, and b is melting depth. $S1$ is melting height area, and $S2$ is melting depth area). Figure 5 shows the transverse cross section of the claddings using a scanning electron microscope. The relative geometric dimensions of the cladding layer were measured with AutoCAD software. It was found that there were no obvious cracks, pores, inclusions nor other defects in the cladding layer section.

The machining data are shown in Table 2, where $D\%$ can be obtained according to Equation (2) [17]. PU is calculated via Equation (3), and ME is obtained by Equation (4).

$$D\% = \frac{S2}{S1 + S2} \tag{2}$$

$$PU = \frac{S1 \times V \times 0.00819 \times 60}{0.28815 \times Fv} \tag{3}$$

$$ME = S1 \times V \tag{4}$$

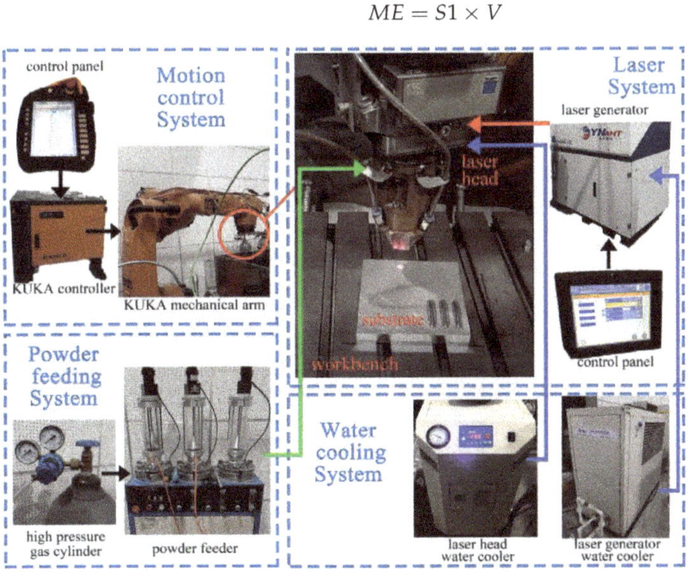

Figure 3. Laser cladding system.

Table 1. Chemical composition of clad powder and substrate materials.

Material	Form	Elements wt (%)								
		C	V	Mn	Cr	Mo	Co	Si	W	Fe
42CrMo	Plate	0.4	-	0.63	0.99	0.19	-	0.21	-	Rem
T15	Powder	1.6	4.7	0.45	4.5	-	5.4	0.48	11.7	Rem

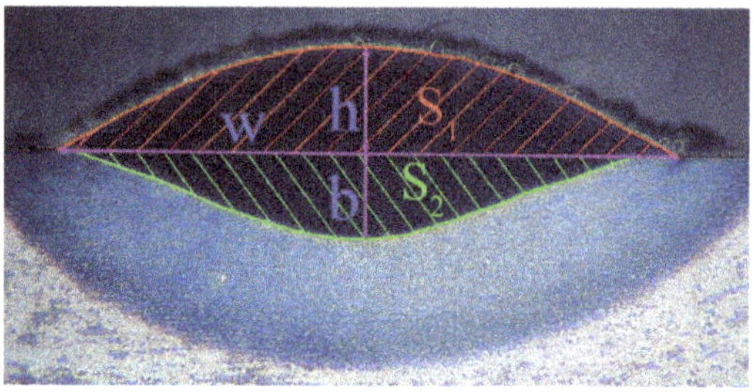

Figure 4. Geometric shape diagrammatic sketch of the cladding layer.

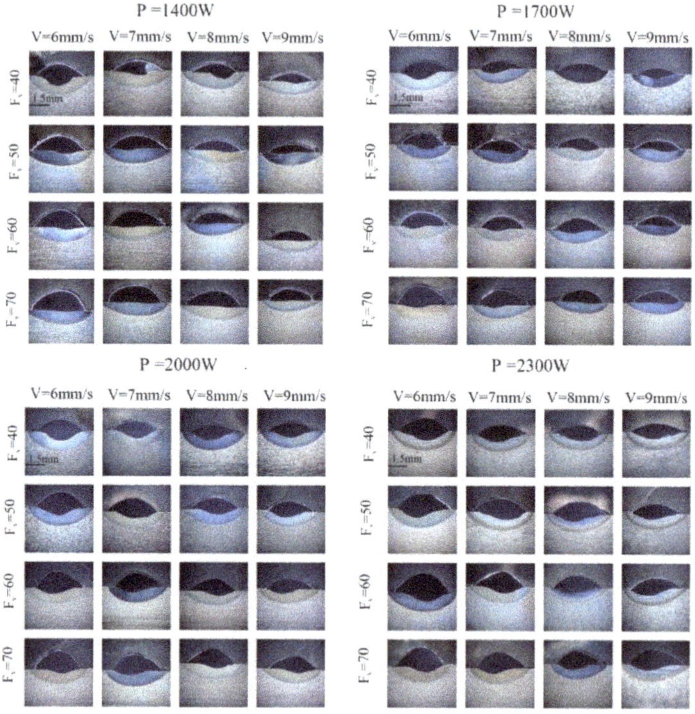

Figure 5. Cross section optimal micrographs of the single clad tracks for different process parameters.

With the machining data in Table 2, the proposed MOSMA-SVR-POLC was used to search the optimized process parameters. The parameter settings of MOSMA-SVR-POLC are shown in Table 3. The machining data is regarded as the training set, and X is taken as the test set. The prediction $D\%$, $1/PU$ and $1/ME$ are obtained via these approaches. The final Pareto optimal solution and front of MOSMA-SVR-POLC are revealed in Table 4. The corresponding graphics of MOSMA-SVR-POLC are shown in Figure 6. From the figures, the relationship between $D\%$, PU and ME is extremely complex.

Table 2. Machining parameters and measurement results for each single clad.

ID	*P* (W)	*Fv* (V)	*V* (mm/s)	*D*%	*PU*	*ME* (mm^3/s)
1	1400	40	6	37.51%	0.5082	11.921
2	1400	40	7	21.71%	0.432	10.132
3	1400	40	8	31.95%	0.4636	10.874
4	1400	40	9	18.71%	0.3966	9.3024
5	1400	50	6	11.18%	0.4654	13.65
6	1400	50	7	48.71%	0.4509	13.223
7	1400	50	8	7.74%	0.4035	11.835
8	1400	50	9	7.87%	0.3697	10.842
9	1400	60	6	5.73%	0.4497	15.826
10	1400	60	7	3.46%	0.4446	15.644
11	1400	60	8	3.38%	0.3939	13.86
12	1400	60	9	3.31%	0.3648	12.837
13	1400	70	6	2.90%	0.4675	19.193
14	1400	70	7	1.55%	0.439	18.02
15	1400	70	8	0.69%	0.4514	18.531
16	1400	70	9	0.49%	0.3968	16.288
17	1700	40	6	29.19%	0.4797	11.251
18	1700	40	7	26.65%	0.4791	11.237
19	1700	40	8	28.27%	0.4815	11.294
20	1700	40	9	34.03%	0.4673	10.961
21	1700	50	6	17.46%	0.5342	15.667
22	1700	50	7	18.50%	0.4816	14.123
23	1700	50	8	16.66%	0.4893	14.35
24	1700	50	9	17.86%	0.4348	12.753
25	1700	60	6	8.81%	0.4919	17.311
26	1700	60	7	11.61%	0.4923	17.324
27	1700	60	8	11.31%	0.4934	17.362
28	1700	60	9	11.37%	0.4495	15.819
29	1700	70	6	7.03%	0.5152	21.148
30	1700	70	7	7.07%	0.4809	19.741
31	1700	70	8	4.98%	0.455	18.679
32	1700	70	9	7.43%	0.4282	17.576
33	2000	40	6	36.10%	0.588	13.791
34	2000	40	7	37.95%	0.549	12.878
35	2000	40	8	38.59%	0.5437	12.754
36	2000	40	9	39.37%	0.5456	12.798
37	2000	50	6	26.31%	0.6026	17.674
38	2000	50	7	27.72%	0.6127	17.968
39	2000	50	8	29.92%	0.5654	16.582
40	2000	50	9	31.49%	0.5601	16.427
41	2000	60	6	19.79%	0.5891	20.73
42	2000	60	7	21.59%	0.5807	20.435
43	2000	60	8	21.60%	0.5501	19.358
44	2000	60	9	21.75%	0.5057	17.795
45	2000	70	6	12.90%	0.5743	23.573
46	2000	70	7	14.43%	0.5379	22.082
47	2000	70	8	14.92%	0.525	21.55
48	2000	70	9	15.71%	0.5068	20.805
49	2300	40	6	42.39%	0.6426	15.072
50	2300	40	7	39.94%	0.6557	15.379
51	2300	40	8	42.53%	0.6237	14.629
52	2300	40	9	41.97%	0.6167	14.465

Table 2. *Cont.*

ID	P (W)	Fv (V)	V (mm/s)	$D\%$	PU	ME (mm^3/s)
53	2300	50	6	30.71%	0.6159	18.064
54	2300	50	7	34.84%	0.5758	16.888
55	2300	50	8	36.04%	0.5794	16.993
56	2300	50	9	39.57%	0.5202	15.256
57	2300	60	6	25.77%	0.6217	21.876
58	2300	60	7	27.57%	0.5631	19.816
59	2300	60	8	29.52%	0.5669	19.949
60	2300	60	9	30.35%	0.5241	18.443
61	2300	70	6	18.37%	0.6274	25.756
62	2300	70	7	18.88%	0.5917	24.289
63	2300	70	8	22.34%	0.5601	22.991
64	2300	70	9	22.47%	0.5741	23.568

Table 3. Parameter setting of MOSMA-SVR-POLC and other approaches.

Approach	Parameter Item	Value
All	Number of search agents	200
	Number of maximum iterations	200
	Kernel type of ϵ-SVR	RBF
	Penalty factor of ϵ-SVR	0.7
	Coefficient γ of ϵ-SVR	1/3
	Coefficient ϵ of ϵ-SVR	0
MOSMA-SVR-POLC	Parameter z	0.03
MODA-SVR-POLC	Enemy distraction weight	0.096
MOEA/D-SVR-POLC	Probability of selecting parents	0.9
	Maximal copies of a new child	2
	Number of neighbours	13
MOPSO-SVR-POLC	Personal learning coefficient	1.5
	Global learning coefficient	2
	Inertia weight	1
	Inertia weight damping ratio	0.99
NSGAII-SVR-POLC	Crossover rate	0.5
	Mutation rate	0.02
MOGWO-SVR-POLC	Number of grids	10
	Leading wolf selection pressure coefficient	4
	Noncracking selected coefficient	2

Many optimized process parameters were obtained, which is a challenge for practical application: how to select the optimal one for practical machining? The TOPSIS method [18] was used to obtain the best process parameter. The key lies in the weight of the three indicators: $D\%$, $1/PU$ and $1/ME$. The Saaty weight method was used to calculate the weight. The problem becomes to determine how much users value the $D\%$, $1/PU$ and $1/ME$. In this work, $D\%$ is considered to be slightly more important than the other two indicators. Therefore, the weight of $D\%$, $1/PU$ and $1/ME$ is 0.6:0.2:0.2 using the Saaty weight method. The optimal process parameters under this weight are shown in Table 5, based on Table 4. The relative error between MOSMA-SVR-POLC and the actual laser cladding is also calculated and shown in Table 5. From Table 5, it can be seen that only the last error of ME is relatively large, and the others are small. The feasibility of this method is verified to a great extent.

Table 4. Pareto optimal solutions and fronts of MOSMA-SVR-POLC.

ID	P	Fv	V	$D\%$	$1/PU$	$1/ME$
1	1400	67.9	6	16.86%	1.9885	0.0596
2	1400	68.3	6	14.08%	2.0169	0.0589
3	1400	68.4	6	13.27%	2.0251	0.0587
4	1400	68.5	6	12.43%	2.0342	0.0585
5	1400	68.6	6	11.56%	2.0433	0.0583
6	1400	68.7	6	10.67%	2.053	0.0581
7	1400	68.8	6	9.77%	2.0627	0.0578
8	1400	68.9	6	8.87%	2.0725	0.0576
9	1400	69.2	6	6.34%	2.1004	0.057
10	1400	69.3	6	5.59%	2.1088	0.0568
11	1400	69.9	6	2.93%	2.1395	0.0562
12	1400	70	6	2.87%	2.1404	0.0562
13	1400	70	6.1	2.5%	2.1687	0.056
14	1400	70	6.2	2.2%	2.1954	0.0559
15	1400	70	6.3	1.96%	2.2198	0.0558
16	1400	70	6.4	1.78%	2.2411	0.0557
17	1400	70	6.6	1.57%	2.2707	0.0556
18	1400	70	6.7	1.53%	2.2789	0.0555
19	1400	70	6.9	1.51%	2.2815	0.0556
20	1400	70	7.1	1.52%	2.2691	0.0557
21	1400	70	7.2	1.51%	2.2589	0.0559
22	1400	70	7.3	1.49%	2.2477	0.056
23	1400	70	7.4	1.45%	2.2361	0.0562
24	1400	70	7.5	1.37%	2.2257	0.0564
25	1400	70	7.6	1.28%	2.2168	0.0566
26	1400	70	7.7	1.15%	2.2109	0.0568
27	1400	70	8.2	0.3%	2.2462	0.0583
28	1999	70	9	17.21%	1.9662	0.0572
29	2300	40	6	42.43%	1.5557	0.064
30	2300	40	6.2	42.01%	1.5349	0.0642
31	2300	40	6.4	41.4%	1.5216	0.0644
32	2300	40	6.5	41.07%	1.5177	0.0645
33	2300	40	6.6	40.75%	1.5156	0.0646
34	2300	40	6.7	40.46%	1.5156	0.0647
35	2300	40	6.8	40.22%	1.5175	0.0648
36	2300	40	6.9	40.04%	1.5209	0.0649
37	2300	40	7	39.94%	1.5256	0.065
38	2300	40	7.1	39.91%	1.5316	0.0651
39	2300	40.4	6	41.33%	1.5718	0.0639
40	2300	40.6	6	40.04%	1.5916	0.0637
41	2300	70	6	18.35%	1.5944	0.0559
42	2300	70	6.2	18.1%	1.601	0.0556
43	2300	70	6.4	18%	1.6152	0.0552
44	2300	70	6.5	18.03%	1.6252	0.0551
45	2300	70	6.6	18.1%	1.6364	0.0549
46	2300	70	6.8	18.4%	1.6622	0.0547
47	2300	70	6.9	18.63%	1.6764	0.0546
48	2300	70	7.4	20.28%	1.7452	0.0543

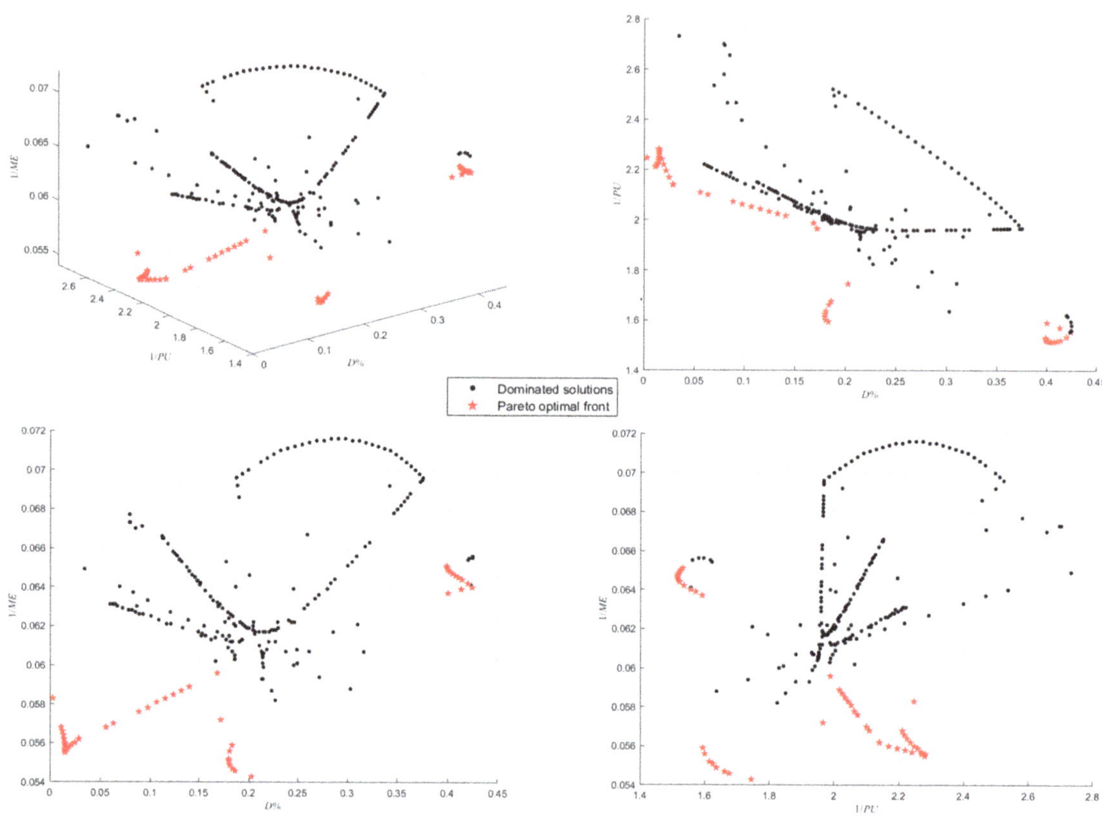

Figure 6. Pareto front of MOSMA-SVR-POLC (3D and 2D).

Table 5. Optimal process parameters and relative errors (*RE*) of MOSMA-SVR-POLC.

ID in Table 4	(P, Fv, V)	RE
12	(1400, 70, 6)	(1.03%, 0.06%, 7.29%)

3.2. Comparative Experiment

The experimental conditions are consistent with those in the feasibility experiment. The proposed MOSMA-SVR-POLC is compared with the mainstream methods: DFM [5] and RSM [11], and some variants of MOSMA-SVR-POLC, such as MODA-SVR-POLC [12], MOEA/D-SVR-POLC [13], MOPSO-SVR-POLC [19], NSGAII-SVR-POLC [20], MOGWO-SVR-POLC [21]. The parameter settings of these approaches are also shown in Table 3. The TOPSIS method [18] is also used to obtain the best process parameter. The weight of $D\%$, $1/PU$ and $1/ME$ is also set to 0.6:0.2:0.2. The best process parameters and relative errors (*RE*) of these approaches are shown in Table 6. Figure 7 shows the Pareto optimal front obtained by MOSMA-SVR-POLC and its similar algorithms.

From Tables 5 and 6, it can be seen that the proposed MOSMA-SVR-POLC had the same optimal process parameters as DFM and MOGWO-SVR-POLC. In terms of *RE*, it was much better than DFM and the same as MOPSO-SVR-POLC and MOGWO-SVR-POLC. DFM performed extremely badly in the *RE* of $D\%$. RSM, MODA-SVR-POLC, MOEA/D-SVR-POLC and NSGAII-SVR-POLC obtained different optimal process parameters. RSM lagged behind MOSMA-SVR-POLC. MODA-SVR-POLC and MOEA/D-SVR-POLC also had a good performance, except for the *RE* of *ME*. NSGAII-SVR-POLC did not perform well in the *RE* of $D\%$. From Figure 7, MOPSO-SVR-POLC, NSGAII-SVR-POLC and

MOGWO-SVR-POLC had a good performance in area C. MOSMA-SVR-POLC performed well in regions A and B. The above results confirm the capacity of MOSMA-SVR-POLC. The proposed approach proves to be very competitive.

Table 6. Optimal process parameters and relative errors (*RE*) of the comparison methods.

Approach	(P, Fv, V)	RE
DFM	(1400, 70, 6)	(180.61%, 8.10%, 6.98%)
RSM	(1400, 70, 9)	(51.81%, 6.51%, 24.02%)
MODA-SVR-POLC	(1700, 70, 6.2)	(0.14%, 0.04%, 14.42%)
MOEA/D-SVR-POLC	(1700, 70, 6)	(0.14%, 0.04%, 15.56%)
MOPSO-SVR-POLC	(1700, 70, 6)	(1.03%, 0.06%, 7.29%)
NSGAII-SVR-POLC	(1402, 68.3, 6)	(81.73%, 10.69%, 3.93%)
MOGWO-SVR-POLC	(1400, 70, 6)	(1.03%, 0.06%, 7.29%)

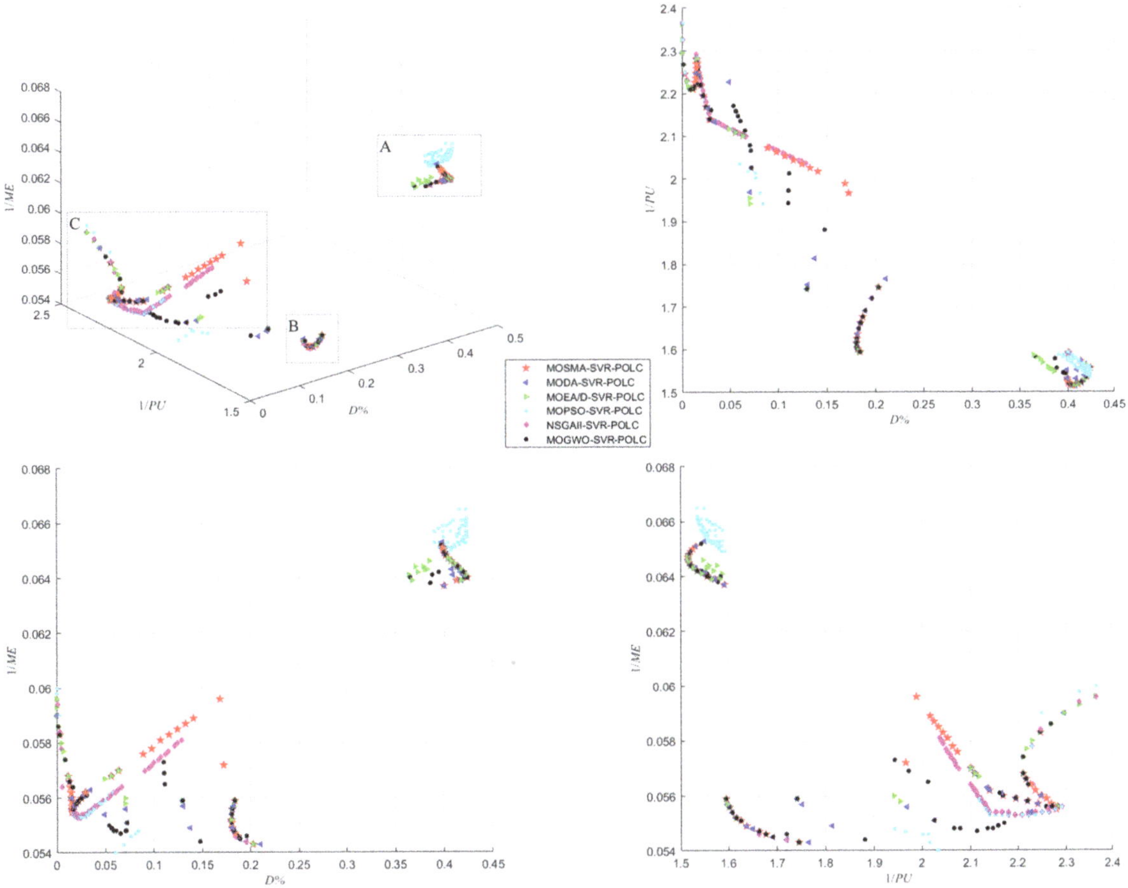

Figure 7. Pareto optimal front obtained by MOSMA-SVR-POLC and its similar algorithms.

4. Discussion

In the feasibility experiment, the TOPSIS method was used to obtain the optimal process parameters. However, the difference in weight ratios between optimization objectives affects the selection of optimal process parameters. In this section, the different weight ratios were selected to study the output of the optimal process parameters, and the prediction accuracy of the proposed approach will be further discussed. The results are shown in Table 7.

From Table 7, it can be seen that the different weight ratios do define the different optimal process parameters. This meets the needs of different users for objectives. No matter which weight ratio, the prediction accuracy of the first two targets, $D\%$ and $1/PU$, are higher than $1/ME$. Especially in the weight ratios 1/3:1/3:1/3, 0.2:0.6:0.2 and 0.2:0.2:0.6, this situation is more obvious. The improvement ofthe prediction accuracy of $1/ME$ may be the next important research topic.

Table 7. Optimal process parameters and relative errors (RE) of the proposed approach under different weight ratios.

Objective Weight Ratio	ID in Table 4	(P, Fv, V)	RE
1/3:1/3:1/3	44	(2300, 70, 6.4)	(8.57%, 5.26%, 32.90%)
0.6:0.2:0.2	12	(1400, 70, 6)	(1.03%, 0.06%, 7.29%)
0.2:0.6:0.2	42	(2300, 70, 6)	(0.11%, 0.03%, 43.98%)
0.2:0.2:0.6	47	(2300, 70, 6.8)	(11.76%, 3.30%, 39.54%)

In addition, the big oh notation is used to calculate the computation cost of the proposed approach. The computation cost of MOSMA and SVR is $O(MI(n \cdot d + cof \cdot n))$ and $O(l \cdot tk)$, respectively, cof is the consumption time of objective calculation, d is the attribute number of individuals, l represents the sample size, and tk represents the time consumed for the calculation of the SVR kernel function. The proposed approach is composed of MOSMA and SVR. Therefore, its computation cost is $O(MI(n \cdot d + l \cdot tk \cdot n))$.

5. Conclusions

This work proposed a new hybrid approach for the process parameter optimization of laser cladding using a multiobjective slime mould algorithm and support vector regression. The proposed MOSMA-SVR-POLC realizes objectives prediction and process parameter optimization in laser cladding and provides users with the most valuable parameters for different objectives. Furthermore, ϵ-SVR is applied to predict the $D\%$, $1/PU$ and $1/ME$. The training data are from the actual laser cladding. MOSMA is used to obtain the Pareto optimal solutions and fronts. The feasibility experiment and comparative experiment were carried out to test the performance of MOSMA-SVR-POLC. The experimental results reveal that MOSMA-SVR-POLC achieves a competitive predictive performance compared with other well established approaches. The study confirms the feasibility and effectiveness of MOSMA-SVR-POLC in the field of process parameter optimization in laser cladding.

For future work, the improvement of the setting parameters and prediction accuracy of ϵ-SVR should be further studied. Secondly, more kinds of laser cladding experiments, such as multichannel laser cladding, should be added to verify and modify the method in this paper.

Author Contributions: Conceptualization, Y.Z. and W.C.; methodology, Y.Z. and W.C.; software, B.G.; validation, B.G. and Z.T.; formal analysis, Z.T.; investigation, B.G.; resources, B.G.; data curation, B.G.; writing—original draft preparation, Y.Z. and B.G.; writing—review and editing, W.C.; visualization, W.C.; supervision, Y.Z.; project administration, Y.Z.; funding acquisition, Y.Z. and W.C. All authors have read and agreed to the published version of the manuscript.

Funding: This research was funded by the National Natural Science Foundation of China (Grant No. 51905148) and Fundamental Research Funds for the Central Universities (Grant No. B200202218 and BC210202088).

Institutional Review Board Statement: Not applicable.

Informed Consent Statement: Not applicable.

Data Availability Statement: The data is presented in Table 2.

Acknowledgments: The authors want to thank Tsinghua University for providing the laser cladding equipment to support the research of the authors.

Conflicts of Interest: The authors declare no conflict of interest. The funders had no role in the design of the study; in the collection, analyses, or interpretation of data; in the writing of the manuscript; or in the decision to publish the results.

References

1. Lian, G.; Yao, M.; Liu, Z.; Yang, S.; Chen, C.; Wang, H.; Xiang, Y.; Cong, W. Near-net shaping control of triangular stacking in laser cladding process. *Procedia Manuf.* **2019**, *34*, 233–238. [CrossRef]
2. Carcel, B.; Serrano, A.; Zambrano, J.; Amigo, V.; Carcel, A.C. Laser cladding of TiAl intermetallic alloy on Ti6Al4V—Process optimization and properties. *Phys. Procedia* **2014**, *56*, 284–293. [CrossRef]
3. Nie, Z.; Wang, G.; Mcguffin, J.; Narayanan, B.; Zhang, S.; Schwam, D. Experimental study and modeling of H13 steel deposition using laser hot-wire additive manufacturing. *J. Mater. Process. Technol.* **2016**, *235*, 171–186. [CrossRef]
4. Fatoba, O.S.; Akinlabi, E.T.; Akinlabi, S.A.; Obiegbu, M.C. Data related to optimized process parameters influence on hardness, microstructural evolution and wear resistance performance of Al-Si-Sn-Cu/Ti-6Al-4V composite coatings. *Data Brief* **2019**, *23*, 103724. [CrossRef]
5. Shayanfar, P.; Daneshmanesh, H.; Janghorban, K. Parameters optimization for laser cladding of inconel 625 on ASTM A592 steel. *J. Mater. Res. Technol.* **2020**, *9*, 8258–8265. [CrossRef]
6. Wang, X.J.; Su, S.C. Modeling and parameter calculation for laser cladding silicon films. *Opt. Precis. Eng.* **2011**, *19*, 60–63.
7. Nenadl, O.; Ocelik, V.; Palavra, A.; Hosson, J. The prediction of coating geometry from main processing parameters in laser cladding. *Phys. Procedia* **2014**, *56*, 220–227. [CrossRef]
8. Alouane, C.; Kasser, A. Consolidation by atmospheric pressure of T15 tool steel powder. *Powder Technol.* **2019**, *352*, 331–339. [CrossRef]
9. Li, S.; Chen, H.; Wang, M.; Heidari, A.A.; Mirjalili, S. Slime mould algorithm: A new method for stochastic optimization. *Future Gener. Comput. Syst.* **2020**, *11*, 300–323. [CrossRef]
10. Manoharan, P.; Jangir, P.; Ravichandran, S.; Alhelou, H.H.; Chen, H. MOSMA: Multi-objective slime mould algorithm based on elitist non-dominated sorting. *IEEE Access* **2021**, *9*, 3229–3248.
11. Alam, M.K.; Urbanic, R.J.; Nazemi, N.; Edrisy, A. Predictive modeling and the effect of process parameters on the hardness and bead characteristics for laser-cladded stainless steel. *Int. J. Adv. Manuf. Technol.* **2017**, *94*, 397–413. [CrossRef]
12. Mirjalili, S. Dragonfly algorithm: A new meta-heuristic optimization technique for solving single-objective, discrete, and multi-objective problems. *Neural Comput. Appl.* **2016**, *27*, 1053–1073. [CrossRef]
13. Zhang, Q.; Li, H. MOEA/D: A multiobjective evolutionary algorithm based on decomposition. *IEEE Trans. Evol. Comput.* **2007**, *11*, 712–731. [CrossRef]
14. Vapnik, V.N. An overview of statistical learning theory. *IEEE Trans. Neural Netw.* **1999**, *10*, 988–999. [CrossRef] [PubMed]
15. Chang, C.C.; Lin, C.J. LIBSVM: A library for support vector machines. *ACM Trans. Intell. Syst. Technol.* **2011**, *2*, 1–27. [CrossRef]
16. Cao, W.; Liu, X.; Ni, J. Parameter optimization of support vector regression using henry gas solubility optimization algorithm. *IEEE Access* **2020**, *8*, 88633–88642. [CrossRef]
17. Reddy, A.L.; Preston, S.P.; Shipway, P.H.; Davis, C.; Hussaina, T. Process parameter optimisation of laser clad iron based alloy: Predictive models of deposition efficiency. *Surf. Coat. Technol.* **2018**, *349*, 198–207. [CrossRef]
18. Peng, T.; Zhou, J.; Chu, Z.; Sun, N. Modeling and combined application of orthogonal chaotic NSGA-II and improved TOPSIS to optimize a conceptual hydrological model. *Water Resour. Manag.* **2018**, *32*, 3781–3799. [CrossRef]
19. Coello, C.A.C.; Pulido, G.T.; Lechuga, M.S. Handling multiple objectives with particle swarm optimization. *IEEE Trans. Evol. Comput.* **2004**, *8*, 256–279. [CrossRef]
20. Deb, K.; Pratap, A.; Agarwal, S.; Meyarivan, T. A fast and elitist multiobjective genetic algorithm: NSGA-II. *IEEE Trans. Evol. Comput.* **2002**, *6*, 182–197. [CrossRef]
21. Mirjalili, S.; Saremi, S.; Mirjalili, S.M.; Coelho, L.D.S. Multi-objective grey wolf optimizer: A novel algorithm for multi-criterion optimization. *Expert Syst. Appl.* **2015**, *47*, 106–119. [CrossRef]

machines

MDPI

Article

Obstacle Modeling and Structural Optimization of Four-Track Twin-Rocker Rescue Robot

Xiaobin Xu [1,2,*], Wen Wang [1,2], Guangyu Su [1,2], Cong Liu [1,2], Wei Cai [1,2], Haojie Zhang [1,2], Yingying Ran [1,2], Zhiying Tan [1,2] and Minzhou Luo [1,2]

1 College of Mechanical & Electrical Engineering, Hohai University, Changzhou 213022, China; 211619010135@hhu.edu.cn (W.W.); suguangyuhhu@hhu.edu.cn (G.S.); 191319010018@hhu.edu.cn (C.L.); wei_cai@hhu.edu.cn (W.C.); 211619010102@hhu.edu.cn (H.Z.); 200219030004@hhu.edu.cn (Y.R.); zytan@hhu.edu.cn (Z.T.); lmz@hhuc.edu.cn (M.L.)
2 Jiangsu Key Laboratory of Special Robot Technology, Hohai University, Changzhou 213022, China
* Correspondence: xxbtc@hhu.edu.cn

Abstract: In order to achieve the best obstacle surmounting performance of a mobile robot in the rescue environment, a four-track twin-rocker bionic rescue robot with an inner and outer concentric shaft was designed in this paper. From the viewpoint of dynamics, the motion process of the mass center of the robot when climbing steps forward and backward was studied. The maximum obstacle height of the robot was calculated. The relationship between the elevation angle of the car body, the swing angle of the rocker arm and the height of the steps was analyzed by simulation. The simulation results show that the maximum forward and reverse obstacle crossing heights were 92.99 mm and 155.82 mm, respectively. Obstacle climbing experiments of the designed robot prototype were carried out. It was found that the measured maximum height of the step was 95 mm, and the measured maximum height of the reverse obstacle was 165 mm. Finally, bionic particle swarm optimization was used to optimize the structural parameters of the rocker arm with an optimal length of 315.2 mm. The study of this paper can be referenced for the design and analysis of obstacle surmounting rescue robots with similar structures.

Keywords: bionic tracked rescue robot; double rocker arm; obstacle crossing; centroid; bionic particle swarm optimization algorithm

Citation: Xu, X.; Wang, W.; Su, G.; Liu, C.; Cai, W.; Zhang, H.; Ran, Y.; Tan, Z.; Luo, M. Obstacle Modeling and Structural Optimization of Four-Track Twin-Rocker Rescue Robot. *Machines* **2022**, *10*, 365. https://doi.org/10.3390/machines10050365

Academic Editor: Raffaele Di Gregorio

Received: 11 April 2022
Accepted: 9 May 2022
Published: 10 May 2022

1. Introduction

Large-scale natural disasters can cause the collapse of buildings and significant casualties. Rescue workers in harsh environments and the limited time to carry out large-scale search and rescue are the current concerns of people. Using a rescue robot to assist in rescue is an effective means to improve the search area and shorten the rescue time.

At present, rescue robots include wheeled rescue robots, bionic foot rescue robots, hybrid aerial/terrestrial robots and caterpillar rescue robots. Wheeled rescue robots move fast, and the structure and control are simple, but the ability to surmount obstacles is poor, which limits their application to relatively flat terrains. Bionic foot rescue robots can be used in very complicated terrains, but their mechanical structure is complex and their control is tedious. Hybrid aerial/terrestrial robots [1,2] can fly over obstacles and drive on the ground to improve energy efficiency. The ability to autonomously explore complex environments is improved, but the structure and control are more complex. Crawler rescue robots are suitable for surmounting obstacles, with a simpler mechanical structure and control compared with bionic foot robots. Crawler mobile robots can move at a relatively high speed with stability on rugged ground such as steps and slopes. Takemori [3] proposed a multi-functional tracked rescue robot, FUHGA2. The main track covers the main body and carries four sub-tracks, with longer six-axis arms and parallel grippers at the top, giving it high dexterity, maneuverability and high search capability. Cho [4] proposed a rescue robot with a chain double-track mechanism with triangular and square track hinges. The relative rotation of the front and rear body enables the robot to better adapt to the terrain and reduce energy consumption on rugged ground. However, the roof is relatively low from the ground, and it is easy to get stuck in complex terrain, resulting in a low driving speed and efficiency. Kim [5] proposed a single-track crawler rescue robot adapted to obstacles. With different shapes of obstacles, the robot's mobile planetary wheel structure could change the orbit shape to enhance its ability to overcome different obstacles, but the energy consumption was too large. Li [6] proposed a W-shaped rocker crawler robot through the combination of a four-wheel rocker structure and a crawler. It can avoid the situation where the legs of the W-shaped rocker four-wheel robot are stuck on obstacles, and it can improve the adaptability to chaotic terrain and the stability of obstacle crossing.

In this paper, considering the stability in complex terrains and the ability to jump over obstacles, the crawler movement mode was adopted to accomplish obstacle crossing for a rescue robot. Based on the two-tracked mobile robot, the double rocker arm mechanism was added. To date, four-crawler double rocker rescue robots have received a lot of research attention. Liu [7] proposed a robot with a variable configuration of the rocker arm. A triangular wheel structure in the rocker mechanism and application of the elliptical form principle were used to improve the crawler tension, reduce track deformation and improve the ability to overcome obstacles. The authors of [8–12] proposed four-track twin-rocker rescue robots with good adaptabilities which can climb over rugged terrains, such as convex platforms and gullies. The tracks on the rocker arm are tightly attached to the rigid body, which enables the robot to climb over convex platforms that have good support. However, the rocker arm structure in the above literature is shorter than that of the car body, so the robot can only overcome obstacles in the forward direction, resulting in a small range of forwarding movement of the robot's center of mass. In the face of higher steps, the robot cannot reverse the obstacle by rotating the rocker arm to support the lifting of the car body. Therefore, through the research and design of the rocker arm structure, the rocker arm of the rescue robot is enough to support the lifting of the car body to reverse the obstacle. In the literature [13], the rocker arm structure was designed based on forward obstacle crossing, so that it can also lift the car body for reverse obstacle crossing by rotating clockwise. However, due to the short length of the rocker arm, the lifting angle of the car body is small, and the height of the reverse obstacle crossing is limited.

Thus, the length of the rocker arm has a great influence on the obstacle surmounting ability, but few studies have looked for the optimal arm length to achieve the maximum obstacle surmounting ability. Given the above obstacle crossing modeling problems of robots, static models [14–17] with different motion states are usually adopted to realize static analysis, and the obstacle crossing performance of robots is analyzed through the variation rule of the centroid position [18,19] during the robot's movement.

The innovations of this paper are as follows:

(1) Through the research and design of the bionic leg-type rocker arm structure, the robot can surmount obstacles upward and downward;
(2) The bionic particle swarm optimization algorithm is used to optimize the structural parameters of the robot, and the optimal length of the rocker arm is obtained to achieve the maximum obstacle crossing capability of the robot.

The rest of this article is organized as follows: In Section 2, the structural design of a four-track twin-rocker rescue robot is described. In Section 3, the mathematical model is established from the perspective of dynamics, and the maximum height that the robot can surmount and the pose when it achieves the best performance of obstacle crossing are deduced according to the position of the robot's centroid. In Section 4, the simulation and experimental verification of the rescue robot in this study are carried out, and the bionic particle swarm optimization algorithm is used to optimize the structural parameters of the rocker arm. Finally, Section 5 summarizes the research results of this paper.

2. Structure Design of Four-Track Twin-Rocker Arm Robot

A four-track twin-rocker mobile rescue robot was designed, as shown in Figure 1. The robot consists of a chassis, leg rocker system, track system, drive system and external sensor system. The structural design of the double rocker arms of the obstacle surmounting robot mimics the climbing movement of legs. The internal and external concentric shaft design is adopted to satisfy the independent driving of the off-road wheel and rocker arm, and four sets of driving equipment are placed on the same axis to reduce the volume of the car body. The center of gravity of the vehicle is set at the front wheel of the vehicle, which is conducive to crossing rugged terrain and obstacles such as gullies and steps.

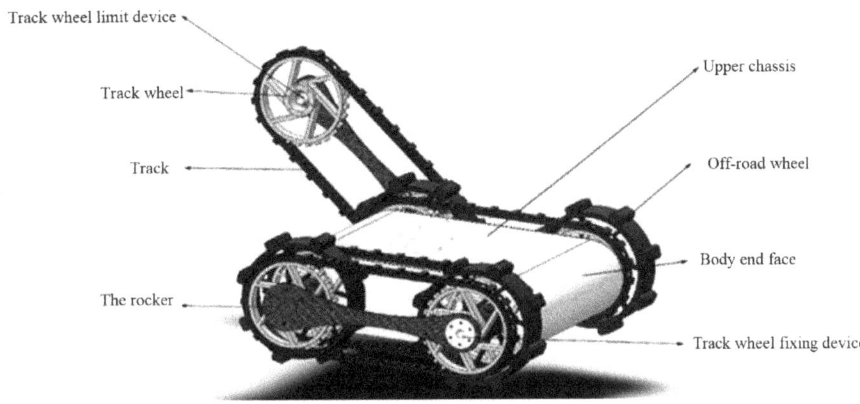

Figure 1. Four-track twin-rocker arm mobile robot object.

The system control diagram is shown in Figure 2. In the figure, STM32F427 is used as the control board, and the whole system is powered by a 24 V mobile power supply. During the operation, it receives the speed instruction issued by the remote control through the serial port and sends the data to the "C610" electrical adjustment. Then, the corresponding PWM signal is calculated and sent to the "M3508" deceleration DC motor and "57AIM30" servo motor to provide power for the track drive system and rocker arm drive system, respectively. The rocker arm drive system guarantees 360° rotation of the rocker arm.

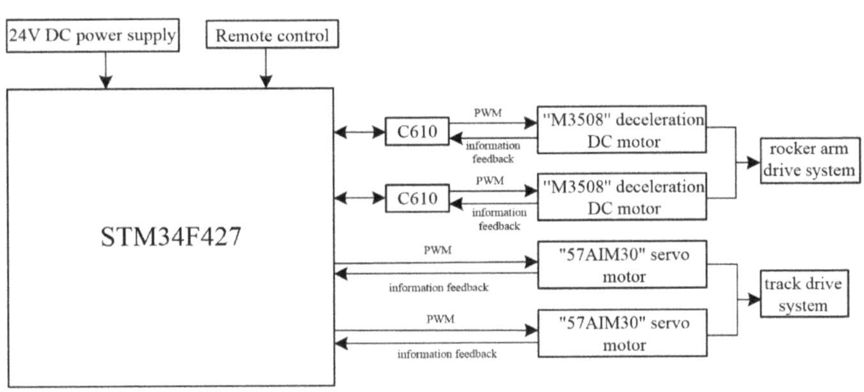

Figure 2. The system control diagram.

To reduce the weight of the car, the whole vehicle is made of carbon fiber. The car adopts a harmonic reducer and a spiral bevel gear to increase the torque on the rocker up to 120 N·m. The structural parameters are presented in Table 1.

Table 1. Structural parameters of a four-track twin-rocker mobile robot.

Indicators	Parameter
Size/mm × mm × mm	378 × 300 × 136.5
Diameter of track wheel/mm	173
Cross-country wheel diameter/mm	143.5
Rocker arm mass/kg	0.189
Car body quality/kg	12.94

3. Obstacle Crossing Analysis of a Four-Track Twin-Rocker Rescue Robot

In rescue work, the four-track twin-rocker robot faces various terrains, which can be simplified into a combination of typical obstacles such as slopes, steps and ditches. Among these obstacles, steps are often used to analyze the obstacle negotiating capability of the rescue robot.

The obstacle crossing function of the rescue robot is to use a walking mechanism to drive the robot to move so that its center of mass can cross the boundary line of the obstacle. During this process, it should be ensured that the robot does not flip over and remains relatively stable. The obstacle crossing process can be divided into the following two types:

1. The robot's forward obstacle surmounting process is shown in Figure 3. The robot is driven by its power to move forward, and the rocker arm rises at a certain distance from the step so that its track wheel can hit and be supported by the rectangular corner of the step, as shown in Figure 3a. Then, the rocker arms rotate clockwise to make the robot body tilt up to a certain angle, as shown in Figure 3b. Under the action of the driving force, the robot moves forward until its mass center crosses the boundary of the step, as shown in Figure 3c. Finally, the robot will be pulled up the step with the force of gravity in the first half of the robot body, as shown in Figure 3d.

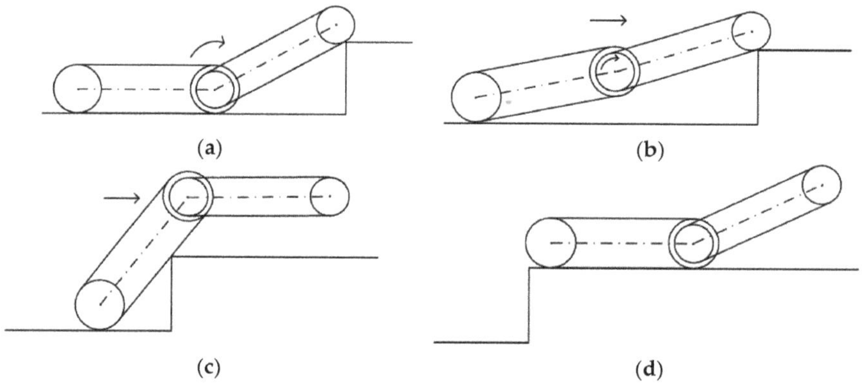

Figure 3. The process of the robot climbing the steps forward: (**a**) lift the rocker arm; (**b**) lift the body with the support of the rocker arm; (**c**) the center of mass crosses the step boundary; (**d**) the body crosses the step.

2. The reverse obstacle surmounting process of the robot is shown in Figure 4. The first four steps are identical to the forward obstacle surmounting process in Figure 3a–c. When the driving wheel fails to cross the step boundary only by rotating the double rocker arm clockwise to support the front part of the car body, the rocker arm is rotated backward until it hits and is supported by the ground, causing the center of mass to rise and move forward, as shown in Figure 4e. When the robot is lifted to a certain height, it climbs the step under the joint action of the driving force, friction and support force of the track on the ground, as shown in Figure 4f.

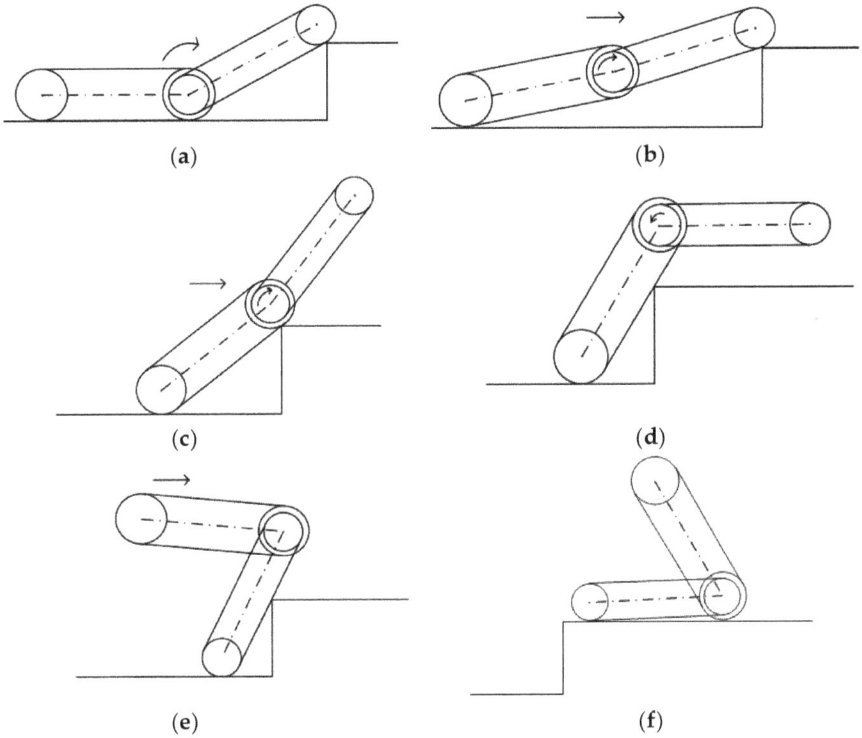

Figure 4. Reverse climbing process of the robot: (**a**) lift the rocker arm; (**b**) lift the body with the support of the rocker arm; (**c**) off-road wheel withstands step angle; (**d**) car body crawler withstands step edges; (**e**) the center of mass crosses the step boundary; (**f**) the body crosses the step.

3.1. Centroid Distribution of Four-Track Twin-Rocker Rescue Robot

When climbing stairs, the gesture of the rocker arm needs to be constantly adjusted according to the height of the step. When the angle between the rocker arm and the robot body changes, the position of the robot's centroid changes accordingly, thus affecting the robot's obstacle crossing performance.

The robot's center of mass trajectory is shown in Figure 5. The coordinate system XO_1Y is established with the center O_1 of the rescue robot's rear cross-country wheel as the origin, O_1O_2 as the abscissa and O_1O_2 as the vertical. $G_1(L,h)$ is the center of mass of the robot body, and G_2 is the center of mass of the robot rocker arm. The variation rule of the robot's center of mass $G(X,Y)$ with the swing arm movement is as follows:

$$\begin{cases} X = \frac{m_1}{m_1+m_2}L + \frac{m_2(L_1+L_2\cos\theta)}{m_1+m_2} \\ Y = \frac{m_1}{m_1+m_2}h + \frac{m_2L_2\sin\theta}{m_1+m_2} \end{cases} \tag{1}$$

$$(X - \frac{m_1L + m_2L_1}{m_1+m_2})^2 + (Y - \frac{m_1h}{m_1+m_2})^2 = \frac{m_2^2L_2^2}{(m_1+m_2)^2} \tag{2}$$

where m_1 is the mass of the robot body, m_2 is the mass of the robot swing arm, L_1 is the center distance between the two driving wheels, L_2 is the distance between the center of the robot swing arm from G_2 to O_2, L is the abscissa of the robot body's center of mass, h is the ordinate of the robot body's center of mass and θ is the swing angle of the robot swing arm.

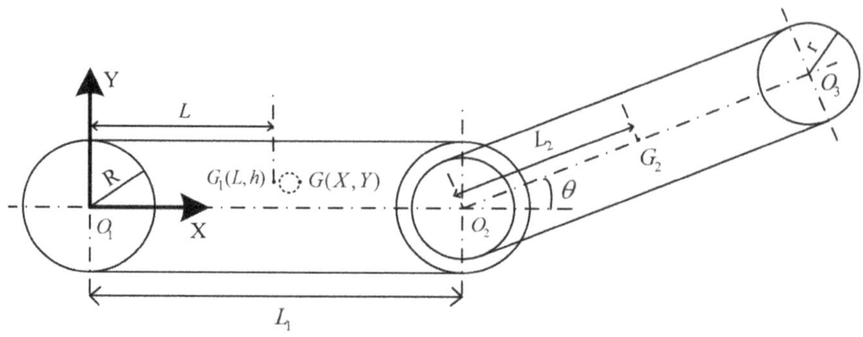

Figure 5. The trajectory of the robot's center of mass.

Therefore, the trajectory of the robot's center of mass is a circle with $(\frac{m_1L+m_2L_1}{m_1+m_2}, \frac{m_1h}{m_1+m_2})$ as the center and $r = \frac{m_2L_2}{m_1+m_2}$ as the radius.

3.2. Forward Obstacle Crossing Analysis of Four-Track Twin-Rocker Rescue Robot

1. When the step height is low (1–2 cm), the force of the robot crossing the step can be obtained by using the static equilibrium equation, as shown in Figure 6.

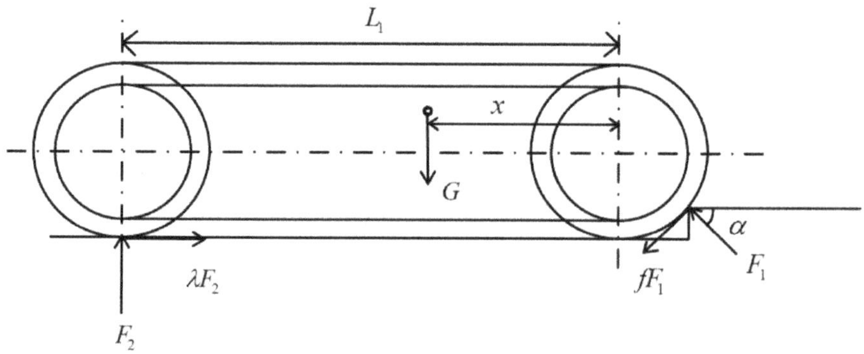

Figure 6. Robot's front wheel's contact with the step.

The function can be established as

$$\begin{cases} F_1 \sin\alpha - fF_1 \cos\alpha - G + F_2 = 0 \\ F_1 \cos\alpha - fF_1 \sin\alpha - \lambda F_2 = 0 \\ fF_1R - GL_1/2 + F_2(L_1 - x) - \lambda F_2 R = 0 \end{cases} \tag{3}$$

According to Equation (3), the function can be expressed as

$$\frac{fR}{L_1} = (\frac{\lambda R - x + \lambda f x - \lambda f L_1}{\lambda L_1}) \cos\alpha + \frac{\lambda(L_1 - x) - fx + \lambda R f}{\lambda L_1} \sin\alpha \tag{4}$$

According to Figure 5, the geometric relationship is

$$\sin\alpha = \frac{R - H}{R} \tag{5}$$

Substituting Equation (5) into Equation (4) and letting f be 0, Equation (4) can be rewritten as

$$\frac{H}{R} = 1 - \sqrt{\frac{1}{1 + (\frac{\frac{L_1}{x} - 1}{1 - \frac{\lambda R}{x}})^2}} \tag{6}$$

where L_1 is the center distance between the front and rear driving wheels of the robot, R is the radius of the driving wheel, G is the gravity of the robot, F_1 is the support force of the step applied on the front wheel, F_2 is the support force of the ground applied on the rear wheel, f is the rolling resistance coefficient, λ is the ground adhesion coefficient, H is the height of the step and x is the distance between the centroid of the robot and the center of the front wheel.

As can be seen from Equation (6), as the parameters $\frac{L_1}{x}$ and $\frac{\lambda R}{x}$ increase, $\frac{H}{R}$ increases. Moreover, it is easier for the front wheel to cross the steps.

2. When the rescue robot crosses the step at a certain height while climbing, the vertical edge line of the step is defined as the key boundary line, as shown in Figure 7. When the centroid of the robot hits the vertical edge line of the step, the rocker arm is kept horizontal. The maximum height of the robot crossing steps forward can be calculated:

$$H(X, Y, \alpha) = R + X\sin\alpha + Y\cos\alpha - \frac{Y + R}{\cos\alpha} \tag{7}$$

where X is the abscissa of the robot's center of mass, Y is the ordinate of the robot's center of mass and α is the angle between the car body and the ground.

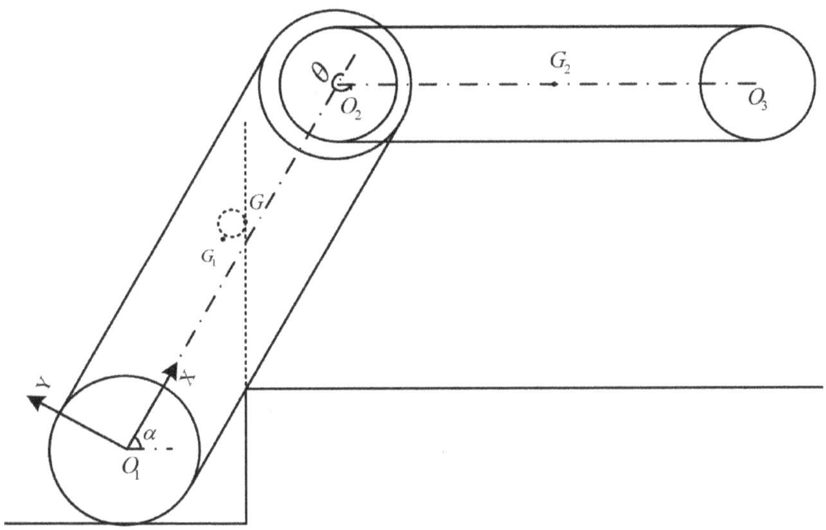

Figure 7. Robot centroid crossing the step boundary in forward obstacle crossing.

To ensure that the rescue robot does not flip over, the elevation angle of the robot needs to satisfy $\alpha \in (0, 90^\circ)$. The relationship between the elevation angle of the robot α and the included angle between the rocker arm and the centerline of the car body θ is

$$\alpha + \theta = 2\pi \tag{8}$$

Take the partial derivative of $H(X,Y,\alpha)$ with respect to the abscissa X and ordinate Y of the robot's centroid and obtain the following formula:

$$\frac{\partial H}{\partial X} = \cos\alpha > 0 \tag{9}$$

$$\frac{\partial H}{\partial Y} = -\sin\alpha - \frac{\sin\alpha}{\cos^2\alpha} < 0 \tag{10}$$

$H(X,Y,\alpha)$ is an increasing function and reduction function of the abscissa and ordinate for the robot centroid. When the center of mass of the robot is closer to the front wheel and step, the maximum height to be crossed is higher, and it is easier for the robot to cross the step.

Take the first partial derivative and the second partial derivative of $H(X,Y,\alpha)$ with respect to elevation α:

$$\frac{\partial H}{\partial \alpha} = X\cos\alpha - Y\sin\alpha - (Y+R)\frac{\sin\alpha}{\cos^2\alpha} \tag{11}$$

$$\frac{\partial^2 H}{\partial \alpha^2} = -X\sin\alpha - Y\cos\alpha - \frac{1+\sin^2\alpha}{\cos^3\alpha}(Y+R) < 0 \tag{12}$$

When $\alpha \in (0, \frac{\pi}{2})$, $\frac{\partial^2 H}{\partial \alpha^2} < 0$, H has a maximum value. When $\frac{\partial H}{\partial \alpha} = 0$, the maximum height that the robot can cross can be obtained as $H_{\max}$.

3.3. *Reverse Obstacle Crossing Analysis of Four-Track Twin-Rocker Rescue Robot*

When the step height is too high, to prevent the robot from flipping over, the forward obstacle crossing cannot be performed by increasing the elevation angle between the robot body and the ground. With O_3 as the support point and O_2 as the rotation center, the rocker arm is rotated counterclockwise to drive the robot's centroid forward movement. The robot's centroid can cross the key boundary of the step under the action of the driving force, as shown in Figure 8. The center coordinate system $X_1O'Y_1$ is established with O_3 as the origin, O_2O_3 as the horizontal axis and O_2O_3 as the vertical axis. The centroid coordinate of the robot $G'(X', Y')$ in the new coordinate system can be expressed as

$$\begin{cases} X' = \frac{m_1(L_1-(L_1-L)\cos\theta_1+h\sin\theta_1)+m_2(L_1-L_2)}{m_1+m_2} \\ Y' = \frac{m_1(h\cos\theta_1+(L_1-L)\sin\theta_1)}{m_1+m_2} \end{cases} \tag{13}$$

where θ_1 is the angle between the centerline O_1O_2 of the car body and the centerline O_2O_3 of the rocker arm.

It can be seen that in the new coordinate system $X_1O'Y_1$, the trajectory of the robot's overall centroid changes with θ_1, which is a circle with $(\frac{m_1L_1+m_2(L_1-L_2)}{m_1+m_2}, 0)$ as the center and $r = \frac{m_1\sqrt{(L_1-L)^2+h^2}}{m_1+m_2}$ as the radius.

The center of mass of the abscissa of the angle θ_1 derivation can obtain $\frac{dX'}{d\theta_1} = \frac{m_1(h\cos\theta_1+(L_1-L)\sin\theta_1)}{m_1+m_2} > 0$, and the centroid abscissa increases with the increase in the angle θ_1, making the center of mass move forward, which favors the surmounting. However, if the included angle θ_1 is too large, the robot will tip forward under the action of inertia forces after crossing the steps, thus causing the car body to overturn. Therefore, as long as $X' \le L_1$, the robot will not flip forward around point O_2. When $X' = L_1$, $\frac{m_1(L_1-(L_1-L)\cos\theta_1+h\sin\theta_1)+\frac{m_2L_1}{2}}{m_1+m_2} = L_1$, the maximum critical value $\theta_{1max} = 71.2°$ can be obtained through calculation.

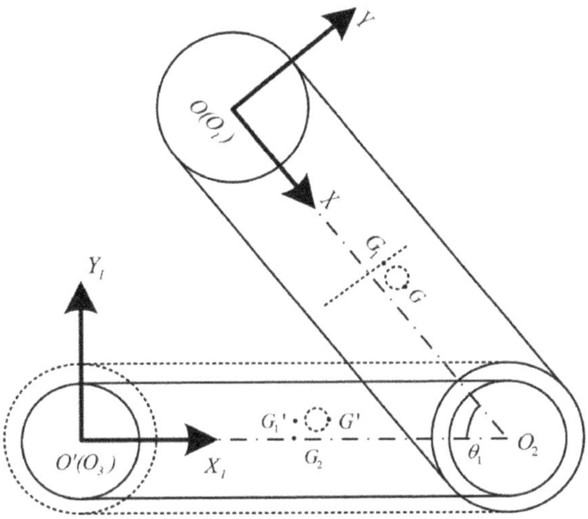

Figure 8. Lifting the state of the car body supported by the rocker arm.

Figure 9 shows the diagram of the robot's reverse obstacle crossing when the centroid of the robot happens to be at the key boundary of the step. According to Formula (7), the height of the step under the new coordinate system $X_1O'Y_1$ has the following relationship with the elevation angle α and included angle θ_1:

$$\begin{aligned} H &= r + X' \sin\alpha + Y' \cos\alpha - \frac{Y' + r}{\cos\alpha} \\ &= r + \frac{m_1(L_1 - (L_1 - L)\cos\theta_1 + h\sin\theta_1) + \frac{m_2 L_1}{2}}{m_1 + m_2} \sin\alpha - \frac{m_1(h\cos\theta_1 + (L_1 - L)\sin\theta_1)\sin^2\alpha}{(m_1 + m_2)\cos\alpha} - r\sec\alpha \end{aligned} \tag{14}$$

where r is the cross-country wheel radius.

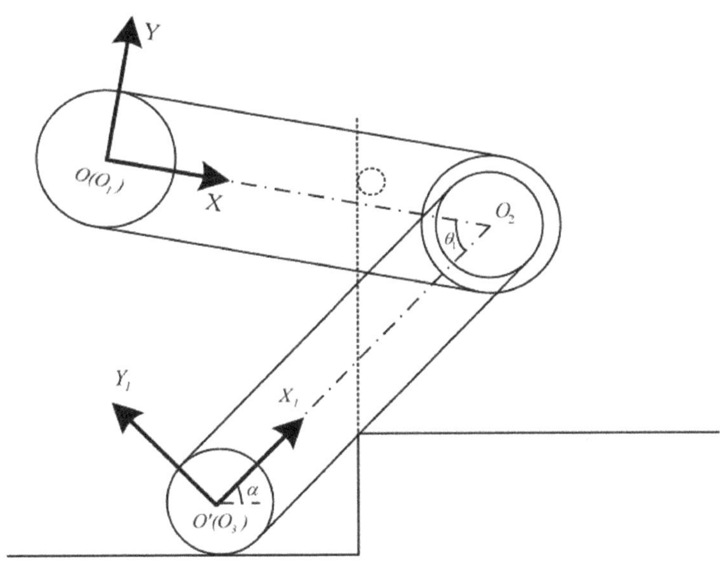

Figure 9. Robot centroid crossing the step boundary in reverse obstacle crossing.

Let the partial derivative of H with respect to angle θ_1 be 0, i.e., $\frac{\partial H}{\partial \theta_1} = 0$; then, Equation (15) can be obtained:

$$\frac{m_1 \sin\alpha}{m_1 + m_2}[((L_1 - L)\tan\alpha + h)\cos\theta_1 + (L_1 - L - h\tan\alpha)\sin\theta_1] = 0 \tag{15}$$

When $\alpha \in (0, 90^\circ)$, $\theta_1 \in (0, 71.2^\circ)$ and the elevation α and included angles θ_1 satisfy Equation (15), H has the maximum value.

4. Simulation and Experiment

4.1. Simulation Value of Obstacle Crossing Performance

1. The rescue robot crosses the barrier.

Substitute the parameters of the rescue robot into Equation (7). According to the calculation, when the pendulum angle is $\theta = 318^\circ$ and the elevation angle is $\alpha = 42^\circ$, the maximum height is achieved at 92.99 mm. The 3D relationship diagram of the step height, the robot body elevation angle and the rocker arm swing angle parameters for the forward obstacle crossing is obtained, as shown in Figure 10a. It is shown in the figure that when the robot is climbing over step obstacles, when the rocker arm is adjusted to the horizontal state, the height H that the robot can cross rises first and then descends with the increase in the elevation angle α of the car body. Therefore, increasing the elevation angle α of the car body within a certain range is suitable for the step crossing performance.

2. The rescue robot surmounts the obstacle in reverse.

When the rescue robot adopts reverse climbing steps, the parameters of the robot are substituted into Equation (13). According to the calculation, when the elevation angle $\alpha = 37.8^\circ$ and the included angle $\theta_1 = 48.6^\circ$, the maximum height of the reverse climbing steps of the robot is 155.82 mm. The 3D relationship diagram of the step height H, robot elevation angle α and rocker arm and car body parameter θ_1 for reverse obstacle crossing was obtained by simulation, as shown in Figure 10b. In the range of $\alpha \in (0, 90^\circ)$ and $\theta_1 \in (0, 71.2^\circ)$, the height H that the robot can cross rises first and then descends with the increase in the angle θ_1. Therefore, when the robot does not flip over, increasing the angle θ_1 between the rocker arm and the car body benefits the step crossing performance.

4.2. Obstacle Crossing Performance Test

In order to verify the maximum obstacle surmounting capability of the four-track twin-rocker rescue robot, a step with a height of 93 mm, as shown in Figure 11, was selected for forward obstacle surmounting, while a step with a height of 156 mm, as shown in Figure 12, was selected for reverse obstacle surmounting. It can be stated that the rescue machine can surmount the step according to the obstacle surmounting mode shown in Figure 3. In the theoretical calculation, it is assumed that the track is rigid, but in the actual measurement, the track is soft, which leads to a drop in the center of mass when the track touches the ground, which is more conducive for the robot to overcome obstacles. However, when the robot crosses the obstacle in the forward direction, the support force, friction force and tension force of the track between the two off-road wheels are less than those of the track on the rocker arm when crossing the obstacle in the reverse direction. The driving force, friction force and edge line support make it easier for the center of mass to overcome obstacles. As a result, the maximum obstacle surmounting ability of the robot in reverse is stronger than that when moving forward. The measured value of the maximum height that can be crossed in the forward direction is 95 mm. The measured maximum height that can be crossed in reverse is 165 mm.

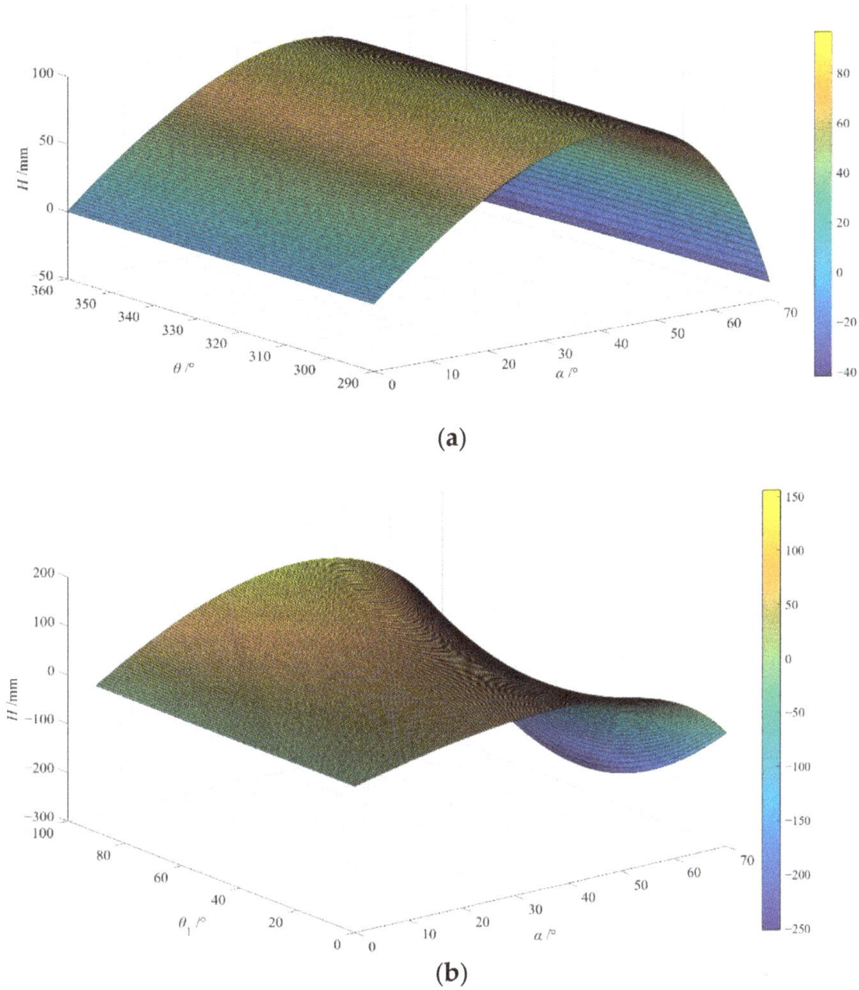

Figure 10. Simulation analysis: (**a**) with a positive obstacle height H, body elevation angle α and radial angle θ of the 3D curved surface; (**b**) reverse with the obstacle height H, body elevation angle α and rocker arm and body angle θ_1 of the 3D surface.

Figure 11. Robot climbing up steps: (**a**) lift the rocker arm; (**b**) track wheels against the step; (**c**) centroid crossing the step; (**d**) car body over the step.

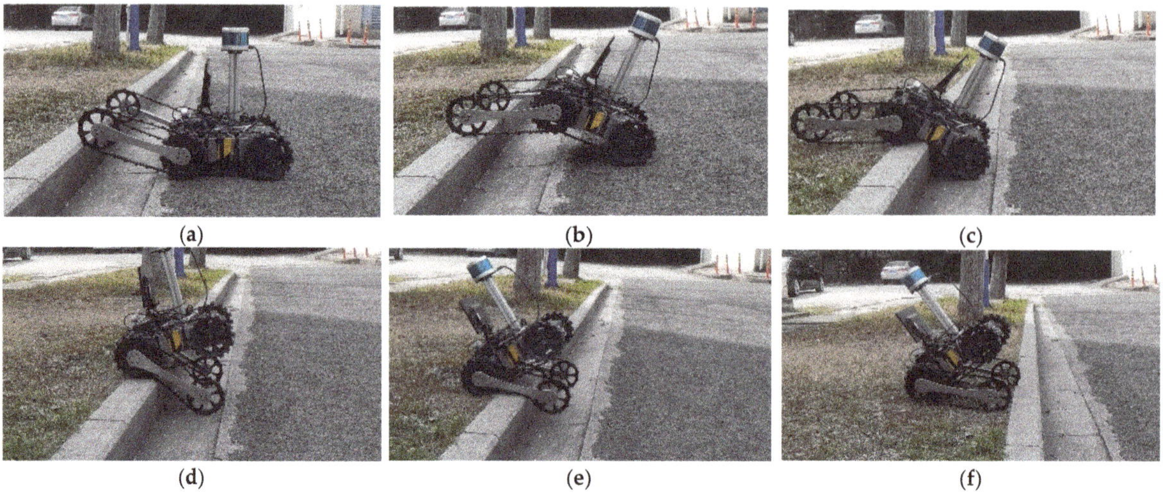

Figure 12. Reverse climbing process of the robot: (**a**) lift the rocker arm; (**b**) track wheels against the step; (**c**) body track against the step; (**d**) radial supporting body; (**e**) centroid crossing the step; (**f**) car body crosses the step.

4.3. Optimization Design of Structural Parameters

Through experiments, it was found that the length of the rocker arm has a great influence on the performance of the reverse obstacle surmounting. In order to further improve the performance of obstacle surmounting, particle swarm optimization was adopted to optimize the structural parameters of the robot's double rocker arm. The bionic particle swarm is initialized by setting the number of particles in the swarm, the maximum number of iterations and the position parameters. The optimal solution of a single particle and the global optimal solution for the population are obtained according to the fitness function. By iterations, if the fitness value of the new-found particle is better than that of the previous one, the particle position and historical optimal solution are updated and compared with the current global optimal solution. If the individual extreme values of all particles are better than the current global optimal solution, the solution is updated, and the position of the particle is recorded until the maximum number of iterations is reached or the optimal result has been found. The specific procedure is shown in Figure 13.

Figure 14 shows that when the number of iterations is 9, the curve gradually converges. When $L_2 = 157.59$ mm, and the optimal length of the rocker arm is 315.18 mm, the maximum obstacle crossing performance of the robot can be achieved. For $L_2 = 157.59$ mm and $L_2 = 189$ mm, the curved surfaces of the crossing height H versus the elevation angle α and the included angle θ_1 are shown in Figure 15. It can be seen that when $L_2 = 157.59$ mm, the maximum height of the reverse climbing step of the robot is 171.58 mm, which is 10.15% higher than the maximum obstacle height that can be crossed when $L_2 = 189$ mm.

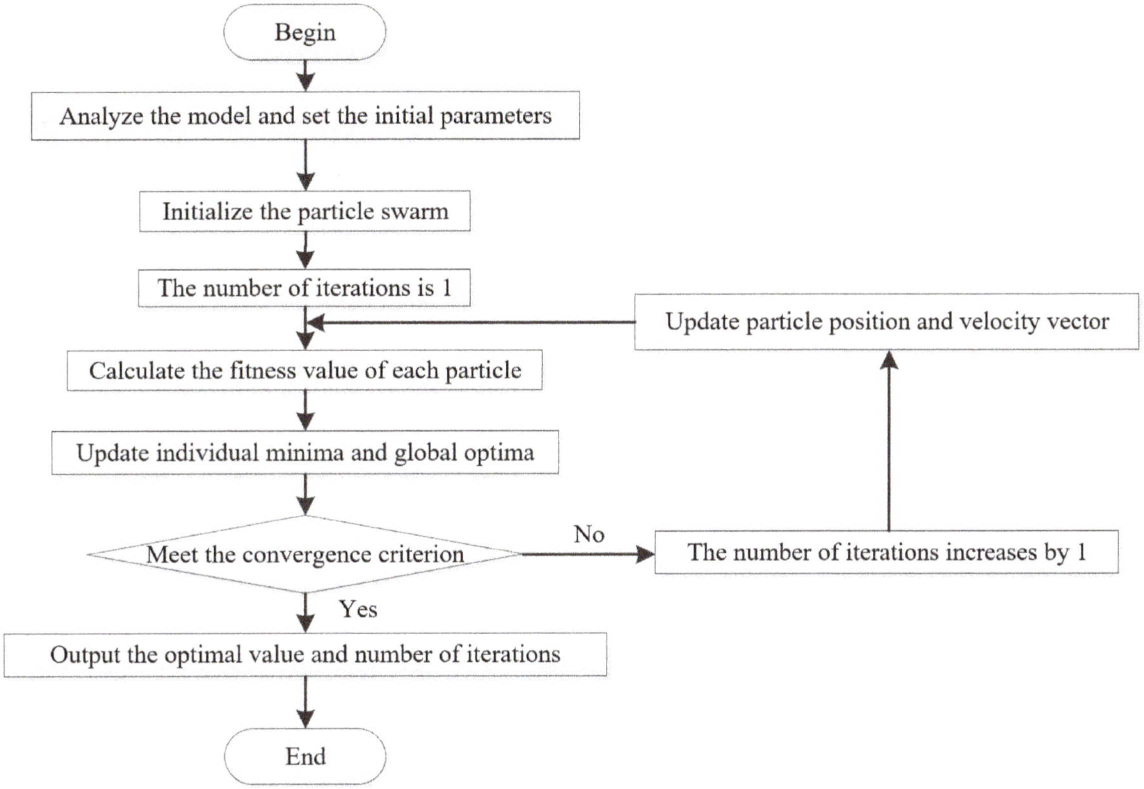

Figure 13. Flow chart of the particle swarm optimization algorithm.

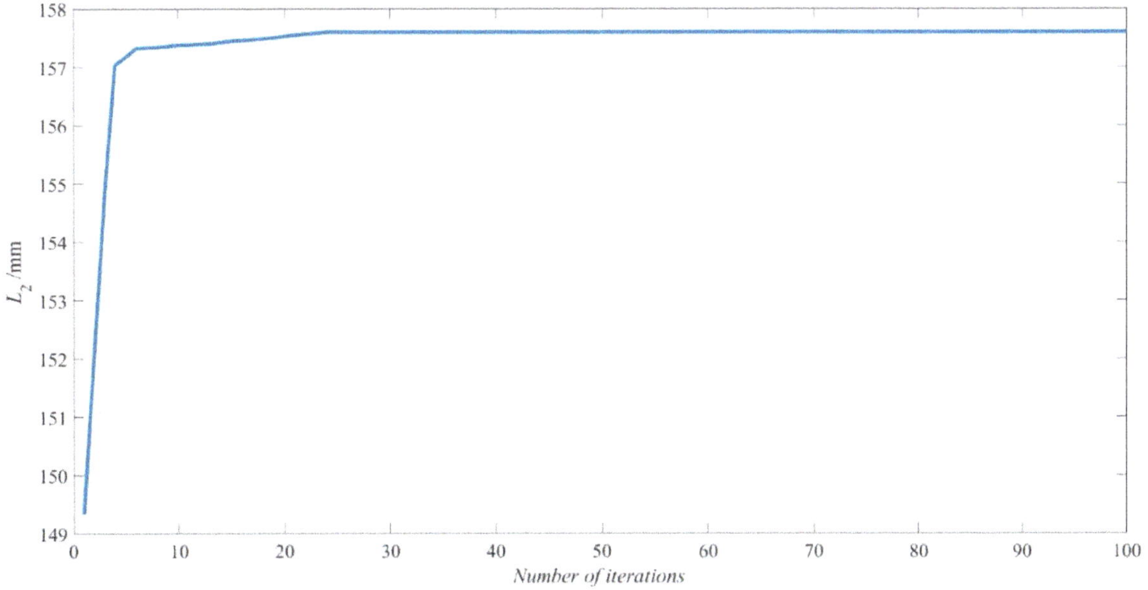

Figure 14. Convergence curve of L_2 with the number of iterations.

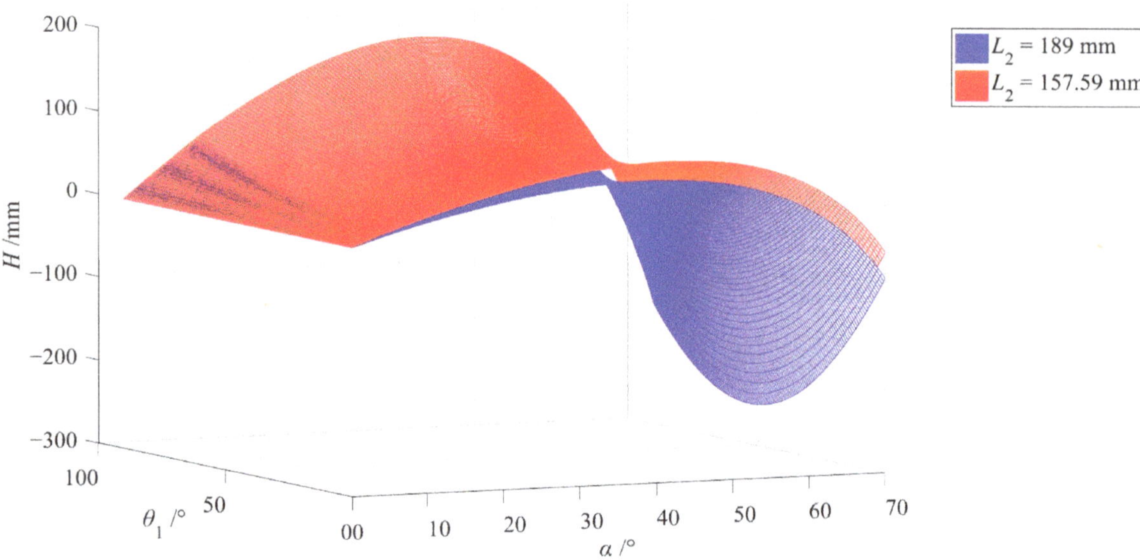

Figure 15. Comparison of obstacle heights with different arm lengths.

5. Discussion

Table 2 provides a brief comparison with previously proposed four-track twin-rocker robots. From the table, we can see that the mass of the double rocker arm in the literature [8,11] is relatively short, and only forward obstacle crossing can be carried out, thus reducing the obstacle crossing performance. Moreover, the obstacle crossing ability is only simulated without actual measurement. In addition, almost all the previous studies emphasized the influence of the arm's length on the robot's obstacle crossing function. The following conclusions are drawn: The increase in the centroid height and the forward shift in position provide more favorable conditions for climbing steps and increase the turning moment required for obstacle crossing. The change of centroid is closely related to the arm length, but few people optimize the arm length through the particle swarm optimization algorithm. Therefore, it is of great significance to combine simulations and experiments to optimize the arm length structure.

Table 2. Comparison of our four-track twin-rocker robot with similar structures.

	Xue [8]	Fang [11]	Li [13]	Wang [17]	This Paper
Body length/mm	100	350	350	400	378
Rocker arm length/mm	45	150	225	350	378
Obstacle was reversed	No	No	Yes	Yes	Yes
Actual measurement was conducted	No	No	Yes	Yes	Yes
Structure was optimized	No	No	No	No	Yes

6. Conclusions

Along with the structure of the four-track twin-rocker rescue robot, the mathematical model of the robot climbing steps forward and backward was established in this paper. The theoretical formula of the pose state of the robot was derived to achieve the maximum obstacle surmounting performance. The obstacle crossing performance of the four-track twin-rocker rescue robot was analyzed by simulation. Particle swarm optimization was used to optimize the structure of the rocker arm, and the following conclusions were obtained:

1. By comparing the theoretical calculation with the experiment, it can be seen that the experimental measurement was larger. The measured maximum forward and reverse crossing heights were 93 mm and 156.1 mm, respectively.
2. Combined with the simulation and experiment, it was found that the length of the rocker arm is critical to the obstacle surmounting process. The particle swarm optimization algorithm was used to optimize the structural parameters of the robot, and the optimal arm length of the robot to achieve the maximum obstacle surmounting ability was found to be 315.18 mm.

Author Contributions: Conceptualization, X.X. and W.W.; methodology, X.X. and W.W.; software, X.X. and G.S.; validation, X.X., W.W. and C.L.; formal analysis, X.X., W.W. and H.Z.; investigation, X.X., H.Z. and C.L.; resources, X.X.; data curation, X.X. and G.S.; writing—original draft preparation, X.X. and W.W.; writing—review and editing, X.X. and W.C.; visualization, X.X.; supervision, Y.R., Z.T. and M.L.; project administration, X.X.; funding acquisition, X.X. All authors have read and agreed to the published version of the manuscript.

Funding: This research was funded by the Changzhou Sci&Tech Program (Grant No. CE20215041), the Fundamental Research Funds for the Central Universities (Grant No. B220202023) and the Jiangsu Key R&D Program (Grant No. BE2020082-1).

Institutional Review Board Statement: Not applicable.

Informed Consent Statement: Not applicable.

Data Availability Statement: Not applicable.

Conflicts of Interest: The authors declare no conflict of interest.

References

1. Premachandra, C.; Otsuka, M.; Gohara, R.; Ninomiya, T.; Kato, K. A study on development of a hybrid aerial terrestrial robot system for avoiding ground obstacles by flight. *IEEE/CAA J. Autom. Sin.* **2018**, *6*, 327–336. [CrossRef]
2. Zhang, D.; Guo, C.; Ren, H.; Zhu, P.; Xu, M.; Lu, H. The Design of an Aerial/Ground Dual-Modal Mobile Robot for Exploring Complex Environments. In Proceedings of the 2021 IEEE International Conference on Real-Time Computing and Robotics (RCAR), Xining, China, 15–19 July 2021; IEEE: Piscataway, NJ, USA; pp. 393–398.
3. Takemori, T.; Miyake, M.; Hirai, T.; Wang, X.; Fukao, Y.; Adachi, M.; Yamaguchi, K.; Tanishige, S.; Nomura, Y.; Matsuno, F.; et al. Development of the multifunctional rescue robot FUHGA2 and evaluation at the world robot summit 2018. *Adv. Robot.* **2019**, *34*, 119–131. [CrossRef]
4. Cho, C.; Lee, W.; Kang, S.; Kim, M.; Song, J.-B. Uneven terrain negotiable mobile platform with passively adaptive double tracks and its application to rescue missions. *Adv. Robot.* **2005**, *19*, 459–475. [CrossRef]
5. Kim, J.; Lee, C.; Kim, G. Study of machine design for a transformable shape single-tracked vehicle system. *Mech. Mach. Theory* **2010**, *45*, 1082–1095. [CrossRef]
6. Li, Y.W.; Ge, S.R.; Zhu, H.; Fang, H.F. Mobile Platform of a Rocker-Type W-Shaped Track Robot. *Key Eng. Mater.* **2010**, *419–420*, 609–612. [CrossRef]
7. Liu, S.; Guo, Y.; Jia, H.; Lin, S.; Zhao, D. Mechanism principle and obstacle-crossing analysis of robot with automatic-strained track and variable main arm configuration. *J. Cent. South Univ.* **2013**, *44*, 2289–2297.
8. Xue, T.; Liu, R.; Zhai, S.M. Structure Design and Analysis of Moving Character for Detect Robot for Underground Coal Mine. *Appl. Mech. Mater.* **2014**, *526*, 205–210. [CrossRef]
9. Han, X.; Lin, M.; Wu, X.; Yang, J. Design of an Articulated-Tracked Mobile Robot with Two Swing Arms. In Proceedings of the 2019 IEEE 4th International Conference on Advanced Robotics and Mechatronics (ICARM), Toyonaka, Japan, 3–5 July 2019; IEEE: Piscataway, NJ, USA; pp. 684–689.
10. Li, C.S.; Cui, G.Q.; Pang, S.H. Dynamic Modeling and Analysis of a Mobile Robot for Multi-Movement States. *J. Hebei Univ. Technol.* **2009**, *38*, 31–35.
11. Fang, H.F.; Ge, S.R.; Li, Y.W. Obstacle Performance Analysis of Four-Track Robot with Compliant Swing Arms. *J. China Univ. Min. Technol.* **2010**, *39*, 682–686.
12. Xie, S.; Bao, S.; Zou, B.; Pu, H.; Luo, J.; Gu, J. The research on obstacle-surmounting capability of six-track robot with four swing arms. In Proceedings of the 2013 IEEE International Conference on Robotics and Biomimetics (ROBIO), Shenzhen, China, 12–14 December 2013; IEEE: Piscataway, NJ, USA, 2013; pp. 2441–2445. [CrossRef]
13. Li, Y.; Ge, S.; Hua, Z.; Zhu, H.; Liu, J. Obstacle-Surmounting Mechanism and Capability of Four-Track Robot with Two Swing Arms. *Robot* **2010**, *32*, 157–165. [CrossRef]

14. Ye, C.; Li, J.; Yu, S.; Ding, G. Movement Analysis of Rotating-Finger Cable Inspection Robot. *Int. Conf. Intell. Robot. Appl.* **2019**, *11744*, 326–337. [CrossRef]
15. Cui, D.; Gao, X.; Guo, W.; Li, J. Multimode obstacle-crossing analysis of a wheel/track mobile robot. In Proceedings of the 2017 IEEE International Conference on Unmanned Systems (ICUS), Beijing, China, 17–29 October 2017; IEEE: Piscataway, NJ, USA, 2017; pp. 197–203. [CrossRef]
16. Ma, J.M.; Li, X.F.; Yao, C.; Wang, Z. Dynamic Modeling and Analysis for Obstacle Negotiation of Ground Mobile Robot. *Robot* **2008**, *30*, 273–278.
17. Wang, W.; Du, Z.; Sun, L. Dynamic Load Effect on Tracked Robot Obstacle Performance. In Proceedings of the 2007 IEEE International Conference on Mechatronics, Kumamoto, Japan, 8–10 May 2007; IEEE: Piscataway, NJ, USA, 2007; pp. 370–375. [CrossRef]
18. Yamada, Y.; Miyagawa, Y.; Yokoto, R.; Endo, G. Development of a Blade-Type Crawler Mechanism for a Fast Deployment Task to Observe Eruptions on Mt. Mihara. *J. Field Robot.* **2015**, *33*, 371–390. [CrossRef]
19. Ye, C.Q.; Shang, W.Y.; Xu, Z.P.; Liu, J.J.; Huang, Q.R.; An, B. Structural Design of Robot Applied in Post-Earthquake Relief. *Appl. Mech. Mater.* **2013**, *437*, 619–622. [CrossRef]

machines

Article

Design and Modeling of a Bio-Inspired Compound Continuum Robot for Minimally Invasive Surgery

Gang Zhang [1,2], Fuxin Du [1,2,3,*], Shaowei Xue [1], Hao Cheng [1], Xingyao Zhang [1], Rui Song [4] and Yibin Li [4]

1 School of Mechanical Engineering, Shandong University, Jinan 250100, China; gangzhang@mail.sdu.edu.cn (G.Z.); shaowei_xue@mail.sdu.edu.cn (S.X.); haocheng@mail.sdu.edu.cn (H.C.); zhangxingyao@mail.sdu.edu.cn (X.Z.)
2 Key Laboratory of High Efficiency and Clean Mechanical Manufacture of Ministry of Education, Shandong University, Jinan 250061, China
3 Beijing Advanced Innovation Center for Intelligent Robots and Systems, Beijing Institute of Technology, Beijing 100811, China
4 School of Control Science and Engineering, Shandong University, Jinan 250100, China; rsong@sdu.edu.cn (R.S.); liyb@sdu.edu.cn (Y.L.)
* Correspondence: dufuxin@sdu.edu.cn

Abstract: The continuum robot is a new type of bionic robot which is widely used in the medical field. However, the current structure of the continuum robot limits its application in the field of minimally invasive surgery. In this paper, a bio-inspired compound continuum robot (CCR) combining the concentric tube continuum robot (CTR) and the notched continuum robot is proposed to design a high-dexterity minimally invasive surgical instrument. Then, a kinematic model, considering the stability of the CTR part, was established. The unstable operation of the CCR is avoided. The simulation of the workspace shows that the introduction of the notched continuum robot expands the workspace of CTR. The dexterity indexes of the robots are proposed. The simulation shows that the dexterity of the CCR is 1.472 times that of the CTR. At last, the length distribution of the CCR is optimized based on the dexterity index by using a fruit fly optimization algorithm. The simulations show that the optimized CCR is more dexterous than before. The dexterity of the CCR is increased by 1.069 times. This paper is critical for the development of high-dexterity minimally invasive surgical instruments such as those for the brain, blood vessels, heart and lungs.

Keywords: compound continuum robot; concentric tube continuum robot; notched continuum robot; dexterity

Citation: Zhang, G.; Du, F.; Xue, S.; Cheng, H.; Zhang, X.; Song, R.; Li, Y. Design and Modeling of a Bio-Inspired Compound Continuum Robot for Minimally Invasive Surgery. *Machines* **2022**, *10*, 468. https://doi.org/10.3390/machines10060468

Academic Editors: Yanjie Wang, Xiaofeng Liu, Aihong Ji, Shichao Niu and Bo Li

Received: 18 April 2022
Accepted: 9 June 2022
Published: 11 June 2022

1. Introduction

The powerful locomotion capability of biologically inspired continuum robots in small spaces has attracted increasing attention from researchers [1–4]. As shown in Figure 1, this robot achieves extreme dexterity through the biological structure of bionic snakes, octopus tentacles and elephant trunks.

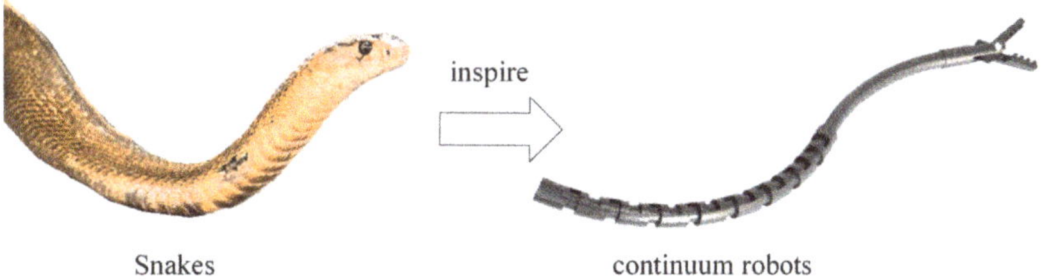

Figure 1. Bio-inspired continuum robot.

The ultra-high dexterity makes the continuum robot widely used in various fields of medicine such as laparoscopy and thoracoscopy [5–7]. However, with the development of minimally invasive surgery, the requirements for surgical instruments have become stricter. The instruments are required to bypass human organs or tissues to perform operations deeper in the human body, such as transseptal puncture, vascular surgery and skull base surgery [8,9]. While maintaining high dexterity, continuum robots with smaller diameters and a larger central access are required in these procedures.

Continuum robots were first built in the 1960s. After the 2000s, with the clarification of the application scenarios of continuum robots, related research has gradually increased. The application in the medical field has greatly promoted the research and development process of continuum robots [10,11]. In the past 10 years, many continuum robots have been proposed; these continuum robots have obvious trade-offs between the size of the center channel and the size of the continuum robot. By imitating the bones of snakes, a continuum robot composed of hinged joints was proposed by Li [12–14]. This robot has excellent dexterity, but the existence of the hinge narrows the central channel and limits the use of end instruments. A continuum robot composed of discrete joints and a central rod was proposed by Simaan [15,16]. Although this structure has unique advantages in terms of accuracy and stiffness, the central rod of this continuum limits the end instruments. To solve the problem of the tiny central channel, Murphy et al. [17–19] proposed a notched continuum robot (NCR). This kind of continuum robot replaces the hinge joints of the hinged-joints continuum robot by machining notches on the Nitinol tube in order to obtain a more significant central channel. NCRs have good stiffness and motion accuracy. In subsequent studies, this notched continuum robot has been used in laryngeal surgery [20], pediatric vascular surgery [21] and other aspects. However, similar to the hinged-joints continuum robot, the diameter of the robot is limited to the millimeter level. In 2009, Webster et al. [22] proposed a concentric tube robot (CTR) made of a nested pre-bent Nitinol tube. By eliminating the drive cable, the diameter of the robot can easily reach sub-millimeters. The surgical instruments can be made smaller in size. This special continuum robot has been used in skull base tumor resection [23], cardiac intervention [24], lung biopsy [25], etc. Although the CTR can achieve a satisfactory diameter, Alfalahi et al. [26] pointed out an apparent trade-off between the stiffness of the CTR and the working space. The stiffness of this robot is mainly determined by the material, which makes it difficult to change. Therefore, the workspace of the CTR with the same configuration is smaller than that of the NCR.

Other kinds of continuum robots have also been further developed. Fluid-driven and magnetic-field-driven continuum robots are a research hotspot. The fluid-driven continuum robot has excellent flexibility and a great turning angle. Such robots have been intensively studied by Greer and Laschi et al. [27,28]. Miniaturization and safety issues are the keys to hindering the application of fluid-driven continuum robots in surgery. There is a risk of rupture of the fluid bag when moving near scalpels. Magnetic-field-driven continuum robots can obtain a satisfactory diameter. However, such continuum robots have limitations in terms of biocompatibility and flexibility [1]. All in all, there are still many problems to be solved in the application of these two continuum robots in surgery.

Mixing different forms of continuum robots to improve the performance of robots has attracted the attention of researchers. The cable-driven continuum robot and CTR can complement each other's shortcomings. The cable drive continuum increases the CTR's workspace. The CTR reduces the size of cable-driven continuum robots. In 2017, the initial concept of a hybrid-actuated continuum robot was proposed by Li [29]. Li proposed the combination of a cable-driven continuum with a CTR. The dexterity of this continuum robot was evaluated. Li's paper is very instructive. However, Li did not consider the application limitations of this continuum robot, especially in terms of the size. In 2021, Abdel-Nasser [30] proposed the use of an articulated continuum robot combined with a CTR for use in minimally invasive surgery. Abdel-Nasserd's research shows that the hybrid-actuated continuum robots have a greater workspace and dexterity. However, not all types of cable-driven continuum robots are suitable for being combined with CTR. The diameter of the articulated continuum robot is too large to be applied in minimally invasive surgery. Moreover, Li and Abdel-Nasser did not consider the stability issues of the CTR part when the cable-driven continuum robot par is in a state of high curvature. It should be noted that instability refers to the rapid jump of the concentric tube robot from one equilibrium position with higher potential energy to another equilibrium point with lower potential energy. This movement is difficult to control and is extremely dangerous for surgical procedures [31,32]. In the design, we need to try to avoid this situation.

This paper proposed a compound continuum robot combining the CTR and the NCR which can achieve a smaller diameter and a larger central cavity. The possible stability issues of the CCR are considered. Compared with CCRs proposed in previous studies, the CCR proposed in this article is more suitable for minimally invasive surgery. The contributions can be summarized as follows:

- A compound continuum robot (CCR) combining the concentric tube continuum robot (CTR) and the notched continuum robot is proposed to design high-dexterity minimally invasive surgical instruments. The simulations show that the CCR's workspace is bigger than that of the CTR and that the dexterity indices of the CCR are 1.231 times larger than those of the CTR.
- Stability issues in the CTR part were considered. The failure boundaries of the workspace are defined. In the newly defined workspace, the CCR can perform stable movements. The CTR section avoids instability.
- The dexterity index of the CCR is proposed. The length distribution of the compound continuum robot is optimized using a fruit fly algorithm based on the dexterity index.

The rest of the paper is as follows. Section 2 elaborates on the critical issues of this paper. Section 3 summarizes the methods and calculations used in this paper. Section 4 simulates the workspace and dexterity of the CCR and discusses the obtained results. Section 5 summarizes the whole paper.

2. Problem Formulation

Miniaturization and dexterity have been the focus of research on microsurgical instruments. The excellent adaptability and dexterity of the continuum robot make it widely used in surgical instruments. In the past decade, continuum robots have been widely adopted in laparoscopy and thoracoscopy. With the advancement of technology, minimally invasive surgery has gradually developed into deeper areas of the human body that are more difficult to reach with traditional minimally invasive surgery, such as the skull base, heart and lungs, as shown in Figure 2. These operations not only require the extremely high dexterity of surgical instruments but also have strict requirements as to the peripheral diameter of the surgical instruments and the diameter of the central cavity. The importance of the diameter is obvious. During the design stage, the importance of the cavity in the center of a surgical instrument is often overlooked. However, in the actual surgical process, it is very unlikely that only one instrument is used, and the existence of the central cavity exists to realize the replacement of these surgical instruments.

However, it is difficult to reduce the diameter and increase the diameter of the central cavity of the cable-driven continuum robots. The emergence of the concentric tube robot is a turning point in terms of the miniaturization of the continuum robot. The design of the cable drive is canceled in CTR. It completes the dexterous movement by advancing and rotating the pre-bent elastic tube. This design provides a large central channel for surgical tools, and the diameter can be easily reduced by using flexible tubes with small diameters. The CTR's invariable curvature cable-driven continuum robot leads to a smaller working space and minimizes the dexterity compared with the cable-driven continuum robot. Some researchers found that the defects of continuum robots can be overcome by combining different continuum robots. However, the current design of the compound continuum robot is still at the theoretical stage. Neither the small diameter nor the large central lumen of the surgical instruments required in actual surgery are discussed. A feasible solution still needs to be devised. As shown in Figure 2, in this paper, a biomimetic excitation composite continuum robot with high dexterity and a small diameter and large central cavity is proposed. This special continuum robot has broad application prospects in surgical robots. Meanwhile, like ordinary continuum mechanisms, the CCR can also provide a solution for the development of dexterous and manual small instruments.

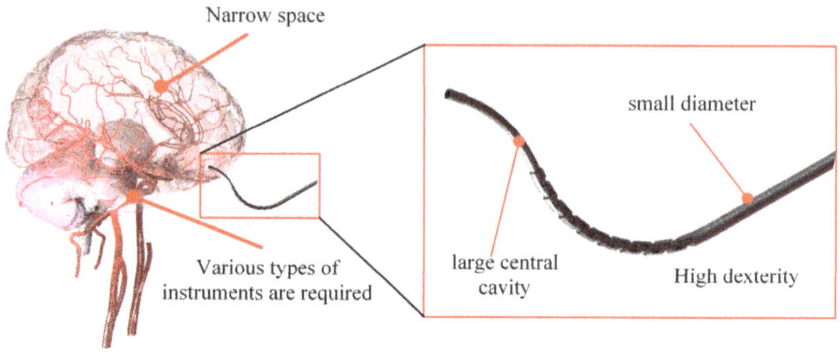

Figure 2. Requirements for surgical instruments in surgery.

3. Methods

This section summarizes the methods and theories used in this paper. The motion mechanism was expounded, and the kinematic model of the CCR was established. The stability condition of the CCR was introduced as the boundary condition of the workspace. Then, a posture dexterity evaluation method was introduced to evaluate the dexterity index of the CCR. Finally, a fruit fly algorithm was used to optimize the length assignments for continuum robots.

3.1. Design and Mechanism of Compound Continuum Robot

Keeping a large central cavity in the small size of the continuum robot is one of the key factors that hinder designers. The traditional processing methods make it difficult for ordinary cable-driven continuum robots to meet the requirements of operations in small spaces such as the skull base and heart. In the process of research, we found that the combination of the NCR and CTR not only circumvents the shortcomings of each but also meets physicians' expectations for a new generation of surgical instruments, as shown in Figure 3.

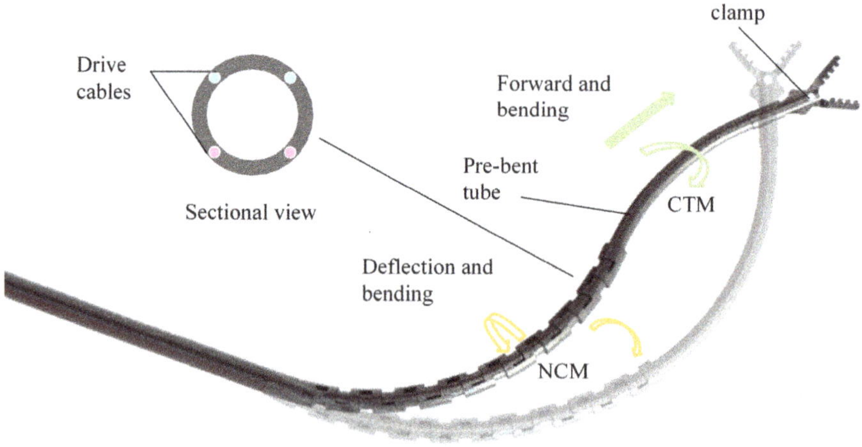

Figure 3. A bio-inspired continuum robot compound composed of a CTR with an NCR.

Femtosecond laser machined Nitinol tubes can be used as NCR parts. The tube wall needs to be pre-machined with four through holes through which the drive cables are to pass. The CTR part is a curved Nitinol tube with a curvature that is set in advance. The 3D printing of elastic materials to process the CCR is also an ideal processing method.

The CCR is divided into three parts: the NCR part, the CTR part and the surgical instrument part. The NCR has degrees of freedom in rotation and bending. The bending degree of freedom of the NCR is achieved by two pairs of antagonistic filaments, and the rotational degree of freedom is achieved by the rotation of the straight tube. The CTR has a unique operating mechanism with degrees of freedom for feed and rotation. The feeding of the CTR releases the elastic potential energy accumulated by the pre-bent tube, thereby enabling bending. The curvature of the CTR, without releasing the elastic potential energy, will be limited to the curvature of the NCR portion. Figure 3 shows the clamps commonly used in surgery. The surgical clamps were passed through the cavity of the CTR section. When an instrument needs to be replaced, the clamp can be pulled out along the cavity, and a new surgical instrument can be inserted through the cavity.

Before the combination, the NCR was able to obtain a large working space, and the size was difficult to reduce. The CTR enables smaller diameters but less flexibility and a smaller workspace. Both have a larger central cavity. After bonding, the larger central cavity is preserved. The robot's dexterity and minimum size constraints are lifted.

3.2. *Kinematics Model and Smooth-Running Workspace*

Based on the piecewise constant curvature hypothesis proposed by Hannan and Walker [33] et al., the kinematic model of the composite continuum can be established. It assumes that the curvatures of all the points of the continuum robot backbone are equal constants. As shown in Figure 4, the parameters of the CCR are defined. The skeleton curves of NCR and CTR can be regarded as two arcs. We stipulate that the constant length of the NCR part is L_1, the bending angle is θ_1 and the rotation angle is φ_1. The variable length of the CTR part is L_2, the bending angle is θ_2, the rotation angle is φ_2 and the fixed curvature is k. The parameter θ_2 is related to the invariant parameter k, and the relationship between θ_2 and k is as follows.

$$\theta_2 = kL_2 \tag{1}$$

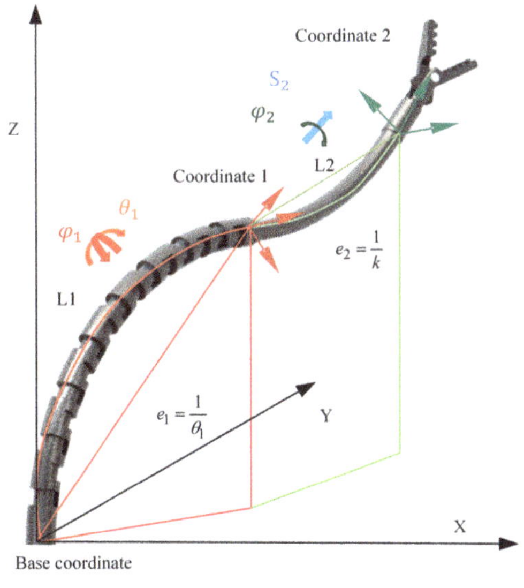

Figure 4. The backbone curve and coordinates of the CCR.

The NCR part can be abstracted into a circular tube, as shown in Figure 5. The effect of incisions on the NCR on the shape of the NCR backbone was not considered when the number of incisions was sufficient. The yellow and blue rods represent the cables that drive the NCR. The amount of change (θ) of each parameter in the joint space of the NCR part is driven by the change in the cable length in the drive space (l).

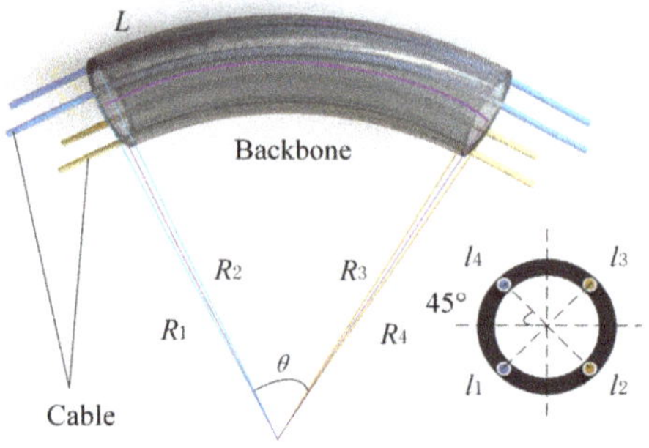

Figure 5. The relationship between the cable length and angle of the NCR.

Set the yellow cable as cable 1 and cable 2, the blue cable as cable 3 and cable 4 and the bending radius of each cable as R_i. i refers to the i-th cable. There is 45° spacing between the cables.

According to Figure 5, the radius of the backbone (R_0) of the NCR can be expressed by:

$$R_0 = L/\theta \tag{2}$$

According to the geometric relationship in Figure 5, the length of each cable can be obtained.

$$l_i = \theta R_i \tag{3}$$

$$R_i = R_N \mp \sqrt{2}d/4 \tag{4}$$

Use minus when i is 1 or 2 and plus when i is 3 or 4.

The length of the CTR parts is L_2 = [0, $L_{2\max}$]. Taking the fixed end of the NCR as the basic coordinate system, the DH parameters of the CCR can be obtained, as shown in Table 1:

Table 1. CCR's D-H parameters.

D-H Parameters	θ	a	α	d
NCR parts	φ_1	$L_1\sin\theta_1/\theta_1$	θ_1	$L_1(1-\cos\theta_1)/\theta_1$
CTR parts	$\pi/2$	0	φ_2	0
	$\sin L_2/k$	kL_2	$(1-\cos kL_2)/k$	$\sin kL_2/k$

Bring it into the following homogeneous change matrix:

$$T_{i-1}^{i} = \begin{vmatrix} c\theta_i & -s\theta_i c\alpha_i & s\theta_i s\alpha_i & a_i c\theta_i \\ s\theta_i & c\theta_i c\alpha_i & -c\theta_i s\alpha_i & a_i s\theta_i \\ 0 & s\alpha_i & c\alpha_i & d_i \\ 0 & 0 & 0 & 1 \end{vmatrix} \tag{5}$$

In Equation (5), $c\theta_i = \cos\theta_i$, $s\theta_i = \sin\theta_i$, $c\alpha_i = \cos\alpha_i$, $s\alpha_i = \sin\alpha_i$.

By multiplying the homogeneous change matrices of each joint in turn, the end pose matrix $\boldsymbol{T_z}$ of the CCR can be solved, as shown in Equation (6).

$$T_z = T_0^1 T_1^2 T_2^3 = \begin{pmatrix} R_z & P_z \\ 0 & 1 \end{pmatrix} = \begin{pmatrix} R_{11} & R_{12} & R_{13} & P_1 \\ R_{21} & R_{22} & R_{23} & P_2 \\ R_{31} & R_{32} & R_{33} & P_3 \\ 0 & 0 & 0 & 1 \end{pmatrix} \quad (6)$$

$R_{11} = -c\theta_2(s\varphi_1 s\varphi_2 - c\theta_1 c\varphi_1 c\varphi_2) - s\theta_2 c\varphi_1 s\theta_1$, $R_{21} = c\theta_2(c\varphi_1 s\varphi_2 - c\theta_1 s\varphi_1 c\varphi_2) - s\theta_2 s\varphi_1 s\theta_1$, $R_{31} = -s\theta_2 c\theta_1 - c\theta_2 c\varphi_2 s\theta_1$, $R_{12} = -c\varphi_1 c\varphi_2 - c\theta_1 c\varphi_1 s\varphi_2$, $R_{22} = c\varphi_1 c\varphi_2 - c\theta_1 s\varphi_1 s\varphi_2$, $R_{32} = s\theta_2(c\varphi_1 s\varphi_2 + c\theta_1 s\varphi_1 c\varphi_2) + c\theta_2 s\varphi_1 s\theta_1$, $R_{13} = c\theta_2 c\varphi_1 s\theta_1 - s\theta_2(s\varphi_1 s\varphi_2 - c\theta_1 c\varphi_1 c\varphi_2)$, $R_{23} = s\theta_2(c\varphi_1 s\varphi_2 + c\theta_1 s\varphi_1 c\varphi_2) - c\theta_2 s\theta_1 s\varphi_1$, $R_{33} = c\theta_2 c\theta_1 - s\theta_2 c\varphi_2 s\theta_1$, $P_1 = \frac{1}{k}((c\theta_2 - 1)(s\varphi_1 s\varphi_2 + c\theta_1 c\varphi_1 c\varphi_2) + s\theta_2 s\theta_1 c\varphi_1) - \frac{1}{\theta_1} L_1 c\varphi_1 (c\theta_1 - 1)$, $P_2 = \frac{-1}{k}((c\theta_2 - 1)(c\varphi_1 s\varphi_2 + c\theta_1 s\varphi_1 c\varphi_2) - s\theta_2 s\theta_1 s\varphi_1) - \frac{1}{\theta_1} L_1 s\varphi_1 (c\theta_1 - 1)$, $P_3 = \frac{1}{k}((c\theta_2 - 1)s\theta_1 c\varphi_2 + s\theta_2 c\theta_1) + \frac{1}{\theta_1} L_1 s\theta_1$, $\theta_2 = kL_2$.

In the same way, the forward kinematics model of the two-segment NCR and the forward kinematics model of the two-segment CTR can be solved. The DH parameters are shown in Tables 2 and 3.

Table 2. NCR's D-H parameters.

D-H Parameters	i	θ	a	α	d
Segment 1	1	φ_1	$L_1 \sin\theta_1/\theta_1$	θ_1	$L_1(1-\cos\theta_1)/\theta_1$
Segment 2	2	$\pi/2$	0	φ_2	0
	3	$-\pi/2$	$L_2 \sin\theta_2/\theta_2$	θ_2	$L_2(1-\cos\theta_2)/\theta_2$

Table 3. CTR's D-H parameters.

D-H Parameters	i	θ	a	α	d
Segment 1	1	φ_1	$\sin k_1 S_1/k_1$	$k_1 S_1$	$(1-\cos k_1 S_1)/k_1$
Segment 2	2	$\pi/2$	0	φ_2	0
	3	$-\pi/2$	$\sin k_2 S_2/k_2$	kS_2	$(1-\cos k_2 S_2)/k_2$

Where φ_1 and φ_2 are the rotation angles of the first segment and the second segment of the NCR, L_1 and L_2 are the lengths of the first and second segments of the NCR and θ_1 and θ_2 are the bending angles of the first and second segments of the NCR.

S_1 and S_2 are the feed movements of the CTR's first and second sections. k_1 and k_2 are the curvatures of the first and second segments of the CTR.

The CTR part is a pre-curved tube with a fixed curvature. When the NCR part bends, the curvature of the CTR part nested in the NCR part is changed. The elastic force caused by the changing curvature of the CTR part acts on the tube wall of the NCR part. When this curvature change is greater than the maximum allowable changing curvature, the elastic force causes stability problems in the CTR part. As shown in Figure 6, the stability problem refers to a sudden change in the movement speed of the CTR part when the CTR part is driven to rotate. The CTR part jumps rapidly from one point (usually a high-potential stabilization point) to another (usually a low-potential stabilization point).

The condition that the CTR is partially stable is introduced to correct the kinematics. The relevant theory proposed by Xu and Dupont [34,35] is used to construct the stability boundary conditions of the CCR. The theory is constructed based on the principle of least potential energy.

$$L_c\sqrt{r} < \arctan\left(\frac{K_1 + K_2}{(K_1 l_2 + K_2 l_1)\sqrt{r}}\right) \quad (7)$$

$$r = (1 + v)\|u_1\|\|k\| \quad (8)$$

$$v = \frac{k_{2xy}}{k_{2z}} - 1 \quad (9)$$

Among them, L_c is the length of the curved part of the concentric tube, K_1 and K_2 are the stiffnesses of each tube and l_1 and l_2 are the lengths of the straight part of each tube. u_1 is the curvature of the current NCR part. k_{2xy} is the bending stiffness, and k_{2z} is the torsional stiffness. The Newton iteration method was used to solve Equation (7).

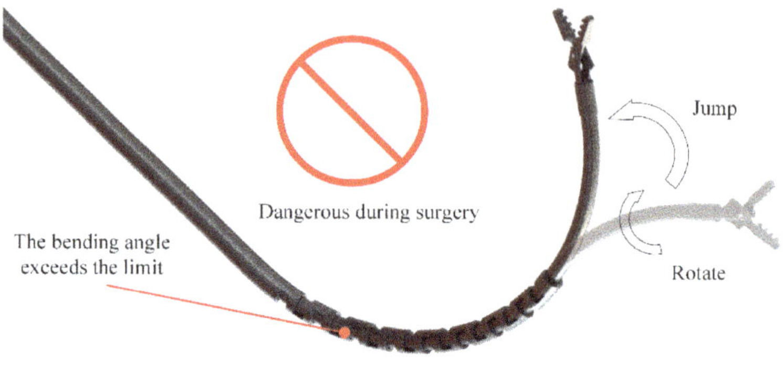

Figure 6. Stability issues of the CTR.

In actual motion control, it is necessary to obtain all of the possible motion trajectories of the robot. All of the kinematic parameters of the points on the trajectory need to be obtained. Because of the redundant nature of the CCR, the CCR can reach the same point with different attitudes. All of the kinematic parameters corresponding to each pose are listed in an array. When solving, it is stipulated that the position information of the point is already known. The kinematic parameters of all the poses that can reach this point need to be solved.

Let the position information of the point at the end of the robot be $\boldsymbol{P_e}$ $[x_e\ y_e\ z_e]^{\mathrm{T}}$. $\boldsymbol{R_e}$ is the pose information of the point. The equation that is to be solved for the inverse kinematics can be expressed as Equation (10).

$$P_e = \begin{bmatrix} ((c\theta_2 - 1)(s\varphi_1 s\varphi_2 + c\theta_1 c\varphi_1 c\varphi_2) + s\theta_2 s\theta_1 c\varphi_1)/k - L_1 c\varphi_1 (c\theta_1 - 1)/\theta_1 \\ ((c\theta_2 - 1)(c\varphi_1 s\varphi_2 + c\theta_1 s\varphi_1 c\varphi_2) - s\theta_2 s\theta_1 s\varphi_1)/k - L_1 s\varphi_1 (c\theta_1 - 1)/\theta_1 \\ ((c\theta_2 - 1)s\theta_1 c\varphi_2 + s\theta_2 c\theta_1)/k + L_1 s\theta_1/\theta_1 \end{bmatrix} \tag{10}$$

Assuming the pose is known, transform the pose into equations related to the kinematic parameters.

$$\boldsymbol{Re}(\varphi_1, \theta_1, \varphi_2, L_2) = \boldsymbol{Rs} \tag{11}$$

To eliminate the redundant nature of Equation (10), Equation (11) is introduced to solve the inverse kinematics of the CCR. Let the known terminal pose be $\boldsymbol{Rs}$ and let $\boldsymbol{Re}$ be the pose expression solved in Equation (7). By traversing all possible $\boldsymbol{Rs}$ values, the inverse kinematics of the setpoint can be solved by the LM algorithm. The iterative equation is shown in Equations (12) and (13):

$$q_{n+1} = q_n + H_n^{-1} g_n \tag{12}$$

$$H_k = J_k^T W_E J_k + W_N \tag{13}$$

The LM algorithm can obtain an inverse kinematics solution of the continuum robot. The geometric properties of the continuum robot show that its inverse kinematics solution is symmetric, as shown in Figure 7. Therefore, another solution can be obtained by exploiting the symmetry of the continuum robot.

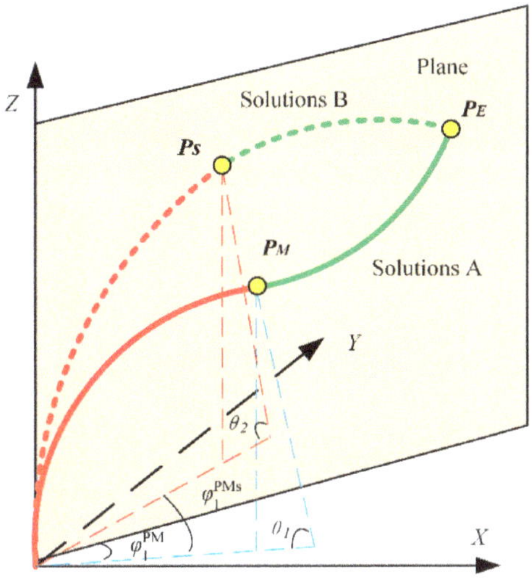

Figure 7. Two symmetric solutions at P_EOZ.

Assuming that one of the solutions A ($\varphi_1{}^{PM}$, θ_1, $\varphi_2{}^{PM}$, $L_2{}^{PM}$) is known, the midpoint of the continuum robot shape corresponding to this solution is P_M. The midpoint refers to the endpoint of the NCR portion. Let the parameter of its symmetrical solution be B ($\varphi_1{}^{Ps}$, θ_2, $\varphi_2{}^{Ps}$, $L_2{}^{Ps}$) and the midpoint of the corresponding continuum robot be P_S. A can be solved by the symmetry of the continuum robot:

$$\varphi_1^{Ps} = 2\varphi_{PE} - \varphi_1^{PM} \tag{14}$$

From Equation (15), other parameters can be obtained:

$$F(\varphi_2^{Ps}, L_2^{Ps}) = \begin{cases} x_e = \left(\left(c(kL_2^{Ps}) - 1\right)\left(s\varphi_1 s\varphi_2^{Ps} + c\theta_1 c\varphi_1 c\varphi_2^{Ps}\right) + s(kL_2^{Ps})s\theta_1 c\varphi_1\right)/k - L_1 c\varphi_1(c\theta_1 - 1)/\theta_1 \\ z_e = \left(\left(c(kL_2^{Ps}) - 1\right)s\theta_1 c\varphi_2^{Ps} + s(kL_2^{Ps})c\theta_1\right)/k + L_1 s\theta_1/\theta_1 \end{cases} \tag{15}$$

3.3. Dexterity Evaluation and Optimization

Dexterity is one of the general evaluation indexes of robots. The commonly used dexterity evaluation methods of continuum robots include the condition number, end attitude angle, etc. The dexterity of the continuum robot was evaluated based on the condition number by Wang et al. [36]. The existing research shows that the evaluation method based on the condition number examines the uniformity of the robot Jacobian transformation matrix in all directions [37]. This method cannot intuitively describe the performance of the robot. In 2016, the continuum robot's dexterity was evaluated based on a posture dexterity evaluation method that uses forward kinematics to compute the dexterity index by Wu [38]. This method can intuitively express the dexterity of the continuum robot. In this paper, the forward kinematics-based pose dexterity evaluation method is adopted.

In the pose dexterity assessment method, the dexterity of a point is defined as the number of poses the robot can achieve at that point, as shown in Figure 8. For statistical convenience, the number of all poses of that point (usually a ball) is taken as the denominator. The greater the number of poses a robot can achieve at a certain point, the more actions the robot can perform.

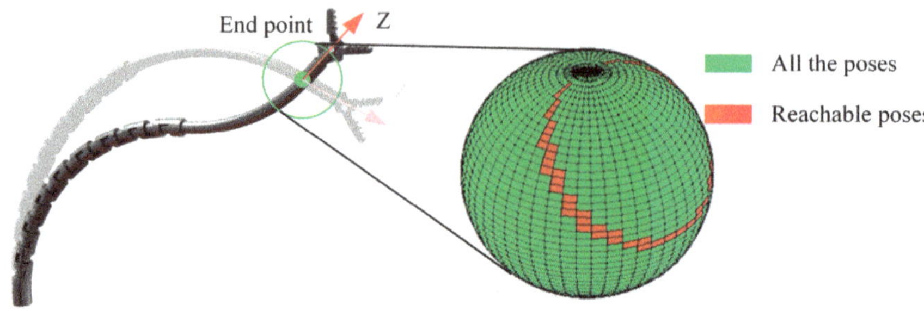

Figure 8. Cross-section of the robot workspace under the current configuration.

The end pose of the robot is represented by θ_0 and φ_0 in the Cartesian coordinate system. The end posture of the CCR can be expressed by Equation (16).

$$R_z = Rotx(\theta_0)Rotz(\varphi_0) \tag{16}$$

The *Rotz* (φ_0) is the rotation matrix around the X-axis. The *Rotz* (φ_0) is the rotation matrix around the Z-axis. The X-axis and Z-axis are divided into n_1 and n_2 parts. The dexterity of the point can be represented by the graph which the X-axis and Z-axis form. Using the graph, the dexterity value of this point can be calculated using Equation (17).

$$a_{\text{point}} = \frac{n_p}{n_1 n_2} \tag{17}$$

In Equation (17), n_p is the number of shares of θ_0 and φ_0 in the dexterity graph obtained by solving the inverse solution at this point. Then, the dexterity value of the entire workspace can be expressed by Equation (18).

$$DI = \frac{\sum_{1}^{N} a_{\text{point}}}{N} \tag{18}$$

In Equation (18), *DI* is called the global dexterity index of the robot in this configuration. The numerical calculation of dexterity here does not consider the moving of the base coordinates. When the moving of the base coordinates is introduced, the dexterity value at this point can be calculated using Equation (19).

$$a^h_{\text{point}} = \frac{n^h_p}{n_1 n_2} \tag{19}$$

where h is the distance of the feed. n^h_p is the number of shares in the dexterity map with all of the CCR's end poses under a given feed distance.

After obtaining the dexterity values of the CCRs with different length configurations, the fruit fly algorithm Toolbox of MATLAB2018b was used to optimize the length configuration of the CCRs.

4. Results and Discussion

This section simulates the properties of the CCR using the method in Section 3. All the results are analyzed and discussed. The simulations in this section were all done in MATLAB 2018b. First, the smooth workspace of the CCR was drawn. The comparison with the other continuum robot workspaces was used to demonstrate the advantages of the CCR in the workspace. Then, the dexterity map of the CCR was drawn. A comparison of the dexterities of different continuum robots was proposed. Finally, using MATLAB's FOA toolbox, the length configuration of the CCR was optimized.

4.1. Workspace

According to Equations (7)–(9), the relationship between the maximum bending angle of the NCR and the length of each part can be shown in Figure 9 when the stability is considered. The maximum angle of the NCR is limited to π due to the mechanical structure limitation.

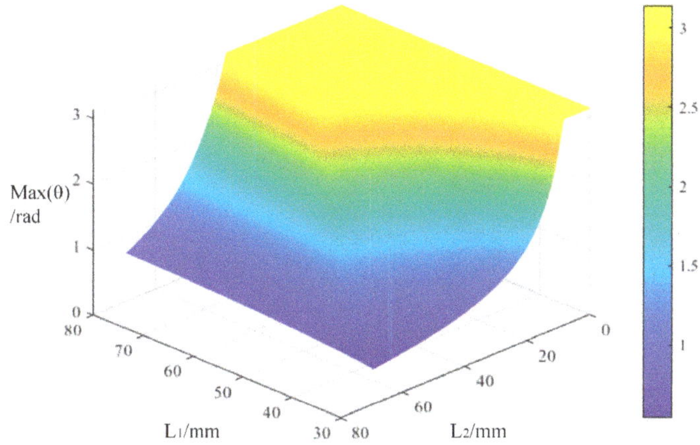

Figure 9. The relationship between the maximum bending angle of the NCR part and the length of each part.

Set the length of the CCR to be 50 mm long for the NCR part and 50 mm for the CTR part. As shown in Figure 10, the workspaces of those robots are drawn by the Monte Carlo method. Figure 10a is the workspace of the CCR considering the stability problem and the workspace without considering the stability problem (CCRF). It can be seen that the curvature of the edge portion of the CCRF exceeds the maximum curvature limit. Movement in this region can cause instability in the CTR portion. As shown in Figure 10b, the workspaces of the NCR are more three-dimensional, and the workspaces of the CTR are flatter. The reachable points of the NCR are concentrated at the far end of the robot. The reachable points of the CCR are concentrated in the middle part, and the reachable points of the CTR are focused on the end. This means that, under the same configuration, all three robots have good motion accuracy at the distal, middle and proximal end, respectively.

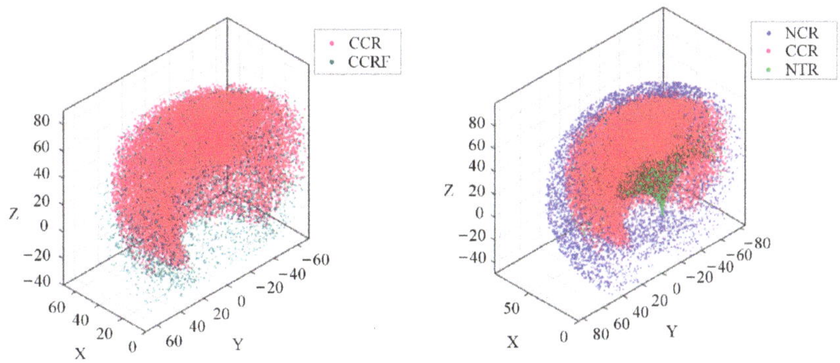

(a) (b)

Figure 10. Workspace comparison of continuum robots. (**a**) The workspace considering the stability problem and the workspace without considering the stability problem, (**b**) The workspaces of different continuum robots.

4.2. Dexterity Comparison between the CTR and CCR

A total of 20 million groups of parameters were used to randomly operate the forward kinematics so that there were enough poses to reach the same point. With enough data, it can be directly considered that the poses obtained at this time are all attainable poses. Calculate the dexterity index of all the points in the workspace. The dexterity distribution graph is drawn using the dexterity values of these points, as shown in Figure 11.

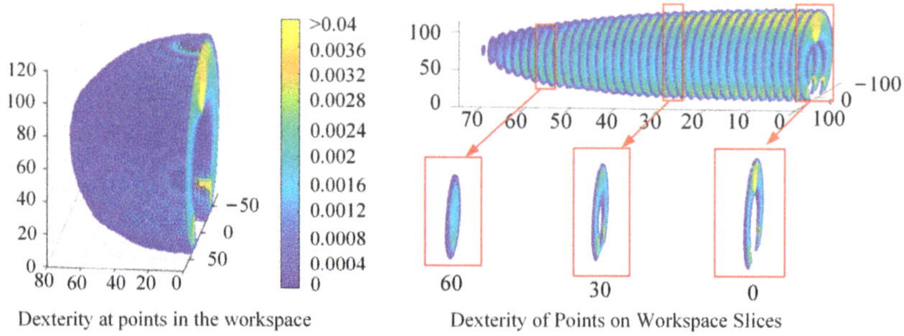

Figure 11. The dexterity distribution chart under the current configuration.

Figure 11 shows that the area with good dexterity is mainly concentrated in the middle of the robot's workspace. The points with poor dexterity are focused on the edge of the workspace. By avoiding the movement of the robot to the edge, the robot has excellent dexterity.

The global dexterity index of the CTR with the same configuration is calculated. The global dexterity index of the CCR and CTR is shown in Table 4.

Table 4. The dexterity index of the CCR and TCM.

	Total Length	Length of the NCR Part	Length of the TCR Part	k of the TCR Part	Global Dexterity
CCR	100	50	50	$\pi/100$	0.0159
CTR	100	-	50–50	$\pi/100$	0.0108

As can be seen in Table 4, the global dexterity index of the CTR is 0.0108. The dexterity index of the CCR is 1.472 times that of the CTR. The results show that the dexterity of the CTR can be improved significantly by the CCR. The dexterity graph of the CCR is shown in Figure 12. It shows that the dexterity of the CCR is improved compared with the CTR. More importantly, the high dexterity points of the CCR are more evenly distributed, and the points inside the workspace have better dexterity. The points of poor dexterity are distributed on the workspace surface.

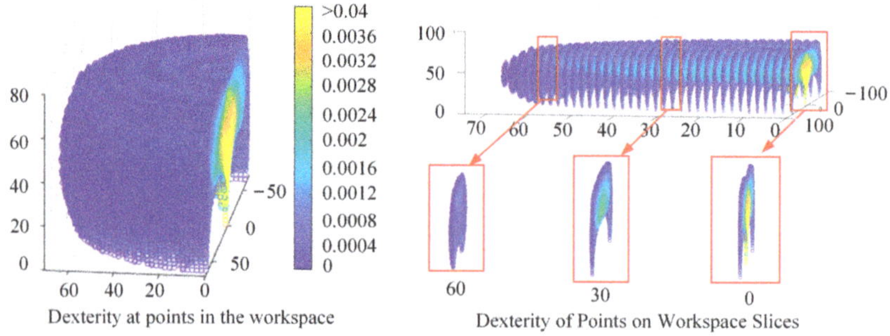

Figure 12. The dexterity performance of the CTR.

4.3. Length Distribution Optimization

The parameters and dexterity indexes for the CCR with different lengths can be found in Table 5. A total of 50 million points were introduced to calculate the dexterity indexes of the CCR under different length configurations.

Table 5. The dexterity indices of the CCR with different length ratios.

	Segment 1			*Segment 2*			*Dexterity Index (DI)*
	θ_1/rad	φ_1/rad	L_1/mm	L_2 (mm)	φ_2/rad	k/(rad/mm)	
Group 1	$[0, \theta_{g1}]$	$[0, 2\pi]$	35	[0, 65]	$[0, 2\pi]$	$\pi/100$	0.0156
Group 2	$[0, \theta_{g2}]$	$[0, 2\pi]$	40	[0, 60]	$[0, 2\pi]$	$\pi/100$	0.0168
Group 3	$[0, \theta_{g3}]$	$[0, 2\pi]$	45	[0, 55]	$[0, 2\pi]$	$\pi/100$	0.0170
Group 4	$[0, \theta_{g4}]$	$[0, 2\pi]$	50	[0, 50]	$[0, 2\pi]$	$\pi/100$	0.0159
Group 5	$[0, \theta_{g5}]$	$[0, 2\pi]$	55	[0, 45]	$[0, 2\pi]$	$\pi/100$	0.0135
Group 6	$[0, \theta_{g6}]$	$[0, 2\pi]$	60	[0, 40]	$[0, 2\pi]$	$\pi/100$	0.0123
Group 7	$[0, \theta_{g7}]$	$[0, 2\pi]$	65	[0, 35]	$[0, 2\pi]$	$\pi/100$	0.0123
Group 8	$[0, \theta_{g8}]$	$[0, 2\pi]$	70	[0, 30]	$[0, 2\pi]$	$\pi/100$	0.0123
Group 9	$[0, \theta_{g9}]$	$[0, 2\pi]$	75	[0, 25]	$[0, 2\pi]$	$\pi/100$	0.0122

The total length of the robot is set as L_m = 100 mm. The curvature of the CTR is set as $\pi/100$, and the maximum bending angle of the NCR is set as π. The global dexterity of the robots with different length distributions is calculated using the method proposed in Section 3.3. The dexterity graphs of CCRs configured with different lengths can be found in Figure 13.

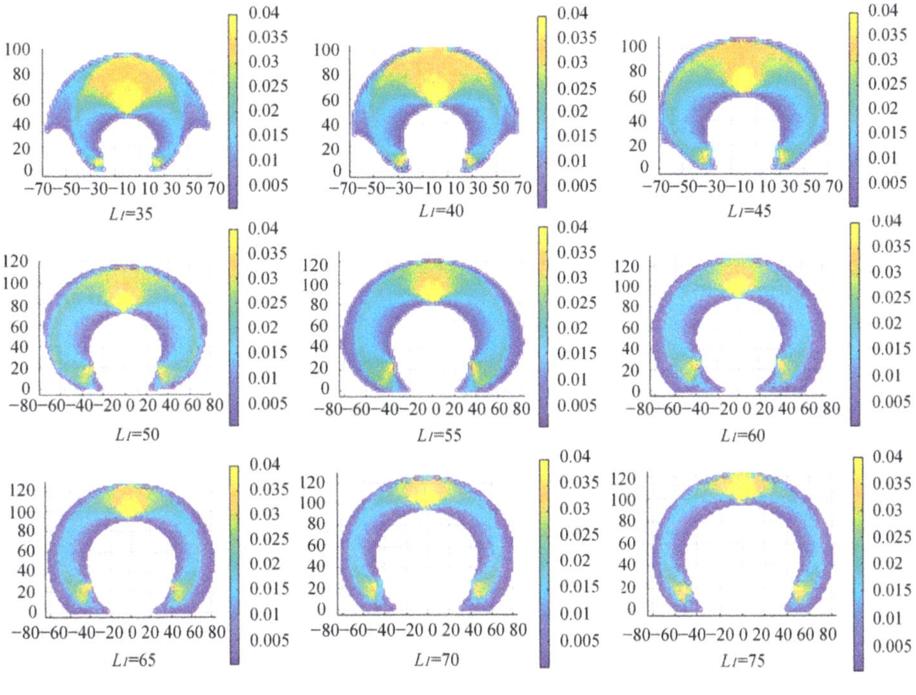

Figure 13. The dexterity of the continuum robot with different length distributions.

In Table 5, $\theta_{gi} = u_{1i}L_{1i}$,$\theta_{gi}$ is the maximum allowable bending angle of the NCR part (segment 1) in the stable operation of the CCR. Due to physical constraints, θ_{gi} is a maximum of π. u_{1i} can be calculated from Equations (7)–(9). i is the ith group of simulations.

The smoothing spline of the MATLAB Fitting Toolbox is used to fit the relationship between L and the dexterity index, as shown in Figure 14. A fruit fly optimization algorithm (FOA) was used to optimize the CCR length assignment for optimal dexterity. The result of the optimization is shown in Figure 15. After optimization, the CCR has the optimal dexterity when the length of the NCR part is 44.3032 mm. To facilitate processing, we select integers, and the length of the NCR part is 44 mm. The global dexterity index is 0.0170.

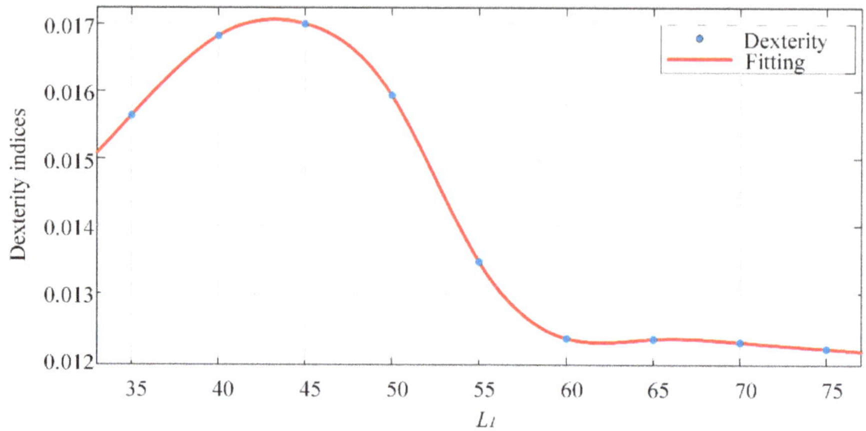

Figure 14. Fitted model.

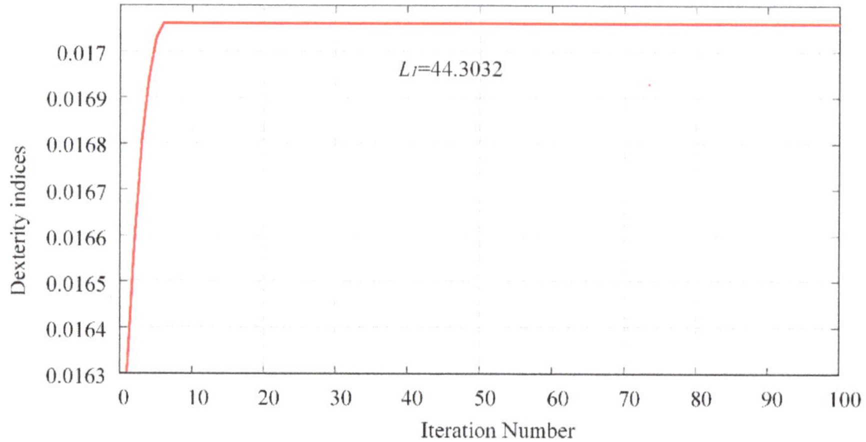

Figure 15. FOA-based dexterity index optimization.

The dexterity distribution diagram of the CCR is shown in Figure 16. Comparing Figures 12 and 16, the improved dexterity index is 7.4% after optimization. The CCR has good dexterity under the originally set length configuration. Compared with the CCR under other configurations, the dexterity index is significantly improved. For example, when the robot is configured with 65 mm (NCR part) and 35 mm (CTR part), DI = 0.0123. Its dexterity is increased by 38.2%.

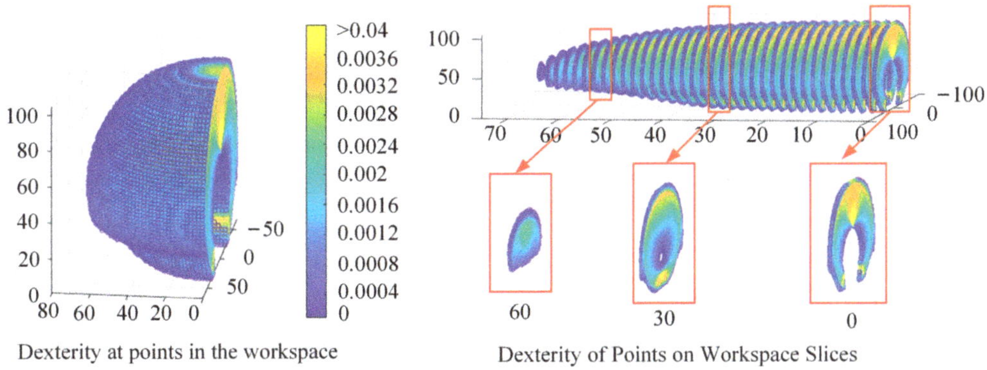

Figure 16. The dexterity after optimization.

5. Conclusions

To meet the requirements of minimally invasive surgery, a bio-inspired composite continuum robot (CCR) that combines the features of the NCR and CTR is proposed in this paper. Unlike other continuum robots, the CCR enables smaller diameters and larger central cavities. Then, to avoid unstable phenomena such as the 'bifurcation' of the CCR during operation, a stability limit is introduced to limit the maximum value of the bending angle of the NCR part. A kinematic model considering the stable motion conditions of the continuum robot is established. The workspace of the CCR is compared to the workspace of the NCR and CTR. The workspace of the CCR is significantly larger than that of the CTR. Then, the dexterity of the CCR is evaluated with an attitude angle-based evaluation method. The simulation shows that the dexterity index of the CCR was 2.32 times higher than that of the CTR. Finally, based on the dexterity index, the length distribution of the CCR was optimized using a fruit fly algorithm. The optimization results show that when the length of the CCR is configured as 44 mm (NCR part) and 56 mm (CTR part), it has the best dexterity. The dexterity value of the optimized CCR is 0.0170. The proposal of the CCR makes it possible to develop small-sized surgical instruments with a large central channel. It is ideal for minimally invasive procedures such as those for the skull base, lungs and heart.

In subsequent research, the authors will study the dynamics and statics of the CCR, study the tremor caused by CTR motion and build a corresponding experimental platform to verify the performance of the proposed model.

Author Contributions: Conceptualization, G.Z. and F.D.; methodology, G.Z.; software, S.X.; formal analysis, H.C. and X.Z.; resources, Y.L., R.S. and F.D.; data curation, G.Z.; writing—original draft preparation, G.Z.; writing—review and editing, F.D.; visualization, G.Z. and S.X.; supervision, Y.L., R.S. and F.D.; project administration, F.D.; funding acquisition, F.D. All authors have read and agreed to the published version of the manuscript.

Funding: This work was supported by the China Postdoctoral Science Foundation funded project (Grant No. 2019M662346), the Shandong Provincial Postdoctoral Innovative Talents Funded Scheme (Grant No. 238226), the Focus on Research and Development Plan in Shandong Province (Grant No. 2022CXGC010503), the Intelligent Robots and Systems High-Precision Innovation Center Open Fund (Grant No. 2019IRS06), the Fundamental Research Funds for the Central Universities and the Young Scholars Program of Shandong University.

Data Availability Statement: Not applicable.

Conflicts of Interest: The authors declare no conflict of interest.

References

1. Da Veiga, T.; Chandler, J.; Lloyd, P.; Pittiglio, G.; Wilkinson, N.J.; Hoshiar, A.K.; Harris, R.A.; Valdastri, P. Challenges of continuum robots in clinical context: A review. *Prog. Biomed. Eng.* **2020**, *2*, 032003. [CrossRef]
2. Kolachalama, S.; Lakshmanan, S. Continuum Robots for Manipulation Applications: A Survey. *J. Robot.* **2020**, *2020*, 4187048. [CrossRef]
3. Li, S.; Hao, G. Current Trends and Prospects in Compliant Continuum Robots: A Survey. *Actuators* **2021**, *10*, 145. [CrossRef]
4. Lu, J.; Du, F.; Yang, F.; Zhang, T.; Lei, Y.; Wang, J. Kinematic modeling of a class of n-tendon continuum robots. *Adv. Robot.* **2020**, *34*, 1254–1271. [CrossRef]
5. Püschel, A.; Schafmayer, C.; Groß, J. Robot-assisted techniques in vascular and endovascular surgery. *Langenbeck's Arch. Surg.* **2022**, 1–7. [CrossRef] [PubMed]
6. Iwasa, T.; Nakadate, R.; Onogi, S.; Okamoto, Y.; Arata, J.; Oguri, S.; Ogino, H.; Ihara, E.; Ohuchida, K.; Akahoshi, T.; et al. A new robotic-assisted flexible endoscope with single-hand control: Endoscopic submucosal dissection in the ex vivo porcine stomach. *Surg. Endosc.* **2018**, *32*, 3386–3392. [CrossRef]
7. Lei, Y.; Li, Y.; Song, R.; Du, F. Development of a novel deployable arm for natural orifice transluminal endoscopic surgery. *Int. J. Med Robot. Comput. Assist. Surg.* **2021**, *17*, e2232. [CrossRef]
8. Almendárez, M.; Alvarez-Velasco, R.; Pascual, I.; Alperi, A.; Moris, C.; Avanzas, P. Transseptal puncture: Review of anatomy, techniques, complications and challenges, a critical view. *Int. J. Cardiol.* **2022**, *351*, 32–38. [CrossRef]
9. Seung, S.; Liu, P.; Park, S.; Park, J.-O.; Ko, S.Y. Single-port robotic manipulator system for brain tumor removal surgery: SiromanS. *Mechatronics* **2015**, *26*, 16–28. [CrossRef]
10. Seetohul, J.; Shafiee, M. Snake Robots for Surgical Applications: A Review. *Robotics* **2022**, *11*, 57. [CrossRef]
11. Burgner-Kahrs, J.; Rucker, D.C.; Choset, H. Continuum Robots for Medical Applications: A Survey. *IEEE Trans. Robot.* **2015**, *31*, 1261–1280. [CrossRef]
12. Li, Z.; Du, R.; Lei, M.C.; Yuan, S.M. Design and analysis of a biomimetic wire-driven robot arm. In Proceedings of the ASME International Mechanical Engineering Congress and Exposition, Denver, CO, USA; 2011; Volume 54938, pp. 91–198.
13. Li, Z.; Du, R. Design and Analysis of a Bio-Inspired Wire-Driven Multi-Section Flexible Robot. *Int. J. Adv. Robot. Syst.* **2013**, *10*, 209. [CrossRef]
14. Li, Z.; Du, R.; Yu, H.; Ren, H. Statics modeling of an underactuated wire-driven flexible robotic arm. In Proceedings of the 5th IEEE RAS/EMBS International Conference on Biomedical Robotics and Biomechatronics, Sao Paulo, Brazil, 12–15 August 2014; pp. 326–331.
15. Xu, K.; Simaan, N. An Investigation of the Intrinsic Force Sensing Capabilities of Continuum Robots. *IEEE Trans. Robot.* **2008**, *24*, 576–587. [CrossRef]
16. Simaan, N.; Xu, K.; Wei, W.; Kapoor, A.; Kazanzides, P.; Taylor, R.; Flint, P. Design and Integration of a Telerobotic System for Minimally Invasive Surgery of the Throat. *Int. J. Robot. Res.* **2009**, *28*, 1134–1153. [CrossRef]
17. Wilkening, P.; Alambeigi, F.; Murphy, R.J.; Taylor, R.H.; Armand, M. Development and Experimental Evaluation of Concurrent Control of a Robotic Arm and Continuum Manipulator for Osteolytic Lesion Treatment. *IEEE Robot. Autom. Lett.* **2017**, *2*, 1625–1631. [CrossRef]
18. Gao, A.; Murphy, R.J.; Liu, H.; Iordachita, I.I.; Armand, M. Mechanical Model of Dexterous Continuum Robots with Compliant Joints and Tendon/External Force Interactions. *IEEE/ASME Trans. Mechatron.* **2017**, *22*, 465–475.
19. Murphy, R.J.; Kutzer, M.D.; Segreti, S.M.; Lucas, B.C.; Armand, M. Design and kinematic characterization of a surgical robot with a focus on treating osteolysis. *Robotica* **2014**, *32*, 835–850. [CrossRef]
20. Wang, H.; Wang, X.; Yang, W.; Du, Z. Design and kinematic modeling of a notch continuum robot for laryngeal surgery. *Int. J. Control. Autom. Syst.* **2020**, *18*, 2966–2973. [CrossRef]
21. Francis, P.; Eastwood, K.W.; Bodani, V.; Looi, T.; Drake, J.M. Design, Modelling and Teleoperation of a 2 mm Diameter Compliant Instrument for the da Vinci Platform. *Ann. Biomed. Eng.* **2018**, *46*, 1437–1449. [CrossRef]
22. Webster, I.R.J.; Romano, J.M.; Cowan, N.J. Mechanics of Precurved-Tube Continuum Robots. *IEEE Trans. Robot.* **2008**, *25*, 67–78. [CrossRef]
23. Bruns, T.L.; Remirez, A.A.; Emerson, M.A.; Lathrop, R.A.; Mahoney, A.W.; Gilbert, H.B.; Liu, C.L.; Russell, P.T.; Labadie, R.F.; Weaver, K.D.; et al. A modular, multi-arm concentric tube robot system with application to transnasal surgery for orbital tumors. *Int. J. Robot. Res.* **2021**, *40*, 521–533. [CrossRef]
24. Bergeles, C.; Gosline, A.H.; Vasilyev, N.V.; Codd, P.J.; Del Nido, P.J.; Dupont, P.E. Concentric Tube Robot Design and Optimization Based on Task and Anatomical Constraints. *IEEE Trans. Robot.* **2015**, *31*, 67–84. [CrossRef] [PubMed]
25. Swaney, P.J.; Mahoney, A.W.; Hartley, B.I.; Remirez, A.A.; Lamers, E.; Feins, R.H.; Alterovitz, R.; Webster, I.R.J. Toward Transoral Peripheral Lung Access: Combining Continuum Robots and Steerable Needles. *J. Med. Robot. Res.* **2017**, *2*, 17500015. [CrossRef]
26. AlFalahi, H.; Renda, F.; Stefanini, C. Concentric Tube Robots for Minimally Invasive Surgery: Current Applications and Future Opportunities. *IEEE Trans. Med Robot. Bionics* **2020**, *2*, 410–424. [CrossRef]
27. Greer, J.D.; Morimoto, T.K.; Okamura, A.M.; Hawkes, E.W. Series pneumatic artificial muscles (sPAMs) and application to a soft continuum robot. In Proceedings of the IEEE International Conference on Robotics and Automation, Singapore, 29 May–3 June 2017. [CrossRef]

28. Laschi, C.; Cianchetti, M.; Mazzolai, B.; Margheri, L.; Follador, M.; Dario, P. Soft Robot Arm Inspired by the Octopus. *Adv. Robot.* **2012**, *26*, 709–727. [CrossRef]
29. Dupont, P.; Simaan, N.; Choset, H.; Rucker, C. Continuum Robots for Medical Interventions. *Proc. IEEE* **2022**, 1–24. [CrossRef]
30. Li, Z.; Wu, L.; Ren, H.; Yu, H. Kinematic comparison of surgical tendon-driven robots and concentric tube robots. *Mech. Mach. Theory* **2017**, *107*, 148–165. [CrossRef]
31. Abdel-Nasser, M.; Salah, O. New continuum surgical robot based on hybrid concentric tube-tendon driven mechanism. *Proc. Inst. Mech. Eng. Part C J. Mech. Eng. Sci.* **2021**, *235*, 7550–7568. [CrossRef]
32. Rucker, C.; Childs, J.; Molaei, P.; Gilbert, H.B. Transverse Anisotropy Stabilizes Concentric Tube Robots. *IEEE Robot. Autom. Lett.* **2022**, *7*, 2407–2414. [CrossRef]
33. Hannan, M.W.; Walker, I.D. Novel kinematics for continuum robots. In *Advances in Robot Kinematics*; Springer: Dordrecht, The Netherlands, 2000; pp. 227–238.
34. Xu, R.; Atashzar, S.F.; Patel, R.V. Kinematic instability in concentric-tube robots: Modeling and analysis. In Proceedings of the 5th IEEE RAS/EMBS International Conference on Biomedical Robotics and Biomechatronics, Sao Paulo, Brazil, 12–15 August 2014; pp. 163–168.
35. Dupont, P.E.; Lock, J.; Itkowitz, B.; Butler, E.J. Design and Control of Concentric-Tube Robots. *IEEE Trans. Robot.* **2010**, *26*, 209–225. [CrossRef]
36. Wang, J.; Lau, H.Y.K. Dexterity Analysis based on Jacobian and Performance Optimization for Multi-segment Continuum Robots. *J. Mech. Robot.* **2021**, *13*, 061012. [CrossRef]
37. Biyun, X.; Jing, Z. Advances in Robotic Kinematic Dexterity and Indices. *Mech. Sci. Technol.* **2011**, *30*, 1386–1393.
38. Wu, L.; Crawford, R.; Roberts, J. Dexterity Analysis of Three 6-DOF Continuum Robots Combining Concentric Tube Mechanisms and Cable-Driven Mechanisms. *IEEE Robot. Autom. Lett.* **2016**, *2*, 514–521. [CrossRef]

machines

MDPI

Article

Reducing Helicopter Vibration Loads by Individual Blade Control with Genetic Algorithm

Renguo Yang, Yadong Gao *, Huaming Wang and Xianping Ni

National Key Laboratory of Rotorcraft Aeromechanics, Nanjing University of Aeronautics and Astronautics, Nanjing 210016, China; nuaayrg@nuaa.edu.cn (R.Y.); hm_wang@nuaa.edu.cn (H.W.); nxp201301@163.com (X.N.)
* Correspondence: gydae@nuaa.edu.cn

Abstract: A rotor that can realize individual blade pitch control was designed. This paper focuses on finding the trend of helicopter vibration loads after applying multiple high-order harmonic control. The Glauert inflow model was introduced to calculate the induced velocity of rotor blades in a rotor disk plane, and the Leishman Beddoes (L-B) unsteady dynamic model was employed to calculate the aerodynamic forces of each section of a rotor blade. It was found that the influence of each high-order harmonic control on individual blade vibration load reduction is similar in different advanced ratios. After these calculations, the genetic algorithm was used to calculate the best combination of amplitude and phase of the higher order harmonic under a specific flight state. Under the effect of high harmonic input, the vibration loads of the hub could be reduced by about 65%. These results can be theoretically applied to design control law to reduce helicopter vibration loads.

Keywords: helicopter; individual blade control; vibration control; optimal state; genetic algorithm

Citation: Yang, R.; Gao, Y.; Wang, H.; Ni, X. Reducing Helicopter Vibration Loads by Individual Blade Control with Genetic Algorithm. *Machines* **2022**, *10*, 479. https://doi.org/10.3390/machines10060479

Academic Editors: Davide Astolfi and Jan Awrejcewicz

Received: 2 May 2022
Accepted: 13 June 2022
Published: 15 June 2022

1. Introduction

Dynamic loads at helicopter rotor hubs are one of the main vibration sources of helicopters. Reducing rotor vibration loads is an important approach to helicopter vibration suppression. Centrifugal pendulums, double-wire pendulums and other passive dynamic vibration absorbers were installed on helicopters in the past to reduce rotor vibration loads [1]. With the continuous development of computer technology, sensing technology and control technology, active control methods have become a research hotspot. These methods mainly include high-order harmonic control (HHC), individual blade control (IBC), active flap control (AFC) [2], active torsion control (ATC) [3] and active control of structural response (ACSR) [4].

Individual blade control used to reduce helicopter vibration loads has been investigated and proven successful in recent literatures, which is a new method developed from HHC. HHC reduces vibration load components corresponding to the passing frequency of the rotor system by applying an excitation to the non-rotating ring of the swashplate [5]. However, IBC is more precise than HHC in theory, which can individually apply HHC to targeted blades. It can replace the traditional helicopter swashplate on the structure, as shown in Figure 1.

Between 1977 and 1985, the Ham team at the Massachusetts Institute of Technology (MIT) conducted an early experimental study of IBC [6,7]. In 2001, Sikorsky and NASA carried out a full-scale wind tunnel test to find the influence of IBC on the vibration level of a UH-60 helicopter [8]. In 2008, a group of German scientists headed by Fuerst applied practically on the basis of single blade control and proposed the concept of an electro-mechanical-actuator (EMA) rotor system [9]. In an EMA, a servo motor is installed in each arm of the rotor system to reduce rotor vibration loads. An EMA system implements the first-order periodic pitch control of the blades and high-order harmonic control of a single blade through the servo motor motion control. In September 2015, the DLR completed the first wind tunnel test of its Multiple Swashplate System. During these tests, the potential of this new active rotor control system to effectively reduce noise, vibrations and power consumption using several IBC strategies was successfully demonstrated on two different model rotors without using actuators in the rotating frame [10].

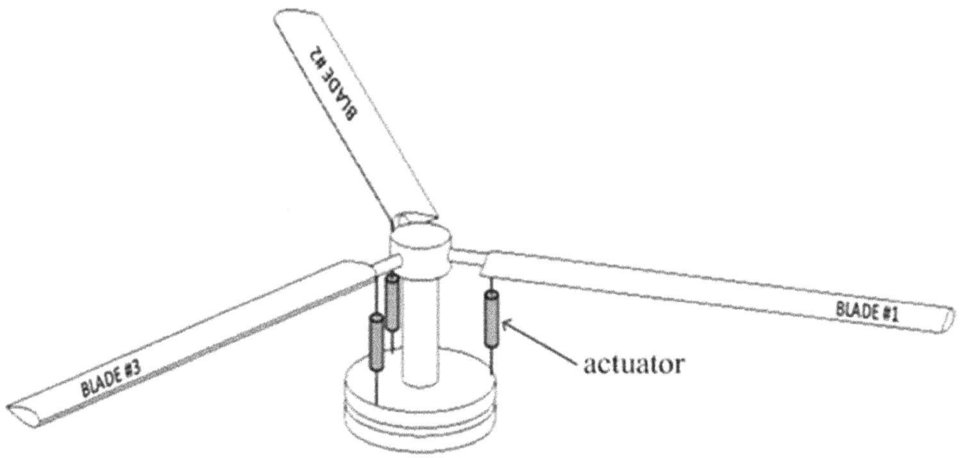

Figure 1. Helicopter individual blade control.

Genetic algorithm (GA) is a random search method derived from the evolutionary law of "survival of the fittest, survival of the fittest" in the biological world [11]. It has been widely used in combinatorial optimization, signal processing, machine learning and other fields [12–14]. The genetic algorithm regards the solution set of the problem as a population, and continuously uses genetic operators to combine individuals in the population to generate a new generation of candidate solution sets and selects the best solution from the population according to certain criteria during the iteration process, until the convergence condition is satisfied [15,16].

Although the researchers have conducted much experimentation on IBC, the number of tests that can be performed at that time is limited and finding the parameters of optimal amplitudes and phases of multiple harmonics to reduce vibration loads in this limited number of trials is difficult and expensive. The purpose of the study is to analyze the influence of IBC on helicopter hub vibration loads reduction. This study gives some analysis details on how to calculate the influence of IBC on hub vibration loads. This study presents the optimum control of selective order harmonics and their combinations in order to reduce helicopter hub vibration loads.

In order to analyze the effect of higher-order harmonics on the hub load, a new rotor is designed, as shown in Figure 2. The new rotor does not change the basic structure of the original rotor. An actuator is added to the variable-pitch tie rod. The actuator includes a motor, a spring and a rod, which can be seen in Figure 3. The lower surface of the rod is in the shape of a sine curve, so the high-order harmonic pitch control can be applied to the blades when the motor rotates at a fixed speed. In this paper, the optimal states are obtained through the GA method, which could be used to guide the design of actuator parameters.

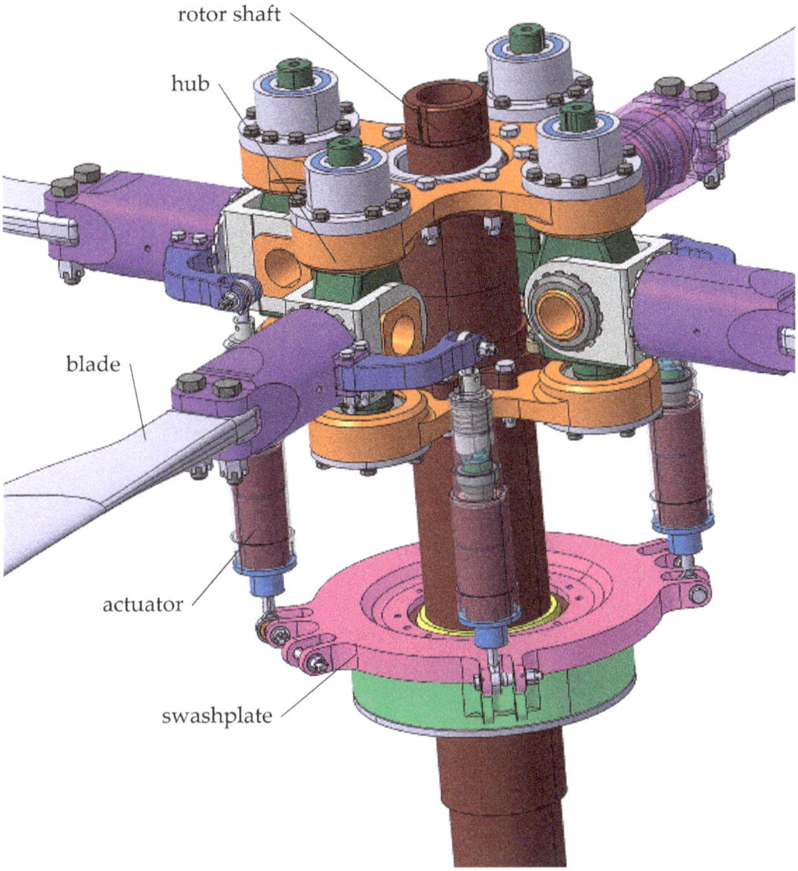

Figure 2. The rotor, controlled by IBC.

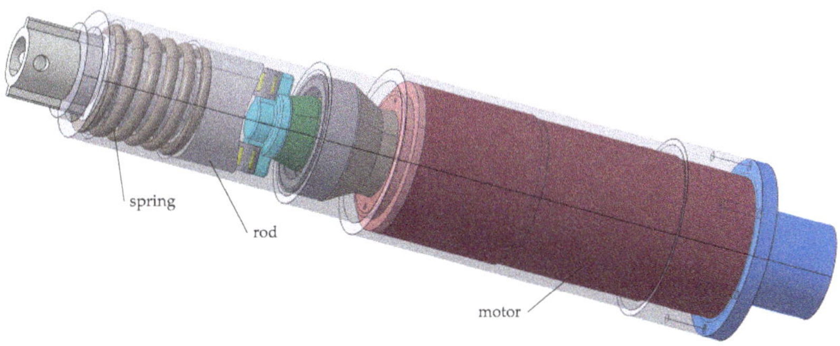

Figure 3. The actuator.

2. Calculation Method

2.1. Rotor Blade Airfoil Aerodynamics

This paper ignores the impact of IBC on original trimming of the helicopter. The individual pitch control superimposes the higher-order harmonic pitch on the basis of the rotor collective pitch control and cyclic pitch control. When a helicopter is flying forward steadily, the angle of attack (AoA) of the rotor blade airfoil is determined by the manipulated variable, the blade twist, the flapping adjustment coefficient and the higher-order harmonic pitch, shown as Equation (1).

$$\theta(t,\bar{r}) = \theta_0 + \theta_{1c}\cos(\Omega t) + \theta_{1s}\sin(\Omega t) + \theta_w\bar{r} + A_2\cos(2\Omega t + \varphi_2) + A_3\cos(3\Omega t + \varphi_3) + \ldots \quad (1)$$

where θ is the pitch angle of the blade. $\bar{r}$ is relative radius of blade section position. A_2 and A_3 represent the amplitudes of the 2nd and 3rd harmonic. φ_2, φ_3 are the phases of the 2nd and 3rd harmonic.

The flow angle of the blade airfoil depends on the relative air flow velocity and direction, as shown in Figure 4.

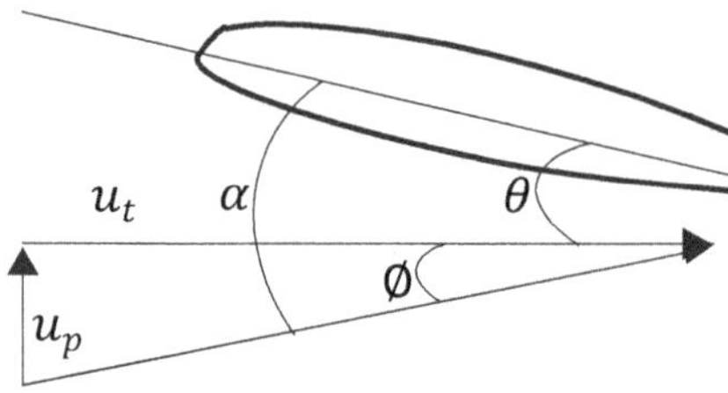

Figure 4. The flow angle of the blade airfoil.

Parameters shown in the figure can be obtained from flowing formulas.

$$\varnothing = \arctan(u_P/u_t) \quad (2)$$

$$u_P = V\sin\alpha_s - v - r\dot{\beta} - V\cos\alpha_s\cos\psi\sin\beta \quad (3)$$

$$u_t = \Omega r + V\cos\alpha_s\sin\psi \quad (4)$$

where $\varnothing$ is the angle of the wind speed and ψ is the azimuth of the blade. u_P and u_t are the vertical and lateral components of wind speed.

The flow field of a helicopter rotor is very complex. The method, proposed by Glauert, was used to calculate the inflow air velocity in the hub plane [17]. The induced velocity is expressed as a superposition of the uniform term and the periodic variation of the radius along the radial direction, shown as Equation (5).

$$v(\bar{r},\psi) = v_0(1 + K_x\bar{r}cos\psi) \quad (5)$$

In the formula, the inducing speed v_0 is the non-uniform coefficient which can be calculated from uniform inflow model (v_0—induced velocity at the rotor disc center, calculated by the momentum theory). K_x is the uneven coefficient, which is different in various theories. This article is based on the recommended values of Coleman, Feingol and Stempin, as shown in Equation (6).

$$K_x = \sqrt{1 + (\lambda/\mu)^2} - |\lambda/\mu| \quad (6)$$

The aerodynamic model proposed by Leishman and Beddoes is used to calculate the airfoil aerodynamics [18]. This model takes into account the delay effect of aerodynamic load response caused by dynamic stalls, and it also takes into account the different conditions of the aerodynamic environment (attachment flow, trailing edge separation and leading-edge separation), which accounts for the compressibility of the air flow, the lift loss caused by separation of the airfoil trailing edge and the leading edge.

2.2. Response Solution

The rotor blade is analyzed as an elastic beam [19–21]. Hamilton's variational principle is used to derive the system equations of motion, which can be expressed as

$$\delta\Pi = \int_{t1}^{t2} (\delta U - \delta T - \delta W)dt = 0 \tag{7}$$

where δU, δT, and δW are the virtual variation of strain energy, kinetic energy and the virtual work done by external forces. The variations can be written as

$$\delta U_i = \int_0^R \iint (E\varepsilon_{xx}\delta\varepsilon_{xx} + G\varepsilon_{x\eta}\delta\varepsilon_{x\eta} + G\varepsilon_{x\zeta}\delta\varepsilon_{x\zeta})d\eta d\zeta dx \tag{8}$$

$$\delta T_i = \int_0^R \iint \rho_s \vec{V}_i \cdot \delta\vec{V}_i d\eta d\zeta dx \tag{9}$$

$$\delta W_i = \int_0^R \left(L_u^A \delta u + L_v^A \delta v + L_w^A \delta w + L_\phi^A \delta\phi\right)dx \tag{10}$$

where ε_{xx} is axial strain, and $\varepsilon_{x\eta}$ and $\varepsilon_{x\zeta}$ are engineering shear strains. $\Xi\eta\zeta$ is rotating deformed blade coordinate system. L_u^A, L_v^A, and L_w^A are the distributed airloads in the x, y, z directions, respectively, and L_ϕ^A is the aerodynamic pitching moment about the undeformed elastic axis.

For the i-th blade, the virtual energy expression in Equation (7) is written in the discretized form such that

$$\delta\Pi_i = \int_{\psi I}^{\psi F} [\sum_{j=1}^{N}(\delta U_j - \delta T_j - \delta W_j)]d\psi = 0 \tag{11}$$

Using the notation

$$\Delta_j = \delta U_j - \delta T_j - \delta W_j \tag{12}$$

The blade is discretized into a number of beam elements. There are six degrees of freedom at each element boundary node. The elemental nodal displacement vector is defined as

$$q_j^T = [u_1, v_1, v_1', w_1, w_1', \phi_1, u_2, v_2, v_2', w_2, w_2', \phi_2] \tag{13}$$

Using appropriate shape functions, the elemental variation in energy Δ_j can be written in the following matrix form as

$$\Delta_j = \delta q_j^T \left([M]_j \ddot{q}_j + [C]_j \dot{q}_j + [K]_j q_j - \{F\}_j\right) \tag{14}$$

where $[M]_j$, $[C]_j$ $[K]_j$ and $\{F\}_j$ are (blade) elemental mass, damping, stiffness and load matrices.

By assembling elemental matrices, the total energy can be expressed as

$$\delta\Pi_i = \int_{\psi I}^{\psi F} \delta q^T([M]\ddot{q} + [C]\dot{q} + [K]q - \{F\})d\psi = 0 \tag{15}$$

The above formula can be transformed into the following finite element equation of the motion of the blade.

$$[M]\ddot{q} + [C]\dot{q} + [K]q = \{F\} \tag{16}$$

The modal superposition method is used in the case study to calculate the blade response. The modal superposition method refers to expressing the response of the structural system as a linear superposition of the natural modes of each order according to the inherent characteristics of the structural system [22–24], as shown in Equation (17).

$$Y(t) = \sum_{i=1}^{N} y_i \gamma_i(t) = \phi\gamma \tag{17}$$

where ϕ is modal matrix of the blade. ϕ was derived from the $[M]$, $[C]$, $[K]$ in Equation (16) by using the EIG function in matlab.

The blade system has a large degree of freedom, and it is impossible to superimpose the modes of all orders. It is necessary to perform modal truncation, and so a certain degree of calculation accuracy will be lost. Since the rotor vibration loads are mainly concentrated in the low frequency part and magnitudes of high frequency components are low, the high-order modal response amplitudes are small, and the modal truncation has little effect on the response results, so the method can satisfy the calculation needs.

2.3. Hub Loads

The hub loads can be obtained by the superposition of the loads at the root of each blade, as shown in Equation (18). Figure 5 shows the positional relationship of each force direction.

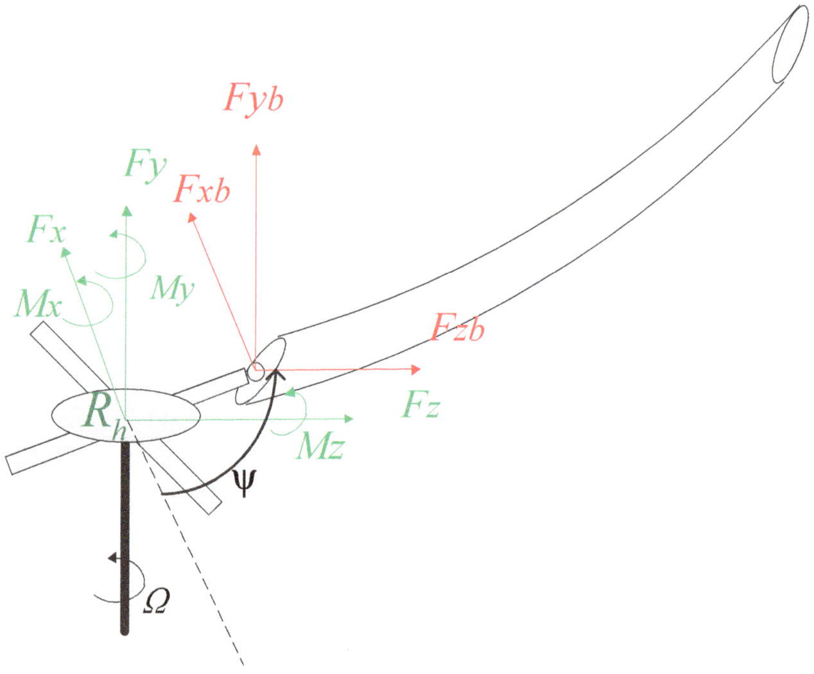

Figure 5. The relationship of each force direction.

$$
\begin{cases}
F_x(\psi) = \sum_{i=0}^{N-1} \left[F_{xb}(\psi + \frac{2\pi}{N} \times i) \sin(\psi + \frac{2\pi}{N} \times i) - F_{zb}(\psi + \frac{2\pi}{N} \times i) \cos(\psi + \frac{2\pi}{N} \times i)\right] \\
F_y(\psi) = \sum_{i=0}^{N-1} F_{yb}(\psi + \frac{2\pi}{N} \times i) \\
F_z(\psi) = \sum_{i=0}^{N-1} \left[F_{zb}(\psi + \frac{2\pi}{N} \times i) \sin(\psi + \frac{2\pi}{N} \times i) + F_{xb}(\psi + \frac{2\pi}{N} \times i) \cos(\psi + \frac{2\pi}{N} \times i)\right] \\
M_x(\psi) = e \cdot \sum_{i=0}^{N-1} \left[-F_{yb}(\psi + \frac{2\pi}{N} \times i) \sin(\psi + \frac{2\pi}{N} \times i)\right] \\
M_y(\psi) = e \cdot \sum_{i=0}^{N-1} \left[F_{xb}(\psi + \frac{2\pi}{N} \times i)\right] \\
M_z(\psi) = e \cdot \sum_{i=0}^{N-1} \left[-F_{yb}(\psi + \frac{2\pi}{N} \times i) \cos(\psi + \frac{2\pi}{N} \times i)\right]
\end{cases} \tag{18}
$$

where e is flap and lag hinge offset of the blade. The loads of the blade are obtained by integrating the loads of each blade profile in the spanwise direction, as shown in Equation (19).

$$
\begin{bmatrix} F_{xb} \\ F_{yb} \\ F_{zb} \end{bmatrix} = \int_e^R \mathrm{Fb}dr \tag{19}
$$

The loads Fb can be written as Equation (20).

$$
\mathrm{Fb} = \mathrm{Fb}^{\mathrm{A}} + \mathrm{Fb}^{\mathrm{I}} \tag{20}
$$

where Fb^{A} is the aerodynamic load which can be written as Equation (21), and Fb^{I} is inertial loads, which refers to [14].

$$
\mathrm{Fb}^{\mathrm{A}} = \begin{bmatrix} -dQ \\ dL \cos \omega' \\ dL \sin \omega' \end{bmatrix} \tag{21}
$$

where dQ and dL are the drag and lift of blade profile, and ω' is the angular displacement of the profile.

3. Case Study

The model established in Section 1 can be used to calculate the hub loads of traditional single-rotor helicopters and is not applicable to new configuration helicopters such as coaxial helicopters. A helicopter with a 3-bladed main rotor is adopted in the case study as an example [25]. The rotor parameters of the helicopter are shown in Table 1 and blade properties of each section are listed in Table 2, in which r means radial station, $\mathrm{EI_f}$ is flap stiffness, $\mathrm{EI_l}$ is lag stiffness, GJ is torsional stiffness, M is blade sectional mass and YG is chordwise blade c.g. location. In the studied case, the influence of different amplitudes and phases of different harmonic components of IBC and their combinations was analyzed.

Table 1. Parameters of the main rotor.

Number of Blades	Radius	Blade Chord	Rotor Rotational Speed	Shaft Angle of Attack	Flap and Lag Hinge Offset
3	5.25 m	0.35 m	40.5 rad/s	−4°	0.475 m

Table 2. Properties of the blade section.

r m	EIf $\times 10^3$ N.kg	EIl $\times 10^3$ N.kg	GJ $\times 10^3$ N.m^2	M kg/m	YG m	r m	EIf $\times 10^3$ N.kg	EIl $\times 10^3$ N.kg	GJ $\times 10^3$ N.m^2	M kg/m	YG m
0.445	4500	1100.0	250	25	0	3.59	7.95	420.5	11.1	5.02	−0.0038
0.475	400	350.0	1000	25	0	3.634	7.98	414.0	11.2	6.68	0.112
0.505	770	490.0	1250	15	0	3.754	7.98	414.0	11.2	6.58	0.112
0.525	9350	885.0	2200	105	0	4.173	7.91	399.6	11.2	7.59	0.204
0.545	64	320.0	2000	100	0	4.35	7.85	395.6	12.8	8.36	0.237
0.565	66	330.0	34	38.65	0	4.39	8.50	477.4	12.8	8.67	0.181
0.585	55.6	250.0	27.3	16.15	0	4.41	7.71	482.3	12.8	8.59	0.0183
0.625	51.6	186.5	21.7	8.55	0	4.44	10.04	530.8	12.8	9.64	0.0093
0.665	50.7	134.5	19.3	8.15	0	4.49	17.75	623.7	12.8	15.07	0.0096
0.705	52.00	134.5	19.3	7.75	0	4.61	17.75	623.7	12.8	14.36	0.0133
0.745	52.10	96.9	21.2	10.10	0	4.623	17.75	623.7	12.8	14.92	0.0096
0.795	47.00	104.5	22	7.15	0	4.64	21.43	820.7	19	20.16	−0.0021
0.815	48.50	115.4	22.1	6.80	−0.0017	4.665	13.45	622.5	19	21.26	0.033
0.885	38.64	132.9	18	6.25	−0.0021	4.705	14.49	1020.9	19	21.09	−0.002
0.955	21.77	329.9	17.9	6.82	0.007	4.77	22.60	1288.6	19	17.74	−0.0172
1.025	10.21	555.2	8.2	6.73	−0.0148	4.8	23.53	2301.2	25.7	19.60	−0.0215
1.165	10.31	552.8	8.2	8.50	−0.314	4.855	7.57	730.4	9.9	6.11	0.0054
1.235	7.72	646.1	6.8	8.22	−0.175	4.875	7.80	791.7	9.9	11.55	0.0026
1.305	6.49	703.3	7.9	7.73	−0.189	4.89	8.20	995.3	6.6	5.33	−0.0188
1.375	5.92	585.0	8.8	6.99	−0.144	4.95	8.20	995.3	6.6	5.48	−0.016
1.515	7.81	560.4	9.8	6.95	−0.0067	4.99	8.50	998.7	6.6	6.05	−0.015
1.605	7.71	447.5	10.6	6.51	0.0052	5.03	8.19	966.0	5.1	4.20	−0.0141
1.635	7.61	406.2	11.1	6.41	0.0102	5.07	6.46	809.5	4.9	3.84	−0.088
3.276	7.61	406.2	11.1	6.67	0.0102	5.15	6.46	809.5	4.9	3.69	−0.125
3.396	7.58	413.2	11.1	4.90	−0.0056	5.25	6.46	809.5	4.9	3.69	−0.125

3.1. Model Validation

According to the above rotor parameters and sectional blade properties, the modal frequency of the blade during rotating state was obtained and compared with the calculation results in [16]. The results are shown in Table 3.

Table 3. The mode frequencies.

Mode	Calculation/Ω	Reference/Ω
1st flap	1.03	1.03
2nd flap	2.90	2.81
3rd flap	5.32	5.24
1st lag	0.54	0.58
2nd lag	4.56	4.76
1st torsion	3.47	4.13
2nd torsion	11.26	12.81

Then, the parameters were substituted into the model established in Section 1 to obtain the hub loads. The result is subjected to FFT transformation, and the 3/rev vibration loads obtained is compared with the results calculated by Heffernan using uniform inflow when the helicopter flew at a speed of 56.4 m/s [26,27]. The comparison result is shown in Figure 6. The calculation results differ by about 10%. The calculation model is credible.

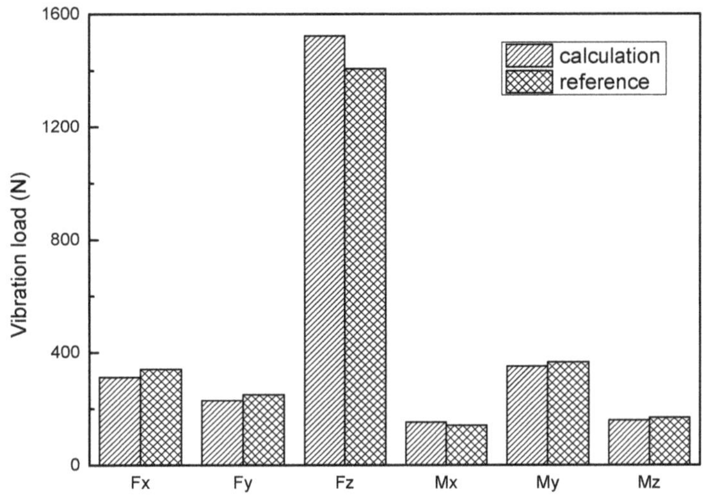

Figure 6. A comparison of calculated vibration load values with reference values.

3.2. Influence of Harmonic Phases and Amplitudes

In order to analyze the influence of amplitudes and phases of high-order harmonics on the vibration loads of the hub, the vibration load parameter *Fvh* is defined. *Fvh* is given by

$$\begin{aligned} Fvh = \; & (Max(Fx(\psi)) - Min(Fx(\psi))) + (Max(Fy(\psi)) - Min(Fy(\psi))) \\ & + (Max(Fz(\psi)) - Min(Fz(\psi))) \\ & + (Max(Mx(\psi)) - Min(Mx(\psi))) + (Max(My(\psi)) \\ & - Min(My(\psi))) + (Max(Mz(\psi)) - Min(Mz(\psi))) \end{aligned} \quad (22)$$

Fvh0 is used to represent the baseline vibration loads without IBC. *Fv*, which is the ratio of *Fvb* to *Fvb0*, can represent the impact of IBC on the vibration loads of the hub. *Fv* is given by

$$Fv = \frac{Fvh}{Fvh0}$$

If *Fv* is less than 1, it means the IBC algorithm with the selected amplitudes and phases are beneficial to vibration loads reduction; otherwise, vibration loads will become greater than those of the baseline model and IBC algorithm fails, which is adverse.

At first, some specific amplitudes of the harmonic were selected and kept unchanged in the analysis. The phases were changed from 0° to 360° with 30° spacing. Results for when the helicopter was flying under the speed of μ = 0.26 (advance ratio) are shown in Figure 7.

It can be seen in Figure 7 that the vibration load ratio changed significantly with the variety of phases of second order harmonics. They all decreased firstly, then increased as the phases increased. The vibration loads became minimum when phase angles of the second order harmonics reached about 150°. It can be seen from Figure 7 that under the action of the second order harmonic with an amplitude of 0.3° and a phase of 150°, the vibration load was less than 50% of the original.

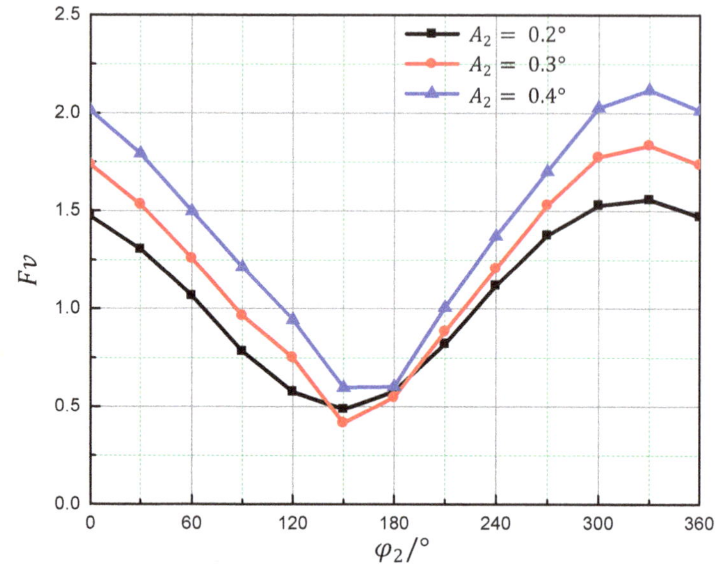

Figure 7. The influence of the phase of second order harmonicon vibration loads.

Similarly, some specific phases of the harmonics were selected and kept unchanged in the analysis in order to investigate the impact of harmonic magnitudes on vibration loads reduction under IBC.

The phases investigated in the analysis changed from 0° to 0.3°. When the helicopter was flying at the advance ratio μ = 0.26, the vibration loads were calculated, the results of which are shown in Figure 8.

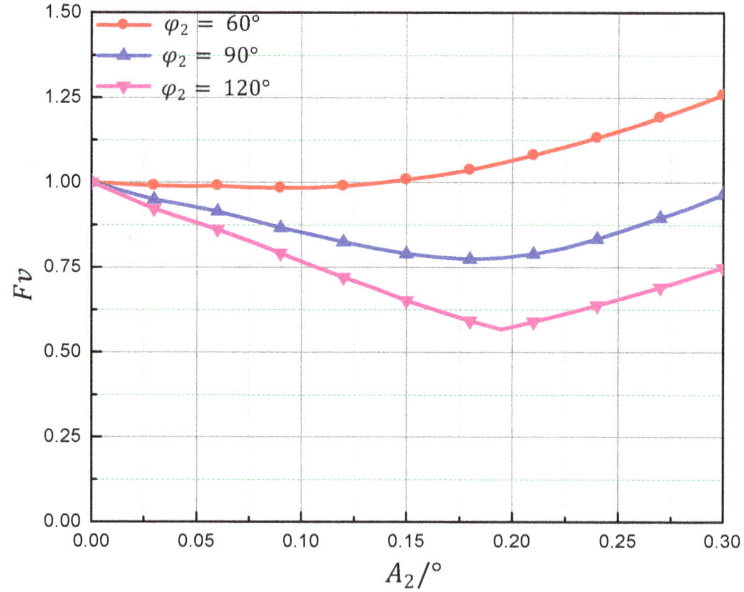

Figure 8. The influence of amplitudes of second order harmonics on vibration loads.

It can be seen in the figure that the vibration loads decrease first and then increase as the amplitudes of the second and third harmonics increase. Also, the figures show that the minimum vibration loads with IBC vary with both the amplitudes and phases of the harmonics applied.

3.3. Optimal States

Due to the complexity of the calculation of the helicopter hub loads, it is difficult to directly obtain the relationship between the vibration loads and the amplitudes and phases of each harmonic, so it is impossible to directly obtain the optimal parameters. So, the genetic algorithm is used to find the harmonic state when the vibration load is minimal [28,29]. The calculation process is shown in Figure 9.

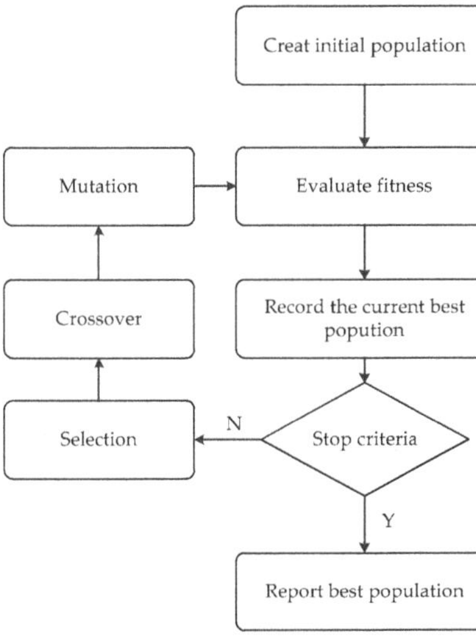

Figure 9. The process flow of a genetic algorithm.

At first, each individual is encoded as a 22-bit string. The first 10 bits represent the amplitude of the high-order harmonic, and the last 12 bits represent the phase of the high-order harmonic. The variation range of amplitude is 0–1°. The phase variation range is 0–360°. The amplitude calculation accuracy is $\frac{1}{2^{10}-1} = 0.00097 < 0.001$, and the phase calculation accuracy is $\frac{360}{2^{12}-1} = 0.0879 < 0.1$.

80 randomly generated individuals form the initial population, which is shown in Figure 10. Then, the fitness evaluation of each individual is evaluated. The fitness is reciprocal of Fv; that is, the smaller the vibration, the higher the fitness. Then, the top 50% in fitness are selected to generate offspring for the next generation by crossing their string. Some individuals also have mutations. After that, a new population is formed which will be used in the next round of calculation. The GA parameters used in the calculation are listed in Table 4.

Table 4. Parameters of the genetic algorithm.

Parameter	Value
Number of chromosomes	22
Population size	80
Number of interactions	100
Crossover probability	0.8
Mutatin probability	0.1
selection rate	0.5

The population after 100 generations of reproduction is shown in Figure 11. Figure 12 shows the variation of the optimal individuals in each generation. The trend chart of *Fv* under each generation of optimal individuals is shown in Figure 13.

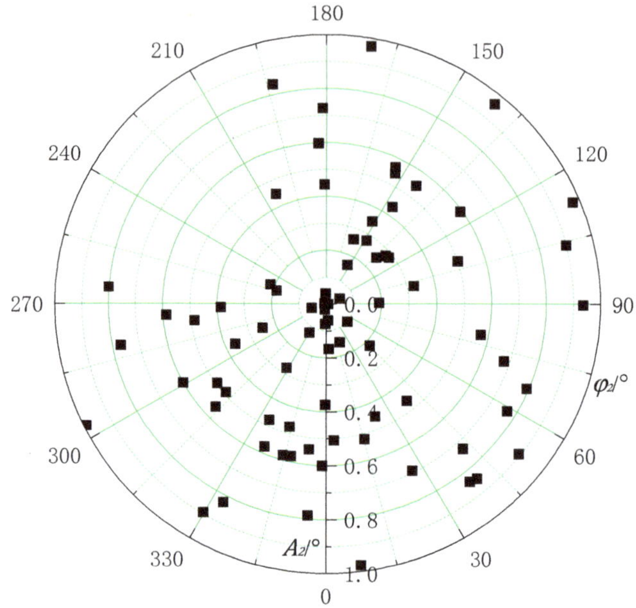

Figure 10. The initial population.

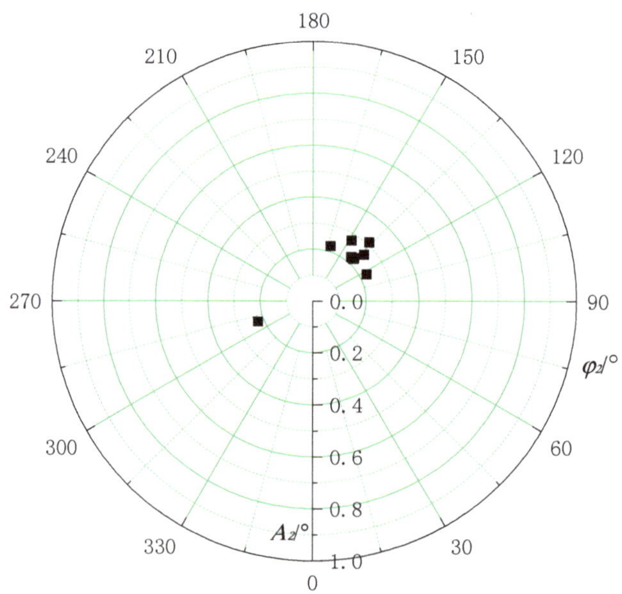

Figure 11. The final population.

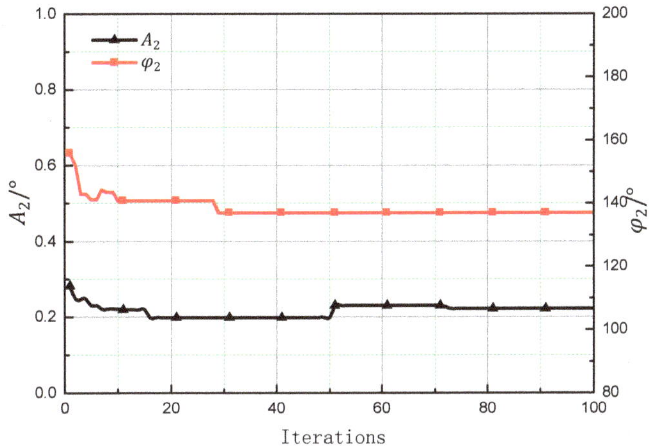

Figure 12. Variations of the optimal individual.

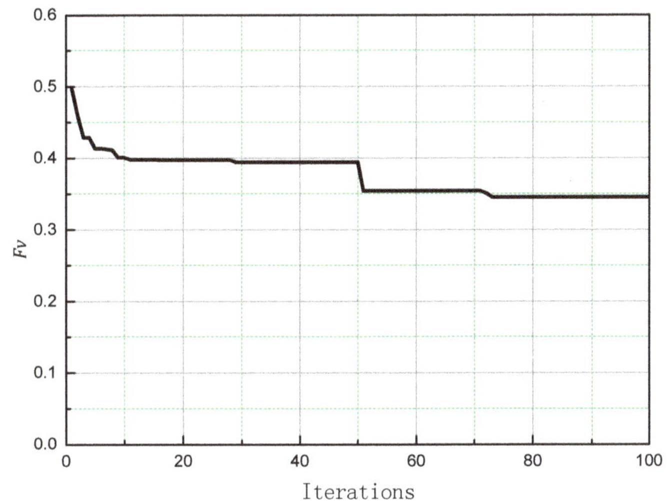

Figure 13. The trend chart of *Fv* under each generation of optimal individuals.

It can be seen in Figures 10 and 11 that the initial population is randomly distributed, the distribution is relatively scattered and the final population distribution is concentrated. The population appears to be small in Figure 10 because the optimal individual data overlap.

Figure 12 shows that after 50 generations of evolution, the optimal individual in the population is close to the final optimal value, indicating that the genetic algorithm can quickly solve the optimal state in vibration control by IBC. As can be seen in Figure 13, 60% of the vibration reduction effect can be achieved after 30 evolutions, and the final vibration reduction effect can reach more than 65%.

The initial population is widely distributed, and the probability of new optimal individuals generated by crossover is high. Therefore, in the first 10 iterations, the optimal individual changes greatly, and the vibration load decreases rapidly. At the 50th generation, with the generation of mutation, the first 10 bits of new individuals changed greatly, which was manifested as a sudden change in the amplitude of the harmonic. Because of this, the vibration load decreased rapidly once again. After that, the optimal individual did not change much, and had little effect on the vibration load of the hub. After 10 iterations of data updates, the population tended to be stable, although the crossover probability was relatively large, and it was difficult to generate new optimal individuals. Therefore, the changes in amplitude and phase were very small or even unchanged. Correspondingly, the vibration load of the hub changed little. The phase and amplitude of the optimal individual in the population changed greatly at the 30th and 50th iteration, but the vibration load of the hub decreased more at the 50th iteration compared with that at the 30th iteration, indicating that the parameter sensitivity of amplitude was greater than the phase during optimization, so the appropriate surface height difference (i.e., the amplitude of the second order harmonic) is particularly important in the structural design of the actuator.

4. Conclusions

This article investigated the influence of IBC on helicopter hub vibration load reduction. Not all states of amplitudes and phases could be used for reducing hub vibration loads. The vibration loads could become even greater than those of a baseline system without IBC for some control laws.

The parameters of the blade were simplified compared with reference. The blade segments with similar parameters were merged together, which reduced the blade segments and effectively shortened the calculation times during a large number of repeated iterative calculations. By comparing the results of the characteristics of the blade in the original text, it was found that the errors were very small, indicating that reducing the number of segments of the blade was accurate.

The GA can quickly and effectively realize the solution of the optimal state. By choosing appropriate parameters of amplitude and the phase of the second harmonic, calculated by GA, the vibration load of the hub could be reduced by 65%.

For future work, the accuracy of the proposed method should be verified via experimentation. It is necessary to analyze the influence of other harmonics—such as the third harmonic—on the vibration load of the hub.

Author Contributions: Conceptualization, H.W.; methodology, X.N.; software and validation, Y.G.; investigation and writing—original draft preparation, R.Y. R.Y. and Y.G. contributed equally to this paper. All authors have read and agreed to the published version of the manuscript.

Funding: This research was funded by the Fundamental Research Funds for the Central Universities, grant number NS2015013 and the Priority Academic Program Development of Jiangsu Higher Education Institutions.

Institutional Review Board Statement: Not applicable.

Informed Consent Statement: Not applicable.

Data Availability Statement: Not applicable.

Conflicts of Interest: The authors declare no conflict of interest.

References

1. Ellis, C.W.; Jones, R. Application of an absorber to reduce helicopter vibration levels. *J. Am. Helicopter Soc.* **1963**, *8*, 30–42. [CrossRef]
2. Lu, Y. Research on Electronically Controlled Rotor System. Ph.D. Thesis, Nanjing University of Aeronautics and Astronautics, Nanjing, China, 2004.
3. Cesnik, C.E.S.; Shin, S.J.; Wilbur, M.L. Dynamic response of active twist rotor blades. *Smart Mater. Struct.* **2001**, *10*, 62–76. [CrossRef]
4. Welsh, W.; Fredrickson, C.; Rauch, C.; Lyndon, I. Flight test of an active vibration control system on theUH-60 black hawk helicopter. In Proceedings of the 51th Annual Forum of the American Helicopter Society, Fort Worth, TX, USA, 8–11 May1995.
5. Miao, W.; Kottapalli, S.B.R.; Frye, H.M. Flight demonstration of higher harmonic control (HHC) on S-76. In Proceedings of the 39th Annual Forum of the American Helicopter Society, Washington, DC, USA, 9–11 May 1983.
6. Ham, N.D. A simple system for helicopter individual-blade control and its application to gust alleviation. In Proceedings of the 16th European Rotorcraft Forum, Bristol, UK, 18–21 September 1990.
7. Ham, N.D. Helicopter individual-blade-control: Promising technology for the future helicopter. In Proceedings of the 21th European Rotorcraft Forum, Saint Petersburg, Russia, 30 August–1 September 1995.

8. Jacklin, S.A.; Haber, A.; Simone, G.; De Norman, T.R.; Kitaplioglu, C.; Shinoda, P. Full-scale wind tunnel test of an individual blade control system for a UH-60 helicopter. In Proceedings of the 58th Annual Forumof American Helicopter Society, Montréal, QC, Canada, 11–13 June 2002.
9. Fuerst, D.; Hausberg, A. Experimental verification of an Electro-Mechanical-Actuatorfor a swashplateless primary and individual helicopter blade control system. In Proceedings of the 64th Annual Forum of American Helicopter Society, Montréal, QC, Canada, 29 April–1 May 2008.
10. Kuefmann, P.; Bartels, R.; van der Wall, B.G. The first wind-tunnel test of the DLR's multiple swashplate system: Test procedure and preliminary results. In Proceedings of the 72nd Annual Forum of the American Helicopter Society, West Palm Beach, FL, USA, 17–19 May 2016.
11. Ma, Y.J.; Yun, W.X. Research progress of genetic algorithm. *Appl. Res. Comput.* **2012**, *29*, 1201–1210.
12. Ullah, Z.; khan, M.; Raza Naqvi, S.; Farooq, W.; Yang, H.; Wang, S.; Vo, D.V.N. A comparative study of machine learning methods for bio-oil yield prediction—A genetic algorithm-based features selection. *Bioresour. Technol.* **2021**, *335*, 125292. [CrossRef] [PubMed]
13. Naranjo-Pérez, J.; Jiménez-Manfredi, J.; Jiménez-Alonso, J.; Sáez, A. Motion-based design of passive damping devices to mitigate wind-induced vibrations in stay cables. *Vibration* **2018**, *1*, 269–289. [CrossRef]
14. Zhang, Y.; He, L.; Yang, J.; Zhu, G.; Jia, X.; Yan, W. Multi-objective optimization design of a novel integral squeeze film bearing damper. *Machines* **2021**, *9*, 206. [CrossRef]
15. Liu, Q.; Zha, Y.; Liu, T.; Lu, C. Research on adaptive control of air-borne bolting rigs based on genetic algorithm optimization. *Machines* **2021**, *9*, 240. [CrossRef]
16. Omar, H.M. Optimal geno-fuzzy lateral control of powered parachute flying vehicles. *Aerospace* **2021**, *8*, 400. [CrossRef]
17. Johnson, W. *Helicopter Theory*; Dover Publications: New York, NY, USA, 1980; p. 139.
18. Leishman, J.G.; Beddoes, T.S. A semi-empirical model for dynamic stall. In Proceedings of the 42nd Annual Forum of American Helicopter Society, Washington, DC, USA, 2–4 June 1986.
19. Bir, G.; Chopra, I. *University of Maryland Advanced Rotorcraft Code Theory Manual*; Center for Rotorcraft Education and Research University of Maryland: College Park, MD, USA, 1994.
20. Zhou, J. Simulation and Analysis of Smart Rotor with Dual Trailing-Edge Flaps. Master's Thesis, Nanjing University of Aeronautics and Astronautics, Nanjing, China, 2015.
21. Zhou, H. Research on Vibration Suppression and Realization of Smart Rotor based on Dual Trailing-Edge Flaps. Master's Thesis, Nanjing University of Aeronautics and Astronautics, Nanjing, China, 2020.
22. Hodges, D.; Dowell, H. *Nonlinear Equations of Motion for the Elastic Bending and Torsion of Twisted Nonuniform Rotor Blades*; NASA: Washington, DC, USA, 1974.
23. Huiping, S. An improved mode superposition method for linear damped systems. *Commun. Appl. Numer. Methods* **1991**, *7*, 579–580. [CrossRef]
24. John, M.; Newman, S. *Basic Helicopter Aerodynamics*, 4th ed.; Wiley-Blackwell: Hoboken, NJ, USA, 2011.
25. Heffernan, R.; Gaubert, M. *Structural and Aerodynamic Loads and Performance Measurements of SA349/2 Helicopter with an Advanced Geometry Rotor*; NASA: Washington, DC, USA, 1986; pp. 15–103.
26. Heffernan, R.M.; Yamauchi, G.K.; Gaubert, M.; Johnson, W. *Hub Loads Analysis of the SA349/2 Helicopter*; NASA: Washington, DC, USA, 1988.
27. Liu, Q.; Hu, G.; Lei, W. Parameter identification and application of rotor airfoil dynamic stall model. *J. Nav. Aeronaut. Astronaut. Univ.* **2015**, *30*, 129–133.
28. Chen, B.; Liu, W. SAC model based improved genetic algorithm for solving TSP. *J. Front. Comput. Sci. Technol.* **2021**, *15*, 1680–1693.
29. Pan, J.; Qian, Q.; Fu, Y. Multi-population genetic algorithm based on optimal weight dynamic control learning mechanism. *J. Front. Comput. Sci. Technol.* **2021**, *15*, 2421–2437.

machines

MDPI

Article

Mechanical Deformation Analysis of a Flexible Finger in Terms of an Improved ANCF Plate Element

Yu Xing [1,2,†], Lei Liu [3,4,†], Chao Liu [2], Bo Li [2], Zishen Wang [2], Pengfei Li [2,*] and Erhu Zhang [2]

1 School of Mechanical and Electrical Engineering, Xi'an Polytechnic University, Xi'an 710048, China; reevesxing@hotmail.com
2 School of Mechanical and Precision Instrument Engineering, Xi'an University of Technology, Xi'an 710048, China; sheersix@xaut.edu.cn (C.L.); yunfenglibo@126.com (B.L.); wang_zi_shen@163.com (Z.W.); eh-zhang@xaut.edu.cn (E.Z.)
3 School of Mechanical Engineering, Northwestern Polytechnical University, Xi'an 710072, China; liulei2022@nwpu.edu.cn
4 State Key Laboratory for Manufacturing Systems Engineering, Xi'an Jiaotong University, Xi'an 710049, China
* Correspondence: lipengfeinew@xaut.edu.cn
† These authors contributed equally to this work.

Abstract: In recent years, flexible continuum robots have been substantially developed. Absolute nodal coordinates formulation (ANCF) gives a feasible path for simulating the behavior of flexible robots. However, the model of finger-shaped robots is often regarded as a cylinder and characterized by a beam element. Obviously, this is short of characterizing the geometrical feature of fingers in detail, especially under bending conditions. Additionally, for the lower-order plate element, it is hard to characterize the bending behavior of the flexible finger due to fewer nodes; a higher-order plate element often requires an extremely long computing time. In this work, an improved ANCF lower-order plate element is used to increase the accuracy of the Yeoh model and characterize the geometrical structure of silicone rubber fingers, taking into particular consideration the effect of volume locks and multi-body system constraints. Since it is a kind of lower-order plate element, essentially, the computing time is nearly the same as that of conventional lower-order plate elements. The validity of this model was verified by comparing it with the results of the published reference. The flexible finger, manufactured using silicone rubber, is characterized by the novel ANCF lower-order plate element, whereby its mechanical deformation and bending behavior are simulated both efficiently and accurately. Compared to the ANCF beam element, conventional lower-order plate element, and higher-order plate element, the novel plate element in this paper characterizes the external contour of the finger better, reflects bending behavior more realistically, and converges in less computing time.

Keywords: absolute nodal coordinates formulation (ANCF); plate element; flexible finger; silicone rubber; mechanical deformation; hyper-elastic constitutive model

Citation: Xing, Y.; Liu, L.; Liu, C.; Li, B.; Wang, Z.; Li, P.; Zhang, E. Mechanical Deformation Analysis of a Flexible Finger in Terms of an Improved ANCF Plate Element. *Machines* **2022**, *10*, 518. https://doi.org/10.3390/machines10070518

Academic Editors: Yanjie Wang, Xiaofeng Liu, Shichao Niu, Bo Li and Aihong Ji

Received: 19 March 2022
Accepted: 2 June 2022
Published: 27 June 2022

1. Introduction

As an indispensable part of mechanical grippers and robots, the structure of silicone rubber flexible fingers has taken many forms and changed in complexity under the effort of many scholars. Its origin can be traced back to previously pneumatic artificial muscles, which were designed to simulate human muscles and consisted of silicone rubber cylinders and outer woven nets. Since then, the property of silicone rubber has improved, and the volume of artificial muscles has shrunk gradually and has moved towards soft biological tissue because soft biological tissue has unique advantages compared to hard structures, such as strong environmental adaptability, flexibility, and being harmless to the human body. The silicone rubber flexible finger has such characteristics and uses them for the grasping process; its excellent packing skill is superior to traditional rigid mechanical grippers in food handling, medical service robots, and so on.

Silicon rubber materials are classified as hyper-elastic materials in terms of mechanical properties, the attribution of which should be characterized by hyper-elastic constitutive equations. The Neo-Hookean model is one of the most commonly used constitutive models of hyper-elastic materials; however, it is a simple model that expresses stain energy, and it only has good agreement under small strains; the computation will be less accurate as the strain energy increases. Mooney researched the deformation of hyper-elastic materials theoretically and experimentally and then raised the stain energy model of the first and second invariants for the right Cauchy–Green deformation tensor [1]. The Yeoh model is a branch of Mooney's model; its advantage is being more suitable for calculating larger stress and strain problems as it can characterize the deformation behavior of silicone rubber material more accurately [2].

Absolute nodal coordinates formulation (ANCF) is an emerging finite element method (FEM). Its node coordinates include slope coordinates, the element mass matrix is constant, and the stiffness matrix and nodal coordinates are highly correlated. The theory of the ANCF method does not contain the hypothesis of small turning angle motions [3–7], which is the most significant advantage compared to the classic FEM; hence, it is suitable for calculating large deformation problems such as modern flexible fingers or other flexible continuum robots [8–14].

Jung discussed the accuracy of three hyper-elastic constitutive models under ANCF beam elements in modeling silicone rubber, and the result shows that the Yeoh model is more accurate [15]. Pappalardo et al. proposed a new fully parametrized plate finite element based on ANCF kinematic description that fills the theoretical gap of the large rotation vector formulations [16]. Melly et al. obtained the residual strain energy density from equibiaxial loading and fitted it to a term with dependence on the second invariant; then, they added the term to the original expression, whereby modifying Yeoh's model for improving the prediction of vulcanized rubber and thermoplastic elastomers [17]. A rational absolute nodal coordinate formulation (RANCF) thin plate element was developed by Pappalardo; it is used in the accurate geometric modeling and analysis of flexible continuum bodies with complex geometrical shapes [18]. The flexible body may have an inaccurate rigid characteristic when the incompressible material model encounters bending deformation. Namely, the model of the hyper-elastic incompressible material is unable to converge to a correct result, the phenomenon of which is named volume lock. In order to solve this problem, Bayat et al. presented a locking-free element formulation based on reduced integration; physically-based hourglass stabilization (Q1SP) was coupled for the first time with the DG framework, leading to a DG variant with very good convergence properties [19]. Orzechowski et al. established an efficient method of modeling nonlinear nearly incompressible materials with polynomial Mooney–Rivlin models and a volumetric energy penalty function in the ANCF framework [20]. This method was dedicated to the examination of several ANCF fully parameterized beam elements under an incompressible regime. Xu et al. proposed a higher-order plate element formulation with quadratic interpolation in the transverse direction, which can not only alleviate volumetric locking but also improve accuracy in the simulation of large bending deformations compared to improved lower-order plate elements with the selectively reduced integration method [21].

In the published references, models of finger-shaped robots are often regarded as a cylinder and characterized by a beam element. Obviously, this is short of characterizing the geometrical feature of the fingers in detail, especially under bending conditions. Additionally, for lower-order plate elements, it is hard to characterize the bending behavior of the flexible finger due to fewer nodes; higher-order plate elements often require extremely long computing times.

In this work, an improved ANCF lower-order plate element is used to increase the accuracy of the Yeoh model and characterize the geometrical structure of the silicone rubber finger, taking into particular consideration the effect of volume locks and multi-body system constraints. Since it is a kind of lower-order plate element, essentially, the computing time is nearly the same as that of conventional lower-order plate elements. The validity of this model was verified by comparing it with the results of the published reference. The flexible finger, manufactured using silicone rubber, is characterized by the novel ANCF lower-order plate element, whereby its mechanical deformation and bending behavior are simulated both efficiently and accurately.

This paper is organized as follows. After the introduction, an improved Yeoh model based on the ANCF plate element is introduced in Section 2. In Section 3, the relevant examples' verification is conducted and analyzed. On the basis of the above-mentioned theory, the computation of a silicone rubber finger is presented in Section 4. In the last section, we give a short conclusion.

2. Theoretical Foundation and Modeling

2.1. ANCF Plate Element Theory

2.1.1. Kinematics Characterization

A three-dimensional ANCF plate element with four nodes is shown in Figure 1, in which a, b, and t represent the length, width, and depth, respectively. Each node contains 12 node coordinates and, at any point, includes 12 degrees of freedom (DOFs), which are written in the form of a vector $\mathbf{e}_i$,

$$\mathbf{e}_i = \begin{bmatrix} \mathbf{r}_i^{\mathrm{T}} & \frac{\partial \mathbf{r}_i^{\mathrm{T}}}{\partial x} & \frac{\partial \mathbf{r}_i^{\mathrm{T}}}{\partial y} & \frac{\partial \mathbf{r}_i^{\mathrm{T}}}{\partial z} \end{bmatrix}^{\mathrm{T}} \tag{1}$$

where $\mathbf{r}_i$ is the global coordinate vector of the node (A, B, C, or D); $\partial \mathbf{r}_i / \partial \alpha$ $(\alpha = x, y, z)$ is the partial derivative of the global coordinate vector to the element coordinate.

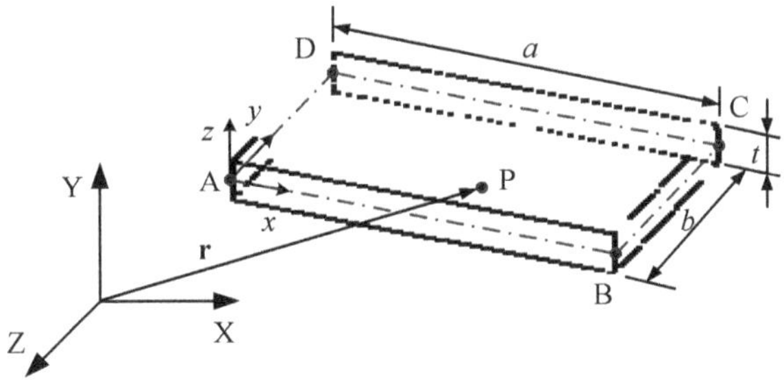

Figure 1. ANCF plate element with 3 dimensions and 4 nodes.

The coordinates of a single element are given in the form of a vector:

$$\mathbf{e} = \begin{bmatrix} \mathbf{e}_A^{\mathrm{T}} & \mathbf{e}_B^{\mathrm{T}} & \mathbf{e}_C^{\mathrm{T}} & \mathbf{e}_D^{\mathrm{T}} \end{bmatrix}^{\mathrm{T}} \tag{2}$$

The coordinates of any point in the element coordinate system are x, y, z, and its position coordinate $\mathbf{r} = [\mathrm{X}, \mathrm{Y}, \mathrm{Z}]^{\mathrm{T}}$ in the global coordinate system is given by the following position interpolation function:

$$\mathbf{r} = \mathbf{S}(x, y, z)\mathbf{e} \tag{3}$$

where $\mathbf{r}$ is the global coordinate vector of any point in the plate element, $\mathbf{S}$ is the shape function matrix, and $\mathbf{e}$ is the coordinate vector of the element, which is given by Formula (2).

S is only related to the element coordinate. Different shape function matrices and node coordinate vectors are multiplied to obtain different global coordinate interpolation functions; the shape function of a typical three-dimension four-node plate element is as follows,

$$\begin{array}{c} S = [\ S_1\mathbf{I} \quad S_2\mathbf{I} \quad S_3\mathbf{I} \quad S_4\mathbf{I} \quad S_5\mathbf{I} \quad S_6\mathbf{I} \quad S_7\mathbf{I} \quad S_8\mathbf{I} \\ S_9\mathbf{I} \quad S_{10}\mathbf{I} \quad S_{11}\mathbf{I} \quad S_{12}\mathbf{I} \quad S_{13}\mathbf{I} \quad S_{14}\mathbf{I} \quad S_{15}\mathbf{I} \quad S_{16}\mathbf{I}] \end{array} \tag{4}$$

2.1.2. The Mass Matrix, Elastic Force, and External Force of an Element

The cell mass matrix can be obtained by the kinetic energy formula

$$T = \frac{1}{2}\int_V \rho_0 \dot{\mathbf{r}}^{\mathrm{T}} \dot{\mathbf{r}} dV \tag{5}$$

where ρ_0 is the density at the beginning of the structure; V is the volume at the initial moment. We substitute Formula (3) into the above formula to obtain the element mass matrix.

$$\mathrm{M} = \int_V \rho_0 S^{\mathrm{T}} S dV \tag{6}$$

Because the shape function of the element is not related to time, in the beginning, the density and size of the element are constant, so the result of Formula (6) is a constant matrix that does not change by time. The node coordinate of the plate element is a 48×1 column vector, the shape function is a 3×48 matrix, and the dimension of the mass matrix of the element is 48×48.

The elastic force can be obtained from stain energy, which can be characterized as:

$$U = \frac{1}{2}\int_V \boldsymbol{\varepsilon}^{\mathrm{T}} \mathbf{E} \boldsymbol{\varepsilon} dV \tag{7}$$

where $\boldsymbol{\varepsilon}$ is the vector form of the strain tensor, and $\mathbf{E}$ is the elastic coefficient matrix.

The strain tensor $\boldsymbol{\varepsilon}_m$ is obtained by using the right Cauchy–Green deformation tensor [22],

$$\boldsymbol{\varepsilon}_m = \frac{1}{2}(\mathbf{J}^{\mathrm{T}}\mathbf{J} - \mathbf{I}) \tag{8}$$

$\mathbf{J}$ is the displacement gradient matrix.

$$\mathbf{J} = \frac{\partial \mathbf{r}}{\partial \mathbf{r}_0} \tag{9}$$

$\mathbf{r}$ and $\mathbf{r}_0$ represent the coordinate vector of the element at this moment and the coordinate matrix of the element at the initial moment, respectively.

Formula (7) means that the elastic force $\mathbf{Q}_s$ is computed by the partial differential of the nodal coordinate vector due to stain energy.

$$\mathbf{Q}_s = \left(\frac{\partial U}{\partial \mathbf{e}}\right)^{\mathrm{T}} \tag{10}$$

Any point on the element is subjected to a concentrated force $\mathbf{F}$, and the virtual work is done by this force,

$$\delta W_e = \mathbf{F}^{\mathrm{T}} \delta \mathbf{r} = \mathbf{F}^{\mathrm{T}} \mathbf{S} \delta \mathbf{e} = \mathbf{Q}_e^{\mathrm{T}} \delta \mathbf{e} \tag{11}$$

where $\mathbf{r}$ is the global coordinate vector, which needs to be substituted into the coordinate values x, y, z of the element. $\mathbf{Q}_e$ is the generalized external force corresponding to the concentrated force of the element

$$\mathbf{Q}_e = \mathbf{S}^{\mathrm{T}} \mathbf{F} \tag{12}$$

2.1.3. Dynamics Equations

The previous parts are the theory of a single board element according to the virtual work; the dynamic equation of the entire system that can be gained under the ANCF undamped and unconstrained dynamic equation is

$$\mathbf{M}\ddot{\mathbf{e}}_s + \mathbf{Q}_s - \mathbf{Q}_e = 0 \tag{13}$$

where $\mathbf{M}$ is assembled from the mass matrix of the element and is not changed with time; $\mathbf{e}_s$ is the system coordinates, which are assembled from element coordinates; $\mathbf{Q}_s$ is the elastic force vector of the system, which is assembled from the elastic force of the element. The elastic force and the node coordinates show extreme nonlinearity with time change; $\mathbf{Q}_e$ is the generalized external force vector of the system [23].

For a system containing constraints and damping, the dynamic equation is

$$\mathbf{M}\ddot{\mathbf{e}}_s + \mathbf{C}\dot{\mathbf{e}}_s + \Phi_{\mathbf{e}_s}^{\mathrm{T}}\lambda + \mathbf{Q}_s - \mathbf{Q}_e = 0 \tag{14}$$

where $\mathbf{C}$ is the system damping matrix; Φ is the constraint equation; λ is the Lagrangian multiplier vector.

2.2. *An Improved Hyper-Elastic Constitutive Model Based on ANCF*

2.2.1. A Constitutive Model Based on ANCF

In the theory of continuum mechanics, the three invariants of the right Cauchy–Green deformation tensor $\mathbf{C}_r = \mathbf{J}^{\mathrm{T}}\mathbf{J}$ are, respectively,

$$I_1 = \mathrm{tr}(\mathbf{C}_r) \tag{15}$$

$$I_2 = \frac{1}{2}\left[\mathrm{tr}(\mathbf{C}_r)^2 - \mathrm{tr}(\mathbf{C}_r^2)\right] \tag{16}$$

$$I_3 = \det(\mathbf{C}_r) \tag{17}$$

In the above three formulas, I_1, I_2 and I_3 are, respectively, the first invariant, the second invariant, and the third invariant of the right Cauchy–Green deformation tensor. When the material is incompressible, $J = \det(\mathbf{J}) = 1$, or equivalently, $I_3 = J^2 = 1$. Therefore, the strain energy of incompressible materials depends only on I_1 and I_2.

The Yeoh model is a function of the first invariant of the right Cauchy–Green deformation tensor, and its model form is:

$$U_y = \sum_{i=1}^{3} \mu_{i0}(I_1 - 3)^i \tag{18}$$

In the above formula, μ_{i0} is the material constant evaluated by experiments. Although the first invariant was used in the Yeoh model, its order reaches the third order. At the same time, there are three material constants that characterize material properties that have been included in this model, and the content of expression is richer.

The model expressed by Equation (18) assumes that the material is completely incompressible, which satisfies $J = \det(\mathbf{J}) = 1$. This condition needs to be added to the equation of the system; the usual methods include the Lagrangian multiplier method and the penalty method. The Lagrangian multiplier method is a constraint method that multiplies the condition of volume incompressibility by a Lagrangian multiplier and substitutes it into the system equation to achieve volume control. Compared with the Lagrangian multiplier method, the penalty method is simpler and more effective here, so the penalty method was chosen for this article to achieve volume constraints. Under the penalty method, additional strain energy density will be added to the constitutive model,

$$U_p = \frac{1}{2}k(J-1)^2 \tag{19}$$

where U_p—the volume penalty equation; k—the volume penalty number, which can represent the bulk modulus of the materials. A reasonable value of k is usually selected to ensure the incompressibility of the volume and set to 1000 MPa. The reason is that an excessively large value of k may lead to numerical problems during the process of computation. It should be pointed out that the volumetric energy penalty equation was used to make the materials nearly incomprehensible rather than absolutely incomprehensible.

The full formula of the Yeoh model, using the penalty equation, is

$$\overline{U}_y = U_y + U_p = \sum_{i=1}^{3} \mu_{i0}(I_1 - 3)^i + \frac{1}{2}k(J-1)^2 \tag{20}$$

According to the conclusion obtained by Orzechowski [20], volume behavior was introduced to the above equation, which may result in a numerical problem during the process of computation, so it is necessary to separate the volume and partial derivative components. In order to decouple, the invariant of the right Cauchy–Green deformation tensor was not directly used in this constitutive equation, which was replaced by the following form:

$$\overline{I}_1 = J^{-2/3} I_1 \tag{21}$$

The above variation was used in this article, which avoided the numerical problem during the process of finding a solution. The new constitutive model after the replacement was as follows:

$$\overline{U}_y = U_y + U_p = \sum_{i=1}^{3} \mu_{i0}(\overline{I}_1 - 3)^i + \frac{1}{2}k(J-1)^2 \tag{22}$$

The above formula is a general strain energy density formula; the volume integration of each element is needed to obtain the strain energy of the element. Therefore, with the improved Yeoh model, the strain energy of the element can be expressed as

$$\overline{U}_Y = \int_V (U_y + U_p)dV = \int_V \left(\sum_{i=1}^{3} \mu_{i0}(\overline{I}_1 - 3)^i + \frac{1}{2}k(J-1)^2 \right) dV \tag{23}$$

Hence, the immutable volume of the Yeoh model was established through the penalty equation, and it was further improved by separating the volume and partial derivation components of deformation.

According to Formula (10), the elastic force of the improved Yeoh model with ANCF is shown as follows:

$$\bar{\mathbf{Q}}_{KY} = \left(\frac{\partial \overline{U}_Y}{\partial \mathbf{e}} \right)^{\mathrm{T}} = \int_V \left(\frac{\partial U_y}{\partial \mathbf{e}} + \frac{\partial U_p}{\partial \mathbf{e}} \right)^{\mathrm{T}} dV \tag{24}$$

2.2.2. The Elimination of Volume Lock

Many researchers have mentioned that ANCF is affected by the lock phenomenon of the element. The lock phenomenon of the continuum mechanics method is particularly obvious, based on which the wrong rigidity characteristics are shown in the bending process of flexible bodies. The influence of this phenomenon is even greater when it comes to the incompressible material model. The super-elastic incompressible material model cannot converge to the correct result. This problem is called the volume lock phenomenon in the reference.

Numerical integration is usually used in the above elastic force formula derived from the Yeoh model. In finite element analysis, the Gaussian integral is the most commonly used numerical integration algorithm. In the ANCF elastic material model and many other finite element methods, selective reduction integration (short for SRI) is often used to prevent the occurrence of volume lock. The integration of strain energy density is divided into two parts with the point of the above method. The first part does not consider the influence of volume locks and performs a complete integration. The second part takes the impact of the rate of volume locks into consideration when using SRI.

From Formula (23), the equation of strain energy under the Yeoh model can simply be split. By removing the volumetric energy penalty equation that imposes constraints on the volume, the first part does not consider the behavior of the volume; hence, it can be completely integrated. The influence of volume is introduced to the second part through the volume penalty equation; under the plate element, it is assumed that there is no thickness, and the thickness is directly multiplied into the equation for the plane integral. The computation of Formula (23) of the strain energy equation is calculated as follows:

$$\overline{U}_Y^{SRI} = \int_V U_y dV + t\int_S U_p dS \tag{25}$$

In the above two formulas, SRI represents selective reduction integration, and t and S represent the thickness and area of the plate element, respectively. The above formula shows that the volumetric energy penalty equation is only evaluated on the neutral plane of the plate element.

2.3. *The Modeling of Length Constraint*

2.3.1. The Constraints of a Multi-Body System

The above system dynamic equations are unconstrained equations. There will be some constraints in the actual calculation process. The Lagrange multiplier method is usually used to compute the constraint equations of system equations. At this time, the system dynamic equations become the following form,

$$\begin{cases} \mathbf{M}\ddot{\mathbf{e}}_s + \Phi_{\mathbf{e}_s}^{\mathrm{T}}\boldsymbol{\lambda} + \mathbf{Q}_s - \mathbf{Q}_e = 0 \\ \Phi(\mathbf{e}_s, t) = 0 \end{cases} \tag{26}$$

where Φ is the constraints equation, $\Phi_{\mathbf{e}_s}$ is the partial derivative matrix of the constraints equations with respect to the system coordinates, and λ is the Lagrange multiplier.

Take the second derivation of the constraint equation system with respect to time t and in conjunction with the first formula in the above formula.

$$\begin{bmatrix} \mathbf{M} & \Phi_{\mathbf{e}_s} \\ \Phi_{\mathbf{e}_s} & 0 \end{bmatrix} \begin{bmatrix} \ddot{\mathbf{e}}_s \\ \boldsymbol{\lambda} \end{bmatrix} = \begin{bmatrix} \mathbf{Q}_e - \mathbf{Q}_s \\ \mathbf{Q}_d \end{bmatrix} \tag{27}$$

where $\mathbf{Q}_d$ is the result of the second partial derivation of the constraint equation Φ with respect to time, which has the following form:

$$\mathbf{Q}_d = \Phi_{\mathbf{e}_s}\ddot{\mathbf{e}}_s = -\Phi_{tt} - 2\Phi_{\mathbf{e}_s t}\dot{\mathbf{e}}_s - \left(\Phi_{\mathbf{e}_s}\dot{\mathbf{e}}_s\right)_{\mathbf{e}_s}\dot{\mathbf{e}}_s \tag{28}$$

Hence, the conventional ordinary differential solution to Equation (27) can calculate the system response with constraints. The explicit solution process will not be discussed in detail here.

2.3.2. Length Constraint

Certain dimensionalities need to be constrained in order to be consistent with the actual condition, such as the skeleton of a flexible finger. In this subsection, a length constraint method is proposed to be applied to the plate element and proven to be both reliable and feasible in the section of theoretical verification.

As plotted in Figure 2, the length, width, and thickness of the initially undeformed single plate element are a, b, and t. Taking the AB edge as an example, if the length of the edge is unchanged during the deformation process, l_{AB} should always be equal to a. In the process of motion deformation, the position vector of any point on the AB edge in the global coordinate system can be obtained according to Formula (3):

$$\mathbf{r} = \mathbf{S}(x, y = 0, z = 0)\mathbf{e} \tag{29}$$

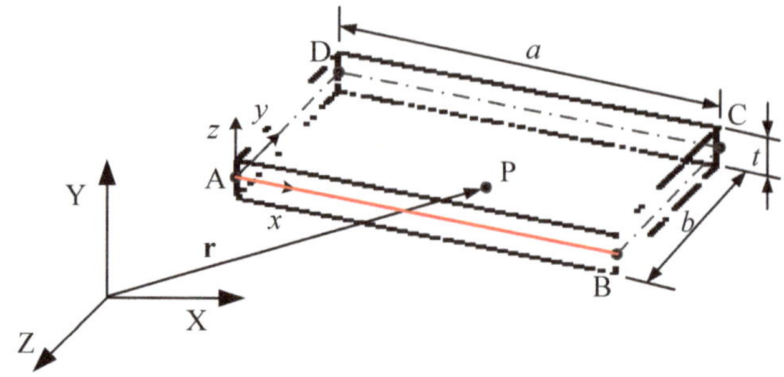

Figure 2. Single plate element.

According to the above formula, at the current moment, X, Y, and Z coordinates of any point on the AB edge in the global coordinate system are related to the element coordinate x, namely, $X = X(x)$, $Y = Y(x)$, $Z = Z(x)$. The AB curve calculated by interpolation is smooth and has continuous derivation due to ANCF. According to geometric theory, if a smooth or piecewise smooth curve has a continuous derivation, the length of the curve can be obtained by integrating the curve of the length of the arc.

$$l_{AB} = \int_L ds = \int_0^a \sqrt{[X'(x)]^2 + [Y'(x)]^2 + [Z'(x)]^2} dx \tag{30}$$

Therefore, the constraint equation at this time can be written as:

$$\Phi(\mathbf{e}, t) = l_{AB} - a = \int_0^a \sqrt{[X'(x)]^2 + [Y'(x)]^2 + [Z'(x)]^2} dx - a = 0 \tag{31}$$

The above equation is substituted into the system of dynamics equation through Lagrangian multipliers as a constraint equation, which can make the AB edge inextensible. However, in the actual calculation process, the content of the root number in Formula (31) is too complicated to open the root number, so it is impossible to calculate the AB length at the current moment. In order to solve this problem, it is unavoidable to seek the AB length.

According to Formula (31), the coordinate of the material point of the AB segment has been determined by the parameter equation: $X = X(x)$ $Y = Y(x)$ $Z = Z(x)$. Hence, the length of AB is the value of the integral of the square root of the formula $[X'(x)]^2 + [Y'(x)]^2 + [Z'(x)]^2$. When the length of AB is a fixed value, the integration of $[X'(x)]^2 + [Y'(x)]^2 + [Z'(x)]^2$ will also be fixed. Taking $\int_0^a [X'(x)]^2 + [Y'(x)]^2 + [Z'(x)]^2 dx$ as a fixed value has the same effect as setting $\int_0^a [X'(x)]^2 + [Y'(x)]^2 + [Z'(x)]^2 dx$ as a fixed value. Additionally, the former has the capacity to integrate.

The constraint equation can be calculated directly by removing the root sign, whereby the length of the plate element is unchanged.

$$\Phi(\mathbf{e}, t) = l_{AB} - a = \int_0^a [X'(x)]^2 + [Y'(x)]^2 + [Z'(x)]^2 dx - a = 0 \tag{32}$$

3. Theoretical Verification

In order to verify the accuracy of the hyper-elastic constitutive model and the validity of volume lock and length constraint, some comparing calculations were conducted in this section.

3.1. Verification of Volume Lock

The simulation example is a silicone rubber plate; the structure is shown in Figure 2: one side of the plate is fixed and the other side is set free. The relevant computing parameters are listed in Table 1. A dynamics simulation is carried out to calculate the movement process of the flat pendulum under the action of gravity (along the negative Z-axis), and the total calculation time is 0.4 s. The dynamics equation is established according to Equation (14), in which the damping matrix is distributed according to Rayleigh damping. The detailed explanation of each parameter is referred to in Xu's work [21]. It is noted that μ_{10}, μ_{20}, and μ_{30} are material constants used to establish strain energy function, which requires μ_{20} or μ_{30} to be negative to meet the condition that the strain energy density is equal to or greater than zero and increases with deformation monotonically [24]. These three parameters can be obtained by tensile test [15].

Table 1. Geometrical and material parameters of a silicone rubber plate in this example.

Name	Symbol	Value	Unit
Length	a	0.18	m
Width	b	0.16	m
Thickness	t	0.04	m
Damping coefficient	c	1.5	N·s/m
Bulk modulus	k	1000	MPa
Material constant	μ_{10}	0.2712	MPa
Material constant	μ_{20}	0.03053	MPa
Material constant	μ_{30}	−0.0004013	MPa

In order to verify whether the SRI method eliminates the volume lock effectively, Formula (23) is used to calculate the model that does not include SRI in Figure 3; SRI is calculated using Formula (25). In the model, P is the midpoint of its side; the changes in X and Z coordinates of point P over time are shown in Figure 4 due to gravity.

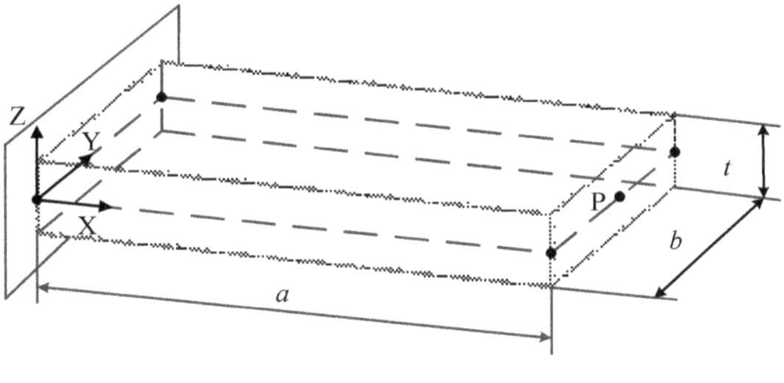

Figure 3. The silicone rubber cantilever plate.

It can be seen from Figure 4 that for the unreduced integration model, point P on the silicone rubber plate has no significant displacement in the X coordinate, and the maximum displacement on the Z-axis does not have the capacity to reduce half of the integration model, which moves to its limit in the vicinity of 0.15 s and then rebounds upward. Figure 5 shows the Yeoh model's non-selectively reduced internal plate at t = 0.2 s. It can be seen that under the influence of the volume lock, the edge of the plate is tilted upward, and the local edge is even higher than the *XY* horizontal plane; thereby, the deformation is incorrect. After adopting SRI, the displacement of point P is significantly increased in both the X-axis and Z-axis directions, and the results of motion and unreduced integration are completely different. From the above comparison, it can be concluded that under the influence of volume locks, the silicone rubber plate cannot be swung freely, and the SRI method can effectively eliminate volume locks and avoid obtaining incorrect results.

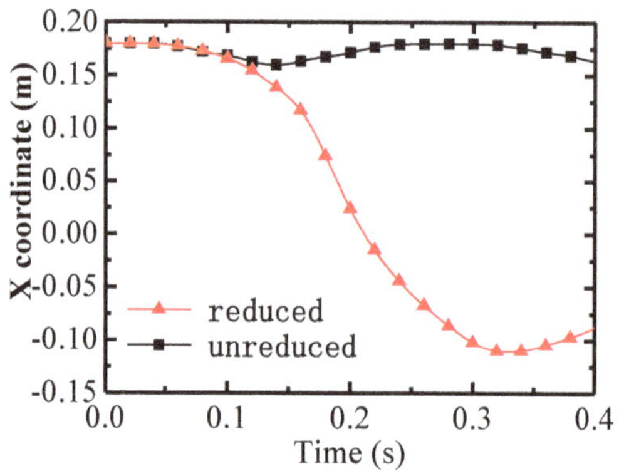

(a)

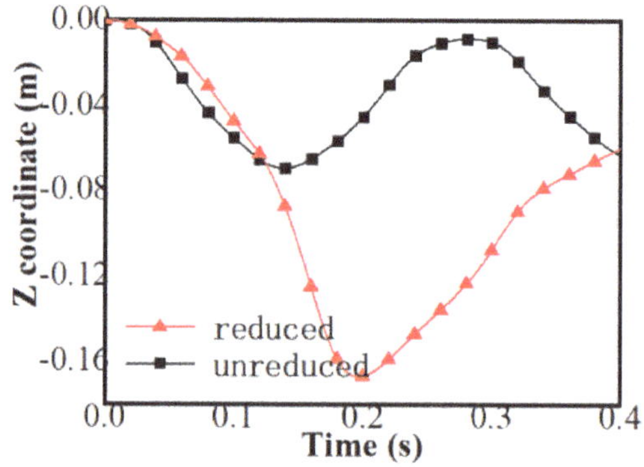

(b)

Figure 4. The coordinate of point P based on the Yeoh model. (**a**) X coordinate. (**b**) Z coordinate.

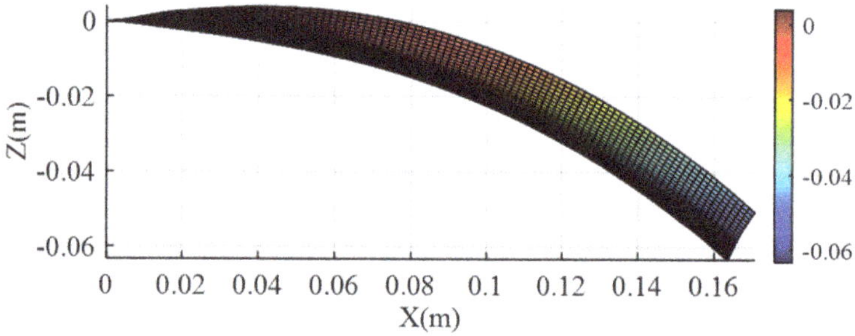

Figure 5. The deformation of a silicone cantilever based on the Yeoh model.

3.2. Verification of an Improved Yeoh Model

On the basis of eliminating the effect of volume lock, the accuracy of the constitutive model is verified further. A comparison is conducted between the results of the improved Yeoh model and the experimental results in the published reference [21]. It needs to be pointed out that the comparison here is the result of improving the low-order plate element (conventional plate element by using the SRI method) rather than the results of the high-order plate element. The simulating parameters are the same as those in Section 3.1.

The comparison among the simulation results of point P in both X and Z coordinates, the experimental results, and conventional Yeoh model results is shown in Figures 6 and 7.

In Figure 6, the two models in the first 0.25 s have good agreement with the experimental results. In the interval from 0.25 s to 0.4 s, there is a deviation between the two models, and the result obtained by the improved Yeoh model is slightly worse.

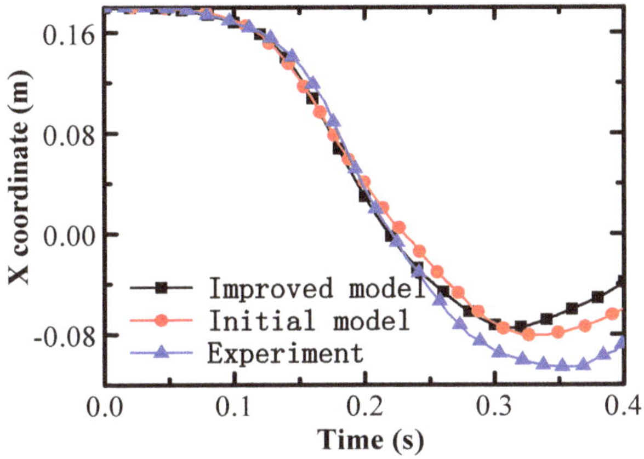

Figure 6. The X coordinate comparison of point P (experimental data are cited from reference [21]).

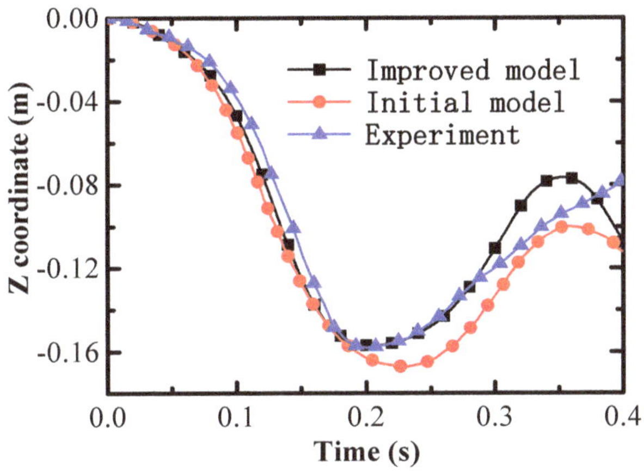

Figure 7. The Z coordinate comparison of point P (experimental data are cited from reference [21]).

In Figure 7, in the first 0.18 s, the results of the two models are close to the experimental results. To be specific, the improved Yeoh model seems better than the initial model. In the interval from 0.18 s to 0.3 s, the results of the improved model coincide with those obtained by the experiments. Their motion's lowest points coincide with each other and are smaller than that of the initial Yeoh model. In the last 0.1 s, the results in terms of the two models fail to match the experimental results. What is more, the improved model shows a larger fluctuation than the others at around 0.35 s. On the whole, the improved Yeoh model that separates the volume components is in good agreement with the experiment.

Figure 8 shows the shape of the silicone rubber plate under the improved Yeoh model at various times. It can be seen from Figures 6 and 7 that the plate is still swinging freely at 0.1 s, and it drops to the lowest point at 0.2 s, which is the smallest value in the Z coordinate in Figure 7. At this moment, a certain amount of elastic and kinetic energy has been accumulated on the plate, and it then continues to swing backward. In the 0.2 to 0.3 s movement, the improved Yeoh model moves faster and then slows down smoothly after

0.3 s. It can be seen from the whole process that the law of motion and shape is in accord with the theoretical expectations.

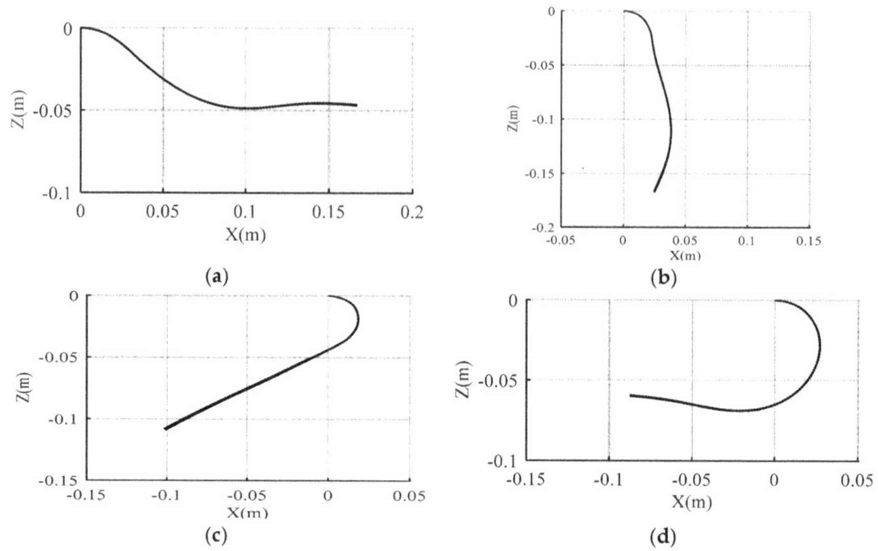

Figure 8. The shapes of the silicone plate at different moments. (**a**) t = 0.1 s, (**b**) t = 0.2 s, (**c**) t = 0.3 s, (**d**) t = 0.4 s.

According to the whole process of the free swing of the silicone rubber plate, the Yeoh model, without any improvement, has a large deviation from the actual results. The method proposed by the published reference [20] is used to remove the coupling behavior of the volume component and the partial conductance component during the deformation process and replace the first invariant in the constitutive equation with $\overline{I}_1 = J^{-2/3}I_1$. The result calculated by this improved method is very close to the experimental result, which is better than the Yeoh model. Meanwhile, computing accuracy can also be guaranteed.

3.3. Verification of Length Constraint

In this subsection, the verification calculation example is designed to illustrate the feasibility of the length constraint. As plotted in Figure 9, there is a hollow thin-walled cylinder with the specific dimensions shown in Table 2. In order to observe the deformation of the thin-walled cylinder with great flexibility, both ends of the cylinder are fixed. It should be noted that the purpose of this subsection is to verify the feasibility of the length constraint; therefore, the parameters in this case (Table 2) are totally different from those in Table 1.

Table 2. Parameters of the hollow cylinder.

Name	Symbol	Value	Unit
Radius	r	0.2	m
Length	L	0.4	m
Thickness	t	0.02	m
Density	ρ	7200	kg/m^3
Elastic modulus	E	1×10^6	MPa
Poisson's ratio	μ	0.3	-

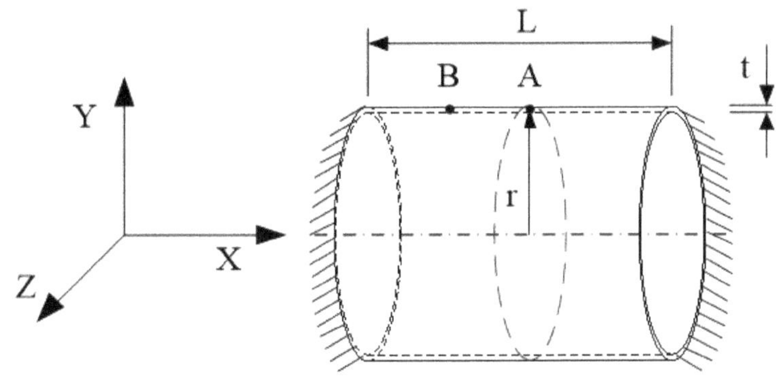

Figure 9. Hollow cylinder.

Air pressure is applied to the interior to make it expand. A constraint condition of constant length is applied to the middle line where point A is located; thus, it only suffers the air load; without expansion, the left and right parts expand and deform separately. The undamped dynamic equation is adopted to establish the dynamic equation according to Formula (26), and the pressure load is applied under ANCF. The applied air pressure increases linearly with time: $P = 100 \cdot t$ kPa; the loading time is 0.5 s, which means the final value of the air pressure is 50 kPa. The deformation effect of a hollow cylinder at different moments is shown in Figure 10.

It can be seen from Figure 10 that the expanded deformation of the hollow thin-walled cylinder structure increases with air pressure. The air pressure is slight at first, and the deformation is not obvious. Significant deformation can be found at 30 kPa; due to the constraint, there is no bulge in the middle section, and symmetrical expansion and deformation only occur on the left and right sides of the constrained part; this deformation is in line with the expected deformation effect. In order to illustrate the effectiveness of the constraints in detail, the change curves in the Z-axis coordinates of points A and B in Figure 9 over time are shown in Figure 11.

In Figure 11, it is clear that the Z-axis coordinates of points A and B both become larger. At 0.5 s, when 50 kPa air pressure is loaded, the expanded deformation of point A and point B is 0.0025 and 0.035, respectively. The expanded deformation of point A does not mean that the constraint is not effective; it can be explained that the system dynamics equation is solved by a numerical algorithm, which will cause errors in the calculation process, and the cylinder itself has an increasing tendency to deform under the driving force. The errors, with an increasing trend, are accumulated during the iterative process, so the obtained results augment gradually. Comparing the small expansion of point A with the displacement of point B, the displacement of the two points is not in the same order of magnitude, which implies that the constraint effect is very good. It is feasible to use Equation (32) instead of (31) as the constraint equation, and the inextensible constraint is realized using this method. It should be pointed out that the result of Figure 11 shows a wave-like rising trend; this is because the system adopts an undamped dynamic equation without energy dissipation, and each moment is not precisely static; the external forces that augment stepwise cannot balance the elastic force of the system immediately, and it is reasonable to expect a resulting antagonistic rise.

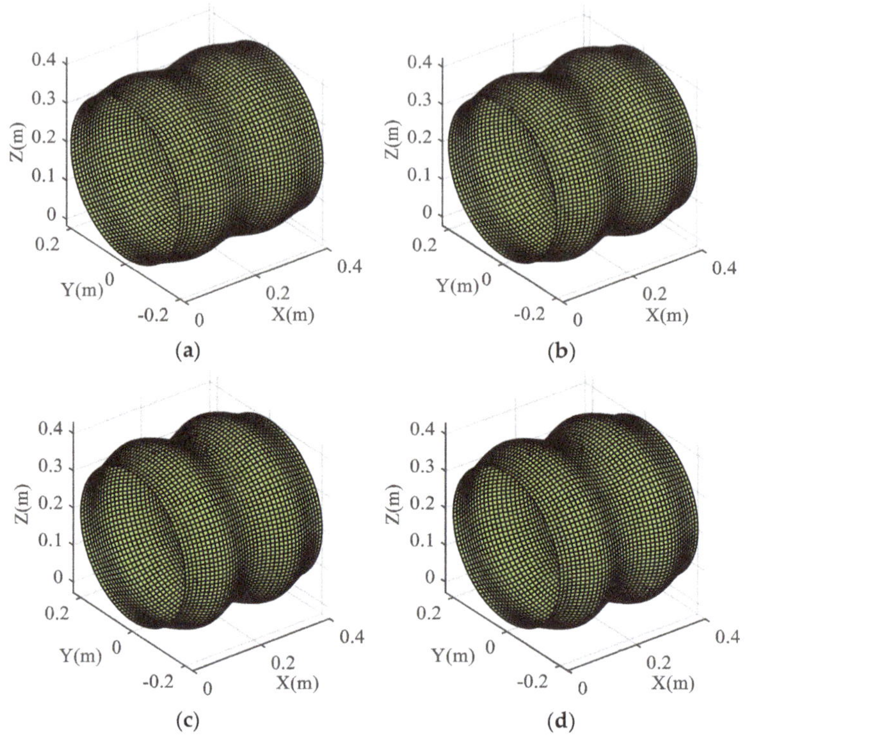

Figure 10. The shape of the cylinder at different moments. (**a**) $t = 0.2$ s, (**b**) $t = 0.3$ s, (**c**) $t = 0.4$ s, and (**d**) $t = 0.5$ s.

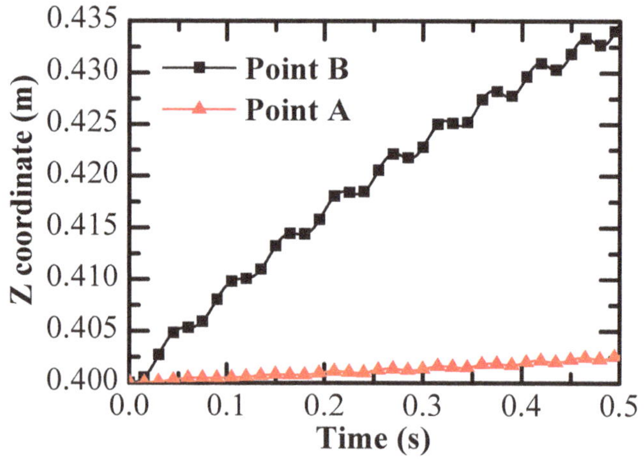

Figure 11. Z coordinate comparison of points A and B.

4. The Calculation of a Silicone Rubber Finger

4.1. The Structure of a Silicone Rubber Finger

In this section, a silicone rubber flexible finger is designed, and its structure is shown in Figure 12. The flexible finger is divided into four sections. The bottom and middle sections are the same flexible cavities made of silicone rubber material. Between the two flexible cavities is a rigid bone ring supported by a more rigid material. The same silicon rubber material is used for physical packaging at the top. Except for the top, the entire flexible finger is hollow, and the thickness of the surface is uniform. The pneumatic flexible finger can be given different sizes, according to actual conditions, to meet the needs of different purposes. Some additional constraints can be added to improve its performance in the future. The deformation of the flexible finger under air pressure will be computed in this section. It should be noted that in order to improve the overall simulation rate, the overall size of the flexible finger is deliberately enlarged. The relevant deformation is still proportional to its size, so the larger size result can be used to derive the actual deformation result. The dimensions of a, b, r, and t in Figure 12 are listed in Table 3, and the relevant material parameters are the same as in Table 1.

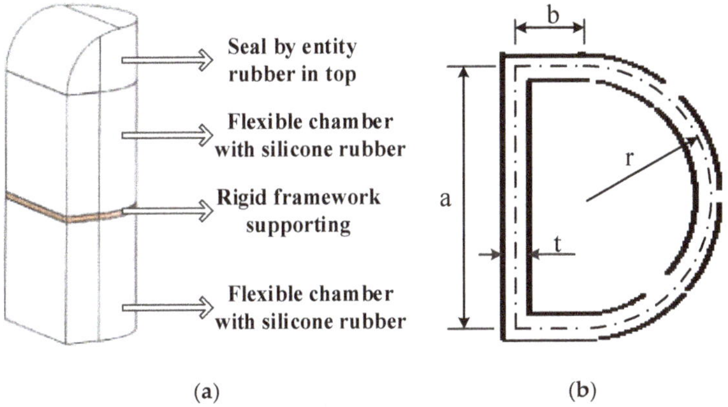

Figure 12. Structure of silicone rubber flexible finger. (**a**) Entire structure; (**b**) cross-section shape.

Table 3. Parameters in Figure 12.

Name	Symbol	Value	Unit
Radius	r	0.1	m
Thickness	t	0.02	m
Length	a	0.2	m
Width	b	0.05	m

4.2. Single Cavity Deformation Analysis

As mentioned above, this pneumatic flexible finger consists of three similar sections, and the internal cavities are connected. The advantage of this structure is that under ideal conditions, each chamber under the same air pressure has the same deformation; thereby, the deformation is repeatable. The research of a single structure is representative, so the deformation simulation of a single cavity is carried out. The constitutive model of the silicone rubber material here adopts the above-mentioned improved Yeoh's constitutive model. The parameters are detailed in the previous section, and the deformation effect of a single cavity after being loaded by air pressure is programmed to be calculated. Some fixed constraints are used in the upper and lower ends of a single cavity, and air pressure is applied inside. The pressure increases linearly with time. The undamped dynamic equation is computed and simulated from 0 to 0.5 s. Then, the pressure is increased to 12 kPa at the end of 0.5 s.

Figure 13 shows the displacement of the cavity in the Z-axis direction. From the comparison of the results, it can be seen that the expansion effect of a single cavity, calculated based on ANCF and the improved Yeoh model, is good, and the original shape of the cavity is symmetrical. After evenly distributing the air pressure, the deformed structure presents a symmetrical shape, which meets the expected expectation and verifies the correctness of the deformation of a single cavity after being fixedly restrained.

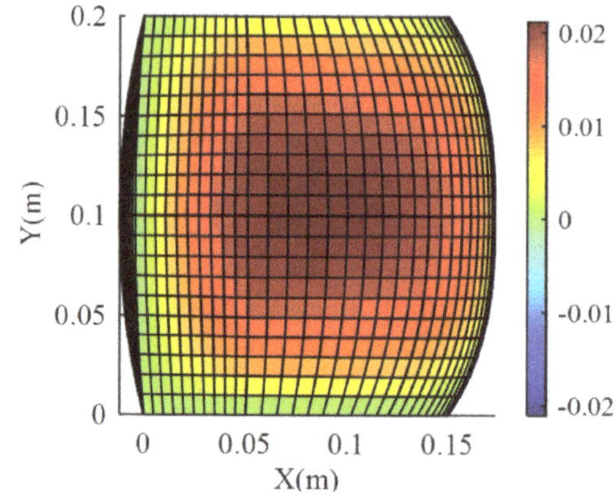

Figure 13. Z-axis displacement of single finger chamber.

4.3. Analysis of the Overall Deformation of a Flexible Finger

At the top of the finger is a solid encapsulation; during the process of applying air pressure, only the displacement of the top solid section emerges, rather than the deformation. Hence, the air pressure at the top is regarded as an equivalent force applied to the top of the middle of the flexible cavity. The rigid skeleton between the two cavities is simulated by a material with a large elastic modulus. The bottom edge of the finger is fixedly constrained. The calculation method is the same as the above calculation of the single-section cavity. The air pressure is increased linearly with time. The computing time is 0.5 s. At the end of the 0.5 s, the air pressure increases to 12 kPa. The improved Yeoh model is used as the constitutive model. When the air pressure is 8, 10, or 12 kPa, the displacements of the entire finger on the Z-axis are as shown in Figures 14–16, respectively.

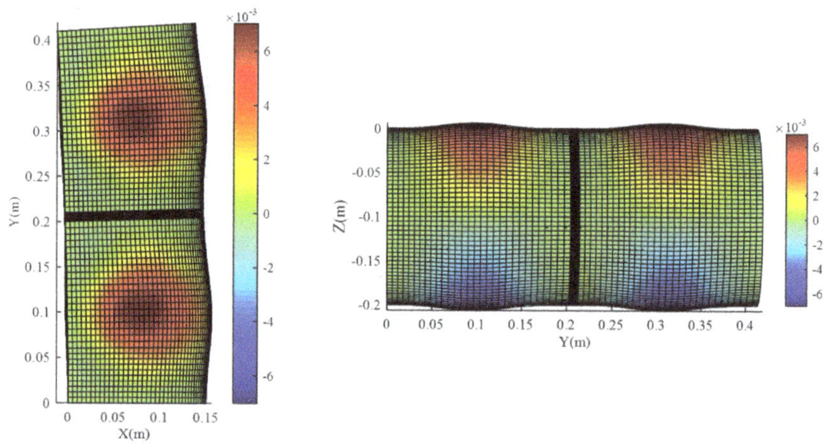

Figure 14. The deformation of two chambers under the pressure of 8 kPa. (**a**) Front view; (**b**) right view.

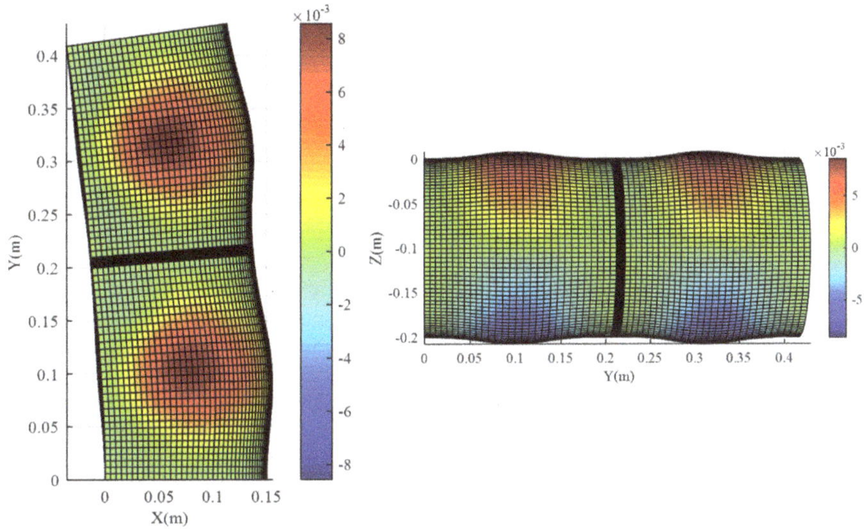

(a) **(b)**

Figure 15. The deformation of two chambers under the pressure of 10 kPa. (**a**) Front view; (**b**) right view.

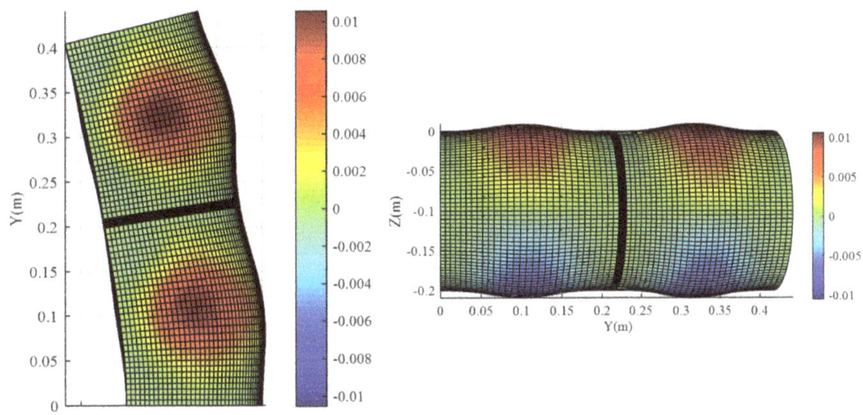

(a) **(b)**

Figure 16. The deformation of two chambers under the pressure of 12 kPa. (**a**) Front view, (**b**) right view.

It can be seen that the overall structure of the flexible finger is effectively bent and deformed under the pneumatic drive. When the air pressure is low, the expansion and deformation, as well as the bending deformation, are not obvious. For instance, the Z-axis expansion displacement is only 7 mm under a pressure of 8 kPa. Additionally, there is nearly no bending. The Z-axis expansion displacement of the finger is increased to 8.2 mm with 10 kPa of pressure. The bending performance can be clearly seen. When the air pressure is up to 12 kPa, the maximum Z-axis expansion displacement is 10 mm. At this moment, the bending angle is larger than at any time. In addition, the expansion and deformation of the flexible finger are symmetrical, which agrees with the predicted tendency. In conclusion, the motion performance of the flexible finger can be characterized by the proposed model effectively, including the expansion, deformation, bending status, and so on, under different air pressures.

5. Conclusions

The ANCF method is known to be well suited for the large deformation analysis of flexible bodies such as flexible fingers. This work presents an improved ANCF low-order plate element based on the hyper-elastic constitutive model, whereby the bending behavior of a silicon rubber flexible finger can be simulated in detail. Compared with the ANCF beam element, conventional lower-order plate element, and higher-order plate element, the novel plate element in this paper characterizes the external contour of the finger better, reflects the bending behavior more realistically, and converges in less computing time.

The effects of volume lock and length constraint were also considered. The former is used to eliminate the incorrect rigidity characteristics of bending deformation for a flexible finger, which is another important factor in realizing large deformation analysis. The skeleton of the flexible finger is simulated by the method of length constraint, which is easy to implement and characterizes the skeleton of the finger.

As a direction of future research, it is desirable to apply the presented method to other ANCF elements, such as solid elements. Furthermore, the method to eliminate the effect of volume lock also needs to be further researched.

Author Contributions: Formal analysis, C.L.; Funding acquisition, L.L.; Investigation, B.L. and E.Z.; Project administration, P.L.; Software, Z.W.; Writing—original draft, Y.X.; Writing—review & editing, Y.X. All authors have read and agreed to the published version of the manuscript.

Funding: This work was supported by the China Postdoctoral Science Foundation (Grant No. 2020M683682XB), the National Natural Science Foundation of China (Grant No. 51805413), the Science and Technology Program of Xi'an, China (Grant No. 2019217814GXRC014CG015-GXYD14.22), the Natural Science Foundation of Shaanxi Province (Grant No. 2020JQ-826), the State Key Laboratory for Manufacturing Systems Engineering (Grant No. sklms2021014), and the State Key Laboratory of Smart Manufacturing for Special Vehicles and Transmission Systems (Grant No. GZ2019KF011).

Data Availability Statement: The datasets generated and/or analyzed during the current study are available from the corresponding author upon reasonable request.

Conflicts of Interest: The authors declare no conflict of interest.

References

1. Mooney, M. A Theory of Large Elastic Deformation. *J. Appl. Phys.* **1940**, *11*, 582–592. [CrossRef]
2. Renaud, C.; Cros, J.M.; Feng, Z.Q.; Yang, B. The Yeoh model applied to the modeling of large deformation contact/impact problems. *Int. J. Impact Eng.* **2009**, *36*, 659–666. [CrossRef]
3. Shabana, A.A. *An Absolute Nodal Coordinate Formulation for the Large Rotation and Deformation Analysis of Flexible Bodies*; Technical Report; Department of Mechanical Engineering, University of Illinois at Chicago: Chicago, IL, USA, 1996.
4. Shabana, A.A. Definition of the Slopes and the Finite Element Absolute Nodal Coordinate Formulation. *Multibody Syst. Dyn.* **1997**, *1*, 339–348. [CrossRef]
5. Berzeri, M.; Campanelli, M.; Shabana, A.A. Definition of the Elastic Forces in the Finite-Element Absolute Nodal Coordinate Formulation and the Floating Frame of Reference Formulation. *Multibody Syst. Dyn.* **2001**, *5*, 21–54. [CrossRef]
6. Omar, M.A.; Shabana, A.A. A two-dimensional shear deformable beam for large rotation and deformation problems. *J. Sound Vib.* **2001**, *243*, 565–576. [CrossRef]
7. Kerkkaenen, K.S.; Sopanen, J.T.; Mikkola, A.M. A Linear Beam Finite Element Based on the Absolute Nodal Coordinate Formulation. *J. Mech. Des.* **2005**, *127*, 621–630. [CrossRef]
8. Russo, M.; Dong, X.J.M.; Theory, M. A calibration procedure for reconfigurable Gough-Stewart manipulators. *Mech. Mach. Theory* **2020**, *152*, 103920. [CrossRef]
9. Ba, W.; Dong, X.; Ahmad-Mohammad, A.; Wang, M.; Axinte, D.; Norton, A. Design and validation of a novel fuzzy-logic-based static feedback controller for tendon-driven continuum robots. *IEEE/ASME Trans. Mechatron.* **2021**, *26*, 3010–3021. [CrossRef]
10. Yu, J.J.; Dong, X.; Pei, X.; Zong, G.H.; Qiu, Q. Mobility and Singularity Analysis of a Class of 2-DOF Rotational Parallel Mechanisms Using a Visual Graphic Approach. In Proceedings of the Asme International Design Engineering Technical Conferences & Computers & Information in Engineering Conference, Washington, DC, USA, 28–31 August 2011.
11. Ma, N.; Dong, X.; Axinte, D. Modeling and Experimental Validation of a Compliant Underactuated Parallel Kinematic Manipulator. *IEEE/ASME Trans. Mechatron.* **2020**, *25*, 1409–1421. [CrossRef]

12. Ma, N.; Dong, X.; Palmer, D.; Arreguin, J.C.; Liao, Z.; Wang, M.; Axinte, D. Parametric vibration analysis and validation for a novel portable hexapod machine tool attached to surfaces with unequal stiffness. *J. Manuf. Process.* **2019**, *47*, 192–201. [CrossRef]
13. Barrientos-Díez, J.; Dong, X.; Axinte, D.; Kell, J.J.R.; Manufacturing, C.-I. Real-Time Kinematics of Continuum Robots: Modelling and Validation. *Robot. Comput.-Integr. Manuf.* **2020**, *67*, 102019. [CrossRef]
14. Russo, M.; Raimondi, L.; Dong, X.; Axinte, D.; Kell, J.J.R.; Manufacturing, C.-I. Task-oriented optimal dimensional synthesis of robotic manipulators with limited mobility. *Robot. Comput.-Integr. Manuf.* **2021**, *69*, 102096. [CrossRef]
15. Jung, S.P.; Park, T.W.; Chung, W.S. Dynamic analysis of rubber-like material using absolute nodal coordinate formulation based on the non-linear constitutive law. *Nonlinear Dyn.* **2011**, *63*, 149–157. [CrossRef]
16. Pappalardo, C.M.; Wallin, M.; Shabana, A.A.; Dynamics, N. A New ANCF/CRBF Fully Parameterized Plate Finite Element. *J. Comput. Nonlinear Dynam.* **2017**, *12*, 031008. [CrossRef]
17. Melly, S.K.; Liu, L.; Liu, Y.; Leng, J. Modified Yeoh model with improved equibiaxial loading predictions. *Acta Mech.* **2022**, *233*, 437–453. [CrossRef]
18. Pappalardo, C.M.; Yu, Z.; Zhang, X.; Shabana, A.A.; Dynamics, N. Rational ANCF thin plate finite element. *J. Comput. Nonlinear Dyn.* **2016**, *11*, 051009. [CrossRef]
19. Bayat, H.R.; Wulfinghoff, S.; Kastian, S.; Reese, S. On the use of reduced integration in combination with discontinuous Galerkin discretization: Application to volumetric and shear locking problems. *Adv. Model. Simul. Eng. Sci.* **2018**, *5*, 10. [CrossRef]
20. Orzechowski, G.; Frczek, J. Nearly incompressible nonlinear material models in the large deformation analysis of beams using ANCF. *Nonlinear Dyn.* **2015**, *82*, 451–464. [CrossRef]
21. Xu, Q.; Liu, J.; Qu, L. A Higher-Order Plate Element Formulation for Dynamic Analysis of Hyperelastic Silicone Plate. *J. Mech.* **2019**, *35*, 795–808. [CrossRef]
22. Mikkola, A.M.; Shabana, A.A. A Non-Incremental Finite Element Procedure for the Analysis of Large Deformation of Plates and Shells in Mechanical System Applications. *Multibody Syst. Dyn.* **2003**, *9*, 283–309. [CrossRef]
23. Dufva, K.; Shabana, A.A. Analysis of thin plate structures using the absolute nodal coordinate formulation. *J. Multi-Body Dyn.* **2005**, *219*, 345–355. [CrossRef]
24. Yeoh, O.H. Some forms of the strain energy function for rubber. *Rubber Chem. Technol.* **1993**, *66*, 754–771. [CrossRef]

machines

MDPI

Article

The Feature Extraction of Impact Response and Load Reconstruction Based on Impulse Response Theory

Dawei Huang †, Yadong Gao *,†, Xinyu Yu and Likun Chen

College of Aerospace Engineering, Nanjing University of Aeronautics and Astronautics, Nanjing 210001, China; damereywong@nuaa.edu.cn (D.H.); y1215225@nuaa.edu.cn (X.Y.); my2001@nuaa.edu.cn (L.C.)
* Correspondence: gydae@nuaa.edu.cn
† These authors contributed equally to this work.

Abstract: Impact load is a kind of aperiodic excitation with a short action time and large amplitude, it had more significant effect on the structure than static load. The reconstruction (or identification namely) of impact load is of great importance for validating the structural strength. The aim of this article was to reconstruct the impact load accurately. An impact load identification method based on impulse response theory (IRT) and BP (Back Propagation) neural network is proposed. The excitation and response signals were transformed to the same length by extracting the peak value (amplitude of sine wave) in the rising oscillation period of the response. First, we deduced that there was an approximate linear relationship between the discrete-time integral of impact load and the amplitude of the oscillation period of the response. Secondly, a BP neural network was used to establish a linear relationship between the discrete-time integral of the impact load and the peak value in the rising oscillation period of the response. Thirdly, the network was trained and verified. The error between the actual maximum amplitude of impact load and the identification value was 2.22%. The error between the actual equivalent impulse and the identification value was 0.67%. The results showed that this method had high accuracy and application potential.

Keywords: structural dynamics; impulse response theory; impact load identification; dynamic inverse problem; BP neural network; signal processing method

Citation: Huang, D.; Gao, Y.; Yu, X.; Chen, L. The Feature Extraction of Impact Response and Load Reconstruction Based on Impulse Response Theory. *Machines* **2022**, *10*, 524. https://doi.org/10.3390/machines10070524

Academic Editor: César M. A. Vasques

Received: 5 June 2022
Accepted: 26 June 2022
Published: 28 June 2022

1. Introduction

Impact load was a kind of aperiodic load that cannot be ignored in engineering practice. Impact load acted on the component or structure at a high amplitude in a short time (the action time was less than half of the fundamental free vibration period of the structure, always tenths of a millisecond).

Identifying (or reconstruction namely) the impact load on the structure accurately has significant meanings, for example, it can verify the structural strength and improve the structural design. This article focuses on the identification of impacts that do not cause damage: for example, the launch of missiles on a helicopter or the waves on an ocean platform.

The main focus of this article is the dynamic response of the structure under impact load. When a structure is subjected to impact load, stress waves and dynamic response of the structure are usually considered. The research of stress waves mainly has focused on the local disturbance of objects and its propagation; the research of structural dynamic response ignored the propagation process and directly studied the deformation, fracture and its relationship with time. Due to the complexity of the shock form, researchers typically focused on only one aspect. As an effective method in establishing linear models, artificial neural networks have been used widely in the research of load identification. He Wei presented a brief review for the application of neural networks in load identification areas.

Tian Yan et al., (2004) used the RBF network to research the load on gearbox and performed load identification of the random excitation and bearing excitation [1]. Tian Yan et al., (2006) also used the improved Elman network to identify the electrical-mechanical-fluid coupling shock excitation of the gearbox [2]. Staszewski et al., (2000) used a neural network model to identify the load acting on the composite box panel and used a genetic algorithm to optimize the sensor parameters [3]. Ghajari et al., (2013) applied a neural network model to composite plate, identified loads that took both linear and nonlinear deformation into consideration [4].

Wang et al., (2015) used the SVM network to conduct multi-point random dynamic load identification in the frequency domain, and it performed well in the low-frequency range [5]. Fang et al., (2018) applied a pseudo-linear neural network (PNN) to the vibration load identification of the primary spur gear box under four working conditions of no-load and loaded [6]. Cheng et al., (2018) used the improved neural network and BP neural network to optimize GA-PSO to identify the excitation load of the double-span sub-rotation system [7].

Zheng Shijie et al., (2009) identified the single point loads of composite plate-shell structures based on a genetic algorithm and BP neural network [8]. Mitsui et al., (1998) proposed the idea of using neural network technology to identify the load of the cantilever beam by using neural network technology [9]. Cooper and Dimaio (2018) used the feed-forward neural network to realize the wing load identification of large rib loads; however, they only identified the static loads [10].

Ren et al., (2018) proposed a method of using deep learning technology to perform load parameter identification and structural failure analysis based on the deformation and damage characteristics of the structure [11]. Chen et al., (2019) adopted the deep neural network (DNN) to realize the impact load identification of the rigid body to the hemispherical shell structure. The results showed that the trained DNN network had high accuracy for various characteristic parameters of impact loads [12].

Zhang Zhihong and Zhang Hong (2022) used a BP neural network to study the driving dynamic load recognition of crawler system [13]. Yang Te (2022) used a deep neural network to extract and identify the time-frequency domain features of stationary random signals [14]. Xia Peng (2022) introduced the long short-term memory neural network into the research of dynamic load identification, combined with the "memory" characteristic of a time-delay neural network and dynamic system [15].

The identification methods above based on neural network had a common feature: the excitation signal and the response signal were synchronized, that meant the length of excitation was equal to the response. The correspondence provided an advantage for the application of neural networks.

However, the excitation and response signal of impact load did not have this advantage. The action time of impact load was too short (10 ms or less) while the response was long (10 s or more); therefore, two signals could not correspond one by one. How to extract the representative information from the response signal was the key point of this article.

The main innovation of this paper was that the peak value (amplitude of sine wave) in the oscillation rising period of the response was extracted as the input, and the discrete time integral of the impact load amplitude was extracted as the output. In doing so, two signals were transformed to the same length, making it possible to be used for training and validation of a neural network.

The main content of this article includes an impact load identification method based on impulse response theory (IRT) and BP neural network. The excitation and response signals were transformed to the same. First, we deduced that there was an approximate linear relationship between the discrete-time integral of impact load and the amplitude of the oscillation period of the response. Secondly, a BP neural network was used to establish the linear relationship between the excitation and response. Thirdly, the network was trained and verified.

The main content of the following article is as follows. Section 2 introduces the theoretical basis of the method. Sections 2.1.1 and 2.1.2 introduce the linear relationship between the discrete-time integral of impact load and the peak value (amplitude of sine wave) in the rising oscillation period of the response in both the single-degree-of-freedom (SDOF) system and multi-degree-of-freedom (MDOF) system. Section 2.2 introduces the basic theory of neural network. Section 3 presents the results. Section 3.1 introduces the configuration of experiment, Section 3.2 introduces the signal processing, and Sections 3.3 and 3.4 introduce the training of the neural network and the identification results.

2. Materials and Methods

This chapter includes the detailed description of the impulse response theory and the basic process of the neural network.

2.1. Impulse Response Theory

The Fourier transform theory and the impulse response theory are two methods to calculate the response under aperiodic excitation. The Fourier transform theory was suitable for signals with a certain time length. The impulse response theory regards any load as the superposition of a series of unit impulses, determines the response of unit impulse and then uses the superposition principle to superpose a series of impulse responses one by one (or integral form) to obtain the dynamic response of the system.

The impact load has a short acting time and was not suitable for frequency domain method. Thus, this paper conducted identification research mainly based on the impulse response method.

2.1.1. Single Degree-of-Freedom (SDOF) System

The aperiodic load excitation $F(t)$ can be regarded as the weighted sum of several impulse excitation acting at different moments. The time width of each pulse tends to 0, and the value of $F(t)$ at the corresponding moments can be taken as the weighted coefficient of each impulse, and thus $F(t)$ can be regarded as the weighted sum of each impulse.

Figure 1 shows the conception of the impulse response theory (IRT).

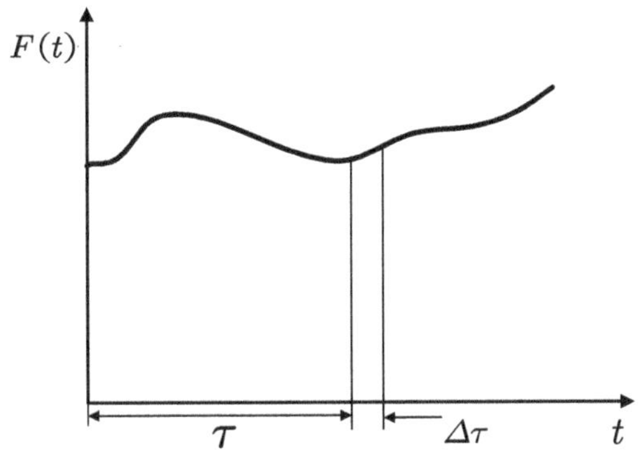

Figure 1. $F(t)$ can be decomposed into discrete impulse excitations.

As it was assumed that the dynamic system was a linear system, if the response of the system under the unit impulse excitation was solved, then the whole response can be solved according to the linear superposition principle. The response of a SDOF system with unit impulse was discussed under zero initial condition. The differential equation set of motion of the system was as follows:

$$\begin{gathered} m\ddot{x}(t) + c\dot{x}(t) + kx(t) = \delta(t) \\ x_0 = x(0) = 0 \\ v_0 = \dot{x}(0) = 0 \end{gathered} \tag{1}$$

In the above equation: $\delta(t)$ was the unit impulse function, also known as the *Dirac* δ function, which was defined as follows:

$$\delta(t-a) = \begin{cases} 0 & t \neq a \\ \infty & t = a \end{cases} ; \int_{-\infty}^{+\infty} \delta(t-a)dt = 1 \tag{2}$$

In the above equation, the impulse width tended to be zero, and the amplitude tended to be ∞; however, the impulse was equal to 1.

The zero-initial condition in Equation (1) mean that the system was in static condition at the beginning and was suddenly affected by $\delta(t)$ at the moment $t = 0$. As the time of $\delta(t)$ action was extremely short, the instant time after the action was over can be written as $t = 0^+$. According to the theorem of momentum, the increase of momentum was equal to the impulse acting on the body; therefore: $m\dot{x}(0^+) - m\dot{x}(0) = 1$.

Assuming that the initial condition was $\dot{x}(0) = 0$, then the system obtained an initial velocity: $\dot{x}(0^+) = v_0^+ = 1/m$. According to this assumption, the impact load can be decomposed into the superposition of several initial excitations. The problem expressed in Equation (1) can be converted to an initial value problem expressed as follows:

$$\begin{gathered} m\ddot{x}(t) + c\dot{x}(t) + kx(t) = 0 \\ x(0^+) = 0 \\ \dot{x}(0^+) = 1/m \end{gathered} \tag{3}$$

The parameters were set as follows:

$$\begin{gathered} \omega_n = \sqrt{k/m} \\ \xi = \frac{c}{2m\omega_n} = \frac{c}{2\sqrt{mk}} \\ \omega_d = \sqrt{1-\xi^2}\omega_n \end{gathered} \tag{4}$$

It was assumed that the SDOF system was in the case of small damping ratio ($0 < \xi < 1$), the solution of Equation (3) was as follows:

$$x(t) = \frac{1}{m\omega_d} e^{-\xi\omega_n t} sin\ \omega_d t \tag{5}$$

The response of the system to unit impulse excitation was denoted as $h(t)$, and then the unit impulse response of the system can be expressed as follows:

$$h(t) = \frac{u(t)}{m\omega_d} e^{-\xi\omega_n t} sin\ \omega_d t \quad u(t) = \begin{cases} 0 & (t < 0) \\ 1 & (t \geq 0) \end{cases} \tag{6}$$

For the impact load excitation $F(t)$, it can be regarded as a combination of a series of impulse excitation, and the impulse force at any time τ was $F(\tau)$. The system responded to the impulse was $F(\tau) \cdot \Delta\tau \cdot h(t-\tau)$. According to the superposition principle of linear system, the whole response of the system was as follows:

$$\sum_{\tau=0}^{t} F(\tau)\Delta\tau h(t-\tau) \tag{7}$$

When $\Delta\tau$ tended to be 0, the summation sign would become into integral; thus, the response of the system to the impact load $F(t)$ can be expressed as follows:

$$\begin{gathered} x(t) = \int_0^t F(\tau)h(t-\tau)d\tau \\ x(t) = \frac{u(t)}{m\omega_d} \int_0^t F(\tau)e^{-\xi\omega_n(t-\tau)} \sin\omega_d(t-\tau)d\tau \end{gathered} \tag{8}$$

In actual conditions, the impact load signal was discrete signal, such as $F(t) = F(t_i)$ $(i = 1, 2, \cdots, n)$. If the sampling interval of the signal was fixed, the time width of each pulse was also fixed, denoted as $\Delta\tau$.

According to Equation (8), the integral form of continuous function can be transformed into the sum form of discrete function, as follows:

$$\begin{aligned} x(t) &= \sum_{\tau=0}^{t} F(\tau)\Delta\tau h(t-\tau) = \sum_{i=1}^{n} F(t_i)\Delta\tau h(t-t_i) \\ &= \frac{u(t)\Delta\tau}{m\omega_d} \sum_{i=1}^{n} F(t_i) e^{-\xi\omega_n(t-t_i)} \sin\omega_d(t-t_i) \end{aligned} \tag{9}$$

The discrete time integral of discrete impact load within the action time $[t_1, t_n]$ can be expressed as:

$$I_F(t_i) = \Delta\tau \sum_{i=1}^{n} F(t_i) \tag{10}$$

For a single excitation, the attenuation of the response within a time step $\Delta\tau$ can be expressed by the relative attenuation coefficient: $e^{-\xi\omega_n\Delta\tau} = e^{-\xi\omega_n\frac{1}{f}}$. The damping ratio of steel structure was between 0.03 and 0.08, and the sampling frequency was 4096 Hz. Thus the attenuation coefficient corresponding to the first 100 Hz mode was about 0.9954, and the attenuation coefficient within 10 sample $10\Delta\tau$ was about 0.9550.

As the attenuation coefficient was very close to 1, the amplitude of response can be regarded as constant in rising oscillation period. The amplitude in the rising oscillation period of each response was as follows:

$$I_x \approx \frac{u(t)\Delta\tau}{m\omega_d} \sum_{i=1}^{n} F(t_i) \approx \frac{u(t)\Delta\tau}{m\omega_d} \sum_{i=1}^{n} F(t_i) e^{-\xi\omega_n(t-t_i)} \approx \frac{u(t)}{m\omega_d} \cdot \Delta\tau \sum_{i=1}^{n} F(t_i) \approx \frac{u(t)}{m\omega_d} \cdot I_F(t_i) \tag{11}$$

The I_x was the amplitude in rising oscillation period of each response (peak value), and the $I_F(t_i)$ was the discrete time integral of the impact load. Equation (12) shows an approximate linear relationship between the two expressions:

$$I_F(t_i) \propto I_x \ (i = 1, 2, \cdots, n) \tag{12}$$

To sum up, if the system parameters of the SDOF system are known, the response can be calculated according to the excitation. Reversely, the excitation can also be solved from the response.

2.1.2. Multiple Degree-of-Freedom (MDOF) System

The same method can be extended to a MDOF system. A constant coefficient differential equation set with n equations was used to describe the system with n degrees of freedom, shown as the matrix equation:

$$[m]\{\ddot{x}(t)\} + [c]\{\dot{x}(t)\} + [k]\{x(t)\} = \{F(t)\} \tag{13}$$

In the above equation, $[m], [c], [k]$ are the mass matrix, damping matrix and stiffness matrix, respectively. $\{\ddot{x}(t)\}$, $\{\dot{x}(t)\}$, $\{x(t)\}$ refer to the generalized displacement, generalized speed and generalized acceleration vector, respectively. $\{F(t)\}$ is the excitation force. The equation could also be expressed as follows:

$$[R] \cdot \left[\dot{Z}\right] + [S] \cdot [Z] = \{Q(t)\} \tag{14}$$

The solution of the equation above can be written as $x_k = x_{k1} + x_{k2}$, where x_{k1} is the general solution and x_{k2} is the particular solution.

- Particular solution

The particular solution of Equation (14) is as follows:

$$x_{k2} = \sum_{j=1}^{2n} \Phi_k^{(j)} \cdot \xi_{j2}(t) = \sum_{j=1}^{2n} \frac{1}{R_j} \Phi_k^{(j)} \cdot \int_0^t e^{\alpha_j (t-\tau)} \cdot \overline{Q}_j(\tau) d\tau \;\; (k = n, n+1, \cdots, 2n) \tag{15}$$

- General solution

According to the theory of vibration mechanics, the general solution of Equation (14) is as follows:

$$x_{k1} = \sum_{j=1}^{2n} \Phi_k^{(j)} \cdot \xi_{j1}(t) = \sum_{j=1}^{2n} \Phi_k^{(j)} \cdot \xi_{0j} \cdot e^{\alpha_j t} \;\; (k = n, n+1, \cdots, 2n) \tag{16}$$

The solution of Equation (14) can be obtained by combining two solutions as follows:

$$x_k = x_{k1} + x_{k2} = \sum_{j=1}^{2n} \Phi_k^{(j)} \cdot (\xi_{0j} e^{\alpha_j t} + \frac{1}{R_j} \int_0^t e^{\alpha_j (t-\tau)} \overline{Q}_j(\tau) d\tau) \;\; (k = n, n+1, \cdots, 2n) \tag{17}$$

Since the system had viscous damping, the free vibration would decay quickly; therefore, only the influence of the particular solution was considered in practice.

If the excitation was the discrete impact load, the expression of the response can be written as follows:

$$x = \sum_{m=1}^{n} \frac{2}{|R_m|} \cdot \sum_{i=1}^{n_t} e^{-\eta_m (t-\Delta\tau)} \cdot |F(t_i)| \cdot \left|\Phi_k^{(m)}\right| \cdot cos(\omega_m (t - \Delta\tau) - \theta_m) \Delta\tau \tag{18}$$

For each natural frequency ω_m, it can be regarded as a combination of SDOF systems.

The response with a frequency of ω_p within a sampling period can be approximated expressed as follows:

$$x_{\omega_p} \approx \frac{2\Delta\tau \left|\Phi_k^{(\omega_p)}\right|}{\left|R_{\omega_p}\right|} \sum_{i=1}^{n_t} |F(t_i)| cos\left(\omega_{\omega_p} (t - \Delta\tau) - \theta_{\omega_p}\right) \tag{19}$$

The amplitude in rising oscillation period of response can be expressed as follows:

$$I_{xd} \approx \frac{2\Delta\tau \left|\Phi_k^{(\omega_p)}\right|}{\left|R_{\omega_p}\right|} \sum_{i=1}^{n_t} |F(t_i)| \tag{20}$$

Similar to the SDOF system, it was still an approximate linear relationship:

$$I_F(t_i) \propto I_{xd} \tag{21}$$

The above conclusions showed that the time integral of discrete impulse load was approximately linear with the peak value (amplitude of sine wave) in the rising oscillation period of response in MDOF system.

2.1.3. Conclusion

Within a given parameter range, the discrete time integral of the impact load had an approximate linear relationship with the peak value (amplitude of sine wave) in the rising oscillation period of response in both SDOF and MDOF systems. Theoretically, this linear relationship can be used to calculate the corresponding load according to the response given.

The accuracy of this method mainly depended on the accuracy of system parameter identification and the attenuation coefficient. If the identification of system parameters was more accurate, and the attenuation coefficient was closer to 1, the identification accuracy was higher.

The key point of the load identification method was to determine the system parameters, that is $\left|\Phi_k^{(\omega_p)}\right|, \left|R_{\omega_p}\right|$. Traditional methods typically determine each parameter by dynamic model or experiment, which is complicated and inaccurate. This paper used the neural network to determine the parameters of the system and establish a linear relationship between two signals.

2.2. BP Neural Network

The linear model of a neural network was a practical method to determine the unknown linear relationship. Theoretically, the mapping between two groups of data with linear relationship can be determined by a neural network. The discrete time integral of the impact load had an approximate linear relationship with the amplitude in the oscillation period of the response. The former was taken as the output and the latter as the input, and the neural network was used to establish the linear model between them.

An Artificial Neural Network (ANN) was established by imitating the structure of the human neuron system, which had a high adaptive ability and learning ability [10]. A typical BP neural network was composed of an input layer, a hidden layer and an output layer. The signals transmitted forward and the errors transmitted reversely. When the result was inconsistent with the reference result, the error was propagated back, and the connection weights and thresholds of the neural network were adjusted according to the learning algorithm until the reference value was reached.

The structure of the neural network is as in Figure 2.

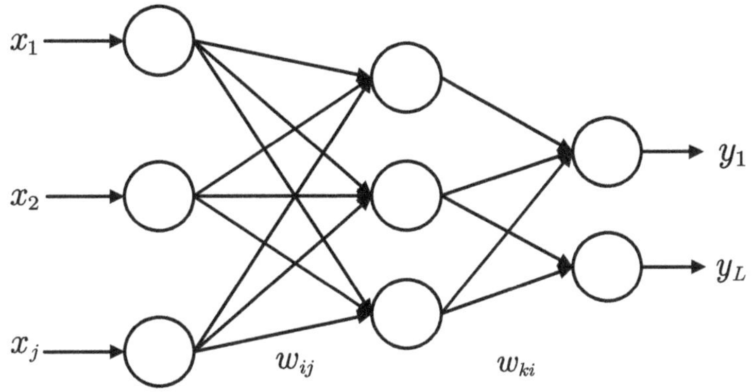

Figure 2. Topological structure of a typical back propagation neural network.

BP neural network consisted of input variables $X = (x_1, x_2, \ldots, x_m)$, hidden layers q and output nodes $Y = (y_1, y_2, \ldots, y_m)$.

The input calculation equation of the node i in the hidden layer is as follows:

$$net_i^p = \sum_{j=1}^{M} \omega_{ij} o_j^p - \theta_i = \sum_{j=1}^{M} \omega_{ij} x_j^p \quad (i = 1, 2, \cdots, q) \tag{22}$$

In the above equations, x_j^p and o_j^p, respectively, represent the input and output of the input layer node under the action of the sample p, ω_{ij} represents the connection weight between the input layer node and the hidden layer node, and θ_i represents the threshold of the hidden layer node.

If the error of output exceeds the set value, the error was fed back from the output layer, and the connection weights between the neuron nodes of each layer were modified until the error was less than the set value. The quadratic error for the output of any sample is expressed as:

$$J_p = \frac{1}{2}\sum_{k=1}^{L}\left(t_k^p - o_k^p\right)^2 \quad (23)$$

The total error expression of the system is:

$$J = \sum_{p=1}^{N} J_p = \frac{1}{2}\sum_{p=1}^{N}\sum_{k=1}^{L}\left(t_k^p - o_k^p\right)^2 \quad (24)$$

The main function of the ANN was the weight correction between the three layers.

- Correction of the connection weight between the output layer and the hidden layer

The connection weights between the neuron nodes of each layer were adjusted in the opposite direction of the gradient of the error function.

The adjustment expression of the connection weight $\Delta\omega_{ki}$ can be obtained as:

$$\Delta\omega_{ij} = \eta\delta_k^p o_i^p = \eta o_k^p\left(1 - o_k^p\right)\left(t_k^p - o_k^p\right)o_i^p \quad (25)$$

Among the equation, t_k^p and o_i^p represent the expected result and the output value of the hidden layer node, respectively.

- Correction of the connection weights between the input layer and the hidden layer

In the same way, the modified expression of the connection weight between the hidden layer and the input layer can be obtained by the gradient of the error function:

$$\Delta\omega_{ij} = \eta\delta_i^p o_i^p = \eta o_i^p\left(1 - o_i^p\right)\left(\sum_{k=1}^{L}\delta_k^p \cdot \omega_{ki}\right)o_j^p \quad (26)$$

Among the equation, o_i^p and o_j^p, respectively, represent the output of the hidden layer neuron node i and the output of the output layer neuron node j under the action of the sample p.

Therefore, the weighting coefficient increment of the output layer node k, and the connection weight increment of the hidden layer node i can be obtained under the action of the sample p:

$$\begin{cases} \omega_{ki}(k+1) = \omega_{ki}(k) + \eta\delta_k^p o_i^p \\ \omega_{ij}(k+1) = \omega_{ij}(k) + \eta\delta_i^p o_j^p \end{cases} \quad (27)$$

The connection weights of each layer of the neural network were adjusted to appropriate values until the error is no more than the set value.

3. Results

This chapter includes four sections, the experimental configuration Section 3.1, signal processing Section 3.2, neural network training and validation Section 3.3 and load identification Section 3.4.

Figure 3 is the flow chart of signal processing and feature extraction. This figure contains the main content of the actual work of this paper.

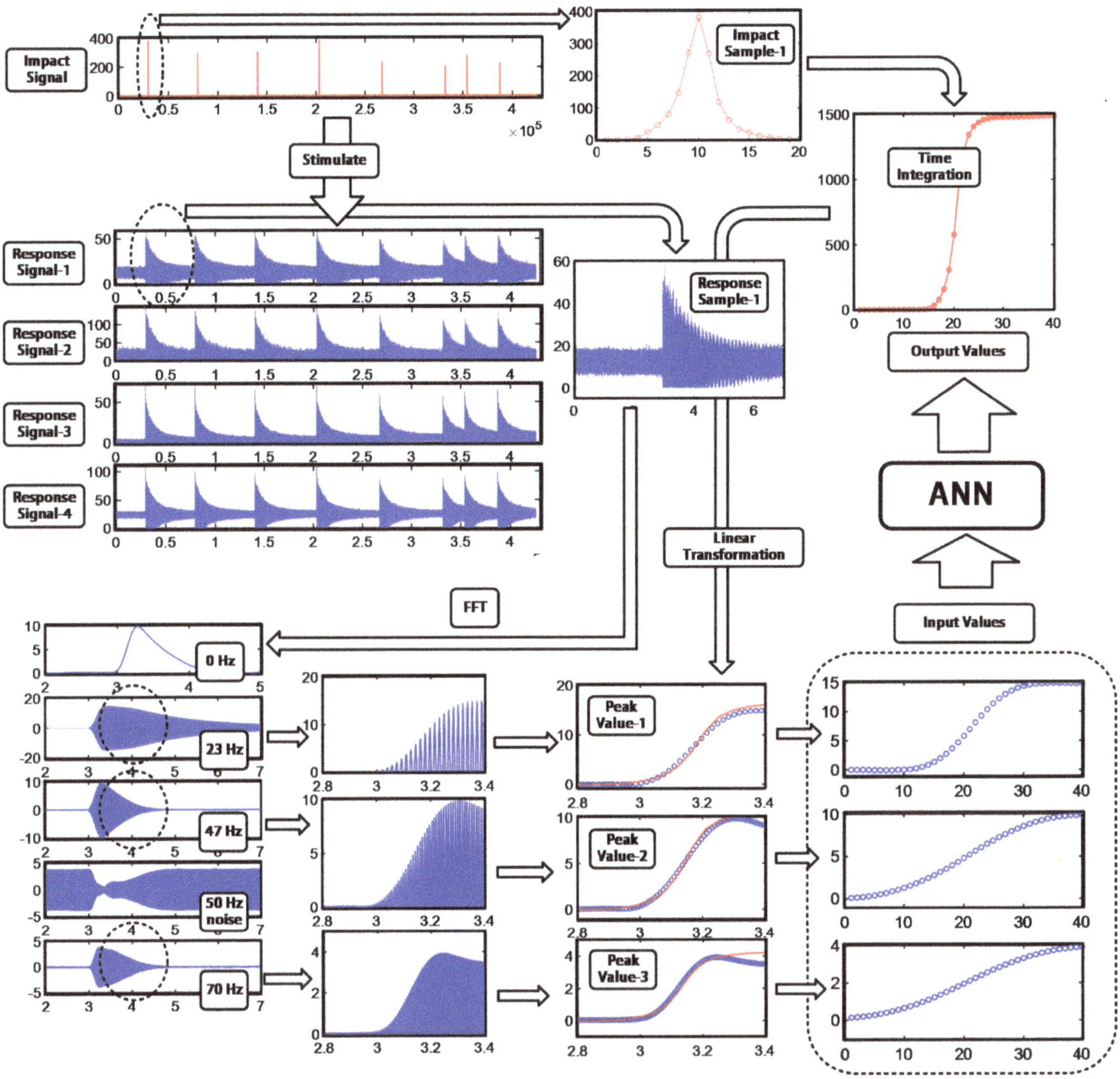

Figure 3. The whole process of the signal processing and featuring extraction.

3.1. Experiment Configuration

In this experiment, the excitation signal of the impact load was acquired by the force sensor on the force hammer, and the response signal was acquired by four sensors (strain rosette). The equipment included a force hammer (PCB086B20), strain rosette, power amplifier, digital acquisition and computer. Before the experiment, the strain rosette was firmly attached to the four measuring points on the steel structure, and the impact load was applied to the structure by the force hammer. The impact load signal was amplified by the power amplifier and then acquired by the digitizer. The response signals were acquired. The sampling frequency was set to 4096 Hz.

The connection diagram of each issue is shown in Figure 4.

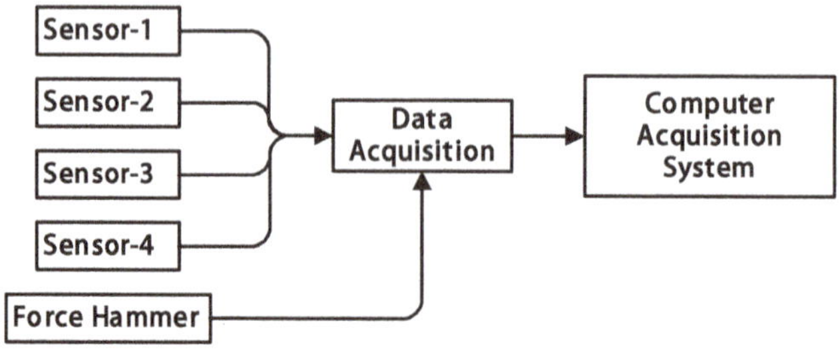

Figure 4. The configuration of the experiment system.

The experimental structure used in this article was a steel cantilever beam with two supporting bars. The density was 7850 kg/m^3, and the elastic modulus was 210 Gpa. The main beam was made of 30 $\times$ 30 mm angle steel with the thickness of 3 mm; the middle is connected by partition frame with the thickness of 3 mm. The actual structure is shown in Figure 5.

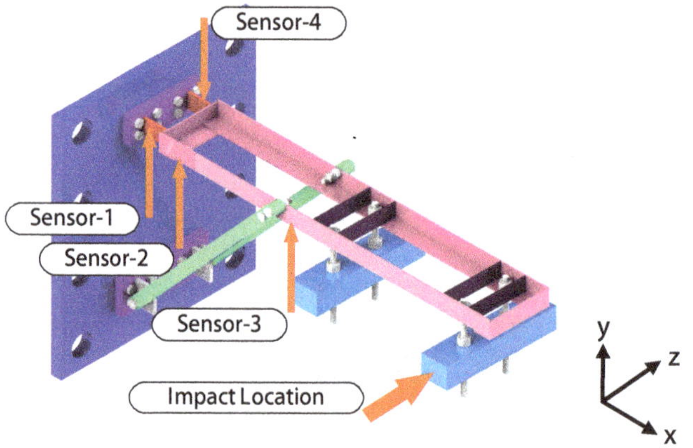

Figure 5. 3D model of the steel frame structure including the location of four sensors and impact load.

Figure 6 shows the details of the strain rosette used in this experiment.

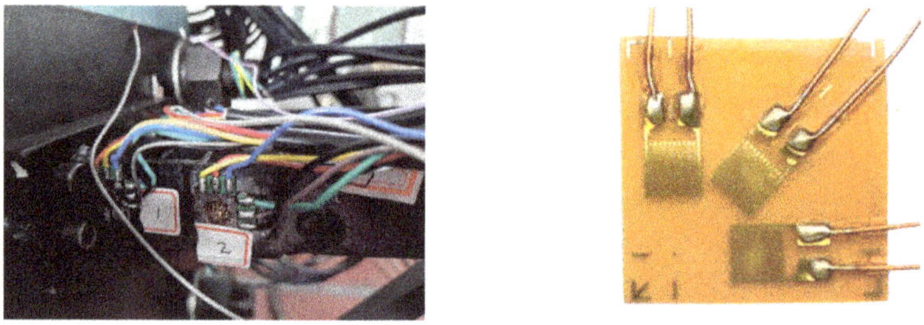

(a) (b)

Figure 6. (**a**) The real configuration of one sensor (strain rosette). (**b**) The specific structure of one strain rosette.

Table 1 listed the exact location of the force hammer and four sensors (strain rosette) in the space coordinate system.

Table 1. Location of four sensors (rosette) and impact load (force hammer).

Item	XYZ Location (mm)
Sensor-1	(0.0, 0.0, 0.0)
Sensor-2	(41.6, 3.2, −13.8)
Sensor-3	(326.7, 3.2, −13.8)
Sensor-4	(0.0, 0.0, 90.2)
Impact	(607.6, −73.8, 156.2)

3.2. Signal Processing

This part includes the design of the digital filter, discrete-time integration of impact load and extraction of peak value (amplitude of sine wave) in the rising oscillation period of response.

Figure 7 shows the impact load records throughout the experiment. The time interval between the last three impact loads was too short for the response to decay completely. Therefore, the last three sets of data were not used in the following signal processing. The signal 1 in Figure 7 is taken as an example.

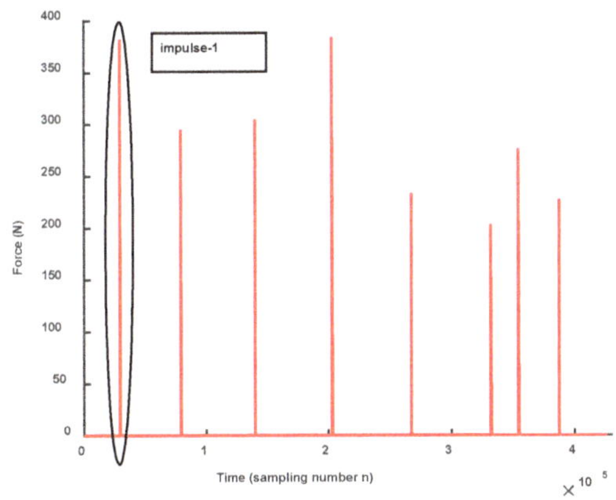

Figure 7. Load-time history of all impact signals.

Figure 8a shows the actual time–force history of the impact load 1, which was a discrete signal. Figure 8b shows the discrete time integral of impact load 1, which was also a set of discrete values. The specific calculation equation of discrete integral value is Equation (10).

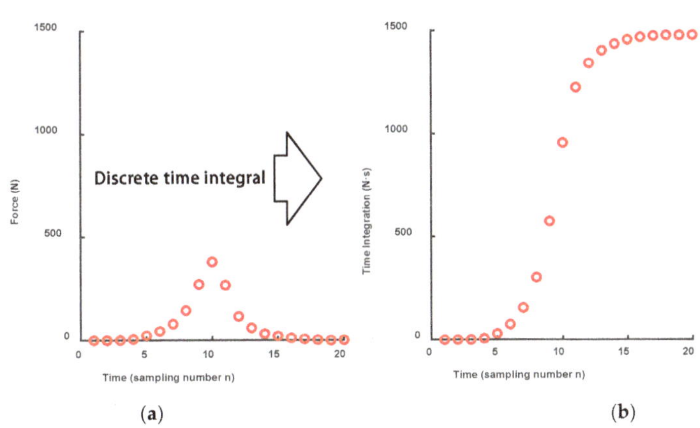

Figure 8. (**a**) Load-time history of impact load 1 (after zoom in). (**b**) Load-time discrete integration of impact load 1.

According to the response solution of the MDOF system, the response signals of each sensor contain multiple natural frequencies. Figure 9 shows the power spectral density (PSD) analysis results of four response from four sensors.

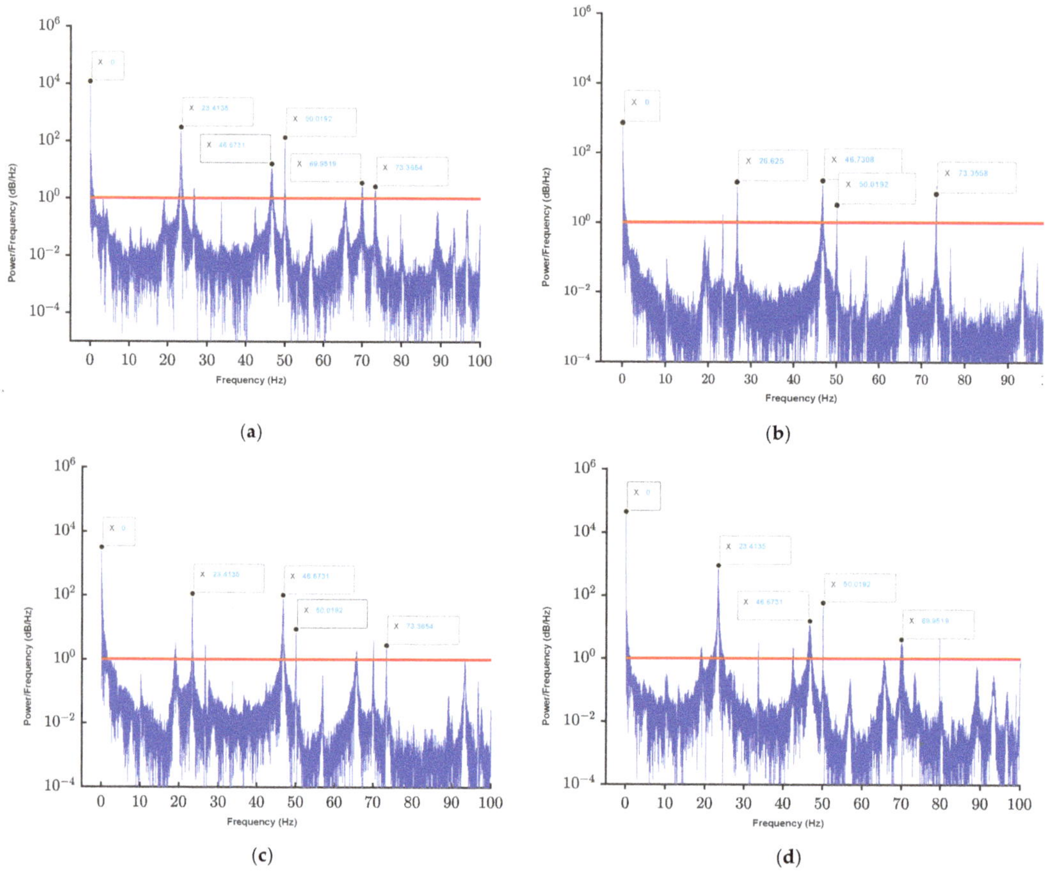

Figure 9. Power Spectral Density (PSD) of four response signals from four strain rosettes. (**a**) Response Signal-1; (**b**) Response Signal-2; (**c**) Response Signal-3; and (**d**) Response Signal-4. The red line in each represented the standard amplitude of the PSD.

The PSD shows that there were several similar frequencies of four sensors. The natural frequencies of the frame structure in Table 1 was obtained from the modal experiment.

According to Figure 9 and Table 2, the primary modes were order 1, order 3 and order 5. Considering that the impact direction was the Z direction, the proportion of the vertical bending mode (order 2) was not significant. There was also a frequency component of 50 Hz. In order to obtain the response of each frequency component individually, a bandpass digital filter was designed as shown in Figure 10.

Table 2. The natural frequency and mode of the actual model.

Order	Natural Frequency (Hz)	Mode of Vibration
1	23.3	Bending (horizontal)
2	33.5	Bending (vertical)
3	45.7	Torsion (*Y* axis)
4	67.3	Torsion (*X* axis)
5	74.3	Torsion (*Z* axis)

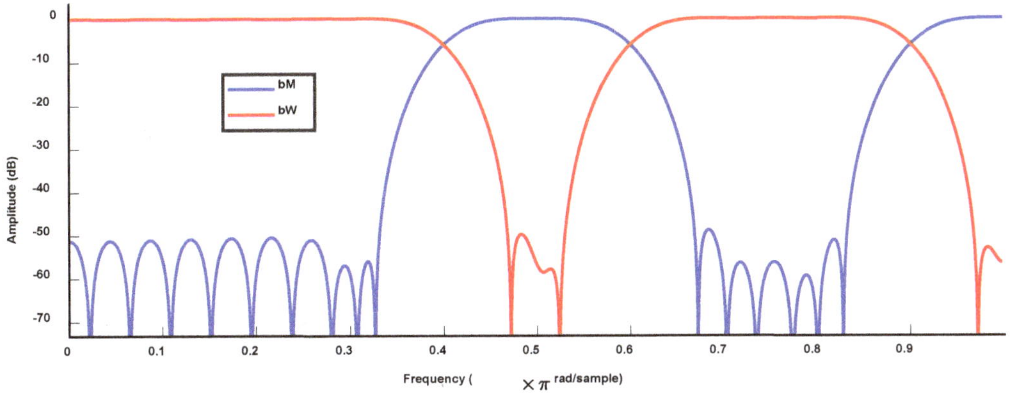

Figure 10. The example amplitude response of the FIR-I filter (Attenuates normalized frequencies below 0.4π rad/sample and between 0.6π and 0.9π rad/sample).

The filter was a FIR1 type bandpass filter. By filtering the primary frequencies individually, the response signals at each frequency were obtained. The parameters of the filter are shown in Table 3.

Table 3. The frequency and the filter band for designing the FIR-I filter.

Frequency (Hz)	Filter Band (Hz)
0	(0, 0.1)
23	(23, 24)
46	(46, 47)
50	(49, 50)
70	(69, 70)

Taking Response 4 as an example, the result is shown in Figure 11.

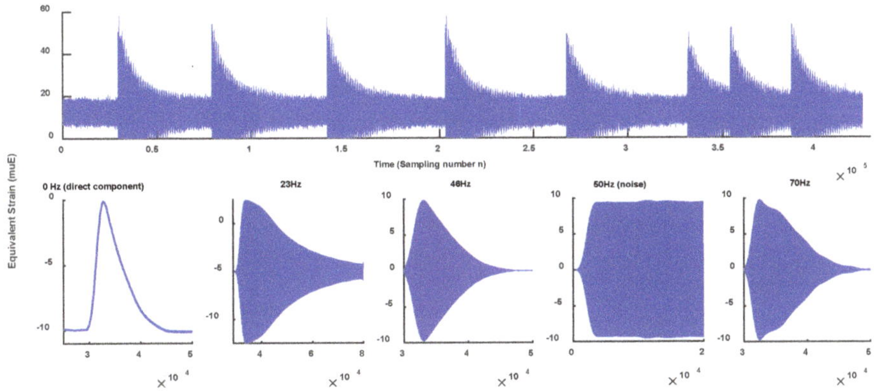

Figure 11. The response signal–4 was decomposed into several frequencies. (0, 23, 46, 50 and 70 Hz).

Figure 11 shows that the 0 Hz component was the fundamental motion without oscillation. The frequency component of 50 Hz did not change with the load; thus, it was noise.

In conclusion, three frequencies of 23, 46 and 70 Hz were selected as identification response signals as shown in Figure 12.

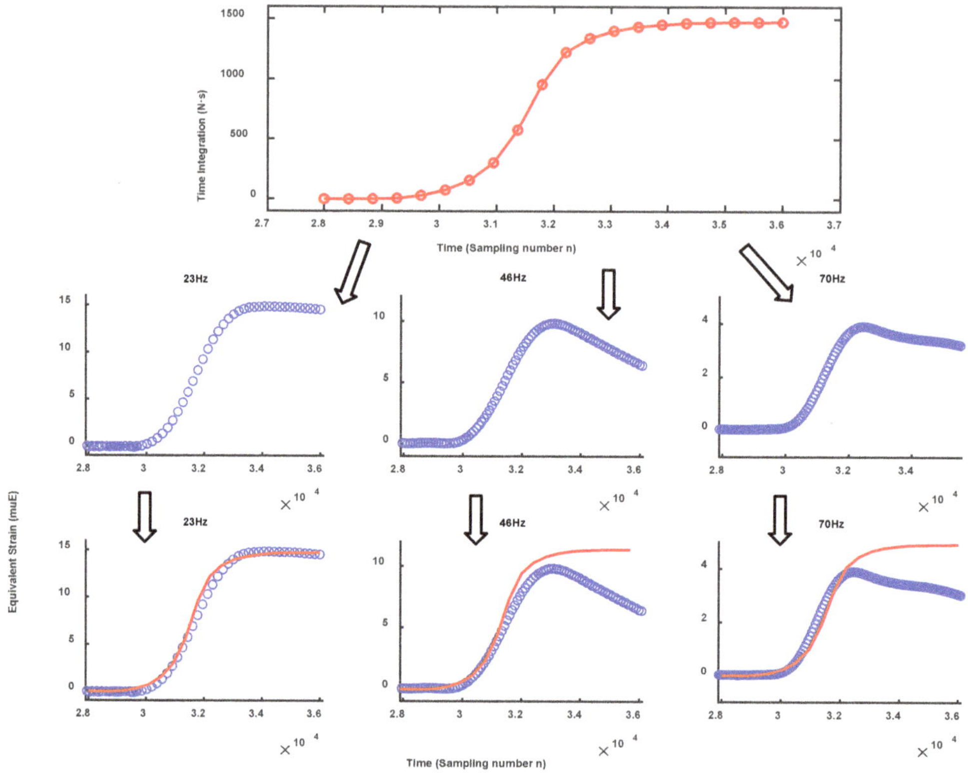

Figure 12. Schematic diagram of linear relation of load–time discrete integration and the peak value (amplitude of sine wave) of the oscillation period.

Figure 12 shows that, through linear transformation, there was indeed a linear relationship between the discrete time integral of the load and the peak value (amplitude of sine wave) in the rising oscillation period of the response. This also confirmed the feasibility of the previous theoretical assumptions. The impact load can be solved according to the linear combination of the response in three frequencies. The peak value (amplitude of sine wave) in the oscillation rising period of the response at three frequencies were extracted as a group of outputs.

The above analysis was given by taking sensor 4 as an example. The situation was similar for other three sensors. A total of 12 sets of data from all four sets of rosettes were extracted to construct the input sample data of the neural network.

To ensure that the maximum value of each group of differential peaks can be extracted, discrete-time differentiation was performed on each peak value (amplitude of sine wave), and Figure 13a was obtained, and the data between the differential maximum point and the left and right 0 points were taken as a set of response samples as shown in Figure 13.

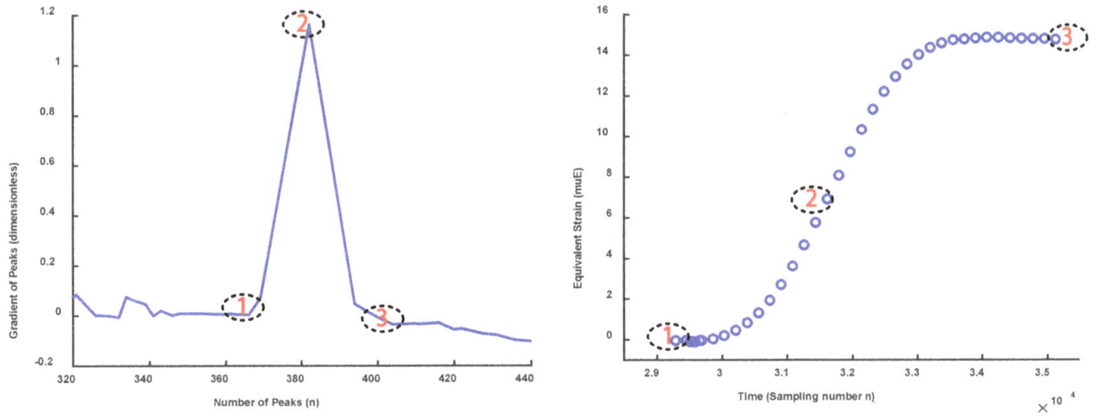

(a) (b)

Figure 13. (**a**) The gradient of the peak value (amplitude of sine wave). (**b**) The value of the oscillation period. The point marked number 1 and 3 represented the point that the gradient equals to zero, the point marked number 2 represented the point that the gradient reached the maximum.

3.3. The Training of the BP Neural Network

The linear relationship between the input and output was constructed through the linear model of the neural network. The structure diagram of the BP neural network used in this paper is shown in Figure 14. It contained 12 inputs, 1 output and 10 hidden layers. The network was trained with the Levenberg–Marquardt backpropagation algorithm.

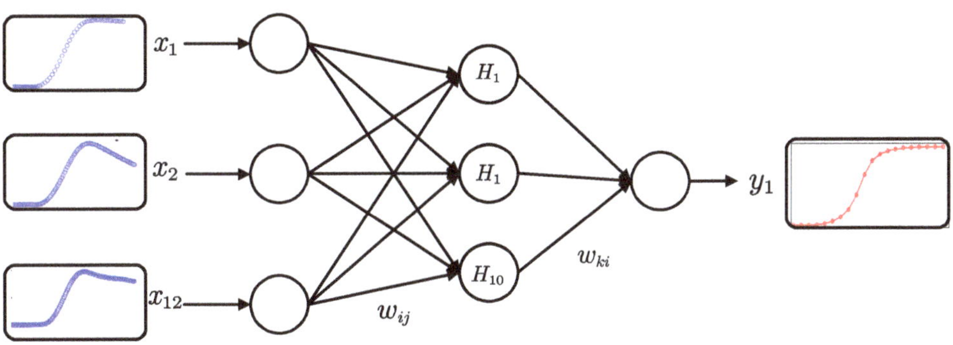

Figure 14. The structure of the BP neural network used in this article.

The performance of the neural network in the training process is shown in Table 4.

Table 4. Performance of the neural network during training and validation.

Situation	Sample Number	MSE [I]	R [II]
Training	84	2.74376	0.999996
Validation	18	7.35701	0.999994
Testing	18	7.62322	0.999992

[I] Regression R Values measure the correlation between outputs and targets. An R value of 1 means a close relationship, and 0 is a random relationship. [II] Mean Squared Error (MSE) is the average squared difference between outputs and targets. Lower values are better. Zero means no error.

As shown in Table 4, the MSE of the three sets of data was small, while the R was close to 1 which meant a strong linear relationship between the input and output. This result also confirmed that the theory proposed in this paper was correct and that the processing of the response signal was feasible.

3.4. Impact Load Identification

This section used the neural network established before to proceed load identification. The recognition results are shown in Figure 15.

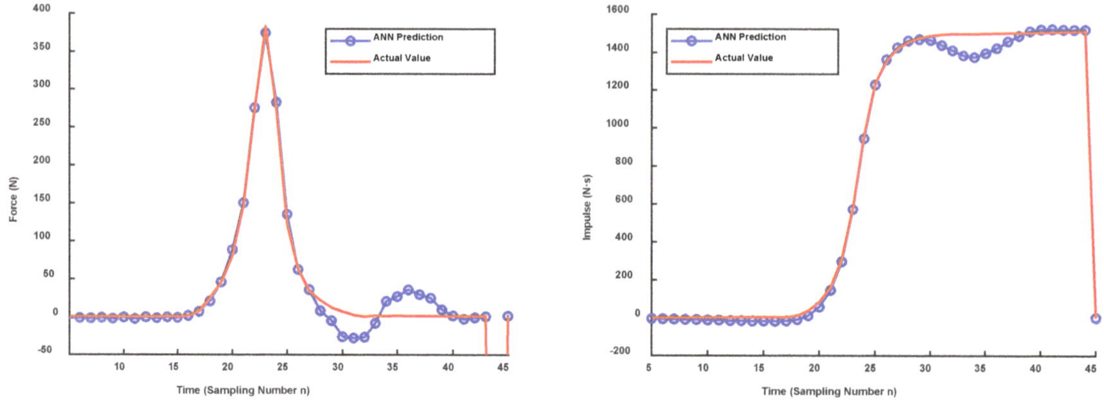

(a) (b)

Figure 15. (**a**) The identification results of the load–time history. (**b**) The identification result of the impulse–time history.

Figure 15a shows the time–force history of load identification and actual value, and Figure 15b shows the time–impulse history of identification and actual value. In practical applications, the maximum amplitude and impulse of the impact load were always given more attention. Table 5 lists the statistics value of the identification results.

Table 5. Relative error of the impact load identification results.

Frequency (Hz)	ANN Prediction	Actual Value	Relative Error
Maximal Force(N)	374.40	382.90	2.22%
Equivalent Impulse (N·s)	1520.00	1510.00	0.67%

Table 2 shows that the identification method had high accuracy. The error between the actual maximum amplitude of impact load and the identification value was 2.22%. The error between the actual equivalent impulse and the identification was 0.67%. These results show that the neural network established in this paper had good performance and that the signal feature extraction method had high credibility.

4. Discussion

In this paper, an impact load identification method based on impulse response theory (IRT) and BP neural network was proposed. By extracting the peak value (amplitude of sine wave) in the rising oscillation period of the response, it transformed the excitation and response signals into the same length. The ANN was used to verify the linear relationship between the time integration of load and peak value of the response, the results showed that there was a strong linear relationship between them. In this article, we did not take the attenuation of response in the rising period into consideration; however, the results were still ideal. That meant the damping ratio of the structure was relatively tiny.

The identification results shown in Figure 15 still had some errors in the end of the signal. The reason might be that the number of training samples was not sufficient. To improve the performance of the ANN, some methods to select the training data should be used. The impulse response theory (IRT) was the basic theory of our manuscript, and the experiment results showed that the theory was effective in this condition. The precondition of this theory was that there was no plastic deformation in structure. If there was plastic deformation, the IRT was not suitable, and the stress wave must be taken into consideration.

Reviewing the research process of this paper, there were still some aspects worth researching in the further. From the experimental perspective: (1) In the response signal, there was a noise mainly at 50 Hz and at the doubling frequencies (100, 150, 200 Hz, etc.). The noise was electric network noise, mainly caused by the absence of an electromagnetic shielding wire. (2) Some intervals of impact were too short, which resulted that the former response signals superposed with the latter ones.

From theory perspective: (1) Only the peak value (amplitude of sine wave) of the response signal was used, which resulted in a certain waste. How to extract more features from other parts of the response signal is worth researching. (2) As the sampling time interval was too short, the actual maximum amplitude of the impact was likely to be missed. Identifying the actual maximum of the impact is a valuable issue.

Though there are some aspects waiting to be explored, the results of this paper have both theoretical and practical value. The neural network tool method was introduced into the field of impact load identification, and a new way to extract signal features was established. A neural network was used to identify the linear parameters of the system, which was more efficient and faster than traditional methods. The validation results showed that the method had high accuracy and practical application potential.

Author Contributions: Conceptualization, methodology, software, writing—original draft preparation and writing—review and editing, D.H.; experiment data, L.C. and X.Y.; supervision, project administration and funding acquisition, Y.G. All authors have read and agreed to the published version of the manuscript.

Funding: This research is a project funded by the Nanjing University of Aeronautics and Astronautics: National Key Laboratory of Rotorcraft Aeromechanics (61422202104) and Priority Academic Program Development of Jiangsu Higher Education (PAPD).

Institutional Review Board Statement: Not applicable.

Informed Consent Statement: Not applicable.

Data Availability Statement: Not applicable.

Conflicts of Interest: The authors declare no conflict of interest.

References

1. Tian, Y.; Xing, S.; Zhang, Z.; Zheng, H. Load Identification of Gearbox Based on Radial Basis Function Neural Networks. *J. Vib. Meas. Diagn.* **2004**, *24*, 16–18+74. [CrossRef]
2. Tian, Y.; Xing, S.; Zheng, H. Load identification of the gearbox based on Elman networks. *J. Vib. Eng.* **2006**, *19*, 114–117. [CrossRef]
3. Staszewski, W.J.; Worden, K.; Wardle, R.; Tomlinson, G.R. Fail-safe sensor distributions for impact detection in composite materials. *Smart Mater. Struct.* **2000**, *9*, 298–303. [CrossRef]
4. Ghajari, M.; Sharif-Khodaei, Z.; Aliabadi, M.H.; Apicella, A. Identification of impact force for smart composite stiffened panels. *Smart Mater. Struct.* **2013**, *22*, 085014. [CrossRef]
5. Wang, J.; Wang, C.; Lai, X. MIMO SVM Based Uncorrelated Multi-source Dynamic Random Load Identification Algorithm in Frequency Domain. *J. Comput. Inf. Syst.* **2015**, *11*, 198–207.
6. Wang, J.G.; Zhang, J.; Wang, Y.J.; Fang, X.Y.; Zhao, Y.X. Nonlinear identification of one-stage spur gearbox based on pseudo-linear neural network. *Neurocomputing* **2018**, *308*, 75–86. [CrossRef]
7. Cheng, G.; Ren, F.; Yang, Z. Study on Load Identification of Double Span Rotor System. *Coal Technol.* **2018**, *37*, 278–280. [CrossRef]
8. Zheng, S.; Guo, T.; Dong, H.; Song, Z. Load identification of piezoelectric structures by using genetic algorithm and finite element analysis. *Chin. J. Comput. Lmechanics* **2009**, *26*, 330–335.
9. Cao, X.; Sugiyama, Y.; Mitsui, Y. Application of artificial neural networks to load identification. *Comput. Struct.* **1998**, *69*, 63–78. [CrossRef]

10. Cooper, S.B.; DiMaio, D. Static load estimation using artificial neural network: Application on a wing rib. *Adv. Eng. Softw.* **2018**, *125*, 113–125. [CrossRef]
11. Ren, S.; Chen, G.; Li, T.; Chen, Q.; Li, S. A Deep Learning-Based Computational Algorithm for Identifying Damage Load Condition: An Artificial Intelligence Inverse Problem Solution for Failure Analysis. *Comput. Modeling Eng. Sci.* **2018**, *117*, 287–307. [CrossRef]
12. Chen, G.R.; Li, T.G.; Chen, Q.J.; Ren, S.F.; Wang, C.; Li, S.F. Application of deep learning neural network to identify collision load conditions based on permanent plastic deformation of shell structures. *Comput. Mech.* **2019**, *64*, 435–449. [CrossRef]
13. Zhang, Z.; Zhang, H.; Chen, Y.; Li, Z.; Fu, Z. Load identification method of track driving system based on genetic neural network. *J. Vib. Shock.* **2022**, *41*, 54–61+89.
14. Yang, T.; Yang, Z.; Liang, S.; Kang, Z.; Jia, Y. Time-frequency feature extraction and identification of stationary random dynamic load using deep neural network. *Acta Aeronaut. Astronaut. Sin.* **2022**, *37*, 98–108.
15. Xia, P.; Yang, T.; Xu, J.; Wang, L.; Yang, Z. Reversed time sequence dynamic load identification method using time delay neural network. *Acta Aeronaut. Et Astronaut. Sin.* **2022**, *42*, 389–397. [CrossRef]

machines

MDPI

Article

Stiffness-Tuneable Segment for Continuum Soft Robots with Vertebrae

Zhipeng Liu [1,2,†], Linsen Xu [3,*,†], Xingcan Liang [2] and Jinfu Liu [4]

1 Hefei Institutes of Physical Science, Chinese Academy of Sciences, Hefei 230031, China; liuzhipeng@mail.ustc.edu.cn
2 University of Science and Technology of China, Hefei 230026, China; lxcan@mail.ustc.edu.cn
3 College of Mechanical and Electrical Engineering, Hohai University, Changzhou 213002, China
4 Changzhou Vocational Institute of Industry Technology, Changzhou 213164, China; liujinfu@mail.ustc.edu.cn
* Correspondence: lsxu@hhu.edu.cn
† These authors contributed equally to this work.

Abstract: In addition to high compliance to unstructured environments, soft robots can be further improved to gain the advantages of rigid robots by increasing stiffness. Indeed, realizing the adjustable stiffness of soft continuum robots can provide safer interactions with objects and greatly expand their application range. To address the above situation, we propose a tubular stiffening segment based on layer jamming. It can temporarily increase the stiffness of the soft robot in a desired configuration. Furthermore, we also present a spine-inspired soft robot that can provide support in tubular segments to prevent buckling. Theoretical analysis was conducted to predict the stiffness variation of the robot at different vacuum levels. Finally, we integrated the spine-inspired soft robot and tubular stiffening segment to obtain the tuneable-stiffness soft continuum robot (TSCR). Experimental tests were performed to evaluate the robot's shape control and stiffness tuning effectiveness. Experimental results showed that the bending stiffness of the initial TSCR increased by more than 15× at 0°, 30× at 90°, and 60× in compressive stiffness.

Keywords: tuneable stiffness; soft continuum robots; stiffening segment; layer jamming; spine-inspired

Citation: Liu, Z.; Xu, L.; Liang, X.; Liu, J. Stiffness-Tuneable Segment for Continuum Soft Robots with Vertebrae. *Machines* **2022**, *10*, 581. https://doi.org/10.3390/machines10070581

Academic Editor: Dan Zhang

Received: 8 June 2022
Accepted: 13 July 2022
Published: 18 July 2022

1. Introduction

Compared to traditional rigid robots, the key advantages of continuum robots are that their weight is lower for the same output force and they are inherently compliant. The structural characteristics of the continuum determine that it has more degrees of freedom, higher dexterity, and can adapt to the shape of various objects, and their ends can be positioned to more 3D space positions. In addition, the simple structure and high compliance of continuum robots are beneficial to improve the safety of human–robot interaction and directly handle sensitive subjects. Continuum robots have also greatly enriched the application field of robots, such as exploration [1], object manipulation [2,3], and surgical applications [4,5]. The most remarkable and common application is invasive surgery. According to the structure, the continuum robots can be divided into two categories: multidisc and soft materials [6].

Although continuum robots have made significant contributions in reducing size and weight and increasing compliance, they suffer from a lack of structural stiffness when manipulating objects compared to rigid robots. More seriously, even small interaction forces can cause large undesired deformations, which are pronounced on long continuum robots. Controllable stiffness provides an opportunity to bridge the gap between soft robots and rigid robots. Therefore, it is a challenge to maintain the trade-off between compliance and stiffness in the design of continuum robots. To deal with this problem, we propose a stiffening segment to increase the current posture stiffness of continuum robots, which are made of soft materials. Our ultimate goal is to develop a stiffness-tuneable SCR that can work closely with humans, such as collaborative robotic arms.

Various methods have been proposed to address this limitation, such as materials with inherent stiffening properties and mechanical mechanisms. In most stiffening material configurations, three types of materials were considered: magnetorheological fluids (MR) [7,8], electro-active phase-change polymers [9,10], or shape memory alloys (SMAs) [11–13]. The stiffness of the stiffening segment employing magnetorheological fluids and electro-active phase-change polymers can be precisely controlled. However, they require strong magnetism and high voltage and current to remain active, so there is a safety risk in human–robot collaboration. Compared with magneto- or electro-active materials, the geometrical arrangement of the stiffening segment with SMAs is simple and the control is easy. The main disadvantage is that they cannot obtain high control precision. Furthermore, the response time of these mentioned materials limits their dynamic performance, which reduces their applicability.

On the contrary, the mechanically driven variable stiffness is more advantageous in obtaining fast reversible stiffening segments. These methods not only meet the requirements of rapid response, but also greatly improve the safety. Because they do not involve high temperatures and high currents, they can meet the basic requirements of human–robot collaboration. One of the most widely used methods is granular jamming, where a stiffening segment is created by packing movable particles into an elastic membrane [8,14]. Furthermore, under the action of negative pressure, the granular particles squeeze each other to form geometric constraints, so the stiffening segment is induced from the unjammed to jammed state. When the negative pressure is removed, the granular particles return to fluidity, so that the stiffening segment also returns to the unjammed state. The level of stiffness that can be achieved depends on the particle-to-particle and particle-to-membrane friction, which are affected by the jammed volume and the applied pressure [15].

Variable-stiffness actuators based on granular jamming have been extensively researched and tested in various robotic applications, including universal robotic grippers [16,17], minimally invasive surgical devices [18], wearable devices [19], and reconfigurable load-carrying structures [20], with some desired results. However, although the granular jamming actuators can resist compressive and shear forces, their resistance to tensile forces is fairly poor. When the tensile stress exceeds the applied pressure of the particle jamming devices, the particles will begin to dissociate and the stiffness of the device will decrease [21]. Due to the inability to effectively resist any significant tensile stress, universal grippers, medical and wearable devices with granular jamming, are designed to be bulky and have low bending loads. Furthermore, when the tensile stress exceeds the range that granular devices can bear in a jammed state, the membrane of the stiffening segment will become the main determinant of the overall stiffness of the actuator [15]. In addition, due to the uneven distribution of particles, buckling is likely to occur during the phase transition from unjammed to jammed state, and the overall stiffness of the device is also unevenly distributed, which greatly weakens the performance of the device. Moreover, the existence of these drawbacks leads to a certain potential risk in the use of granular jamming devices in the medical field.

Another method of structural stiffening that has received extensive attention is the jamming of thin layers of material by negative pressure, whose configurations mainly include planar [22,23], cylindrical [24], and special-shaped structures [25]. This mechanism was first proposed by Kim et al., and its tunable stiffness characteristics were verified by a hollow snake-like manipulator prototype [26]. The layer material of the prototype adopts a flap structure, which is connected by wires to form a tubular shape and then wrapped in a membrane. Compared to granular jamming, to achieve the same stiffness effect, layer jamming is more space-saving [27]. In addition, the tubular shape formed by the layer-jamming stiffening segment is convenient for assembling with the continuum robot. It is beneficial to the miniaturization of the variable-stiffness robot. At the same time, no buckling occurs in tension and compression. However, since the stiffness of the layer jamming depends on the close overlap of the layer materials in the plane, when the cylindrical stiffening segment bends, the deformation of the layer structure results in a decrease in the overall stiffness. In order to deal with this problem, some researches used rigid continuum robots, whose rigid parts can prevent the undesired buckling of the stiffening segment [28–30]. However, there is no related research in the literature that combines SCRs with a support structure and variable-stiffness technology based on layer jamming. It is beneficial to reduce undesired buckling while maintaining the advantages of both.

We combined a support structure and the SCR to propose a unique design, called the flexible drivable spine, which can reduce the undesired buckling of the layer jamming segment. The proposed mechanism was composed of several 3D-printed vertebrae made of acrylonitrile butadiene styrene (ABS) material and an SCR made of super-elastic material (Ecoflex 00-50; Smooth-on, Inc., Macungie, PA, USA), which has the advantage of being highly drivable and not impairing the layer jamming segment. Finally, we integrated the flexible drivable spine with the stiffening segment to obtain a novel tuneable-stiffness soft continuum robot (TSCR), as shown in Figure 1. Based on the design, the TSCR is endowed with bending shape control, as well as variable stiffness capabilities. In this paper, we present the design, performance, and experiments on the TSCR. The rest of the paper is organized as follows. Section 2 provides the details of the TSCR design, including the mechanical design and bio-inspired compliant spine mechanisms. Section 3 proposes an analytical model of a two-layer jamming structure, as well as extending predictions to many-layer jamming structures. Section 4 describes several experiments on the TSCR. Discussion and conclusions are presented in Section 5.

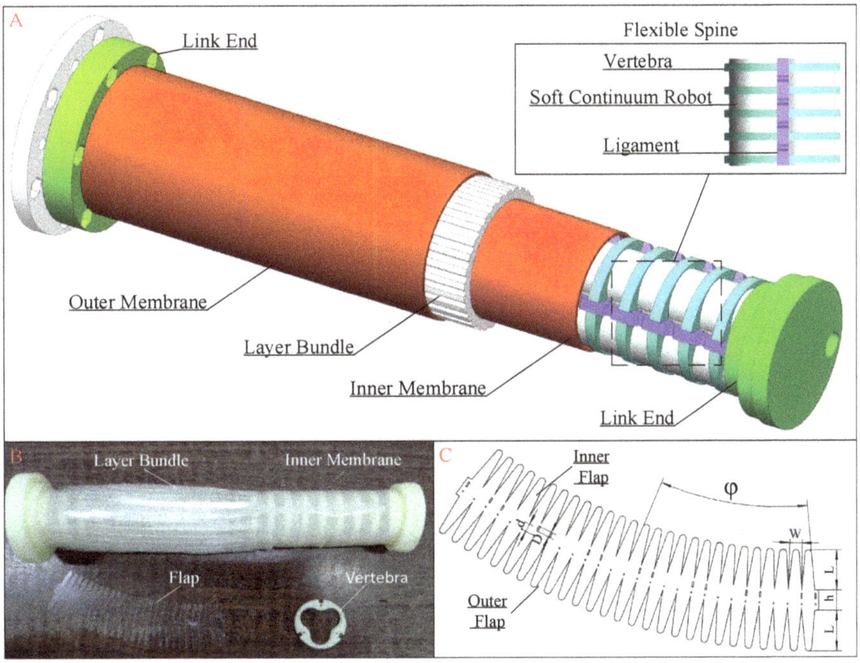

Figure 1. (**A**) Sectional view of a pneumatic-actuated TSCR. (**B**) Structure of TSCR prototype. (**C**) Flap pattern used for stiffening segment based on layer jamming.

2. Robot Design

2.1. The Stiffening Segment Based on Layer Jamming

We employed existing layer jamming techniques to design the stiffening segment to construct the variable-stiffness SCR that can deform both aerodynamically and manually.

The layer design featured the double-sided flap pattern proposed in [4], whose series parameters were the flap length L, the flap width W, the distance between the flaps on both sides h, and the central angle φ, as shown in Figure 1C. The flaps were connected and overlapped by nylon thread to form a conical tube through which the flexible drivable spine can pass. The layer property of the stiffening segment is to provide a frictional overlapping contact area, which depends on the length and width of the flaps, and the stiffening segment hardens through accumulated friction under compression. In addition, the number of layers overlapping each side of the flap is 6.5. The influence parameters on the maximum stiffness can be reflected by the maximum static friction force, which can be expressed as follows:

$$F = \mu n P W L \tag{1}$$

where μ is the friction coefficient of the layer material, n is the number of contact surfaces between the layers at both ends, and P is the applied pressure. The stiffness of the stiffening segment can be tuned by pressure.

The layers were made of polyethylene terephthalate (PET) film, with thickness of 0.23 mm and a friction coefficient of 0.4, which is suitable for large-diameter stiffening segments as PET has high inherent rigidity. Moreover, a high-rigidity material ensures that the layers will not fail and buckle under the external force in the bending state. In order to construct the stiffening segment, the layers were joined together by a needle using 0.108 mm diameter strong nylon thread along the guide holes and slots. Moreover, the overall length and bending capacity of the stiffening segment depend on the length of the guide slot D and the center distance between hole and slot d. The length of the stiffening segment can be calculated as:

$$l = (N-1)(d \pm D/2) + h \tag{2}$$

and the bending angle can be expressed as

$$\theta = (N-1)\arcsin\frac{D}{\phi} \tag{3}$$

where N is the number of the layers, ϕ is the diameter of the stiffening segment. The parameters of the stiffening segment can be seen in the Abbreviations and Parameters section, which also contains the parameters of the flap. According to Equation (2) and the Abbreviations and Parameters section, the adjustable length range of the stiffening segment can be calculated from 363.35 mm to 185.35 mm. The bending angle θ can reach 196°.

To activate the stiffening mechanism, the stiffening segment was wrapped in a sealed tubular membrane, which was made of hyperelastic material (Dragon Skin®10; Smooth-On) with a thickness of 0.3 mm, and then the layers of the segment were compressed together by vacuum. Based on the required parameters, a tubular sealing membrane was made by gluing the cut silicone sheet with glue, and was installed on the 3D-printed link ends with silicone-based adhesive (Sil-Poxy, Smooth-On). Furthermore, the tubular sealing membrane was connected to a vacuum pump through a 6mm PVC pipe on one of the TSCR ends to control the pressure inside the membrane.

2.2. The Flexible Drivable Spine

In order to prevent the potential risk of undesired buckling of the stiffening segment during bending, the flexible drivable spine was fabricated as inspired by the biological spine. It is an SCR with vertebrae that can maintain the constant diameter of the stiffening segment during operation. The spine consists of an SCR, ligaments, and 3D-printed rigid vertebrae, and is defined by the parameters of the gap between vertebrae G_v, vertebra diameter D_v, and vertebra height W_v, as shown in Figure 1. To prevent the motion of the spine from impairing the overall bending performance of the stiffening segment, the ratio of rigid vertebrae to the overall length of the spine should be controlled at 50% or lower to ensure that the spine can have the same or a greater range of tension and compression compared to the stiffening segment. The specifications of spine parameters are presented in the Abbreviations and Parameters section, from which the total length of vertebrae can be calculated as 76 mm. In addition, there are 10 mm rigid supports at both ends. The vertebrae were fitted into the grooves on the SCR in a linear array, and due to the special structural configuration of the SCR, there is no relative rotation with the vertebrae. In addition, a schematic diagram of TSCR fabrication is shown in Figure 2.

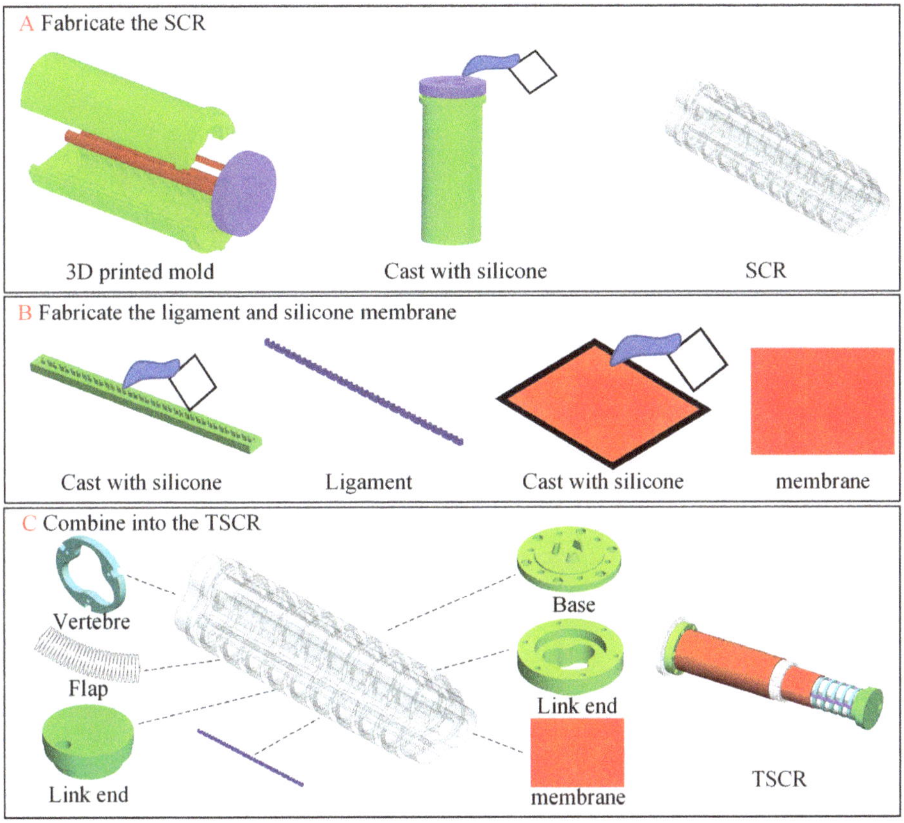

Figure 2. Fabrication of TSCR. (**A**) Fabricate the SCR. Cast uncured silicone (Ecoflex 00-50; Smooth-on, Inc.) into 3D-printed mold. Let the silicone cure to obtain SCR. (**B**) Fabricate the ligament and silicone membrane. Cast uncured silicone (Dragon Skin10®10; Smooth-On) into 3D-printed mold on a flat surface. Let the silicone cure to achieve the ligament and membrane of even thickness. (**C**) Combine all the mentioned components into TSCR.

In this section, we give the design principles, key design parameters, functions of the stiffening segment, and the flexible drivable spine. In addition, we also introduce how the above two parts form the TSCR.

3. Minimal Model of Jamming Mechanism

To illustrate the relationship between the essential parameters and the jammed stiffness of the layer structure, a minimal mechanical model was constructed. There are two reasons for establishing this mechanical model. Fist, it is difficult to manufacture layer jamming specimens to test all possible parameter combinations in the design, so an analytical model helps to describe the general trend of stiffness changes as different parameters are adjusted.

Second, the model can estimate suitable stiffness ranges of layer jamming to accomplish the design task, providing an efficient design tool.

To simplify the analysis, the model only considers two layers of flaps and their interfaces, but at the end, we will further derive the model to include more flaps. Since the Young's modulus of PET is much larger than that of the silicone membrane, we can assume that PET is inextensible. In the normal state, we assume that the initial friction between flaps is zero. Since PET is considered inextensible, the stiffness of the flaps in the tensile direction is only determined by the stiffness of the nylon thread (k_n) used for stitching the flaps and the surrounding silicone membrane (k_m), under the unjammed state. In the jammed state, the flaps are compressed together by external pressure (ΔP). The interfaces form between overlaps of flap–flap and flap–membrane. When pulled, shearing of these interfaces generates extra stiffness (k_{ff} and k_{fm}, respectively). The parameters in the derivation refer to the schematic diagram in Figure 3.

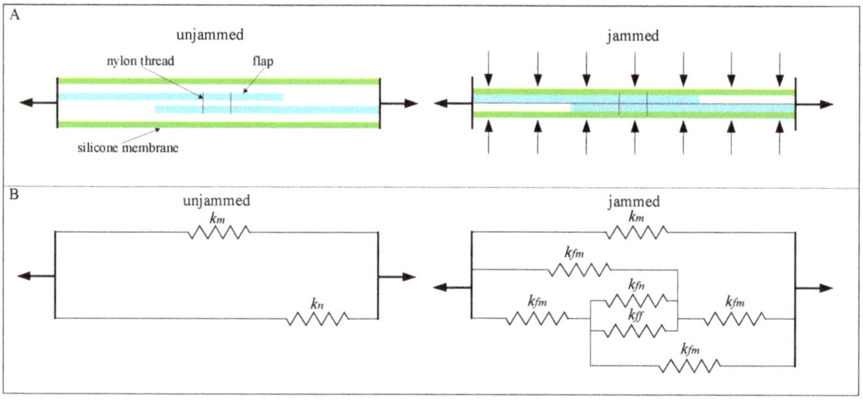

Figure 3. Schematic of analytical model of structural jamming. (**A**) Schematic showing the flap behavior in the unjammed and jammed cases. Only two layers are shown for simplicity. (**B**) Stiffness model of the unjammed and jammed cases of two-layer system.

We use the properties and dimensions of the nylon thread to define the stiffness parameters. The linear model for the tensile stiffness of a single nylon thread is as follows:

$$k_n = \frac{E_n A_n}{L_n} \tag{4}$$

where E_n is the Young's modulus of the nylon thread, L_n is the total length of the nylon thread, A_n is the cross-section area of the nylon thread ($A_n = \pi D_n$, circular-shaped cross-section, D_n is the diameter of nylon thread). The tensile stiffness of a silicone membrane (k_m) can also be established as a linear model as follows:

$$k_m = \frac{E_m A_m}{L_m} \tag{5}$$

where E_m is the Young's modulus of the silicone membrane, L_m is the total length of the silicone membrane, A_m is the cross-section area of the silicone membrane ($A_m = A_{mi} + A_{mo}$, A_{mi} is the cross-sectional area of the inner membrane, and A_{mo} is the cross-sectional area of the outer membrane.). The shear stiffness of the flap–silicone membrane interface can be expressed as:

$$k_f m = \beta \frac{G A_{fm}}{t} \tag{6}$$

where $G = E_m(1 + \vartheta)$ is the shear modulus of silicone (ϑ is the Poisson ratio), A_{fm} is the area between the flap and silicone membrane ($A_{fm} = L_{fm} d$), t is the thickness of the flap–silicone membrane, and β is a parameter to indicate that the contact area between flap and silicone membrane is not all under shear ($\beta \leq 1$).

The mechanical properties of shear interfaces can be divided into a linear region and nonlinear region. In the linear region, it is linear with the displacement until the maximum force is reached. The nonlinear relationship is caused by slip. Therefore, the jammed state of the layer structure also can be divided into three regimes: pre-slip regime, transition regime, and full-slip regime.

In practice, the stiffening segment is subjected to a cantilevered condition, so the Euler–Bernoulli beam theory was employed for analysis. Since the overlapping part of the flaps can be approximated as an isosceles trapezoid, the area of the first moment of the cross-section of the top layer with respect to the interface between the layers can be expressed as $J = \frac{(2WL-Wx)H^2}{4L}$. Furthermore, the area of the second moment of the cross-section can be expressed as $I = \frac{(2WL-Wx)H^3}{6L}$. Specifically, the effective stiffness of the layer structure can be defined as the relationship between the distributed load and the deflection at the free end ($k = \frac{-\partial\omega}{\partial w(x=L)}$).

Substituting I provided into the standard result derived from beam theory, we can obtain the equivalent expression for the pre-slip regime as follows:

$$w_{pre}(x) = -\frac{6\omega L^3}{8EW^*H^3}x^2 + \frac{6\omega L^2}{12EW^*H^3}x^3 - \frac{6\omega L}{48EW^*H^3}x^4 \quad (7)$$

where w is the transverse deflection of the layer structure within the interface, ω is the distributed load, H is the thickness of the flap, W^* is $2WL - Wx$ (W and L are the length and width of the flap, respectively). Substituting Equation (7) into the definition of stiffness, the effective shear stiffness of the flap–flap interface in the pre-slip regime is illustrated as:

$$k_{ff\text{-}pre} = \alpha\frac{EWH^3}{L^4} \quad (8)$$

where α is the effective contact ratio of flaps (a parameter taking incomplete contact into account). Moreover, substituting I and J into the standard result derived from beam theory, the equivalent expression for the transition regime can be written as:

$$w_{tran}(x) = w_{pre}(x) - \frac{81\omega L^5}{128EW^*H^3} + \frac{27\mu pL^3}{16EH^2} - \frac{27\mu pLW^*}{16E\omega H} + \frac{3(\mu p)^3W^{*2}}{4E\omega^2L} - \frac{(\mu p)^4W^{*3}H}{8E\omega^3L^3} \quad (9)$$

Substituting Equation (9) into the definition of stiffness, the effective shear stiffness of the flap–flap interface in the transition regime is illustrated as:

$$k_{ff\text{-}tran} = \frac{-128EWH^3\omega^4}{48(\mu pWH)^4 - 192\omega L(\mu PWH)^3 + 216(\mu P\omega WLH)^2 - 129\omega^4L^5} \quad (10)$$

We modeled collections of flaps in parallel within the inner membrane and the outer membrane. For the system of the stiffening segment in the unjammed state, the only stiffness sources are k_n and k_m. Thus, the unjammed stiffness can be written as:

$$k_u = N_nk_n + k_m \quad (11)$$

where N_n is the number of nylon threads in the stiffening segment. None of the various properties of the silicone membrane are varied; thus, k_m is a constant.

In the jammed state, stiffness is contributed from all interfaces as well as the silicone membrane, flaps, and nylon (see Figure 3, right). The total stiffness can be derived from a combination of series and parallel stiffness. For a two-layer stiffening segment system, we can obtain the following equation:

$$k_{j,N=2} = \frac{k_{fm}^2 + k_{fm}\left(k_{fn} + k_{ff}\right)}{1 + \left(k_{fn} + k_{ff}\right)} + k_m \quad (12)$$

Since k_{ff} and k_{fm} are much larger compared to k_{fn}, they dominate the stiffness of the system. By eliminating the k_{fn} term, we can simplify Equation (12) to:

$$k_{j,N=2} = \frac{k_{fm}^2 + k_{fm}k_{ff}}{1 + k_{ff}} + k_m \quad (13)$$

To illustrate jamming layers with $N > 2$, we observe that for each additional layer, each additional flap increases the number of flap–flap interfaces by one. Since these interfaces are parallel to each other and arrayed along the axis, the k_{ff} term can directly multiply the number of interfaces, $z = N - 1$.

$$k_{j,N>2} = \frac{k_{fm}^2 + k_{fm}zk_{ff}}{1 + k_{ff}} + k_m \quad (14)$$

By observing that k_{ff} and k_{fm} are approximately equal in a certain pressure range, we can further simplify Equation (14) as:

$$k_j = \frac{1+z}{z}k_{ff} + k_m \quad (15)$$

According to the actual situation, the actual values will be used to replace z in the above formula, which can clarify an overall trend. In other words, the stiffness of the flap increases with the increase in the unit length, which is also fully proven by the experimental results.

In the jammed state, the stiffness of the model has been presented before slip. The nonlinear case occurs when the shear force in the interface exceeds the maximum static friction that the interface between adjacent flaps can withstand. The maximum static friction can be written as:

$$F_f^{max}\Delta P = \mu\Delta PA_{ff} \quad (16)$$

where μ is the coefficient of static friction, A_{ff} is the area of shear between flap and flap ($A_{ff} = WL$). After slip occurs, the equivalent expression for the full-slip regime can be written as:

$$w_{slip}(x) = \left(\frac{\mu PL^3}{H^2E} - \frac{3\omega L^2}{EWH^3}\right)x^2 - \left(\frac{2\omega L}{EWH^3} - \frac{3\mu P}{EH^2}\right)x^3 - \frac{\omega}{2EWH^3}x^4 \quad (17)$$

Substituting Equation (17) into the definition of stiffness, the effective shear stiffness of the flap–flap interface in the full-slip regime decreases to:

$$k_{ff-slip} = \frac{2EWH^3}{3L^4} \quad (18)$$

In addition, this stiffness is equal to the effective stiffness of the two-layer structure with no vacuum applied.

All the parameters in the minimal layer jamming model presented in the Abbreviations and Parameters section were used to generate the analytical prediction (dashed line) in Figure 4A. The values of parameters, including α and β, were determined by comparing the similarity between the data and model.

In this section, we construct a minimal mechanical model to illustrate the relationship between essential parameters and the jammed stiffness of the layer structure. The model covers the jammed and unjammed states of the layer structure. Furthermore, the jammed state of the layer structure also can be divided into three regimes: pre-slip regime, transition regime, and full-slip regime. In addition, we also consider the case in which the flaps are not fully in contact.

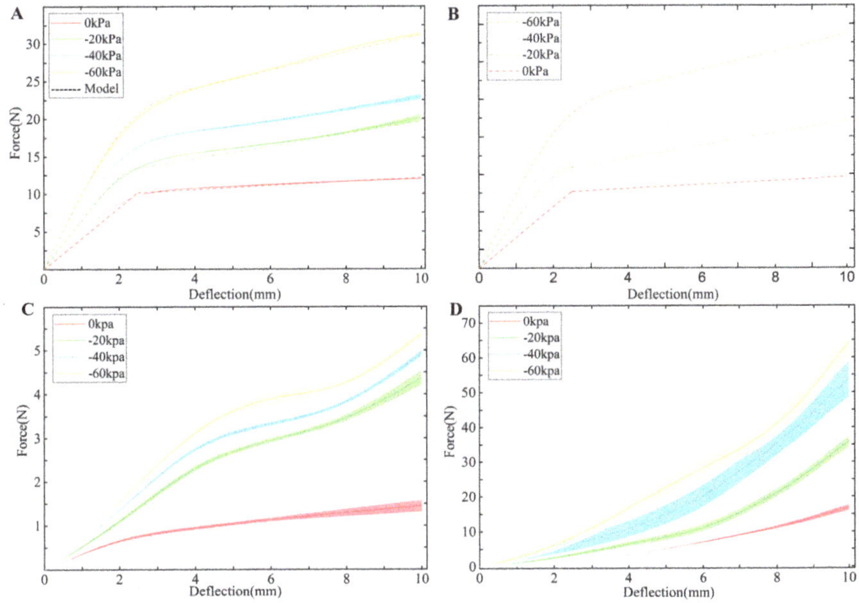

Figure 4. Mechanical characterization of TSCR from 0 kPa to −60 kPa. Each solid line represents the average behavior of the TSCR. Clouds indicate 1 SD from the mean. (**A**) Lateral bending experiment at 90°. Dashed lines are predictions from the analytical model. (**B**) Dashed lines are predictions from the finite element. (**C**) Lateral bending experiment at 0°. (**D**) Compression experiment with straight TSCR.

4. Results and Experimental Setup

Based on the analysis results, the constructed stiffening segment prototype was assembled to the pneumatic the flexible drivable spine and evaluated to demonstrate its applicability. In addition, the stiffness improvement of the flexible spine with the stiffening segment was evaluated.

Having quantified the effect of the essential design parameters on the jamming layer's stiffness, we evaluated the TSCR in three ways. First, to evaluate stiffness and flexibility of the TSCR, we conducted the lateral deflecting experiments at 0° and 90° configurations, as well as the axial compression experiments. Second, in order to evaluate the ability of TSCR to resist abnormal deformation in applications, we manually bent the TSCR from 0° to 180° in 45° increments. The ratio of change in the diameter of the TSCR center at each incremental point can be used as an evaluation criterion. Third, to evaluate the actual performance of the TSCR, we conducted grasping and loading experiments. The experimental platform configuration is shown in Figure 5. It is composed a linear actuator and a force sensor (WD-500, Aidebao Co, Wenzhou, China) mounted on an axially moving slide. The vacuum pressure applied to the stiffening segment was from 0 kPa to −60 kPa with a gradient of −20 kPa (VP 280, 2-stage vacuum pump).

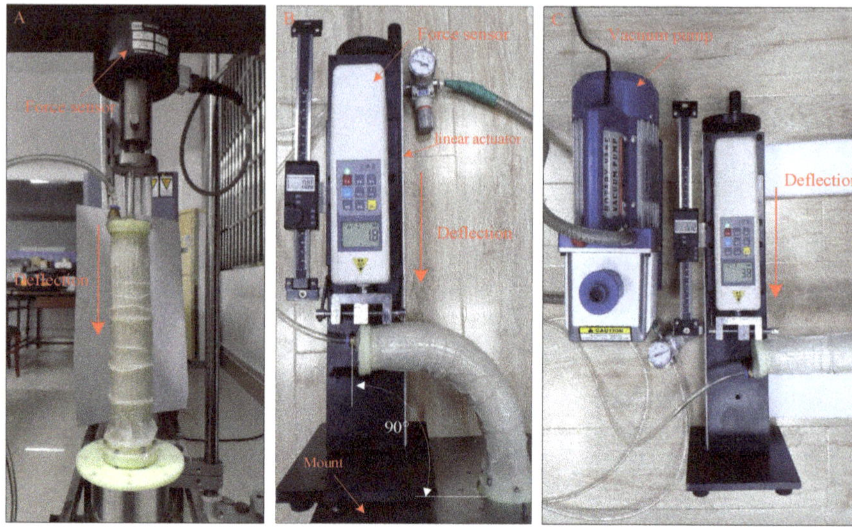

Figure 5. Experimental setups to evaluate the stiffness and flexibility of the TSCR. (**A**) Compression experiment with straight TSCR. (**B**) Lateral bending experiment at 90°. (**C**) Lateral bending experiment at 0°.

4.1. Evaluation of the TSCR's Stiffness and Flexibility

In the experiments, the tested TSCR was mounted on an aluminum plate with a fixed position relative to the linear actuator. The initial relative position of the force sensor to the TSCR is that the force sensor touches the TSCR but no force is detected, and the displacement is 0 mm. During the experiment, the force sensor pushes the TSCR along the axial direction to deflect the required distance. For the lateral deflection experiment, the tested distances for each different position include 2 cm, 4 cm, 8 cm, and 10 cm (Figure 5B,C). For the compression experiment, the axial compression distance is also from 0 cm to 10 cm with a gradient of 2 cm (Figure 5A). We performed 50 cycles of various test protocols in both the unjammed and jammed states. For each test, the required force for the desired deflection needed to be recorded separately.

Here, we adopt a boxplot, which displays the minimum, median, and maximum, the first and third quartile, along with outliers of peak force, for the axial compression experiments as well as lateral deflecting experiments at 0° and 90° configurations, recorded in the measurement; results are summarized in Figure 6. The force–deflection diagrams for the axial compression experiments as well as lateral deflecting experiments at 0° and 90° configurations are shown in Figure 4.

From the results of the lateral deflection experiment in Figure 6, we can observe that the TSCR reached a much higher stiffness in the jammed state, which is up to 60 times higher compared with the initial TSCR. In the 0° configuration, the maximum average deflection force and average axial compression force can reach 5 N and 64 N, respectively. In the 90° configuration, the maximum average deflection force can reach 31 N. Due to the existence of the backbone, the TSCR can maintain high stiffness, instead of decreasing stiffness caused by buckling under the 90° configuration.

It can be seen from Figure 4 that the stiffening segment based on layer jamming bends under compressive force, and the stiffening segment exhibits unstable behavior due to the sudden slip between the flaps. In the unjammed state, the force–displacements of the TSCR are linear, which indicates higher compression for higher stiffness. In Figure 4A, except for some points, the minimum coefficient of determination (R^2) between experimental and model data is 0.9525. Factors that cause discrepancies between the model and the collected results are errors generated during testing, uneven negative pressure distribution, non-standard flap overlap, and small gas leaks. In Figure 4A, the slope of the curve should reflect the change in stiffness. The area where the stiffness changes significantly is located in the transition regime, and the stiffness is a constant in the pre-slip and full-slip regions. In addition, at −60 kPa, the overlap rate between flaps is much higher than at −40 kPa. Therefore, the curve interval between −60 kpa and −40 kpa in Figure 4A is larger than the curve interval between −20 kPa and −40 kPa. Furthermore, the finite element results in Figure 4B are in good agreement with the analytical model results in Figure 4A.

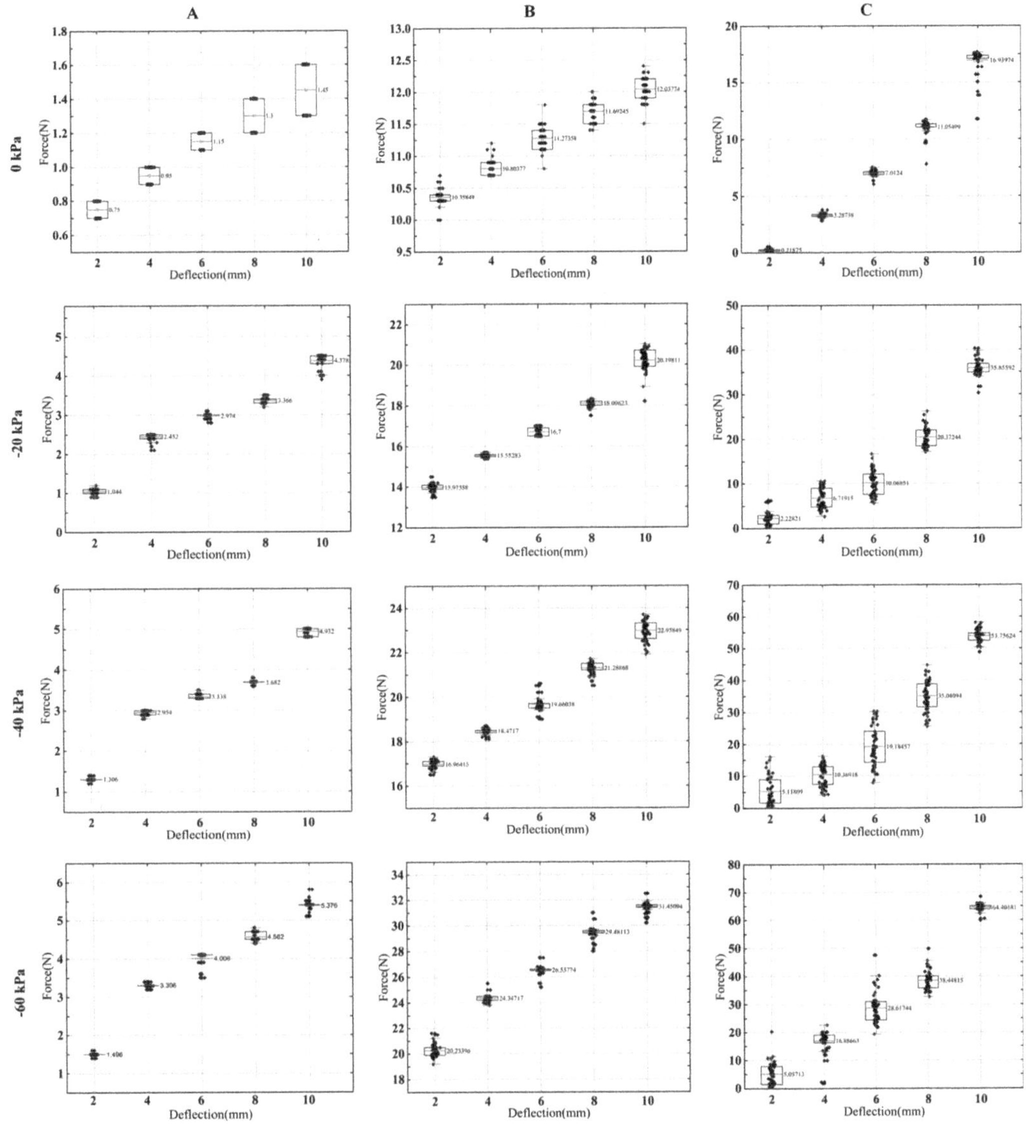

Figure 6. Boxplots of forces displaying the minimum and maximum, the median and the first and third quartile (red line and box) of forces recorded from experimental measurements from 0 kPa to −60 kPa. (**A**) Compression experiment with straight TSCR (left column). (**B**) Lateral bending experiment at 90° (middle column). (**C**) Lateral bending experiment at 0° (right column).

Noticeably, the stiffness increases dramatically in the high-stiffness region, because of the increase in the overlapping area. This also illustrates that the stiffness is positively correlated with the amount of axial compression within a certain range. In order to quantify the stiffening effect of the stiffening segment on the TSCR, a stiffening coefficient γ was proposed as follows:

$$\gamma = \bar{F}_{jammed} / \bar{F}_{unjammed} \tag{19}$$

where $\bar{F}$ represents the average of forces measured in each type of axial compression experiment.

Figure 7 shows the stiffness coefficients for various amounts of axial compression. Compared with the soft continuum robot, the TSCR has comparable stiffness in the unjammed state, which can be indicated by the sudden change in stiffening coefficient in Figure 7. It is clear that the stiffening coefficient and the amount of axial compression within a certain range are positively correlated.

According to the results of these experiments, we can achieve a good balance between stiffness and compliance with the stiffening segment based on layer jamming. Although the stiffening segment is limited to a certain compliance, the stiffness improvement of the SCR is very significant. Sufficient repeatable stiffness of the TSCR was also observed in the examination, and its bending flexibility was not compromised. In addition, at the limit of the length of the stiffening segment, the nylon thread restricts the movement of the flaps. Furthermore, when compressed, the TSCR also showed a tendency to buckle. However, once the vacuum was released, the buckling deformation was reversible.

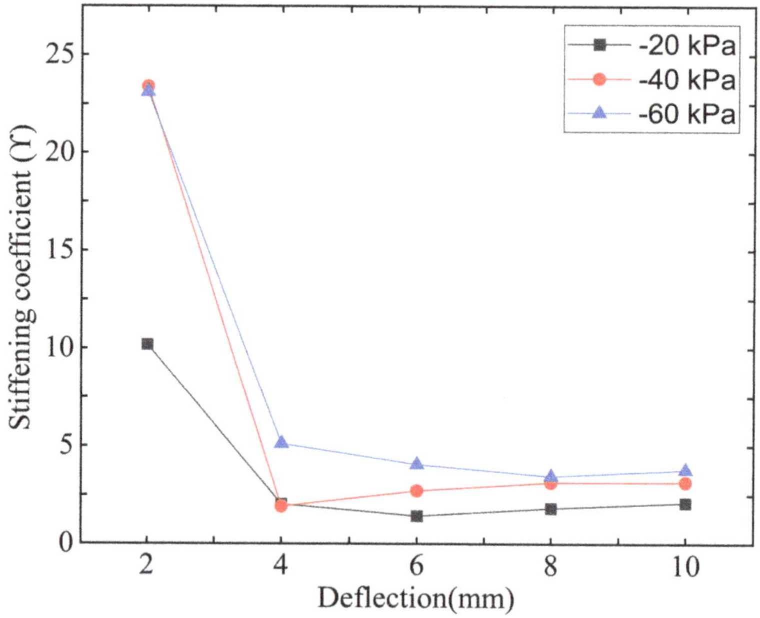

Figure 7. Stiffening coefficient γ for stiffening segment in the axial compression experiment under −20 kPa to −60 kPa state.

4.2. Evaluation of the TSCR Central Diameter

The constant diameter of the soft continuum robot facilitates simplified theoretical analysis, precise position control, and the avoidance of buckling, all of which are important for SCR operation. In particular, under the large structural bending state, the central diameter change is a critical consideration.

Figure 8 shows the effect of the bending angle configuration on the TSCR and SCR diameter. The ratio of the diameter of the TSCR and SCR configurations at each angle to the diameter of the 0° configuration (highlighted green, and highlighted red in 180° configuration on Figure 8) was used as a quantification factor. Then, in order to indicate the relationship between the diameter ratio and the bending angle, the diameter ratio and the bending angle were plotted correspondingly, as shown in Figure 9. The SCR exhibited buckling in a 0.856 ratio at 180°. The TSCR presented the highest diameter ratio in 0.981 at 180°. Therefore, TSCR is significantly better than SCR in center diameter retention. This situation is very beneficial for subsequent control.

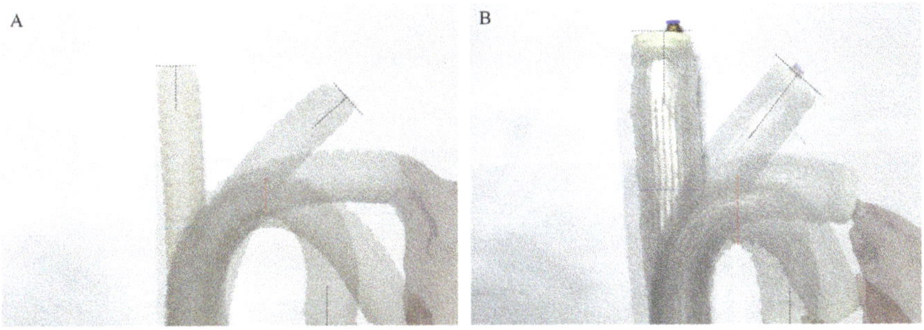

Figure 8. Deformation of centeral diameter under bending. (**A**) Bending experiment of the SCR. (**B**) Bending experiment of the TSCR.

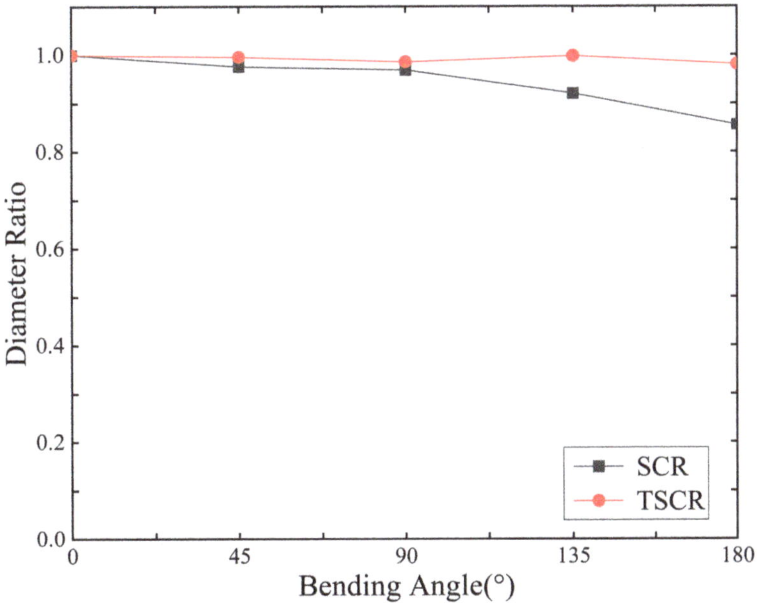

Figure 9. Central diameter ratio: soft continuum robot (SCR) and tuneable-stiffness soft continuum robot (TSCR).

The above experimental results show that the stiffening segment based on layer jamming endows the TSCR with a good balance between compliance and stiffness. Although the axial load capacity of the TSCR is significantly higher than that of other configurations, the minimum load that other configurations can withstand also reaches 5 N. In addition, after repeated testing, the TSCR has shown sufficient repeatable stiffness.

4.3. Evaluation of TSCR by Comparing with Different Actuators

In order to further illustrate the availability of the TSCR, we compare it with representative different actuators from five aspects, such as bending angle, pressure, force, weight, and cost.

From Table 1, we can see that the pressure of most actuators has only a single value, indicating that they can only achieve stiffness changes or other functions. In terms of bending angle and force, the TSCR also has obvious advantages. The continuum robot [28] has the closest performance to the TSCR. However, since the robot is tendon-driven, it cannot extend itself as with TSCR and has an inherent risk of buckling.

Table 1. Multidimensional comparison of different actuators.

Name	Bending Angle [°]	Pressure [kPa]	Force [N]	Weight [g]	Cost
Underactuated robotic hand [2]	160	60	0.24	No data	Medium
Soft gripper [12]	10	70	0.0263	No data	High
Soft robotic gripper [16]	45	20	0.369	66	Medium
Particle phalange [17]	90	−15 and 150	18	460	High
Variable stiffness device [21]	Spiral	−90	0.5	No data	Low
Versatile soft machine [22]	90	−71	10	No data	Low
Kangaroo tail [24]	90	No data	9.8	No data	High
Snake-like manipulator [26]	No data	−101	2	No data	low
Stiffness-tuneable limb segment [27]	187	−60	31.33	No data	Low
Continuum robot [28]	90	No data	16	No data	High
TSCR	196	±60	31.45	440	Low

In this section, we introduce the results of experimental evaluations about the TSCR's stiffness and flexibility. Experiments show that the stiffening segment can indeed greatly improve the stiffness of the SCR. Furthermore, the experimental results are in good agreement with the model. In addition, the experimental results also show that central diameter of the TSCR is well maintained, indicating that the vertebrae can indeed prevent the buckling of the stiffening segment during operation. In addition, the performance of the TSCR is significantly improved compared to other representative actuators. All parameters in this section are shown in the Abbreviations and Parameters section.

5. Conclusions

In this paper, we proposed a stiffness-tuneable segment for soft continuum robots with vertebrae to improve the stiffness in certain desired configurations. The potential undesired buckling of the stiffness-tuneable segment during operation can be prevented by the vertebrae on the SCR. We also have derived an analytical model for jamming structures over three major phases of deformation and effective stiffness, and extracted critical design parameters. We also conducted several experimental evaluations, which illustrated that the effective stiffness improvement by 15× to 60× compared to the initial TSCR could be achieved. In addition, the factors that cause a slight deviation between the jamming model and the experimental results may come from small damage during the production of the silicone membrane, irregular flap stacking, and uneven vacuum distribution.

In future work, miniaturization of the TSCR should be considered for interventional medicine, while the current TSCR can be used for exoskeleton rehabilitation robots. In order to obtain a better compromise between rigidity and flexibility, we can design flaps based on the principle of bionics, and the bionic objects can be scaled objects such as pangolins and fish. In addition, the weaving patterns and materials of flaps can also be used as further research directions. When building the TSCR controller, it is necessary to quantify the hysteresis effect of the loading and unloading of the stiffening segment.

To conclude, we illustrate the mechanics of layer jamming, provide an efficient design tool for jamming structures, and provide references for creating variable robots and mechanisms.

Author Contributions: Conceptualization, Z.L. and L.X.; methodology, Z.L.; software, X.L.; validation, Z.L., L.X. and J.L.; formal analysis, Z.L.; investigation, L.X.; resources, J.L.; data curation, X.L.; writing—original draft preparation, Z.L.; writing—review and editing, L.X.; visualization, X.L.;

supervision, L.X.; project administration, L.X. All authors have read and agreed to the published version of the manuscript.

Funding: This research was funded by a Jiangsu Special Project for Frontier Leading Base Technology, grant number BK20192004; Suzhou Key Industrial Technology Innovation Forward-Looking Application Research Project, grant number SYG202143; Fundamental Research Funds for Central Universities, grant number B220201025; Changzhou Key Research and Development Projects, grant number CJ20210040, and the General Project of the Natural Science Foundation of Colleges and Universities of Jiangsu Province, grant number 21KJB460006.

Institutional Review Board Statement: Not applicable.

Informed Consent Statement: Not applicable.

Data Availability Statement: The data presented in this study are available on request from the authors.

Acknowledgments: The authors thank Hefei Institutes of Physical Science, Chinese Academy of Sciences, for the experimental setup.

Conflicts of Interest: The authors declare no conflict of interest.

Abbreviations and Parameters

The following abbreviations and parameters are used in this manuscript:

Abbreviations	
SCR	Soft continuum robot
TSCR	Tuneable-stiffness soft continuum robot
Parameters	Description
W	Width of flap
L	Length of flap
D	Length of the guide slot
d	Center distance between hole and slot
ϕ	Diameter of stiffening segment
h	Distance between the flaps on both sides
N	No. of layers
n	No. of layers overlapping each side
μ	Friction coefficient
l	The length of stiffening segment
θ	The bending angle of stiffening segment
D_v	Vertebra diameter
W_v	Vertebra height
G_v	Gap between vertebrae
N_v	No. of vertebrae
L_s	Initial length of SCR
L_n	Total length of the nylon thread
k_n	The stiffness of nylon thread
E_n	Young's modulus of nylon thread
A_n	Cross-section area of nylon thread
D_n	Diameter of nylon thread
ΔP	Diameter of nylon thread
k_m	The tensile stiffness of silicone membrane
E_m	Young's modulus of membrane
L_m	Total length of the silicone membrane
A_m	Cross-section area of the silicone membrane
A_{mo}	Cross-sectional area of outer membrane
A_{mi}	Cross-sectional area of inter membrane
E	Young's modulus of PET
G	Shear modulus of silicone
ν	Poisson's ratio of membrane
$k_f m$	Shear stiffness of the flap–silicone membrane interface

A_{fm}	Area between flap and membrane
t	Thickness of flap–silicone membrane
β	Ratio of contact between flap and membrane
k_{ff}	Shear stiffness of the flap–flap interface
A_{ff}	Area between flap and flap
$k_{ff\text{-}pre}$	Shear stiffness of the flap–flap interface in pre-slip regime
$k_{ff\text{-}tran}$	Shear stiffness of the flap–flap interface in transition regime
$k_{ff\text{-}slip}$	Shear stiffness of the flap–flap interface in full-slip regime
W_{pre}	Transverse deflection of the layer structure within the interface in pre-slip regime
W_{tran}	Transverse deflection of the layer structure within the interface in transition regime
W_{slip}	Transverse deflection of the layer structure within the interface in full-slip regime
α	Effective contact ratio of flaps
ω	Distributed load
H	Thickness of the flap
γ	Stiffening coefficient
$\bar{F}_{jammed}$	Average of force at the jammed state
$\bar{F}_{unjammed}$	Average of force at the unjammed state

References

1. Tur, J.M.M.; Garthwaite, W. Robotic devices for water main in-pipe inspection: A survey. *J. Field Robot.* **2010**, *4*, 491–508.
2. Deimel, R.; Brock, O. A novel type of compliant and underactuated robotic hand for dexterous grasping. *Int. J. Robot. Res.* **2016**, *35*, 161–185. [CrossRef]
3. Cheng, N.; Amend, J.; Farrell, T.; Latour, D.; Martinez, C.; Johansson, J.; McNicoll, A.; Wartenberg, M.; Naseef, S.; Hanson, W.; et al. Prosthetic jamming terminal device: A case study of untethered soft robotics. *Soft Robot.* **2016**, *3*, 205–212. [CrossRef] [PubMed]
4. Degani, A.; Choset, H.; Wolf, A.; Ota, T.; Zenati, M.A. Percutaneous intrapericardial interventions using a highly articulated robotic probe. In Proceedings of the First IEEE/RAS-EMBS International Conference on Biomedical Robotics and Biomechatronics, 2006, BioRob 2006, Pisa, Italy, 20–22 February 2006; IEEE: Piscataway, NJ, USA, 2006; pp. 7–12.
5. Burgner-Kahrs, J.; Rucker, D.C.; Choset, H. Continuum robots for medical applications: A survey. *IEEE Trans. Robot.* **2015**, *31*, 1261–1280. [CrossRef]
6. Kolachalama, S.; Lakshmanan, S. Continuum robots for manipulation applications: A survey. *J. Robot.* **2020**, *2020*, 4187048. [CrossRef]
7. Pettersson, A.; Davis, S.; Gray, J.O.; Dodd, T.J.; Ohlsson, T. Design of a magnetorheological robot gripper for handling of delicate food products with varying shapes. *J. Food Eng.* **2010**, *98*, 332–338. [CrossRef]
8. Jaeger, H.M. Celebrating soft matter's 10th anniversary: Toward jamming by design. *Soft Matter* **2015**, *11*, 12–27. [CrossRef]
9. Loeve, A.J.; Bosma, J.H.; Breedveld, P.; Dodou, D.; Dankelman, J. Polymer rigidity control for endoscopic shaft-guide 'Plastolock'—A feasibility study. *J. Med. Devices* **2010**, *4*, 045001. [CrossRef]
10. Lathrop, E.; Adibnazari, I.; Gravish, N.; Tolley, M.T. Shear strengthened granular jamming feet for improved performance over natural terrain. In Proceedings of the 2020 3rd IEEE International Conference on Soft Robotics (RoboSoft), New Haven, CT, USA, 15 May–15 July 2020; IEEE: Piscataway, NJ, USA, 2020; pp. 388–393.
11. Shintake, J.; Schubert, B.; Rosset, S.; Shea, H.; Floreano, D. Variable stiffness actuator for soft robotics using dielectric elastomer and low-melting-point alloy. In Proceedings of the 2015 IEEE/RSJ International Conference on Intelligent Robots and Systems (IROS), Hamburg, Germany, 28 September–2 October 2015; IEEE: Piscataway, NJ, USA, 2015; pp. 1097–1102.
12. Nasab, A.M.; Sabzehzar, A.; Tatari, M.; Majidi, C.; Shan, W. A soft gripper with rigidity tunable elastomer strips as ligaments. *Soft Robot.* **2017**, *4*, 411–420. [CrossRef]
13. Yang, B.; Lin, Q. A latchable microvalve using phase change of paraffin wax. *Sens. Actuators A Phys.* **2007**, *134*, 194–200. [CrossRef]
14. Wang, L.; Yang, Y.; Chen, Y.; Majidi, C.; Iida, F.; Askounis, E.; Pei, Q. Controllable and reversible tuning of material rigidity for robot applications. *Mater. Today* **2018**, *21*, 563–576. [CrossRef]
15. Jiang, A.; Ranzani, T.; Gerboni, G.; Lekstutyte, L.; Althoefer, K.; Dasgupta, P.; Nanayakkara, T. Robotic granular jamming: Does the membrane matter? *Soft Robot.* **2014**, *1*, 192–201. [CrossRef]
16. Li, Y.; Chen, Y.; Yang, Y.; Wei, Y. Passive particle jamming and its stiffening of soft robotic grippers. *IEEE Trans. Robot.* **2017**, *33*, 446–455. [CrossRef]
17. Zhou, J.; Chen, Y.; Hu, Y.; Wang, Z.; Li, Y.; Gu, G.; Liu, Y. Adaptive variable stiffness particle phalange for robust and durable robotic grasping. *Soft Robot.* **2020**, *7*, 743–757. [CrossRef]
18. Cianchetti, M.; Ranzani, T.; Gerboni, G.; Nanayakkara, T.; Althoefer, K.; Dasgupta, P.; Menciassi, A. Soft robotics technologies to address shortcomings in today's minimally invasive surgery: The STIFF-FLOP approach. *Soft Robot.* **2014**, *1*, 122–131. [CrossRef]

19. Choi, J.; Lee, D.Y.; Eo, J.H.; Park, Y.J.; Cho, K.J. Tendon-driven jamming mechanism for configurable variable stiffness. *Soft Robot.* **2021**, *8*, 109–118. [CrossRef]
20. Keller, S.; Jaeger, H.M. Aleatory architectures. *Granul. Matter* **2016**, *18*, 29. [CrossRef]
21. Jadhav, S.; Majit, M.R.A.; Shih, B.; Schulze, J.P.; Tolley, M.T. Variable stiffness devices using fiber jamming for application in soft robotics and wearable haptics. *Soft Robot.* **2022**, *9*, 173–186. [CrossRef]
22. Narang, Y.S.; Vlassak, J.J.; Howe, R.D. Mechanically versatile soft machines through laminar jamming. *Adv. Funct. Mater.* **2018**, *28*, 1707136. [CrossRef]
23. Narang, Y.S.; Degirmenci, A.; Vlassak, J.J.; Howe, R.D. Transforming the dynamic response of robotic structures and systems through laminar jamming. *IEEE Robot. Autom. Lett.* **2017**, *3*, 688–695. [CrossRef]
24. Santiago, J.L.C.; Godage, I.S.; Gonthina, P.; Walker, I.D. Soft robots and kangaroo tails: Modulating compliance in continuum structures through mechanical layer jamming. *Soft Robot.* **2016**, *3*, 54–63. [CrossRef]
25. Ou, J.; Yao, L.; Tauber, D.; Steimle, J.; Niiyama, R.; Ishii, H. jamSheets: Thin interfaces with tunable stiffness enabled by layer jamming. In Proceedings of the 8th International Conference on Tangible, Embedded and Embodied Interaction, Munich, Germany, 16–19 February 2014; pp. 65–72.
26. Kim, Y.J.; Cheng, S.; Kim, S.; Iagnemma, K. Design of a tubular snake-like manipulator with stiffening capability by layer jamming. In Proceedings of the 2012 IEEE/RSJ International Conference on Intelligent Robots and Systems, Vilamoura-Algarve, Portugal, 7–12 October 2012; IEEE: Piscataway, NJ, USA, 2012; pp. 4251–4256.
27. Clark, A.B.; Rojas, N. Stiffness-tuneable limb segment with flexible spine for malleable robots. In Proceedings of the 2019 International Conference on Robotics and Automation (ICRA), Montreal, QC, Canada, 20–24 May 2019; IEEE: Piscataway, NJ, USA, 2019; pp. 3969–3975.
28. Langer, M.; Amanov, E.; Burgner-Kahrs, J. Stiffening sheaths for continuum robots. *Soft Robot.* **2018**, *5*, 291–303. [CrossRef] [PubMed]
29. Kim, Y.J.; Cheng, S.; Kim, S.; Iagnemma, K. A stiffness-adjustable hyperredundant manipulator using a variable neutral-line mechanism for minimally invasive surgery. *IEEE Trans. Robot.* **2013**, *30*, 382–395. [CrossRef]
30. Yoon, H.S.; Yi, B.J. A 4-DOF flexible continuum robot using a spring backbone. In Proceedings of the 2009 International Conference on Mechatronics and Automation, Changchun, China, 9–12 August 2009; IEEE: Piscataway, NJ, USA, 2009; pp. 1249–1254.

machines

Article

The Effects of Unpowered Soft Exoskeletons on Preferred Gait Features and Resonant Walking

Zhengyan Zhang [1], Houcheng Wang [1], Shijie Guo [1,*], Jing Wang [2], Yungang Zhao [2] and Qiang Tian [2]

[1] School of Mechanical Engineering, Hebei University of Technology, Tianjin 300401, China; zzy@hebut.edu.cn (Z.Z.); rijikarrd@sina.com (H.W.)
[2] Tianjin Key Laboratory of Exercise Physiology and Sports Medicine, Institute of Exercise and Health, Tianjin University of Sport, Tianjin 301617, China; xiaojing111213@126.com (J.W.); yungang.zhao@tjus.edu.cn (Y.Z.); jackietian2008@126.com (Q.T.)
* Correspondence: guoshijie@hebut.edu.cn

Abstract: Resonant walking with preferred gait features is a self-optimized consequence of long-term human locomotion. Minimal energy expenditure can be achieved in this resonant condition. This unpowered multi-joint soft exoskeleton is designed to test whether: (1) there is an obvious improvement in preferred speed and other gait features; (2) resonant walking still exists with exoskeleton assistance. Healthy participants (N = 7) were asked to perform the following trials: (1) walking at 1.25 m/s without assistance (normal condition); (2) walking at 1.25 m/s with assistance (general condition); (3) walking at preferred speed with assistance (preferred condition); (4) walking at the speed in trial (3) without assistance (comparison condition). Participants walked at the preferred frequency and ±10% of it. An average 21% increase in preferred speed was observed. The U-shaped oxygen consumption and lower limb muscle activity curve with the minimum at preferred frequency indicated that the resonant condition existed under the preferred condition. Average metabolic reductions of 4.53% and 7.65% were found in the preferred condition compared to the general and comparison condition, respectively. These results demonstrate that the resonant condition in assisted walking could benefit energy expenditure and provide a new perspective for exoskeleton design and evaluation.

Keywords: resonant walking; preferred gait features; unpowered soft exoskeletons; oxygen consumption; lower limb muscle activity

Citation: Zhang, Z.; Wang, H.; Guo, S.; Wang, J.; Zhao, Y.; Tian, Q. The Effects of Unpowered Soft Exoskeletons on Preferred Gait Features and Resonant Walking. *Machines* **2022**, *10*, 585. https://doi.org/10.3390/machines10070585

Academic Editor: Yanjie Wang

Received: 13 May 2022
Accepted: 11 July 2022
Published: 18 July 2022

1. Introduction

Exoskeletons can be divided into three types: rehabilitation exoskeleton, assisting exoskeleton and load-bearing exoskeleton [1,2]. Rehabilitation exoskeletons are mainly used to provide recovery training for people with movement disorders such as stroke patients. Assisting exoskeletons are mainly used for healthy people with declining body functions or the elderly and patients in late rehabilitation. They mainly provide the human body with motion assistance for daily activities, such as level ground walking. Load-bearing exoskeletons are mainly used to improve the user's motor function during weight-bearing activities. Understandably, there has been much interest from the military, where the technology could enhance the performance of soldiers in the battlefield [3]. In addition to whether there is an actuator, the exoskeleton can also be divided into rigid and soft ones according to the structure. The rigid structure is able to transmit large torques, but there are problems such as large volume, heavy mass and joint axis alignment; soft exoskeletons are generally referred to as devices that are made of flexible materials, and their distal end is lightweight and easy to carry and use. The powered rigid exoskeleton can accomplish the above three tasks, but it is not suitable for daily use due to its large volume, heavy mass and short battery life. The powered soft exoskeleton is suitable for daily use as it is light in weight and easy to carry, but its soft structure also means it lacks the ability to bear weight, so it is mainly used for rehabilitation and walking assistance in daily life. Unpowered exoskeletons are also divided into rigid ones and soft ones. The removing of actuator units reduces the mass and volume to make it more suitable for daily use, but this will also lead to lack of assistance ability. As a result, unpowered exoskeletons are mostly used in walking assistance for the reduction of fatigue during daily activities.

Unpowered and soft exoskeletons have been evidenced to improve overall walking efficiency and performance [4–8]. Elastic elements are used to passively store and return energy as one of the main mechanisms in existing unpowered exoskeletons, which are inspired by the deformation of passive structures such as ligaments and tendons during periodic walking [9–11]. Paralleled with the elastic tissues, elastic elements are able to simulate and replace part of their functions, hence the energy savings. During each gait cycle, the energy stored in previous periods will be returned in subsequent movements to propel the body forward [12–17]. As these exoskeletons primarily save energy through the storage and release of elastic elements, the main focus is the spring stiffness (torque profile), the design of devices [18,19] and the assistance timing [6]. When evaluating the assisting ability of exoskeletons, the speed of treadmills was always set in the range of ordinary people's comfortable speed (around 1.25 m/s) to simulate the daily walking condition. In addition, a fix-speed treadmill is useful to constrain speed and isolate the effects of exoskeleton assistance on gait features other than speed.

For unpowered devices, the interaction between gait features and assistance magnitude could be complex, as it should be actuated by users, which will influence the characteristics of human walking. At a certain speed, metabolic rate reaches a minimum at moderate stiffness and then increases rapidly with increasing or decreasing stiffness, with a U-shaped curve. What is more, under the same assisting parameter setting and experimental condition, the metabolic reduction of different subjects varies greatly, and some of them even rise. These results indicate that the effect of the exoskeleton on walking is obviously closely related to the speed and the physical parameters of the individual subjects, and these aspects will greatly affect the exoskeleton assistance. However, previous studies have not paid much attention to the problems above. While for unpowered devices, In the research of Collins [20], participants chose a higher preferred speed with speed-optimized torque obtained from human-in-the-loop optimization rather than torque optimized for energy consumption and normal shoes. This result demonstrated that exoskeleton assistance does have an impact on gait features and can significantly improve the preferred walking speed.

Resonant walking, which appears under preferred gait features, is regarded as the optimal result of human walking, and its periodic behavior could be simulated by the force-driven harmonic oscillator (FDHO) [21,22]. Due to the existence of damping, a periodic forcing function is required to maintain the oscillation of the FDHO. Thus, the minimal force can be reached when its frequency equals the natural frequency of the FDHO. For human walking, lower force (muscle activity) results in a decreased need for oxygen, hence a minimization at resonant frequency. Self-selected preferred speed and frequency have been proven to be almost indistinguishable from the resonant ones. U-shaped oxygen consumption and muscle activity curves were observed with the minimum at the preferred and resonant frequency, increasing away from the preferred frequency [23–25], which also means the frequency range for achieving better metabolic performance is flexible. Local dynamic stability is also an important part of resonant walking, such as maintaining the highest stability of the head at the preferred frequency [24].

Several models have been built to describe resonant walking. Holt has proposed a hybrid mass-spring pendulum model for the swing phase and an escapement-driven, inverted pendulum with a spring model (EDIPS) with viscous damping for the stance phase, based on the force-driven harmonic oscillator [26–28]. Park tuned a model of bipedal walking with damped compliant legs to match human GRFs(ground reaction forces)s at different gait speeds and performed a series of studies based on the bipedal spring-mass model by Geyer [29–34]. The correlation between leg stiffness and center of mass (COM) oscillation behavior during the single-support phase was evaluated by comparing the duration of the damped compliant leg to the duration of the single support phase. The high correlation indicated that COM oscillation behavior takes advantage of the resonance characteristics of leg stiffness during the single support phase, and results also suggested that the leg stiffness increases with speed and load. Contrary to the conventional assumption in gait mechanics that a minimal change in COM excursion can reduce energy cost, previous studies have mentioned that a bouncy gait is more energy-efficient than a flat trajectory [31,35]. This observed bouncy behavior results from the resonant mechanics [30] and has already been used in some assistive devices [36].

As elastic elements are actuated by users during assisted walking, they actually become a part of the human body in the specific period when stretched. At this time, the body and device can be regarded as a new "person" with different global lower limb stiffness. According to a previous study, the preferred step frequency changes with lower limb stiffness and the preferred speed increases with exoskeleton assistance and leg stiffness, thus leading to a new gait feature. It is believed that resonant walking will occur under the new preferred gait feature, and the changes of preferred gait features should be taken into consideration for a further global optimization of assisted walking, as shown in Figure 1.

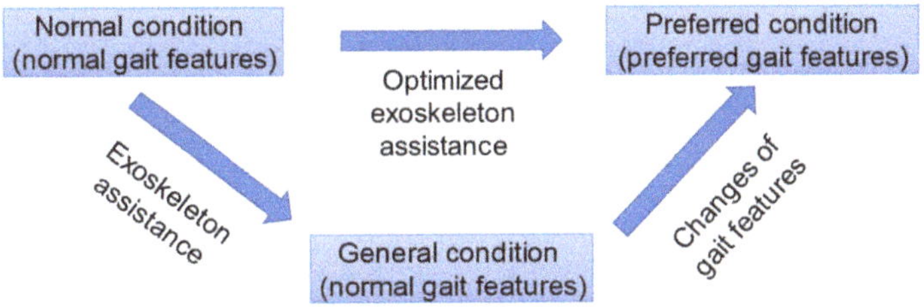

Figure 1. Schematic diagram of assisted walking when considering the changes of the gait feature.

The main purpose of this study was to investigate the questions as follows: whether (1) unpowered exoskeletons can change the preferred speed and gait features; (2) resonant walking occurs under different gait features with exoskeleton assistance. Finally, the net metabolic rate under different conditions was analyzed to evaluate the effect of resonant walking with assistance on walking efficiency. To conduct this research, a multi-joint unpowered soft exoskeleton was designed with two assistance paths, the anterior and posterior of the body. Details of the exoskeleton will be described in the next section.

2. Materials and Methods

2.1. Design of the Exoskeleton

The exoskeleton was made of commercially available resistance bands (TheraBand Professional Latex Resistance Band Gold, TheraBand, Akron, OH, USA), nylon fabric and waist bands, which led to a lightweight device. The three main parts (waistband, hip-ankle assistance path and hip-knee assistance path) and the deformation of springs are shown in Figure 2. The resistance bands were cut into segments with a length of 18 cm (with two 5 cm hook and loop fasteners at each end of both sides) and placed in front and behind the body bilaterally. The waist band is used for an anchor on the upper body to resist the tension of elastic bands. Leg straps 1 are used to connect the anterior spring and foot attachment, while leg straps 2 are used to connect the posterior springs and calf attachment. The connection between the elastic bands and other straps or bands is hook and loop. Before the experiment, elastic bands were first pretensioned by 1.5 cm (initial length) to obtain a preload.

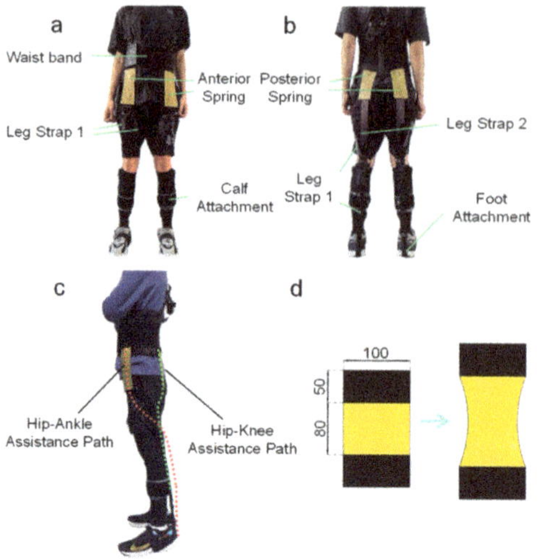

Figure 2. Overview of the soft exoskeleton. (**a**) Front and (**b**) rear view of the device; schematic diagram of the hip-ankle assistance path and the hip-knee assistance path; (**c**) schematic diagram of the geometry of the elastic elements placed anteriorly and posteriorly across the hip; (**d**) dimensions are in millimeters.

The anterior path is used to absorb energy in the stance phase and return when push-off occurs to help plantarflexion and hip flexion. The posterior path is used to absorb energy in the swing phase and return it in the following movements in order to transform the negative power from the knee joint to the hip joint to help hip flexion.

The energy storage and return phase of both anterior and posterior springs are described here (see Figure 3). For the hip-knee assistance path, the elastic energy is stored from the late swing phase and then released in the coming movements. While for the hip-ankle assistance path, the elastic energy is stored from early mid-stance to heel-off, and then released gradually after heel-off. The details of the assistance are not presented here as this is not the main focus of this article.

2.2. Experimental Setup

Participants walked on a treadmill (h/p/cosmos pulsar 4.0, h/p/cosmos, Traunstein, Germany) with a heart rate belt to record heart rates and a gas analysis machine (MetaMax 3B-R2, Cortex, Leipzig, Germany) to measure the oxygen uptake and carbon dioxide exhalation. Electrical muscle activity was measured with 12 wireless EMG (ElectroMyoGraphy) transmitters (Noraxon Ultium EMG, Noraxon, Scottsdale, AZ, USA) at 2000 Hz (Figure 4). The signal was filtered by a low-pass filter at 10 Hz and a high-pass filter at 500 Hz. Surface electrodes were placed bilaterally in a bipolar configuration over six muscles—gluteus maximus (GMAX), rectus femoris (RF), biceps femoris (BF), lateral gastrocnemius medialis (GAS), soleus (SOL) and tibialis anterior (TA). Before placing the electrodes, the body hair was shaved and then the corresponding area was polished with sandpaper and alcohol cotton balls. When placing the transmitters, it was important to notice the interference with the wearable device and overlap with corresponding antagonistic muscles to avoid large measurement errors. All manipulations were performed by experts that specialized in surface EMG measurements. The calf was wrapped with knee-length stockings and the thigh was covered with textiles to prevent the transmitters from being thrown off at fast speeds.

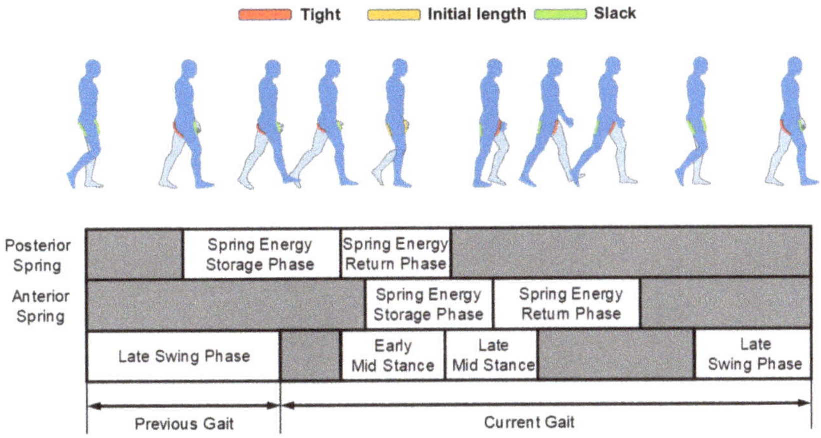

Figure 3. Schematic overview of the functional phases of passive springs located anteriorly and posteriorly across the hip joint in the gait cycle. This begins from the late swing phase in the previous gait to the end of the current gait. Green represents that the spring is slack, yellow represents the initial length of the spring (the spring is pretensioned by 1.5 cm) and red represents that the spring is tight.

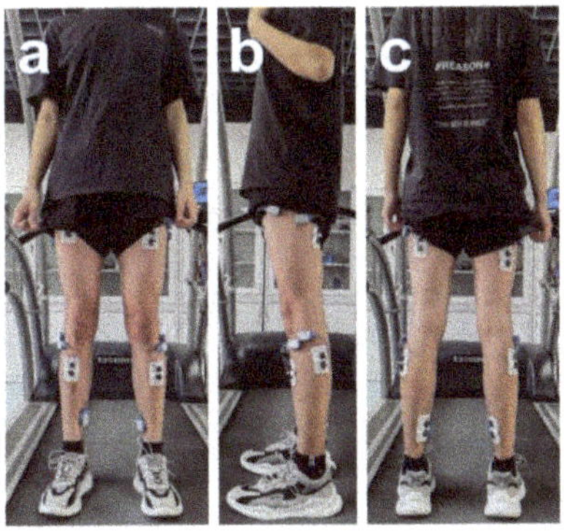

Figure 4. (**a**) Front, (**b**) side and (**c**) rear view of a participant and the placement of electrical electrodes.

Walking kinematics were measured by a motion capture system (QTM, Qualysis, Gothenburg, Sweden) with sixteen lenses (OQUS500&OQUS700, Qualysis, Gothenburg, Sweden) at 300 Hz, tracking a set of 36 reflective markers (Figure 5). Markers were placed symmetrically on both sides of the anterior and posterior iliac muscles, the medial lateral condyle of knee joints, the inner and outer ankles, the heel, the first, second and fifth metatarsal bones, the thighs and the calves. The ground reaction forces (GRFs) and center of pressure (COP) were measured by a force plate (Kistler 9287C, Kistler, Winterthur, Switzerland) at 1000 Hz, which was connected to a computer through an amplifier and a data acquisition system. The physical parameters and pendulum equivalent lengths were measured and calculated according to the previous study [37].

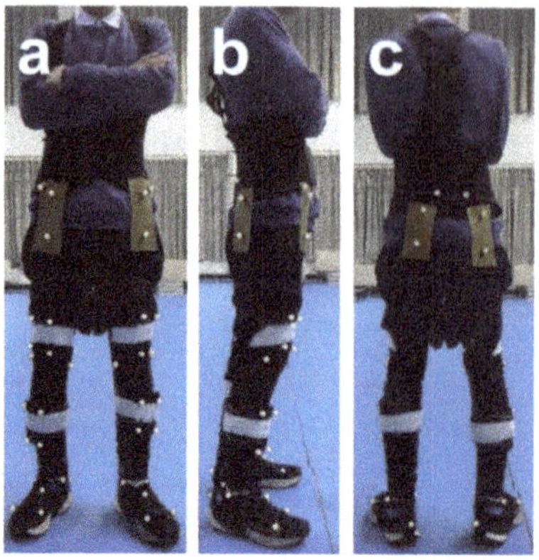

Figure 5. (**a**) Front, (**b**) side and (**c**) rear of a participant and the placement of markers.

2.3. Experimental Protocol

Before the formal experiment, a pre-experiment was conducted to evaluate the rationality of and familiarity with the experimental procedure. The frequencies in the previous study were: preferred frequency, ±15, ±25 and ±35%, but it was found to be difficult to reach ±15%, especially at lower frequencies during assisted walking, so the frequencies were set as preferred and ±10%. In addition, participants performed a preferred speed selection session trial at 4.1, 4.3, 4.5, 4.7, 4.9 and 5.1 km/h (90 s at each speed, higher speeds were eliminated for an obvious increase of oxygen consumption during tests) to decide their preferred speed without exoskeleton assistance. Results showed that 6 out of 7 participants chose 4.5 km/h (1.25 m/s), and only 1 participant chose 4.3 km/h (1.19 m/s). The metabolic rates per distance at these two speeds of the participant who chose 4.3 km/h are almost the same. What is more, there was no statistical significance of differences between the measured step frequencies and the calculated frequencies through paired *t*-test. Based on this, the preferred speed without exoskeleton assistance (normal condition) was set as 1.25 m/s (general speed V_0) in order to simplify the experimental setup slightly.

The experiment consists of three sessions: a familiarization session, a speed selection session and a walking session. In the familiarization session, participants performed an adaptive walking and speed selection protocol; the cardiopulmonary function parameters were also recorded at the same time.

During the speed selection session, a 4-min standing trial was performed at the beginning to collect baseline metabolic power. Then, the participants performed a 15-min speed selection trial at 4.3, 4.7, 5.1, 5.5, 5.9 and 6.3 km/h (90 s at each speed) to decide the preferred speed with assistance. Participants were asked to select a speed that made them feel both comfortable and obviously assisted by the device with the help of rating of perceived exertion (RPE). During the speed-selection trial, participants did not know the real time speed and when the subjects had directed the experimenter to the same speed on two consecutive occasions, it was deemed that this speed was the preferred speed V_1.

The walking session was divided into four conditions: walking at 1.25 m/s without assistance (normal condition) and with assistance (general condition) and walking at preferred speed with assistance (preferred condition) and without assistance (comparison condition). The normal condition represents the free walk under the preferred speed 1.25 m/s. The general condition represents walking with assistance without considering the changes (1.25 m/s) in preferred speed. The preferred condition represents walking with assistance when taking the changes of preferred speeds into consideration. The comparison condition represents the free walk without assistance at the preferred speed of the preferred condition. Sufficient rest intervals were set between every session to eliminate the effect of fatigue.

The treadmill speed was set at the self-selected speed constantly under the same condition in the walking session. At the beginning of a walking session, participants walked on the treadmill under their preferred gait features. The preferred frequency was obtained from the gait cycle duration, calculated by recording 20 steps. Then, participants walked for a 4-min trial following the metronome with ±10% of the preferred step frequency, respectively. Procedures were the same in conditions of walking at 1.25 m/s with and without assistance. The trial for walking at the preferred assisted speed without the device was only performed once at its preferred step frequency as the comparison condition.

As the treadmill is only in the laboratory for measuring the oxygen consumption rather than the laboratory for collecting kinematic and kinetic data, the following method was used to reproduce walking on a treadmill when collecting kinematic and kinetic data. Step lengths were calculated in advance according to the data of the walking session, and simulated by markers placed on the floor. Participants walked under different walking conditions, following the markers and metronome in order to simulate the treadmill speed. Although the difference between walking on a flat ground and walking on a treadmill is unavoidable, the limitations of the equipment can only be offset through extra effort during the experiment. However, since the kinematic parameters only play a limited role in this article and the experimental condition is the same for every participant, the results obtained are still convincing.

2.4. Participants

Seven healthy adults (age: 23 ± 0.93, height: 172.43 ± 1.92 cm, weight: 62.26 ± 4.30 kg) were recruited in the experiments and all of them gave their informed consent before participating. The sample size was chosen based on the previous research [38]. The study and protocol were approved by the Biomedical Ethics Committee of Hebei University of Technology and all experiments were performed in accordance with the relevant guidelines and regulations. Informed consent was obtained from each participant. Detailed information on participants' anthropometric measurements is shown in Table 1.

Table 1. Anthropometric measurements of the participants and equivalent pendulum length (mean and standard deviation).

	Right Tight	**Right Shank**	**Right Foot**	**Equivalent Pendulum Length**
Length (m)	0.42 ± 0.02	0.36 ± 0.03	0.13 ± 0.01	0.55 ± 0.02
Mass (kg)	6.24 ± 0.44	2.90 ± 0.22	0.91 ± 0.06	-

2.5. Data Processing

The added mass of wearable devices was considered negligible with respect to the inertial and gravitational effects on participants' walking. The EMG raw data were first filtered with a 10 Hz high pass filter and a 500 Hz low pass filter. The filtered EMG signals were then rectified and smoothed by calculating the root mean square for a 50 ms moving window. The total activity of each muscle was calculated as the area under this curve for every group of 20 steps in the last 2 min of each 4-min trial.

The total EMG activity of a single muscle for each trial was normalized through dividing it by the total EMG activity of the preferred frequency, and the percentage of activity compared to walking at the preferred frequency was obtained by multiplying these normalized values by 100. The same normalization procedure was performed for V_{O_2}.

The carbon dioxide and oxygen rates were averaged across the last 2 min of each walking condition. Metabolic cost was calculated by using the regression equation of Zuntz based on the thermal equivalent of O_2 for the nonprotein respiratory equivalent [39,40], as in Equation (1):

$$\text{metabolic cost (kcal/min)} = \dot{V}_{O_2} \times (1.2341 \times \text{RER} + 3.8124) \tag{1}$$

where RER is the respiratory exchange ratio. Then, the net metabolic rate per unit distance per unit body mass was calculated, as in Equation (2):

$$\frac{P_{walk} - P_{stand}}{v} \tag{2}$$

where P_{walk} and P_{stand} are the metabolic rate for walking and standing, respectively, and v is the preferred speed under different conditions.

2.6. Statistical Analysis

Wilcoxon test was used for the statistical significance of differences between general speed and preferred speed, as the general speed does not obey a normal distribution. The statistical significance of differences across step frequencies, step lengths, maximum hip angles, maximum plantarflexion angles and maximum dorsiflexion angles was tested using two-way analysis of variance (ANOVA). As VO_2 and muscle activity were normalized to the value obtained for walking at the preferred stride frequency, one-sample t-tests were used to determine the statistical significance of differences of the values between the preferred frequency condition and other conditions. The correlation between the duration of the single support phase and the duration of the gait cycle was also calculated to evaluate the level of participants utilizing the resonance of walking through the square of Pearson's linear correlation coefficient. Since the walking conditions rather than specific parameters were regarded as the variable when evaluating the walking efficiency, one-way analysis of variance (ANOVA) was used to detect the differences across the range of angles and the net metabolic rate per distance under different walking conditions. If a significant effect was found in one-way ANOVA, the Tukey's honestly significant difference (Tukey HSD) test will be used to compare pairs of conditions. A significance level of $\alpha = 0.05$ was used for ANOVA, one-sample t-test Tukey HSD test and Wilcoxon test.

3. Results

3.1. Preferred Gait Features

Four different conditions were evaluated in this study: (1) normal condition: walking at 1.25 m/s without assistance; (2) preferred condition: walking at preferred speed with assistance; (3) general condition: walking at 1.25 m/s with assistance; (4) comparison condition: walking at preferred speed as that mentioned in (3) without assistance.

Exoskeleton assistance is able to substantially increase the self-selected speed (Wilcoxon test, $p = 0.011 < 0.05$). In the speed selection trial, six participants chose 5.5 km/h (1.53 m/s), while one participant chose 5.1 km/h (1.41 m/s), 22.4% and 12.8% higher, respectively, compared to 1.25 m/s (Figure 6).

The frequency and step length were both significantly affected by the speed (ANOVA, $p = 1.51 \times 10^{-4} < 0.05$, $p = 2.15 \times 10^{-8} < 0.05$) (Figure 7). The preferred frequency of the preferred condition was higher than that of the normal condition and general condition by 9.2% and 6.8%, respectively, and frequency increased by 1.7% compared with the comparison condition. Similarly, the step length of the preferred condition was increased by 11.11% and 13.5% compared to the normal condition and general condition.

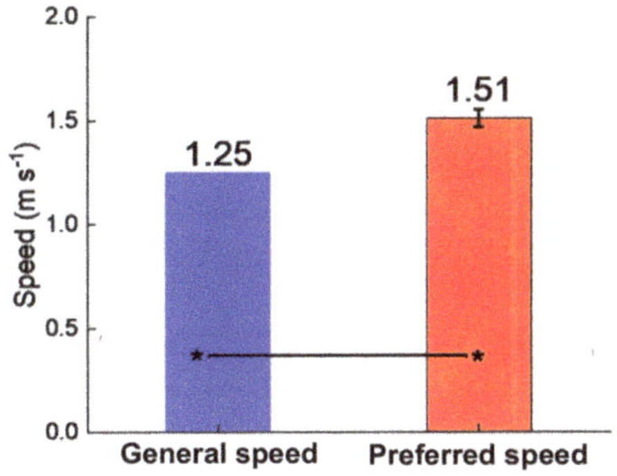

Figure 6. Mean (bars) and standard deviation (whiskers) of preferred walking speed under the normal and preferred condition (general speed and preferred speed); lines with asterisks denote statistically significant differences (Wilcoxon test, $\alpha = 0.05$).

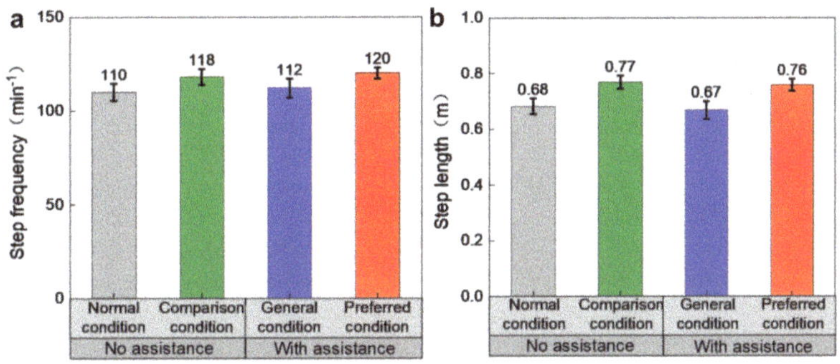

Figure 7. Preferred step frequency and step length under different conditions; (**a**) mean (bars) and standard deviation (whiskers) of step frequency under the normal condition (light grey bars), general condition (blue bars), preferred condition (red bars) and comparison condition (green bars); (**b**) mean (bars) and standard deviation (whiskers) of step length under the normal condition (light grey bars), general condition (blue bars), preferred condition (red bars) and comparison condition (green bars).

Mean joint angles and stick figures are shown to give a direct view of gaits under different conditions (Figure 8). Maximum plantarflexion and dorsiflexion angles were significantly affected by the exoskeleton assistance (ANOVA, $p = 0.01 < 0.05$, $p = 0.02 < 0.05$), as shown in Figure 9. The ranges of ankle, knee and hip angles are shown in Table 2. No significant differences were found in ranges of joint angles between different walking conditions.

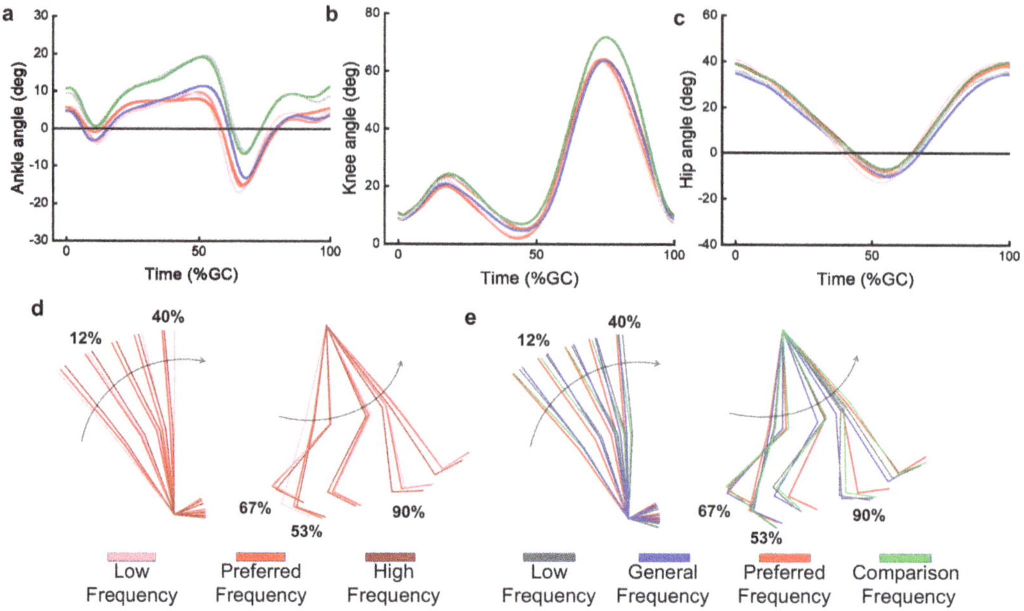

Figure 8. (**a**) Mean ankle angles under different walking conditions; (**b**) mean knee angles under different walking conditions; (**c**) mean hip angles under different walking conditions. (**d**) Stick figures of the mean walking kinematics at different frequencies under preferred speeds. Light red sticks depict walking at lower frequency, red sticks depict walking at preferred frequency and dark red sticks depict walking at higher frequency. Percentages represent different periods of the gait cycle. (**e**) Stick figures of the mean walking kinematics under different conditions. Light grey sticks depict walking under the normal condition, blue sticks depict walking under the general condition, red sticks depict walking under the preferred condition and green sticks depict walking under the comparison condition. Percentages represent different periods of the gait cycle.

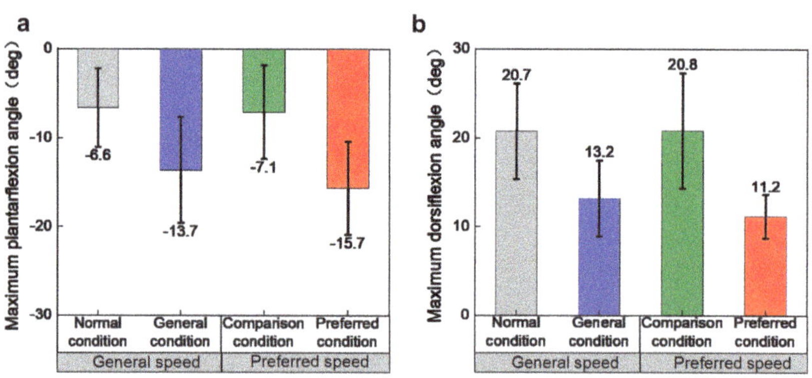

Figure 9. Ankle joint angles under different conditions; (**a**) mean (bars) and standard deviation (whiskers) of maximum plantarflexion angles under the normal condition (light grey bars), general condition (blue bars), preferred condition (red bars) and comparison condition (green bars); (**b**) mean (bars) and standard deviation (whiskers) of maximum dorsiflexion angles under the normal condition (light grey bars), general condition (blue bars), preferred condition (red bars) and comparison condition (green bars).

Table 2. The range of ankle, knee and hip angles under different walking conditions (mean and standard deviation).

	Normal Condition	General Condition	Preferred Condition	Comparison Condition
Ankle angle (deg)	27.34 ± 4.29	26.80 ± 3.76	26.80 ± 3.35	27.88 ± 4.71
Knee angle (deg)	66.28 ± 3.98	61.38 ± 6.15	63.63 ± 4.71	66.22 ± 4.22
Hip angle (deg)	49.63 ± 9.46	50.02 ± 7.27	53.04 ± 6.25	50.86 ± 9.32

3.2. Resonant Walking

Under the general condition (Figure 10a), only 2 of the 7 participants had the least oxygen consumption at the preferred frequency, and the oxygen consumption was only a little better modeled as a quadratic function of $\Delta\omega$ ($R^2 = 0.18$, $p = 0.167$) than a linear function ($R^2 = 0.114$, $p = 0.134$). As in the preferred condition (Figure 10b), the oxygen consumption is the lowest at the preferred frequency among all participants under the preferred condition, and the oxygen consumption was best modeled as a quadratic function of $\Delta\omega$ ($R^2 = 0.236$, $p = 0.089$) over a linear function ($R^2 = 0.05$, $p = 0.328$). There is still a significant trend on the quadratic curve fitting although the fitting degree is not as expected. The oxygen consumption at preferred frequency under the preferred condition was significantly different from lower frequency (One sample *t*-test, $p = 0.004 < 0.05$) and higher frequency (One sample *t*-test, $p = 0.011 < 0.05$). The preferred frequency under the general condition was not significantly different from lower frequency (One sample *t*-test, $p = 0.081$), but was from higher frequency (One sample *t*-test, $p = 0.016$).

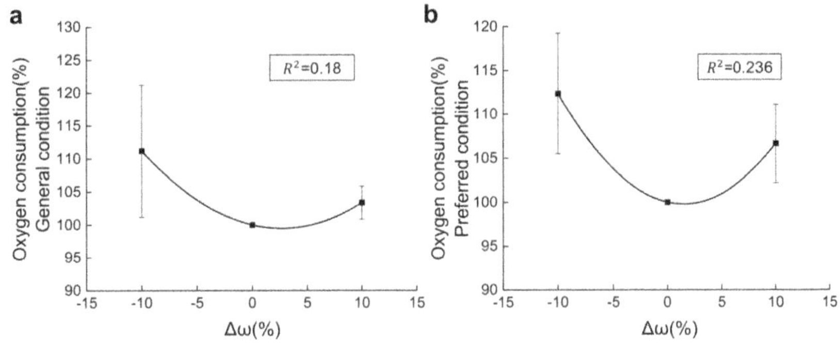

Figure 10. Mean and standard deviation of oxygen consumption (as a percentage of the value for walking at the preferred frequency) for the last two minutes of walking across the three different step frequency conditions. The step frequencies are plotted as the percentage difference from the preferred frequency, $\Delta\omega$. The preferred step frequency is $\Delta\omega = 0\%$. The solid line represents the quadratic curve with error bars of standard deviation. Each subplot is a different condition: (**a**) general condition; (**b**) preferred condition.

Under the general condition (Figure 11), the EMG of all muscles was modeled just a little more accurately as a quadratic function of $\Delta\omega$ (R^2 ranging from 0.041 to 0.375) than linear function (R^2 ranging from 0.16 to 0.272). The EMG of TA, GAS at preferred frequency is significantly different from lower frequencies, while the EMG of RF at preferred frequency is significantly different from higher frequencies. The EMG of SOL, GM and BF at preferred frequency showed no significant difference with either lower or higher frequencies.

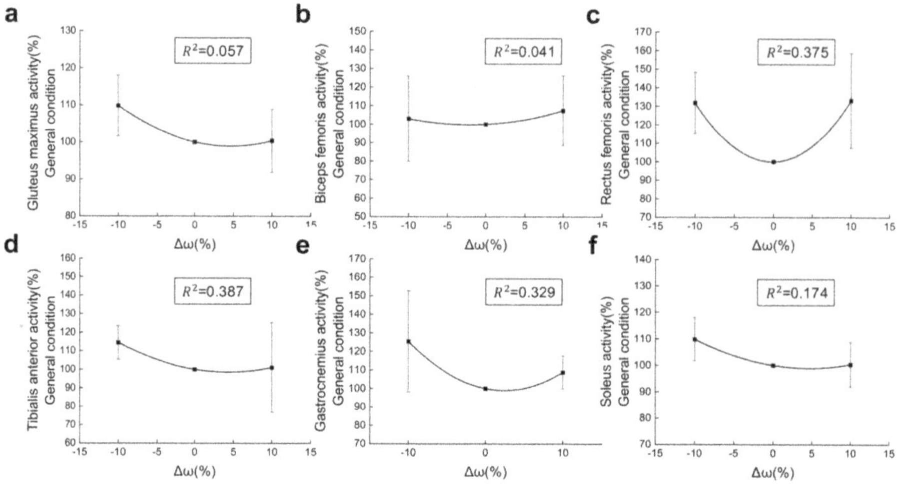

Figure 11. Mean and standard deviation of the muscle activity (as a percentage of the value for walking at the preferred step frequency) for the last two minutes of walking across the three different frequencies under the general condition. Step frequencies are plotted as the percentage difference from the preferred frequency, $\Delta\omega$. The preferred frequency is $\Delta\omega = 0\%$. The solid line represents the quadratic curve with error bars of standard deviation. Each subplot is a different muscle under the general condition: (**a**) gastrocnemius muscle; (**b**) biceps femoris muscle; (**c**) rectus femoris muscle; (**d**) tibialis anterior muscle; (**e**) gastrocnemius muscle; (**f**) soleus muscle.

Under the preferred condition (Figure 12), the EMG of all muscles was best modeled as a quadratic function of $\Delta\omega$ (R^2 ranging from 0.407 to 0.865) over a linear function (R^2 ranging from 0.005 to 0.064). The EMG of TA, GAS, SOL, GM, BF and RF at preferred frequency is significantly different from both lower and higher frequencies. Detailed information on modeling and statistical significance analysis under the preferred and general condition is shown in Tables 3 and 4.

The correlation between the duration of the single support phase and the period of the natural frequency of the spring–mass model under the preferred condition ($R^2 = 0.7$) was better than under the general condition ($R^2 = 0.57$), as shown in Figure 13.

Table 3. Results of modeling analysis for muscle activity and oxygen consumption and the differences between values at the preferred and other frequencies under general condition.

		VO_2	TA	GAS	SOL	GM	BF	RF
R^2	Linear	0.134	0.272	0.135	0.124	0.042	0.016	0.236
	Quadratic	0.167	0.387	0.329	0.174	0.057	0.041	0.375
p	With lower frequency	0.081	0.025	0.037	0.182	0.286	0.774	0.600
	With higher frequency	0.016	0.808	0.106	0.856	0.979	0.216	0.018

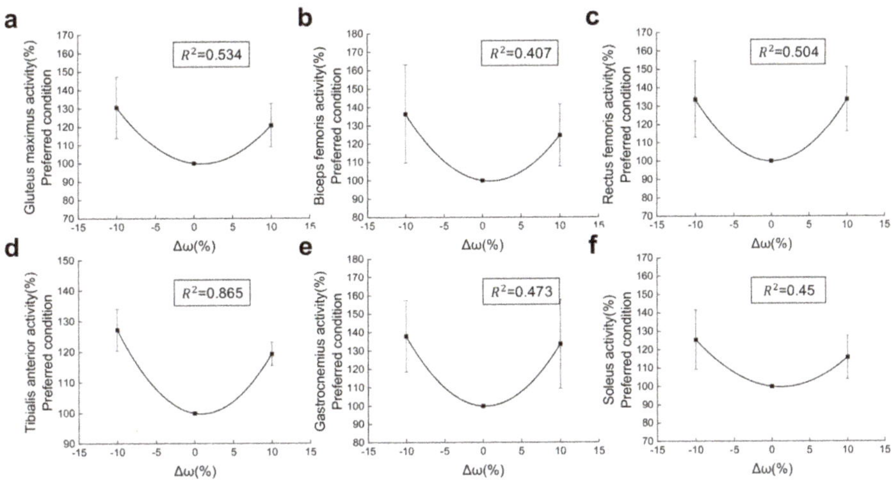

Figure 12. Mean and standard deviation of the muscle activity (as a percentage of the value for walking at the preferred step frequency) for the last two minutes of walking across the three different frequencies under the preferred condition. Step frequencies are plotted as the percentage difference from the preferred frequency, Δω. The preferred frequency is Δω = 0%. The solid line represents the quadratic curve with error bars of standard deviation. Each subplot is a different muscle under the preferred condition: (**a**) gastrocnemius muscle; (**b**) biceps femoris muscle; (**c**) rectus femoris muscle; (**d**) tibialis anterior muscle; (**e**) gastrocnemius muscle; (**f**) soleus muscle.

Table 4. Results of modeling analysis for muscle activity and oxygen consumption and the differences between values at the preferred and other frequencies under preferred condition.

		VO_2	TA	GAS	SOL	GM	BF	RF
R^2	Linear	0.089	0.070	0.005	0.064	0.054	0.041	6.13×10^{-7}
	Quadratic	0.328	0.865	0.473	0.450	0.534	0.407	0.504
p	With lower frequency	0.004	6.6×10^{-5}	0.003	0.009	0.004	0.016	0.007
	With higher frequency	0.011	1.8×10^{-5}	0.014	0.017	0.005	0.012	0.003

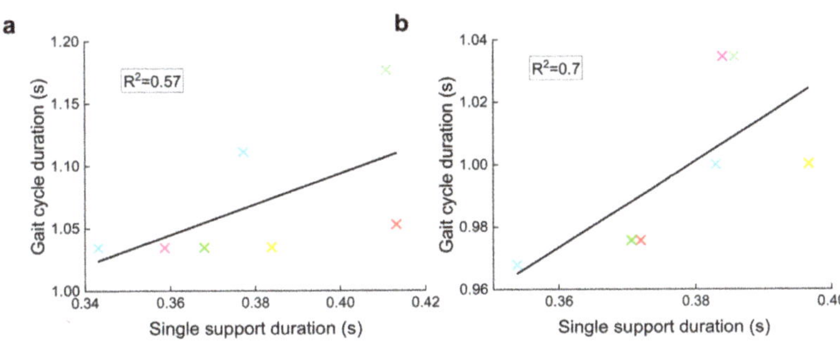

Figure 13. Correlation between the duration of the single support phase and the period of the gait cycle; (**a**) general condition; (**b**) preferred condition. The single support duration was more closely related to the natural frequency of the hybrid spring-mass model under the preferred condition rather than the general condition.

3.3. Net Metabolic Rate

The effects of walking on net metabolic rate varied widely among participants (Figures 14 and 15). Compared with the normal condition, the net metabolic rate changes under the general condition ranged from a 0.59% reduction to a 22.25% increase, with an average increase of 11.62%. Only 1 participant had a reduction in net metabolic rate under the general condition. In the preferred condition, the metabolic rate was reduced by 4.53% compared with the general condition. In total, 4 of 7 participants had a reduction in metabolic rate, 2 of which were reduced by 17.18% and 16.18%; the others were reduced by 5.42% and 7.65%, respectively. The net metabolic rate of the remaining participants increased from 1.57% to 7.52%. Compared with the comparison condition, 6 participants' net metabolic rates decreased, the highest reduction being 16.92%; however, the remaining participants' rates did not decrease by more than 10%. Only 1 participant had an increase in net metabolic rate. There was no significant difference in net metabolic rate between conditions.

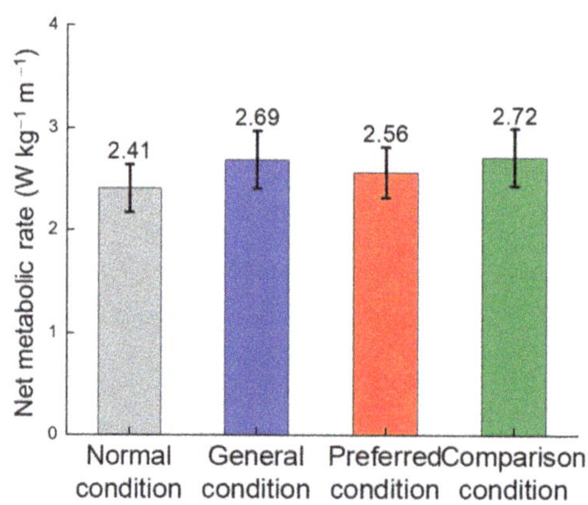

Figure 14. Mean (bars) and standard deviation (whiskers) of the net metabolic rate measured during walking under the normal condition (light grey), general condition (blue), preferred condition (red) and comparison condition (green).

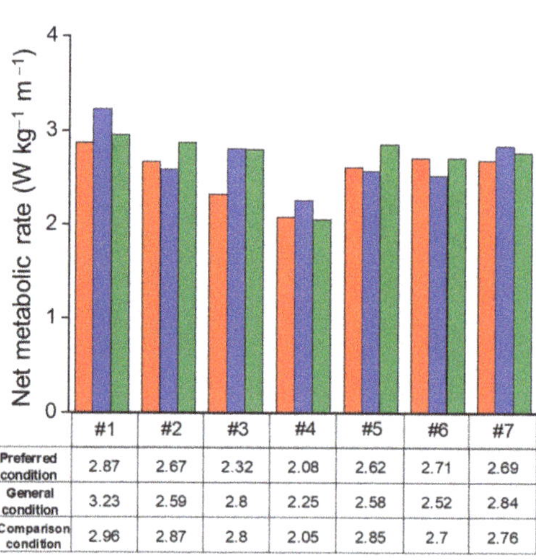

	#1	#2	#3	#4	#5	#6	#7
Preferred condition	2.87	2.67	2.32	2.08	2.62	2.71	2.69
General condition	3.23	2.59	2.8	2.25	2.58	2.52	2.84
Comparison condition	2.96	2.87	2.8	2.05	2.85	2.7	2.76

Figure 15. Net metabolic power for each participant under the preferred (red), general (blue) and comparison (green) condition; the table shows the specific values of net metabolic rate.

4. Discussion

Exoskeleton assistance was found to substantially increase the preferred walking speed. The speed selection session induced participants to walk 0.26 m/s (21%) faster. This is an obvious increase in speed, but not as large as in the previous study of Collins (around 50% of the speed increase of Collins). This is mainly because the assistance device used in this study is unpowered and has effects on multiple joints; a huge increase of speed would cause a corresponding increase in the cost for transportation, as the device should be actuated by users, which is unreasonable behavior for walking efficiency in a subjective perspective. In addition, the formal and pre-experiments showed that the speed of the most comfortable RPE value increased and the RER decreased after several sessions, indicating that the assistance devices have large potential for speed improvement. These results demonstrate that, with appropriate training, unpowered exoskeletons can induce people to walk at faster speeds.

Speed has great effect on the step frequency and step length, and significantly improves them under the same assistance condition. There were only negligible differences between different assistance conditions at the same speed. However, it is still possible that there is a complicated relationship between the exoskeleton assistance and the step frequency and length, as the corresponding changes in preferred gait features are related to the increased speed that is caused by the assistance. Effects of the exoskeleton on the ankle joint angle lasted throughout the gait cycle, especially during the plantarflexion and dorsiflexion phases, and the exoskeleton significantly decreased the dorsiflexion angle and increased the plantarflexion angle. However, the effect on knee and hip angles was not significant, only during the swing phase there was more flexion of the knee angle under assistance. There was no significant difference in the angular range of all the lower extremity joints under different walking conditions. Especially for the ankle joint angle, although the maximum plantarflexion and dorsiflexion angles changed due to the assistance of the exoskeleton, these curves were almost only translated up and down. These results suggested that there were no obvious changes in walking features. These results demonstrated that the effect of exoskeleton on walking kinematics is limited and did not significantly alter walking habits.

In agreement with previous research [25], the preferred step frequency utilizes the least oxygen compared with higher or lower step frequencies, and significant difference between the preferred frequency and others occurred under the preferred condition. As the primary source of metabolic cost in walking, muscles have been proven to have the lowest activity at preferred frequency at a constant speed. In this study, GM, SOL, GAS, TA, BF and RF muscles are minimally activated at the preferred frequency just as in previous reports [25]. As there was a more accurate result from quadratic fitting rather than in linear fitting, it is believed that the resonance did occur during the assisted walking under the preferred condition. However, for the general condition, there was no significant difference between the R^2 of quadratic fitting and linear fitting. This could be the evidence that stable-resonant walking did not exist under the general condition. Yet, there is still a possibility that the adjustment time under the general condition is not enough in this experiment, since the correlation between the single support duration and gait cycle duration is 0.57. This may be the explanation of the above results under the general condition, but some unclear problems still remain. As the lower and higher frequencies are set as $\pm 10\%$ of preferred frequency, it is possible that the small gap and small quantities of frequencies are the main reason for the small difference between the R^2 of the quadratic and linear fitting under the general condition. Moreover, participants had to adjust their cadences in order to coordinate with the metronome, which would impact the breath and EMG during walking. As a result, whether resonant walking did exist under the general condition remains unclear. Unlike in previous studies [25], the muscle activity of BF and RF was higher in lower frequencies with respect to higher frequencies; this may be due to there being more effort conducted in lower frequencies to actuate the device.

A higher correlation between the single support duration and gait cycle duration was shown under the preferred condition (R^2 = 0.7) compared to the general condition

($R^2 = 0.57$), indicating a higher degree of resonance utilization [30]. Walking is more sensitive to resonance under the preferred speed with the exoskeleton. As a result, the proper exoskeleton design and movement parameters will lead to a condition that is much closer to the resonance, thus benefiting the walking efficiency.

Unexpectedly, the net metabolic rate under the general condition showed a totally different result insofar as only one participant had a reduction of 0.59%. Since the gait feature is almost the same, the increase between the normal and general condition can be explained by the device's characteristics such as unpowered and multi-joint-assisted. On the contrary, the net metabolic rate of the preferred condition showed a positive effect on energy saving, with 6 of 7 participants having a reduction on their net metabolic rates compared to the comparison condition. The difference between these two trends suggested the great effect of preferred speeds on assisting results. In addition, a 4.53% reduction was found between the preferred and general condition; 4 of 7 participants under the preferred condition showed a reduction in net metabolic rate with respect to the general condition. The increased net metabolic rate of the general condition is similar to the result in [41] (which had a 23% higher metabolic increase compared to walking without exoskeletons when using the exoskeleton with energy stored from knee extension and released for ankle plantarflexion); this could be explained by the assistance for the whole lower body. As this assistance has a great influence on assisted walking, new gait features will be required to improve the overall walking efficiency. Considering the increase of the net metabolic rate between the normal and general condition, it is believed that the reduction between the preferred and general condition can be explained by the resonant walking. This could be verified by the correlation between the duration of the single support phase and the period of the gait cycle. The higher correlation ($R^2 = 0.7$ for the preferred condition and $R^2 = 0.57$ for the general condition) of the preferred condition indicates that walking under the preferred condition utilizes the resonance of the whole body more efficiently.

The results of this study showed the important impact of resonant walking on exoskeleton devices, especially on unpowered exoskeletons. Preferred speed changed after assistance. When having a greater impact on walking, exoskeletons will also bring greater constraints. If resonant walking cannot be fully utilized, negative results may appear, just as in the general condition in this study. As a result, it is necessary to take the change of gait features into consideration during the design and evaluation of exoskeletons for global and further optimization of assisted walking. As is related to the preferred speed, the introduced concept of resonant walking is able to explain the U-shaped stiffness-metabolic rate curve and the lowest energy consumption at moderate stiffness. What is more, the personalized design is the key to solving the problem of the widely varied metabolic rate. Owing to the close relationship with body parameters, resonant walking is the most representative characteristic of human movement; hence it will certainly make a great contribution to the personalized design of exoskeletons.

Certainly, there are still some limitations of the research. First, the preferred speed could only be closely obtained in the speed-selection session because of the limitation of the treadmill, and this would have some unstable effects when evaluating the resonance condition of walking. Second, muscle activity and energy consumption are only two aspects that affect resonant walking; other aspects such as stability were not evaluated in this research, but muscle activity and oxygen consumption are the only evidence to verify resonant walking in this article, so it is not that comprehensive. More evidence that can prove resonant walking should be considered in a further study. Finally, the reduced number of participants is also a limitation that cannot be ignored, because this may influence the results of statistical analysis. Further studies about assistance performance under different speeds and spring stiffness should be conducted to map out the complicated relationship between exoskeleton assistance and resonant walking, which is consistent with the previous study [42]. In addition, in order to better study resonance in assisted walking, models that are capable of clearly describing resonant walking should be established. The complex relationship between preferred speed, exoskeleton stiffness and body parameters should be delineated in the future study.

5. Conclusions

In this article, the effect of exoskeleton assistance on preferred speeds was tested. Resonant walking was proven to exist in assisted walking, and the higher utilization of resonance with exoskeleton assistance led to lower energy consumption. The relationship of the body parameters, the spring stiffness and corresponding preferred gait features should be the subject of further research in the future. It is believed that changes in preferred speeds should be taken into consideration in further studies, and resonant walking will play an important role in the design and evaluation of unpowered soft exoskeletons.

Author Contributions: Conceptualization, H.W.; methodology, H.W.; software, H.W.; validation, H.W., J.W., Q.T. and Y.Z.; formal analysis, H.W.; resources, Z.Z. and S.G.; data curation, H.W., Z.Z., J.W., Q.T. and Y.Z.; writing—original draft preparation, H.W.; writing—review and editing, S.G. All authors have read and agreed to the published version of the manuscript.

Funding: This research was funded by the National Key Research and Development Program of China under Grant 2019YFB1312500.

Institutional Review Board Statement: The study was conducted in accordance with the Declaration of Helsinki, and approved by the Institutional Review Board of Biomedical Ethics Committee of Hebei University of Technology (NO. HEBUThMEC2022020, 1 November 2021).

Informed Consent Statement: Informed consent was obtained for all subjects involved in the study.

Data Availability Statement: The datasets used and/or analyzed during the current study are available from the corresponding author on reasonable request. Due to the current situation influenced by COVID-19, some portion of the raw data that are stored in Tianjin University of Sport cannot be integrated right now, so we decided to manage the raw data by ourselves and the data are available from the corresponding author Z.Z. (zzy@hebut.edu.cn) upon reasonable request.

Conflicts of Interest: The authors declare no conflict of interest.

References

1. Li, Y.; Sun, H.; Wang, C. Key Technologies of Lower Limb Power-Assisted Exoskeleton Robots: A Review. In Proceedings of the 2021 6th International Conference on Control, Robotics and Cybernetics (CRC), Shanghai, China, 9–11 October 2021; pp. 25–31.
2. Rodríguez-Fernández, A.; Lobo-Prat, J.; Font-Llagunes, J.M. Systematic review on wearable lower-limb exoskeletons for gait training in neuromuscular impairments. *J. Neuroeng. Rehabil.* **2021**, *18*, 22–43. [CrossRef] [PubMed]
3. Proud, J.K.; Lai, D.T.; Mudie, K.L.; Carstairs, G.L.; Billing, D.C.; Garofolini, A.; Begg, R.K. Exoskeleton application to military manual handling tasks. *Hum. Factors* **2022**, *64*, 527–554. [CrossRef] [PubMed]
4. Leclair, J.; Pardoel, S.; Helal, A.; Doumit, M. Development of an unpowered ankle exoskeleton for walking assist. *Disabil. Rehabil. Assist. Technol.* **2018**, *15*, 1–13. [CrossRef] [PubMed]
5. Nasiri, R.; Ahmadi, A.; Ahmadabadi, A.A. Reducing the Energy Cost of Human Running Using an Unpowered Exoskeleton. *IEEE Trans. Neural Syst. Rehabil. Eng.* **2018**, *26*, 2026–2032. [CrossRef]
6. Wang, X.; Guo, S.; Qu, B.; Song, M.; Qu, H. Design of a Passive Gait-based Ankle-foot Exoskeleton with Self-adaptive Capability. *Chin. J. Mech. Eng.* **2020**, *33*, 49. [CrossRef]
7. Zhou, T.; Xiong, C.; Zhang, J.; Hu, D.; Chen, W.; Huang, X. Reducing the metabolic energy of walking and running using an unpowered hip exoskeleton. *Neuroeng. Rehabil.* **2021**, *18*, 95. [CrossRef]
8. Yandell, M.B.; Tacca, J.R.; Zelik, K.E. Design of a Low Profile, Unpowered Ankle Exoskeleton That Fits Under Clothes: Overcoming Practical Barriers to Widespread Societal Adoption. *IEEE Trans. Neural Syst. Rehabil. Eng.* **2019**, *27*, 712–723. [CrossRef]
9. Ishikawa, M.; Komi, P.V.; Grey, M.J.; Lepola, V.; Bruggemann, G.P. Muscle-tendon interaction and elastic energy usage in human walking. *J. Appl. Physiol.* **2005**, *99*, 603. [CrossRef]
10. Sawicki, G.S.; Lewis, C.L.; Ferris, D.P. It Pays to Have a Spring in Your Step. *Exerc. Sport Sci. Rev.* **2009**, *37*, 130–138. [CrossRef]
11. Zelik, K.E.; Kuo, A.D. Human walking isn't all hard work: Evidence of soft tissue contributions to energy dissipation and return. *J. Exp. Biol.* **2010**, *213*, 4257–4264. [CrossRef]
12. Collins, S.H.; Wiggin, M.B.; Sawicki, G.S. Reducing the energy cost of human walking using an unpowered exoskeleton. *Nature* **2015**, *522*, 212–215. [CrossRef]
13. Dijk, W.V.; Kooij, H.; Hekman, E. A passive exoskeleton with artificial tendons: Design and experimental evaluation. In Proceedings of the IEEE International Conference on Rehabilitation Robotics, Zurich, Switzerland, 27 June–1 July 2011; pp. 1–6.

14. Haufe, F.L.; Wolf, P.; Riener, R.; Grimmer, M. Biomechanical Effects of Passive Hip Springs During Walking. *J. Biomech.* **2021**, *98*, 109432. [CrossRef]
15. Yang, J.; Park, J.; Kim, J.; Park, S.; Lee, G. Reducing the energy cost of running using a lightweight, low-profile elastic exosuit. *J. Neuroeng. Rehabil.* **2021**, *18*, 129–140. [CrossRef]
16. Panizzolo, F.A.; Annese, E.; Paoli, A.; Marcolin, G. A Single Assistive Profile Applied by a Passive Hip Flexion Device Can Reduce the Energy Cost of Walking in Older Adults. *Appl. Sci.* **2021**, *11*, 2851. [CrossRef]
17. Panizzolo, F.A.; Bolgiani, C.; Liddo, L.D.; Annese, E.; Marcolin, G. Reducing the energy cost of walking in older adults using a passive hip flexion device. *J. NeuroEng. Rehabil.* **2019**, *16*, 117. [CrossRef]
18. Xiong, C.; Zhou, T.; Zhou, L.; Wei, T.; Chen, W. Multi-articular passive exoskeleton for reducing the metabolic cost during human walking. In Proceedings of the Wearable Robotics Association Conference (WearRAcon), Phoenix, NJ, USA, 25–27 May 2019; pp. 63–67.
19. Zhou, T.; Xiong, C.; Zhang, J.; Zhang, J.; Chen, W.; Huang, X. Regulating Metabolic Energy Among Joints During Human Walking Using a Multiarticular Unpowered Exoskeleton. *IEEE Trans. Neural Syst. Rehabil. Eng.* **2021**, *29*, 662–672. [CrossRef]
20. Song, S.; Collins, S.H. Optimizing Exoskeleton Assistance for Faster Self-Selected Walking. *IEEE Trans. Neural Syst. Rehabil. Eng.* **2021**, *29*, 786–795. [CrossRef]
21. Zarrugh, M.Y.; Radcliffe, C.W. Predicting metabolic cost of level walking. *Eur. J. Appl. Physiol. Occup. Physiol.* **1978**, *38*, 215–223. [CrossRef]
22. Holt, K.G.; Hamill, J.; Andres, R.O. The force-driven harmonic oscillator as a model for human locomotion. *Hum. Mov. Sci.* **1990**, *9*, 55–68. [CrossRef]
23. Holt, K.G.; Hamill, J.; Andres, R.O. Predicting the minimal energy costs of human walking. *Med. Sci. Sports Exerc.* **1991**, *23*, 491–498. [CrossRef]
24. Holt, K.G.; Jeng, S.F.; Ratcliffe, R.; Hamill, J. Energy Cost and Stability During Human Walking at the Prefened Stride Frequency. *J. Mot. Behav.* **1996**, *27*, 164–178. [CrossRef]
25. Russell, D.M.; Haworth, J.L. Walking at the preferred stride frequency maximizes local dynamic stability of knee motion. *J. Biomech.* **2014**, *47*, 102–108. [CrossRef].
26. Obusek, J.P.; Holt, K.G.; Rosenstein, R.M. The hybrid mass-spring pendulum model of human leg swinging: Stiffness in the control of cycle period. *Biol. Cybern.* **1995**, *73*, 139–147. [CrossRef]
27. Fonseca, S.T.; Holt, K.G.; Saltzman, E.; Fetters, L. A dynamical model of locomotion in spastic hemiplegic cerebral palsy: Influence of walking speed. *Clin. Biomech.* **2001**, *16*, 793–805. [CrossRef]
28. Holt, K.G.; Obusek, J.P.; Fonseca, S.T. Constraints on disordered locomotion A dynamical systems perspective on spastic cerebral palsy. *Hum. Mov. Sci.* **1996**, *15*, 177–202. [CrossRef]
29. Hartmut, G.; Andre, S.; Reinhard, B.; Blickhan, R. Compliant leg behaviour explains basic dynamics of walking and running. *Proc. R. Soc. B Biol. Sci.* **2006**, *273*, 2861–2867.
30. Kim, S.; Park, S. Leg stiffness increases with speed to modulate gait frequency and propulsion energy. *J. Biomech.* **2011**, *44*, 1253–1258. [CrossRef]
31. Kim, S.; Park, S. The oscillatory behavior of the CoM facilitates mechanical energy balance between push-off and heel strike. *J. Biomech.* **2012**, *45*, 326–333. [CrossRef]
32. Hong, H.; Kim, S.; Kim, C.; Kim, C.; Lee, S.; Park, S. Spring-like gait mechanics observed during walking in both young and older adults. *J. Biomech.* **2013**, *46*, 77–82. [CrossRef]
33. Lee, M.; Kim, S.; Park, S. Resonance-based oscillations could describe human gait mechanics under various loading conditions. *J. Biomech.* **2014**, *47*, 319–322. [CrossRef]
34. Ryu, H.X.; Park, S. Estimation of unmeasured ground reaction force data based on the oscillatory characteristics of the center of mass during human walking. *J. Biomech.* **2018**, *71*, 135–143. [CrossRef] [PubMed]
35. Kuo, A.D.; Collins, S.H. The six determinants of gait and the inverted pendulum analogy: A dynamic walking perspective. *Hum. Mov. Sci.* **2007**, *26*, 617–656. [CrossRef] [PubMed]
36. Collins, S.H.; Kuo, A.D. Recycling Energy to Restore Impaired Ankle Function during Human Walking. *PLoS ONE* **2010**, *5*, e9307. [CrossRef] [PubMed]
37. Vaughan, C.L.; Davis, B.L.; Oíconnor, J.C. *Dynamics of Human Gait*, 2nd ed.; Hanover: Human Kinetics Pub: Western Cape, South Africa, 1992; pp. 15–22.
38. Malcolm, P.; Galle, S.; Derave, W.; Clercq, D.D. Bi-articular knee-ankle-foot exoskeleton produces higher metabolic cost reduction than weight-matched mono-articular exoskeleton. *Front. Neurosci.* **2018**, *12*, 69–82. [CrossRef]
39. Mcdaniel, J.; Durstine, J.L.; Hand, G.A. Determinants of metabolic cost during submaximal cycling. *J. Appl. Physiol.* **2002**, *93*, 823–828. [CrossRef]
40. Zuntz, N. Ueber die Bedeutung der verschiedenen Nährstoffe als Erzeuger der Muskelkraft. *Arch. Die Gesamte Physiol. Menschen Tiere* **1901**, *83*, 557–571. (In Germany) [CrossRef]
41. Etenzi, E.; Borzuola, R.; Grabowski, A.M. Passive-elastic knee-ankle exoskeleton reduces the metabolic cost of walking. *J. NeuroEng. Rehabil.* **2020**, *17*, 104. [CrossRef]
42. Nuckols, R.W.; Sawicki, G.S. Impact of elastic ankle exoskeleton stiffness on neuromechanics and energetics of human walking across multiple speeds. *J. NeuroEng. Rehabil.* **2020**, *17*, 75–93. [CrossRef]

machines

MDPI

Article

Perching and Grasping Mechanism Inspired by a Bird's Claw

Yongqiang Zhu *, Xiumin He, Pingxia Zhang, Gaozhi Guo and Xiwan Zhang

School of Mechanical and Automotive Engineering, Qingdao University of Technology, Qingdao 266520, China
* Correspondence: zhuyongqiang@qut.edu.cn

Abstract: In nature, birds can freely observe and rest on the surface of objects such as tree branches, mainly due to their flexible claws, thus this paper is inspired by bird perching and shows two imitation bird claw perching grasping mechanisms in the shape of "three in front and one at the back". One is articulated, the other is resilient, the difference being that the former has a pin-articulated claw structure and uses a double fishing line to perform the grasping and resetting action, while the latter uses a resilient linking piece, a single fishing line and resilient linking piece to perform the grasping and resetting action. To verify the grasping effect, experiments were designed to grasp objects of different shapes and maximum grasping weight load. The results show that the two types of perching grasping mechanism can reach a large degree of toe bending, have good passive bending deformation ability, can grasp different types of objects, including the articulated type has a stronger deformation ability, and can grasp branches with a diameter in the range of 12.5–55.8 mm. The elastic reset type is smoother than the articulated type toe bending curve, and the maximum graspable object weight is about three times the overall weight of the grasping mechanism. The maximum gripping weight is about three times the overall weight of the gripping mechanism and the load capacity is about two times that of the articulated type.

Keywords: perching grasping mechanism; fishing line; claw toe; toe bone; toe tip; envelope

Citation: Zhu, Y.; He, X.; Zhang, P.; Guo, G.; Zhang, X. Perching and Grasping Mechanism Inspired by a Bird's Claw. *Machines* **2022**, *10*, 656. https://doi.org/10.3390/machines10080656

Academic Editors: Yanjie Wang, Xiaofeng Liu, Aihong Ji, Shichao Niu and Bo Li

Received: 28 May 2022
Accepted: 3 August 2022
Published: 5 August 2022

1. Introduction

The rapid growth in demand for unmanned aerial vehicles (UAVs) for many civil and military applications has driven the rapid development of UAV technology [1,2]. Aerodynamic inefficiency and high energy consumption lead to limited flight time for UAVs, and UAVs require light weight and cannot carry large-capacity batteries, especially for micro craft. A very effective solution is to design a set of perching and grasping mechanisms that allow UAVs to perch flexibly on the ground or tree branches to perform surveillance and reconnaissance missions, and save energy consumption.

Depending on the UAV perching object, the main categories are planar perching and rod perching, where planar perching implies that the perched object is on a flat surface similar to a wall or ceiling, while rod perching implies perching on a rod-like object such as a tree branch. Drones use embedded perching mechanisms [3–5] and attached perching mechanisms when fixing to planar perches. The embedded perching mechanism mimics the tiny spines on the feet of insects by inserting the spines into the surface of the object being used for perching, often on inclined or vertical rough surfaces. For example, the Aerial Robotics Laboratory at Imperial College London developed a passive adaptive microsatellite perching mechanism weighing only 32 g. The perching mechanism consists of multiple compliant grapple modules. Each module consists of a single plastic linkage with a trapezoidal cross-section and uses cable tension to curl the module. A pair of sharp steel spines protrudes from the lower side of each linkage of the compliant grapple module to increase the perching capacity of the device [5]. In recent years, attachment-based perching mechanisms have become very popular and can be applied to both rough and smooth surfaces. Currently, techniques based on attachment perching include magnetic induction [6,7], dry adhesives [8,9], electrostatic induction adhesives [10,11] and vacuum air pressure [12–14]. For example, Roberts et al. [6] proposed a method to mitigate energy problems in aerospace exploration in indoor environments by relying primarily on ceiling attachments with small ring magnets. Thomas et al. [8] from the GRASP lab at the University of Pennsylvania developed a 60 g weight, gecko-inspired dry-adhesive perching mechanism. Graule et al. [10] proposed a controlled attachment principle for perching small micro air vehicles through electrostatic adhesion. The Mechanical Systems and Vibration Laboratory of Shanghai Jiao Tong University [14] developed a dual elastic combined suction cup (DEC suction cup) to help multi-wing aircraft perch on vertical walls under disturbance. Although embedded and attached perching mechanisms can achieve successful perching in specific scenarios, they are usually demanding on the surface of the perched object and have limited applications. Some of them have limited duration after perching, considering factors such as material adhesion.

In contrast, for rod perching, a gripping mechanism with grasping capability is usually used. Many grippers use simulated human hand designs; for example, Pounds and Dollar mounted a modified version of the SDM hand on a radio-controlled helicopter and used the system to demonstrate grasping objects while hovering [15]. Thomas et al. used the SDM hand and a Festo EXOHand-derived linkage-based hand mounted on a single-degree-of-freedom arm to perform high-speed grasping [16]. A group at Yale University attached an underdriven hand to the underside of a helicopter [17], and the research focused on the effects on object pickup and flight dynamics. The trend in aerial manipulators is always to improve maneuverability and safety by reducing weight and size. Other perching grasping mechanisms mounted on UAVs are inspired by the claws of birds perched on tree branches. Inspired by the perching mechanism of birds, Doyle et al. [18] proposed to use the weight of the UAV to passively apply tendon tension to drive the claws to grasp and passively perch on structures such as tree branches without consuming energy. Bai et al. [19], based on the structure and motion characteristics of bird claws, proposed a UAV perching mechanism with high load-bearing capacity. The mechanism consists of three flexible toes, an inverted crank slider mechanism for achieving opening and closing motions, and a gear mechanism for deforming between the two configurations. Nadan et al. [20] proposed that two opposing, multi-segmented, flexible claw toes can curl and surround objects, such as trees, by folding a four-linked rod when the UAV lands. Although these existing bionic perching mechanisms allow the UAV to perch on targets such as tree branches, they cannot deform to adapt to grasp a wider range of objects and, even for some of the objects that can be successfully grasped, the low envelope fit of the claw toes to the outer surface of the object ultimately results in a poor grasp and affects the stability of the grasp.

In response to the above problems, this paper proposes a new effective solution. Inspired by bird perching grasping, two kinds of bird-like claw perching grasping mechanism are designed, including the driving structure, leg structure and claw structure from top to bottom, which are divided into articulated type and elastic reset type according to the different compositions of the claw structure. The shape of the claws of the two types of perching grasping mechanism is "three in front and one at the back", and two servos actively drive the fishing line to complete the claw toe grasp and release work, so there are no problems with limited perching time and difficult separation due to the sticky pad attachment method. Meanwhile, the use of two drivers, the overall small size and light weight effectively avoid the bulky traditional mechanical gripper. It can finally perch on branches, and the multi-degree-of-freedom claw toe has good adaptability to differently shaped objects, and can grasp objects of certain weight and grasp smoothly.

The imitation bird claw grasping mechanism studied in this paper can be installed in the fuselage of UAVs in the future to realize the transportation of aerial items of UAVs, and they can also perch and stay on tree branches to improve their range and perform special surveillance and search and rescue tasks. Of course, it is not limited to application to UAVs, but also has a wide range of practical application scenarios, improving the structure size as needed and adding auxiliary tools, such as robotic arms and cameras, to be applied in military, medical, aviation, intelligent industrial production and daily life scenarios, as robotic hands instead of humans to complete work with high-volume and high-quality requirements.

The paper is organized as follows: Section 2 explains the structural design of the claw section of the bionic perching grasping mechanism and the actuation method inspired by the perching mechanism of birds in nature. Section 3 designs the structural components of the two perching grasping mechanisms. Section 4 analyzes the articulated claw toe rotation and grasping. Section 5 describes physical experiments to verify the results of the claw toe rotation and grasping analysis and experiments on the grasping object shape adaptability and load capacity of the two perching grasping mechanisms for experimental analysis and comparison, and finally Section 6 concludes the work of the paper and provides an outlook for future research.

2. Bionic Inspiration

A true bird's claw has many functions, such as landing, grasping, perching, scratching and capturing. However, only the perching and grasping functions are of interest for today's flying machines. As shown in Figure 1a, birds in nature show an exceptional ability to adapt and perch efficiently on a wide variety of surfaces, such as branches, cables, flats, etc. The structural morphology of the claw and the actuated grasping mechanism are decisive factors in the bird's ability to achieve a perching grasp.

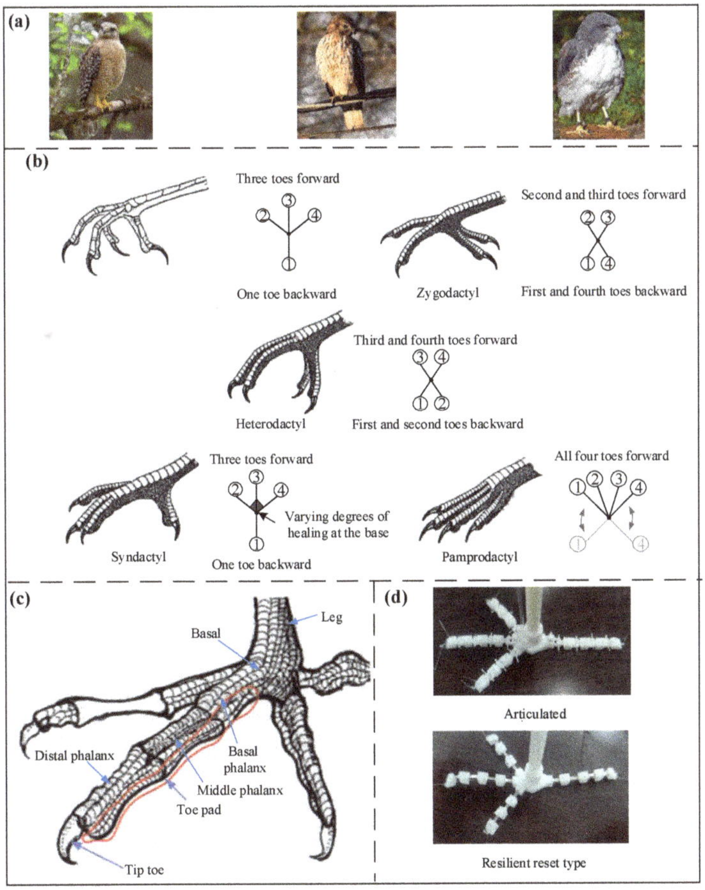

Figure 1. Bird claw bionic principle. (**a**) Bird perching surface; (**b**) claw toe arrangement; (**c**) basic component structure of a bird claw [19]; (**d**) two bionic claw structures.

2.1. *Claw Toe Arrangement Morphology*

Many different toe arrangements exist in the claws of birds, often related to ecology and lifestyle [21–23], with morphology varying from species to species. As shown in Figure 1b, there are five different types of claw toe arrangement [24]: (i) Anisodactyl toe (three toes forward and one toe backward, e.g., sparrows); (ii) zygodactyl toe (first and fourth toes backward, second and third toes forward, e.g., woodpeckers); (iii) heterodactyl toe (first and second toes backward, third and fourth toes forward, e.g., biting cuckoos); (iv) syndactyl toe (all three toes forward and have varying degrees of healing at the base, e.g., kingfishers); and (v) pamprodactyl toe (all four toes forward, e.g., rainbows).

In this design, the anisodactyl toe type was chosen as this arrangement is present in the claws of most birds, and the "three in front and one at the back" design of the claw toe arrangement can be easily adapted to the object being grasped. This four-toed structure uses the middle two long toes for grasping or perching and the outer two short toes to maintain the balance of the grasp and avoid dropping objects when perching or when grasping, making the grasp more stable.

2.2. *The Basic Structure of the Claw Toe*

As shown in Figure 1c, the claw structure consists of a base and four claw toes. The base is used to support its own weight and connect the claw to the leg. The middle claw toe is long and consists of a basal phalanx, a middle phalanx, a terminal phalanx and a toe tip, and the claw toes on either side are short and have only two phalanges, consisting of a basal phalanx, a terminal phalanx and a toe tip. The claws of birds are underdriven, allowing the claws to passively conform to the object that they are attempting to perch on. In analogy to the bird's claw, as shown in Figure 1d, the designed long claw toe has four degrees of freedom and the short claw toe has three degrees of freedom.

At the same time, the toe tip of birds resembles a curved hook and plays an important role as the end of the claw toe, which can grasp larger objects even though the claw toe may not be long enough to wrap around, as the toe tip can generate additional grasping force and even penetrate the surface to which the object is attached [25,26], and thus the design of the toe tip mimics the curved profile of birds.

In addition, claw curvature increases from ground roosting to arboreal, climbing and then predation, with claw curvature and size related to predatory behavior and prey type specialization, respectively, but as revealed by Pike [27] and Fowler [28], there is no clear trend in claw curvature in perching birds. Arbitrary curvature of the claw can thus be considered in the design phase to improve the perching performance of the mechanism.

2.3. *Bird Perching Drive Grasping Mechanism*

A typical bird claw toe mainly consists of the toe bone, toe pad and toe tip, where the toe pad is located at the base of each claw toe, as shown in Figure 1c, which contains tendons and muscles. The muscles contract and the tendons transmit the force that causes each claw toe to bend, and eventually the claw toe grasps the object to achieve a grasp. This inspired the use of fishing line instead of tendons and a servo instead of muscles for actuation.

As shown in Figure 1d, two bionic claw structures are designed in this paper to compare the advantages and disadvantages of each, one is articulated and the other is resilient rest type, the difference being that the former uses a pin articulation between the components of the claw toe and executes a grasp and release using a double fishing line, one line shortening and the other line elongating. Meanwhile, the latter has a resilient connector between the components of the claw toe and executes a grasp using a single fishing line, the line being relaxed and restoring the deformation autonomously under the action of the resilient connector's own elasticity.

3. Design of Two Perching and Grasping Mechanisms

3.1. Overall Structural Design

The overall structural design of the two perching grasping mechanisms is shown in Figure 2a. The articulated perching grasping mechanism consists of a driving structure, a leg structure and a claw structure from top to bottom, with a through-hole connection between the driving structure and the leg structure, and between the leg structure and the claw structure. The difference between the resilient reset perch gripping mechanism and the articulated type lies mainly in the claw structure and the reset method.

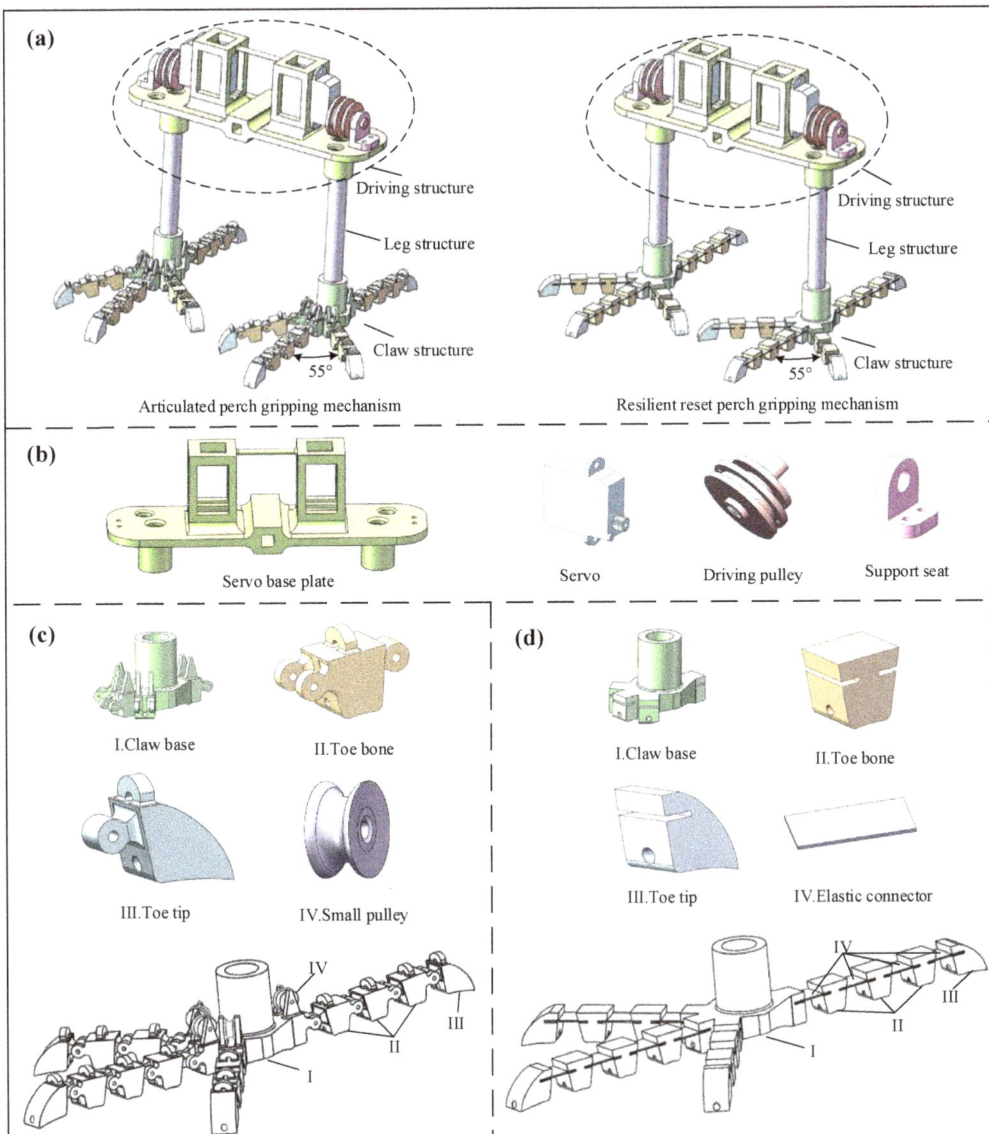

Figure 2. Design of the perch gripping mechanism. (**a**) Overall structural design of the two perching grasping mechanisms. (**b**) Drive structural parts. (**c**) Articulated jaw structural parts. (**d**) Resilient reset jaw structural parts.

3.2. Driving Structure

As shown in Figure 2b, the driving structure includes a servo base plate, a servo, a driving pulley and a support seat. The rectangular section through-hole in the middle of the servo base plate can be nested and connected with the microlight fuselage, and the two through-holes on each side are for internal and external fishing lines to penetrate. The internal teeth in the middle of the driving pulley engage with the teeth of the output gear of the servo. The driving pulley has two slots of different radius sizes, facing the two through-holes in the servo base plate. As the output gear of the servo has a short reach, tightening of the internal and external fishing line will damage the driving pulley, so a support base is designed to support the driving pulley, which is fixed to the servo base.

The resilient repositioning perch gripping mechanism uses only the internal line to perform the grip, so theoretically the servo base plate only needs to be provided with a through-hole for the internal line to pass through and the pulley to be provided with a slide slot. To compare the gripping performance of the two mechanisms, the same servo driving structure is used.

3.3. Claw Structure

The articulated claw structure consists of a claw base, small pulley and four claw toes, as shown in Figure 2a,c. The claw base is provided with a cylindrical tubular projection above the claw base for connection to the leg, surrounded by four projecting branches, the upper side of the branches is hinged to fit the toe bone, the lower through-hole is used to run through the fishing line, the interior is provided with four transitional circular through-holes to change the direction of the internal fishing line. The fishing line run through by the four branches converges at the cylindrical tubular projection above the claw base, the cross-section of the four branches at the root of the claw base is made sloping to allow the claw toe to produce a greater curvature. The bottom of the claw base is curved to better suit the object being grasped. The four small pulleys are connected by pins on the upper surface of the claw base and the external fishing line passes through the bottom of the small pulleys, changing the direction of the external line and reducing friction.

Among the four claw toes, there are two long claw toes in the middle and short claw toes on either side. The angle between the adjacent long and short claw toes is designed to be 55° without interfering with each other, and the long claw toe plays a more obvious gripping role. The clawed toe consists of toe bones and toe tips, with the long claw toe consisting of 3 toe bones and a toe tip, and the short claw toe consisting of 2 toe bones and a toe tip. The toe bones are designed in a trapezoidal shape, the sides of the base are rounded to create a greater angle of curvature between the adjacent parts of the claw toe. The toe bone has a rounded hole underneath for running the internal fishing line through, a protruding earring above for the external fishing line to run through and protruding parts on both sides for connecting adjacent parts. The toe tip is the end of the claw toe, with the same design of protruding earring above and the same design of the left end as the toe bone, with the difference that the fishing line below the through-hole runs in a smooth curve to reduce friction and avoid the toe tip bending upwards to a certain extent, and the end of the toe tip is designed as slender. The toe bone is hinged to the claw base, the toe bone to the toe bone and the toe bone to the toe tip, with the through-hole in the middle of the hinge fitted with a pin gap and the through-holes on both sides fitted with a pin interference, so that the toe bone can be rotated relative to the claw base, the back toe bone relative to the front toe bone and the toe tip relative to the toe bone.

As shown in Figure 2a,d, the resilient reset jaw structure comprises a claw base and four claw toes. The recesses on the upper side of the four protruding branches of the claw base are used to embed the resilient connectors and the lower through-hole is used to allow the fishing line to penetrate. The rest of the structural features are the same as for the claw base of the articulated gripping mechanism. The claw toe includes three main parts: The long claw toe consists of 3 toe bones, 4 elastic connectors and 1 toe tip, the short claw toe consists of 2 toe bones, 3 elastic connectors and 1 toe tip. In order to make the experiment consistent, the number of toe bones and toe tips is the same for both gripping mechanisms; the elastic connectors are flexible and have a high recovery capacity from deformation and are embedded in the grooves of the claw base, toe bones and toe tips. Compared to the articulated ones, there are not protruding earrings above the toe bones and toe tips.

3.4. Tendon Arrangement

The tendon is a fishing line that runs through the entire perching and grasping mechanism. For the articulated perching and grasping mechanism, two fishing lines are used, divided into an external and an internal fishing line, where the external fishing line is exposed outside and runs through the earring above each component of the claw toe, through the claw toe, the small pulley, the servo base plate and the driving pulley. The internal fishing line runs through the through-hole set in the component, through the claw toe, the claw base, the leg structure, the servo base plate and the driving pulley. The internal line, which generates a large amount of displacement, is wound clockwise around the large-radius chute of the driving pulley, with the end point of the line tied to the wall of the chute and the other end point fixed at the toe tip. Meanwhile, the external line, which generates a small amount of displacement, is wound counter-clockwise around the small-radius chute, with the two end points of the line fixed in the same way as the internal line.

For the articulated gripping mechanism, when the claw toe gripping work is carried out, the claw toe is driven and controlled by servos. The servo is deflected counter-clockwise by a certain angle, and synchronously drives the driving pulley rotation, the fishing line wound on the two slots produce different displacement deformation, the internal fishing line is tightened and the external fishing line is loosened, so each claw toe is bent, completing the gripping work. Then, the servo stops deflection. In contrast, when the claw toe is extended, the servo is deflected clockwise by a certain angle and the driving pulley is driven to turn simultaneously, the internal line is released and the external line is tightened, so the claw toe is gradually extended from bending.

The elastic reset perch gripping mechanism uses a single line, which is distributed and fixed in the same way as the internal line of the articulated type. The drive control is the same as that of the articulated type, but the line is pulled tight during grasping, and the elastic connection is deformed by bending. Then, the claw toes grasp the object. While the line is released during extension, the elastic connection restores itself to deformation and finally the claw toes extend.

4. Claw Toe Rotation Gripping Analysis

4.1. Claw Toe Rotation Analysis

The parts of the claw toe have a certain limit range of rotation angle between them and, when greater than the angle constraint range, the parts will show collision interference. Specifically, the angle of rotation of the toe of the articulated perch gripping mechanism relative to the claw base ranges from 0 to 80.34°, the angle of rotation of the latter toe relative to the former toe ranges from 0 to 67.76° and the angle of rotation of the toe tip relative to the toe ranges from 0 to 68.26°.

When the claw toe is bent or extended, a certain amount of deformation of the fishing line occurs and, with reference to the range of angles of rotation between the parts of the claw toe, the deformation of the fishing line through the long and short claw toe is shown in Table 1. When the long and short claw toes are bent from extension, the external line becomes longer and the internal line becomes shorter, and the internal line has approximately twice the amount of deformation as the external line. The design of the driving pulley groove radius is based on the relationship between the internal and external line deformation, so that the large groove radius is also approximately twice as large as the small groove radius.

Table 1. Deformation of articulated long and short claw toe fishing lines.

Claw Toe	Original Length of Outer Wire When Extended (mm)	Original Length of Inner Wire When Extended (mm)	Length of Outer Line after Bending (mm)	Length of Inner Wire after Bending (mm)	Maximum Deformation of the Outer Line (mm)	Maximum Deformation of the Inner Line (mm)
Long claw toe	29.29	33.49	35.64	21.17	6.35	12.32
Short claw toe	21.37	25.75	26.47	15.46	5.10	10.29

In addition, the deformation of both the internal and external line of the long toe is greater than that of the short toe, and the difference between the deformation of the long and short toe is basically constant, with a difference of 1.25 mm for the external line and 2.03 mm for the internal line, the deformation of the long toe is used as a benchmark and the short toe is compensated for with a suitable spring according to the difference in deformation.

The theoretical relationship between the deflection of the line and the angle of deflection of the servo is as follows.

$$L = \frac{n\pi r}{180} - \Delta x \quad (0^\circ \leq n \leq 160^\circ) \tag{1}$$

where L is the amount of line deflection (mm), n is the angle of deflection of the servo (°), r is the radius of the driving pulley slots (mm), Δx is the spring compensation (mm).

The theoretical angle of the servo is 180°, but in practice the maximum operating angle is approximately 160°, which is within the normal operating range of the servo and can achieve the maximum deformation of the internal and external fishing line of the long claw toe. Based on Equation (1), the final radius of the driving pulley is 4.5 mm for the large slot and 2.3 mm for the small slot. r is 2.3 and Δx is 0 when calculating the external line deflection of the long claw toe, r is 4.5 and Δx is 0 when calculating the internal line deflection, r is 3 and Δx is 1.25 when calculating the external line deflection of the short claw toe, r is 4.5 and Δx is 2.03 when calculating the internal line deflection.

Using the long claw toe as an example, the articulated perch gripping mechanism achieves the maximum deformation results, as shown in Figure 3.

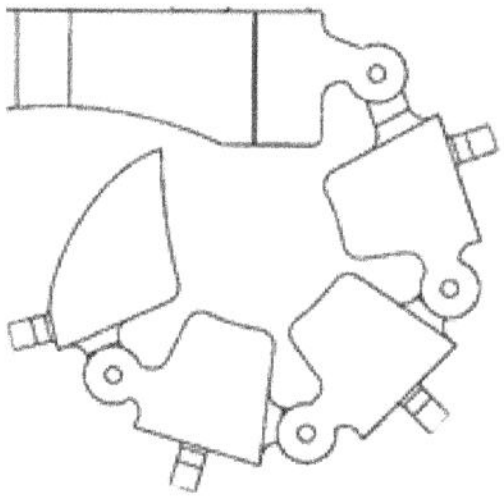

Figure 3. Maximum deformation result for the long claw toe of the articulated perch gripping mechanism.

4.2. *Claw Toe Gripping Analysis*

The articulated perch gripping mechanism was made to grip columnar objects in the diameter range 15–60 mm with a diameter interval of 5 mm and the long claw toe gripping results are shown in Figure 4.

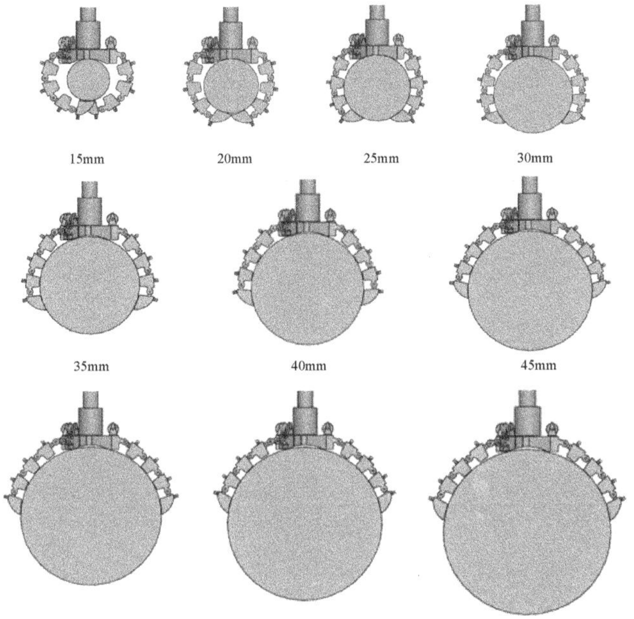

Figure 4. Long claw toe grasp results.

As can be seen from Figure 4, on the one hand, with the increase in the grasping diameter, the degree of bending deformation of the claw toe gradually decreases, and the deformation of the fishing line becomes smaller and smaller; on the other hand, with the increase in the grasping diameter, the envelope angle becomes smaller and smaller. When the diameter is 15 mm and 20 mm, the claw toe can completely envelop the object and leave a residual length, the envelope angle is greater than 360° and the two long claw toe tips appear to interfere with the collision. When the diameter is 25–50 mm, the claw toe can envelop at least 1/2 of the circumference of the object, and the envelope angle is greater than 180°. When the diameter is 55 and 60 mm, the claw toe envelope length is less than 1/2 of the circumference of the object and the envelope angle is less than 180°.

In addition, as the gripping diameter increases, the contact surface between the toe bone and the base of the toe tip and the object becomes larger, which is more stable for gripping. At a gripping radius of 15 mm, only the toe tip touches the object, at a gripping radius of 20 mm, the toe tip and one toe bone touch the object, at a gripping radius of 25–45 mm, the toe tip and two toe bones touch the object and the third toe bone has an increasing contact surface with the object, at a gripping radius of 50–60 mm, the toe tip and three toe bones touch the object, which means that the long claw toe is in close proximity to the object.

As there is a correspondence between the amount of deformation of the external and internal fishing line of the long claw toe, the deformation of the internal fishing line of the long claw toe was used as the measurement standard and a total of 10 sets of data were obtained as shown in Table 2. *d* indicates the diameter of the object, *S* indicates the length of the internal fishing line of the long claw toe after deformation, *L* indicates the amount of deformation of the internal fishing line of the long claw toe and *N* indicates the envelope angle (the angle of the center of the circle corresponding to the envelope arc formed by the claw toe when grasping the object).

Table 2. Corresponding long claw toe internal fissures and envelope angles.

d/mm	15	20	25	30	35	40	45	50	55	60
S/mm	25.98	26.96	27.82	28.47	29.18	29.57	30.22	30.47	30.82	31.12
L/mm	7.51	6.52	5.67	5.02	4.30	3.92	3.27	3.01	2.67	2.37
N/°	557.43	429.33	351.32	297.72	259.87	229.59	207.40	187.82	172.18	158.98

The amount of deformation *L* is known and the corresponding rotation angle *n* of the servo can be calculated by Equation (1). A quadratic polynomial fit is made to the amount of line deformation *L* and the rotation angle *n* of the servo corresponding to the gripping of 15–60 mm diameter objects, and a cubic polynomial fit is made to the envelope centroid angle *N*. The results are shown in Figure 5. It can be seen that as the gripping diameter increases, *L*, *n* and *N* gradually decrease and to an increasingly lesser extent.

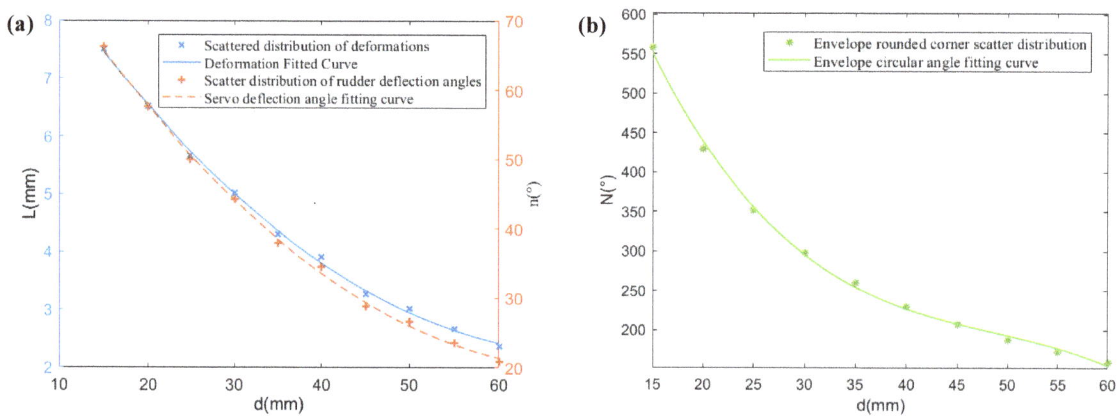

Figure 5. Grab fit curves. (**a**) *L* and *n* curve fit results. (**b**) Envelope angle *N* curve fit results.

When the claw toe grips an object, the claw toe is in contact with the surface of the object and the gripping force of the claw toe on the object is generated by the rudder and transmitted by the line. Theoretically, the greater the weight of the object, the greater the gripping force required, i.e., the greater the force exerted by the line on the lower surface of the line holes in each part of the claw toe, which can easily cause damage to the claw toe. A simplified model was established, with the claw toe in contact with the surface of the object, the finger bones on the upper side of the claw toe articulated with restraint and a force of 5 N applied to the lower surface of the fishing line hole in the direction perpendicular to the toe bone and toe tip. The stress and displacement results are shown in Figure 6, the maximum stress 9.742×10^5 N/m^2 and the maximum displacement 4.465×10^{-4} mm are at the fishtail hole, and the stresses and displacements decrease from the root of the toe to the toe fishtail hole. While the allowable stress of plastic ABS material is 24.5 MPa, the maximum stress in the claw toe is much less than the permissible stress, and the displacement of the claw toe is very small, which shows that the claw toe structure has sufficient resistance to deformation.

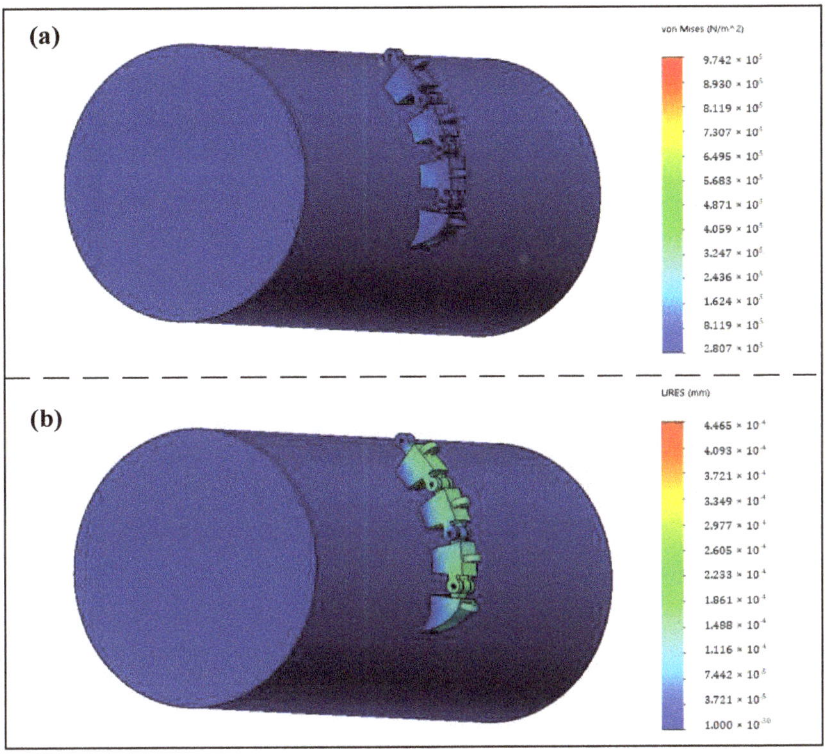

Figure 6. Finite element analysis of claw toe. (**a**) Stress results. (**b**) Displacement results.

5. Physical Experiment Analysis

5.1. Physical Assembly and Experiment Platform Construction

The wire diameter of the pin used in the articulated perch gripping mechanism is 0.7 mm, and the elastic connection of the elastic reset perch gripping mechanism is selected as 0.1 mm 301 stainless steel shrapnel. The weight of the two grasping mechanisms is about 15 g, the claw structure is about 3.5 g, the long claw toe is about 3 cm long and the short claw toe is about 2.5 cm. A 9 g drive servo is chosen according to the rotation angle of the servo required for the grasping work and to meet the requirements of the grasping driving force. A 0.2 mm fishing wire is chosen as the transmission medium of the grasping force for the tendon, because of its toughness, strong tensile capacity and low friction resistance between the parts. A 3D printed resin material was used.

One side of the gripping mechanism was assembled, as shown in Figure 7a. In order to make the gripping structure lighter, the servo type chosen for the first experiment scheme was 2.7 g, but the experiment results showed that the driving torque of the servo was too small, so the servo for the second time was chosen to be 9 g for both habitat gripping mechanisms. The servo was fixed directly to the servo base plate, with the slots facing the holes in the servo base plate. The final assembly of the two perch gripping mechanisms is shown in Figure 7b.

As shown in Figure 7c, the hardware of the experiment platform mainly includes a PC, socket, power supply module, Esp8266, fixing bar, fixing bracket and grasped object, and the software includes Arduino and LabView. After the voltage is reduced to 5 V by the power supply module, the pins of the power module supply power and ground to the servo through the connection line, the signal line of the servo is connected to the pins of the microcontroller Esp8266, the other end of the data line of the microcontroller Esp8266 is connected to the USB port of the PC. The two kinds of perching and grasping mechanism and the fixing bar are fixed together, and the fixing bar is attached to the fixing bracket.

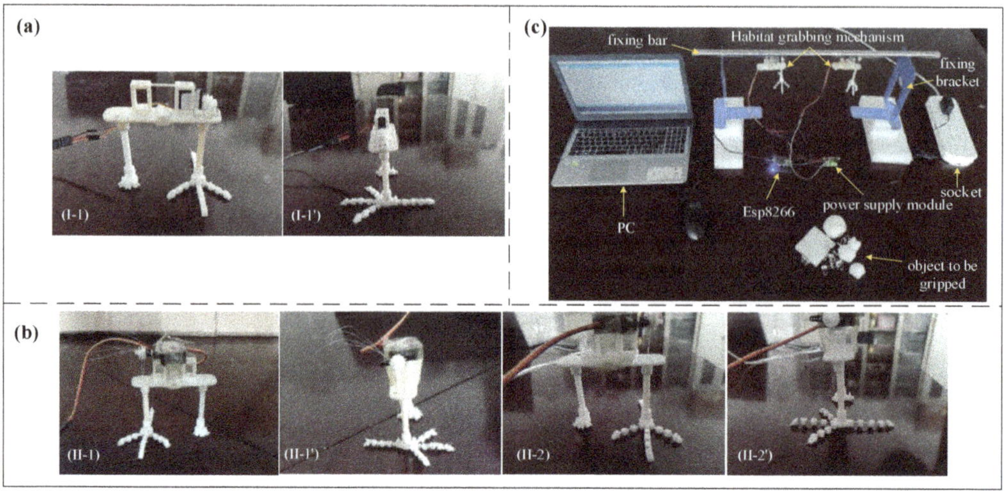

Figure 7. Fabrication of the prototype and the experiment platform. (**a**) First generation of articulated perch gripping mechanism prototype, (I-1) front view, (I-1′) side view. (**b**) Second generation of two perch gripping mechanism prototypes, (II-1) articulated front view, (II-1′) articulated side view, (II-2) resilient reset front view, (II-2′) resilient reset side view. (**c**) The experiment platform.

5.2. Experimental Verification

Performance experiments were carried out for two different perch gripping mechanisms, firstly to verify the claw toe bending deformation capacity of the two perch gripping mechanisms under no-load conditions, as shown in Figure 8.

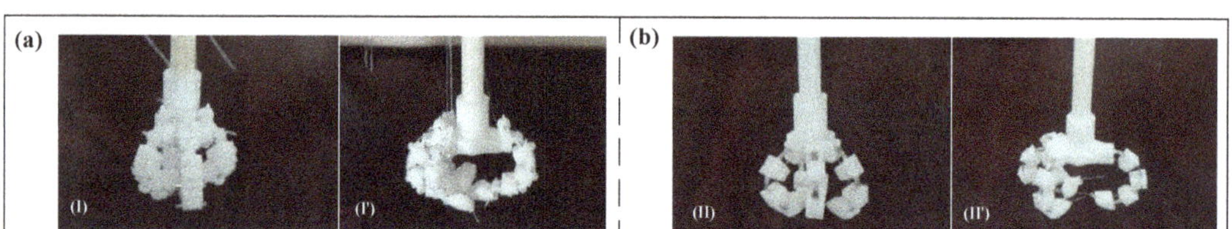

Figure 8. Results of two perch gripping mechanisms with claw toes reaching maximum bending. (**a**) Articulated. (I) Front view, (I′) side view. (**b**) Resilient reset type. (II) Front view, (II′) side view.

The maximum relative rotation angle between the individual parts of the articulated claw toe is consistent with the results of the previous analysis of claw toe rotation. When the claw toe reaches its maximum degree of bending, the articulated servo rotation angle is approximately 158°, which should theoretically be 156.86°, almost reaching the result of the maximum deformation of the fishing line analyzed earlier. The articulated type has a stronger degree of bending deformation compared to the elastically resettable type, but the curve formed after bending is not as smooth as the elastically resettable type.

The claw toe grip results were verified for columnar objects. As the vehicle commonly grips tree branches for perching, tree branches were chosen for the grip experiment in order to make the experiment results more convincing (Video S1). As shown in Figure 9a, the experiment results show that the articulated ones can successfully grasp branches with diameters of 12.5 mm, 21.5 mm, 27.4 mm, 48.6 mm and 55.8 mm, corresponding to servo rotation angles of 145°, 122°, 101°, 60° and 50°, which is approximately twice the result of the n-curve fit, due to the fact that, while ensuring that the branches are wrapped as far as possible, sufficient gripping force is required, thus requiring the servo to rotate further. Both gripping mechanisms can be clearly seen in the side gripping diagrams, where the contact area between the base of the claw toe and the branch becomes larger and the envelope angle becomes smaller as the diameter increases. However, when grasping larger diameter branches of 48.6 mm and 55.8 mm, as shown in Figure 9c, both grasping mechanisms can successfully stand on the branch as the contact area between the base of the claw seat and the top of the branch is sufficiently large. However, in practice, taking into account the natural environment, grasping is also required if the perch is stable, and the grasping results show that the toe bones arch upwards extremely easily and the toe tips play an increasingly prominent role, snapping hard against the surface of the branch.

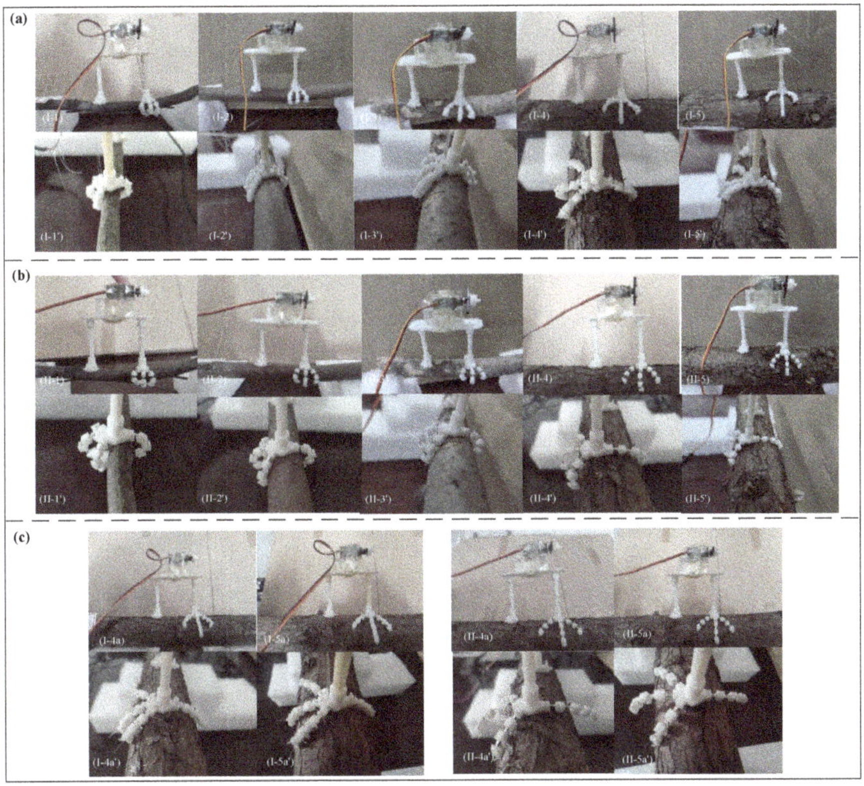

Figure 9. Experiment results of two types of perch gripping mechanism for grasping branches. (**a**) Articulated. (**b**) Resilient reset. (**c**) Two types of perch gripping mechanism for special standing situations. I: Articulated, II: resilient reset, (I-1)–(I-5)/(II-1)–(II-5) branches with diameters of 12.5 mm, 21.5 mm, 27.4 mm, 48.6 mm, 55.8 mm, (I-1′)–(I-5′)/(II-1′)–(II-5′) the corresponding side gripping diagrams, (I-4a)/(II-4a) and (I-5a)/(II-5a) standing on 48.6 mm and 55.8 mm diameter branches, respectively, (I-4a′)/(II-4a′)and (I-5a′)/(II-5a′) corresponding side gripping views.

Compared to the two perch gripping mechanisms, firstly, the resilient reset type does not successfully grip 12.5 mm branches, thus the articulated type has a greater gripping range when gripping branches of different diameters. Secondly, after several gripping experiments, the resilient reset type showed plastic deformation of the elastic joint, which does not fully return to its extended state and will suffer from elastic depletion in long-term use, while the articulated type does not show any effect. Again, when gripping branches of the same diameter, the servo deflection angle required for the resilient repositioning type is greater than that of the articulated type. Finally, the articulated type is not as soft in terms of the flexing effect of the claw toe as the resilient repositioning type.

5.3. Adaptability and Load Capacity Experiment Analysis

To assess the adaptability of the two perching grasping mechanisms for grasping different types of objects, rectangular cardboard and rectangular, spherical and irregularly shaped foam were selected for performance experiments (Video S1), and the grasping results are shown in Figure 10. The results show that both perch gripping mechanisms have good passive bending deformation capability of the claw toes and both can adapt to the experiment object. Specifically, the cardboard was thin and had a smooth surface, and the initial experiments were conducted with a narrow surface as the contact surface, which resulted in a failed grip, mainly due to the small contact area between the bottom surface of the claw structure and the cardboard, the smooth surface of the object and the object's own gravity causing a large relative slip, so a wide surface was chosen as the contact surface instead and a successful grip was achieved. The two gripping mechanisms can easily grip rectangular foam objects with a large contact area and high friction with the foam material, resulting in less relative sliding. The irregular foam object is thick and has a certain slope on the sides, and the contact area of the two gripping mechanisms is very small when gripping, but the four toe tips can be embedded or hooked into the object, which is the key reason for successful gripping. Gripping near spherical objects is easiest, and both types of gripping mechanism have a large contact area between the base of the toe and the object, and have a good envelope for a stable grip.

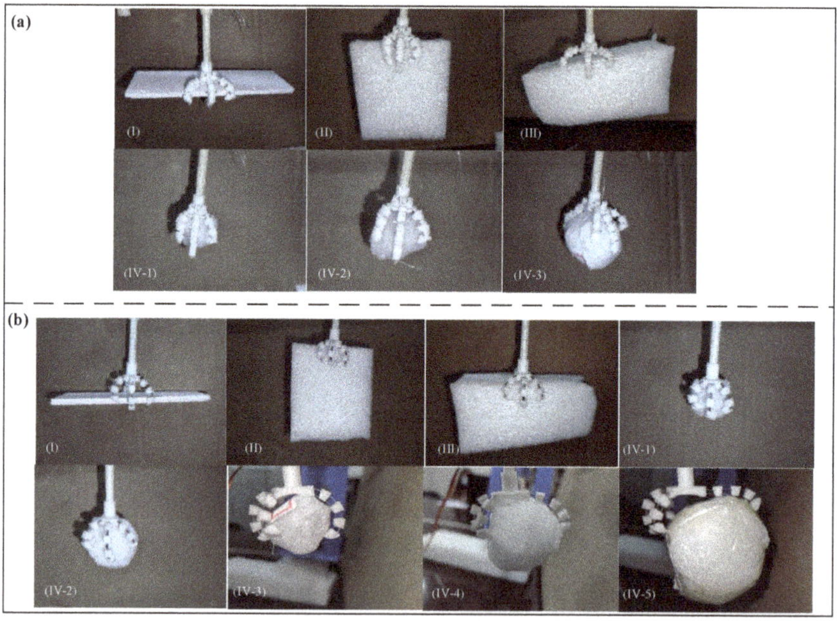

Figure 10. Adaptive performance experiments of two perching grasping mechanisms. (**a**) Articulated. (**b**) Elastic reset type. (I) Rectangular cardboard (thin and smooth), (II) rectangular foam object (thicker), (III) irregular foam object (thicker), (IV) near spherical foam object (IV-1)–(IV-5) (diameters approximately 18 mm, 26 mm, 35 mm, 45 mm, 55 mm in that order).

Therefore, the two perching grasping mechanisms have good adaptability, increasing for cardboard, irregular foam objects, rectangular foam objects and spherical foam objects, in that order. However, the success of the grasp also depends on the point of impact of the grasp, which is theoretically the center of gravity of the object, but in practice tends to make regular grasped objects balanced and symmetrical with respect to both sides of the claw structure. For irregularly shaped and different types of objects, it is particularly important to select the correct contact surface, increase the contact area and give full play to the toe tip.

On the one hand, a standard counterweight carrier is required for the gradual application of forces to the jaw structure of the perching grasping mechanism and, on the other hand, inspired by the different adaptability of the perching grasping mechanism to different objects, two options are proposed for experimenting with the load performance of the two types of perching grasping mechanisms, and a rectangular foam object and a spherical foam object are chosen as counterweight bases, which are counterweighted with clamps and large head pins on the foam object, as shown in Figure 11. The experiment results show that if the rectangular foam object is chosen as the counterweight base, the final maximum applied load force is 0.0435 N for the articulated type and 0.1457 N for the resilient reset type and, if the spherical foam object is chosen as the counterweight base, the final maximum load force is 0.2497 N for the articulated type and 0.4487 N for the resilient reset type, which is about three times the overall weight of the grasping mechanism and about 13 times the weight of the claw structure. Both perching grasping mechanisms were able to restore the claw toe to its original state even after releasing the object, and neither the object nor the claw toe was damaged. It can be seen that the load capacity of the two types of perch gripping mechanism is indeed related to the counterweight base. With the same counterweight base, the resilient reset perch gripping mechanism can withstand approximately twice the load force of the articulated type. As the applied load mass increases, the duration of sustained grip decreases, even though the claw toes can still grip.

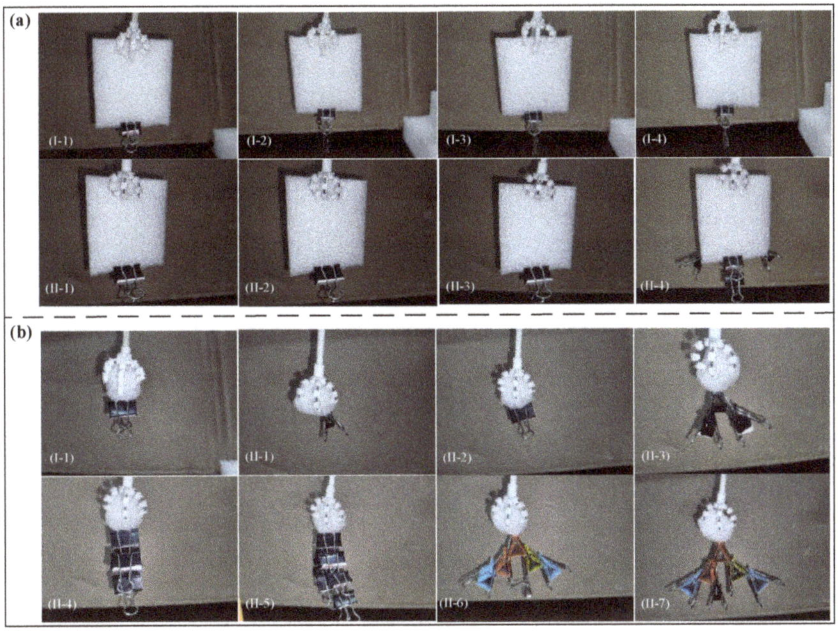

Figure 11. Load performance experiments of the two gripping mechanisms. (**a**) The load was applied to a rectangular base of 1.89 g for counterweight, (I-1)–(I-4) for the articulated gripping mechanism in order of 2.78 g, 3.4 g, 4.03 g, 4.35 g, (II-1)–(II-4) for the elastic repositioning gripping mechanism in order of 2.78 g, 4.97 g, 9.93 g, 14.57 g, 4.97 g, 9.93 g, 14.57 g. (**b**) Applied load to 0.34 g spherical base for counterweighting, I-1: for articulated gripping mechanism to apply object mass of 24.97 g, (II-1)–(II-7) for resilient repositioning gripping mechanism to apply object mass of 2.78 g, 4.97 g, 14.79 g, 24.55 g, 35.69 g, 43.36 g, 44.87 g.

6. Conclusions

This paper proposes two new flexible perching grasping mechanisms inspired by a bird's claw, with a unique shape design of "three in front and one at the back" and multiple degrees of freedom of the claw toe, simple structure and stable grasping. At the same time, the gripping mechanism is driven by two servos, and the fishing line transmits the gripping force, and the speed and gripping force of the servos for the object are controllable. After proposing the mechanical structure design scheme, the articulated perch gripping mechanism claw toe rotation gripping is analyzed, the analysis results are verified through experiments and the adaptability and load-bearing capacity of the two perch gripping mechanisms on objects of different shapes and sizes are tested. The maximum relative rotation angle between the parts of the articulated claw toe under no-load condition was consistent with the results of the claw toe rotation analysis, and the maximum deformation of the fishing line analyzed was almost reached when the claw toe reached the maximum degree of bending. The actual rotation angle of the rudder is about twice the result of the theoretical curve fitting because the claw toe needs sufficient gripping force while ensuring that it wraps around the branch as much as possible. The contact area between the base of the claw toe and the branch becomes larger and the envelope angle becomes smaller as the branch diameter increases for both perching grasping mechanisms. The claw toes of the two perching grasping mechanisms have good passive bending deformation ability and can grasp different types of objects, the toe tips play an important grasping role and the maximum object weight that the elastic reset type can grasp is about three times of its own weight. This imitation bird claw grasping mechanism effectively avoids the use of sticky pad adsorption, perching for a limited time and difficulty in separating the problem, the problems of limited perching time and easy separation that exist with the sticky pad attachment method. compared to the traditional imitation manual robotic perching grasping, and has obvious weight advantage and stable grasp and will not damage the object.

In comparison, the two types of perch gripping mechanism have their own advantages and disadvantages. The articulated type has a stronger degree of bending deformation, grips a larger range of branch diameters and, when gripping branches of the same diameter, the required servo deflection angle is greater for the resilient reset type than for the articulated type. After several gripping experiments, the resilient reset type has plastic deformation of the elastic joint, while the articulated type has no effect. However, the articulated version does not have as smooth and supple a toe bend curve as the resilient repositioning version, and has a significantly greater load capacity than the articulated version. In future research, the structural design of the articulated type will be improved so that the toe bending curve is smoother, and the elastic joint of the resilient repositioning type will be made of a material that is more elastic, more capable of self-recovery from deformation and requires less driving force during deformation and bending. The size of the model of the perching and grasping mechanism can be changed according to the actual needs, and it will be installed in the body of a microlight to improve the range in the future.

Supplementary Materials: The following supporting information can be downloaded at: https://www.mdpi.com/article/10.3390/machines10080656/s1, Video S1: Gripping tests with two perch gripping mechanisms.

Author Contributions: Revision, funding acquisition and conceptualization, Y.Z.; model design, analysis and writing—original draft, X.H.; physical assembly and experimental verification, X.H., G.G. and P.Z.; analysis of experimental results, X.H. and X.Z. All authors have read and agreed to the published version of the manuscript.

Funding: This work was supported by the National Natural Science Foundation of China (51005128), 2021 National Student Innovation and Entrepreneurship Training Program for College Students (No. 202110429008& No. 202110429011).

Institutional Review Board Statement: Not applicable.

Informed Consent Statement: Not applicable.

Data Availability Statement: Not applicable.

Conflicts of Interest: The authors declare no conflict of interest.

References

1. Shakhatreh, H.; Sawalmeh, A.H.; Al-Fuqaha, A.; Dou, Z.; Almaita, E.; Khalil, I.; Othman, N.S.; Khreishah, A.; Guizani, M. Unmanned Aerial Vehicles (UAVs): A Survey on Civil Applications and Key Research Challenges. *IEEE Access* **2019**, *7*, 48572–48634. [CrossRef]
2. Alzahrani, B.; Oubbati, O.S.; Barnawi, A.; Atiquzzaman, M.; Alghazzawi, D. UAV assistance paradigm: State-of-the-art in applications and challenges. *J. Netw. Comput. Appl.* **2020**, *166*, 102706. [CrossRef]
3. Pope, M.T.; Kimes, C.W.; Jiang, H.; Hawkes, E.W.; Estrada, M.A.; Kerst, C.F.; Roderick, W.R.T.; Han, A.K.; Christensen, D.L.; Cutkosky, M.R. A Multimodal Robot for Perching and Climbing on Vertical Outdoor Surfaces. *IEEE Trans. Robot.* **2017**, *33*, 38–48. [CrossRef]
4. Mehanovic, D.; Bass, J.; Courteau, T.; Rancourt, D.; Desbiens, A.L. Autonomous Thrust-Assisted Perching of a Fixed-Wing UAV on Vertical Surfaces. In *Conference on Biomimetic and Biohybrid Systems*; Springer: Cham, Switzerland, 2017; pp. 302–314. [CrossRef]
5. Nguyen, H.-N.; Siddall, R.; Stephens, B.; Navarro-Rubio, A.; Kovač, M. A Passively Adaptive Microspine Grapple for Robust, Controllable Perching. In Proceedings of the 2019 2nd IEEE International Conference on Soft Robotics (RoboSoft) (IEEE), Seoul, Korea, 14–18 April 2019; IEEE: Piscataway, NJ, USA, 2019; pp. 80–87.
6. Roberts, J.F.; Zufferey, J.C.; Floreano, D. Energy management for indoor hovering robots. In Proceedings of the 2008 IEEE/RSJ International Conference on Intelligent Robots and Systems, Nice, France, 22–26 September 2008; IEEE: Piscataway, NJ, USA, 2008; pp. 1242–1247.
7. Garcia-Rubiales, F.J.; Ramon-Soria, P.; Arrue, B.; Ollero, A. Magnetic detaching system for Modular UAVs with perching capabilities in industrial environments. In Proceedings of the 2019 Workshop on Research, Education and Development of Unmanned Aerial Systems (RED UAS), Cranfield, UK, 25–27 November 2019; IEEE: Piscataway, NJ, USA, 2019; pp. 172–176. [CrossRef]
8. Thomas, J.; Pope, M.; Loianno, G.; Hawkes, E.W.; Estrada, M.A.; Jiang, H.; Cutkosky, M.R.; Kumar, V. Aggressive Flight With Quadrotors for Perching on Inclined Surfaces. *J. Mech. Robot.* **2016**, *8*, 051007. [CrossRef]
9. He, Q.; Xu, X.; Yu, Z.; Huo, K.; Wang, Z.; Chen, N.; Sun, X.; Yin, G.; Du, P.; Li, Y.; et al. Optimized Bio-inspired Micro-pillar Dry Adhesive and Its Application for an Unmanned Aerial Vehicle Adhering on and Detaching from a Ceiling. *J. Bionic Eng.* **2020**, *17*, 45–54. [CrossRef]
10. Graule, M.A.; Chirarattananon, P.; Fuller, S.B.; Jafferis, N.T.; Ma, K.Y.; Spenko, M.; Kornbluh, R.; Wood, R.J. Perching and takeoff of a robotic insect on overhangs using switchable electrostatic adhesion. *Science* **2016**, *352*, 978–982. [CrossRef] [PubMed]
11. Park, S.; Drew, D.S.; Follmer, S.; Rivas-Davila, J. Lightweight HighVoltage Generator for Untethered Electroadhesive Perching of Micro Air Vehicles. *IEEE Robot. Autom. Lett.* **2020**, *5*, 4485–4492. [CrossRef]
12. Liu, Y.; Chen, H.P.; Tang, Z.M.; Sun, G.X. A bat-like switched flying and adhesive robot. In Proceedings of the 2012 IEEE International Conference on Cyber Technology in Automation, Control, and Intelligent Systems (CYBER), Bangkok, Thailand, 27–31 May 2012; IEEE: Piscataway, NJ, USA, 2012; pp. 92–97.
13. Huang, C.; Liu, Y.; Ye, X. Design, simulation and experimental study of a force observer for a flying–perching quadrotor. *Robot. Auton. Syst.* **2019**, *120*, 103237. [CrossRef]
14. Liu, S.; Dong, W.; Ma, Z.; Sheng, X. Adaptive Aerial Grasping and Perching with Dual Elasticity Combined Suction Cup. *IEEE Robot. Autom. Lett.* **2020**, *5*, 4766–4773. [CrossRef]
15. Pounds, P.E.I.; Bersak, D.R.; Dollar, A.M. Grasping from the air: Hovering capture and load stability. In Proceedings of the 2011 IEEE International Conference on Robotics and Automation (ICRA), Shanghai, China, 9–13 May 2011; IEEE: Piscataway, NJ, USA, 2011; pp. 2491–2498.
16. Thomas, J.; Polin, J.; Sreenath, K.; Kumar, V. Avian-Inspired Grasping for Quadrotor Micro UAVs. In Proceedings of the ASME 2013 International Design Engineering Technical Conferences and Computers and Information in Engineering Conference, Portland, OR, USA, 4–7 August 2013. [CrossRef]
17. Pounds, P.E.I.; Dollar, A. Hovering Stability of Helicopters with Elastic Constraints. In Proceedings of the ASME 2010 Dynamic Systems and Control Conference, Cambridge, MA, USA, 12–15 September 2010; pp. 781–788. [CrossRef]
18. Doyle, C.E.; Bird, J.J.; Isom, T.A.; Kallman, J.C.; Bareiss, D.F.; Dunlop, D.J.; King, R.J.; Abbott, J.J.; Minor, M.A. An Avian-Inspired Passive Mechanism for Quadrotor Perching. *IEEE/ASME Trans. Mechatron.* **2013**, *18*, 506–517. [CrossRef]
19. Bai, L.; Wang, H.; Chen, X.; Zheng, J.; Xin, L.; Deng, Y.; Sun, Y. Design and Experiment of a Deformable Bird-inspired UAV Perching Mechanism. *J. Bionic Eng.* **2021**, *18*, 1304–1316. [CrossRef]
20. Nadan, P.M.; Anthony, T.M.; Michael, D.M.; Pflueger, J.B.; Sethi, M.S.; Shimazu, K.N.; Tieu, M.; Lee, C.L. A Bird-Inspired Perching Landing Gear System1. *J. Mech. Robot.* **2019**, *11*, 061002. [CrossRef]
21. Proctor, N.S.; Lynch, P.J. *Manual of Ornithology: Avian Structure & Function*; Yale University Press: New Haven, CT, USA, 1993.
22. Bock, W.J.; Miller, W.D. *The Scansorial Foot of the Woodpeckers, with Comments on the Evolution of Perching and Climbing Feet in Birds*; American Museum of Natural History: New York, NY, USA, 1959.

23. Bock, W.J. Functional and evolutionary morphology of woodpeckers. *Ostrich* **1999**, *70*, 23–31. [CrossRef]
24. Roderick, W.R.; Chin, D.D.; Cutkosky, M.R.; Lentink, D. Birds land reliably on complex surfaces by adapting their foot-surface interactions upon contact. *eLife* **2019**, *8*, e46415. [CrossRef] [PubMed]
25. Chi, W.; Low, K.H.; Hoon, K.H.; Tang, J.; Go, T.H. A Bio-Inspired Adaptive Perching Mechanism for Unmanned Aerial Vehicles. *J. Robot. Mechatron.* **2012**, *24*, 642–648. [CrossRef]
26. Xie, C.; Wu, X.; Wang, X. A Three-row Opposed Gripping Mechanism with Bioinspired Spiny Toes for Wall-climbing Robots. *J. Bionic Eng.* **2019**, *16*, 994–1006. [CrossRef]
27. Pike, A.V.L.; Maitland, D.P. Scaling of bird claws. *J. Zool.* **2004**, *262*, 73–81. [CrossRef]
28. Fowler, D.; Freedman, E.A.; Scannella, J.B. Predatory Functional Morphology in Raptors: Interdigital Variation in Talon Size Is Related to Prey Restraint and Immobilisation Technique. *PLoS ONE* **2009**, *4*, e7999. [CrossRef] [PubMed]

machines

Article

DP-Climb: A Hybrid Adhesion Climbing Robot Design and Analysis for Internal Transition

Qingfang Zhang, Xueshan Gao *, Mingkang Li, Yi Wei and Peng Liang

School of Mechatronical Engineering, Beijing Institute of Technology, Beijing 100081, China
* Correspondence: xueshan.gao@bit.edu.cn

Abstract: This paper proposes a double propeller wall-climbing robot (DP-Climb) with a hybrid adhesion system based on the biomimetic design principle to address the problems of single adhesion-powered wall climbing robots (WCRs). Such problems include poor maneuverability and adaptability to orthogonal working surfaces with different roughness and flatness, weak flexibility of ground-wall transition motion, and easy stand stilling of transition. Based on the clinging characteristics of different creatures, the hybrid system combines the rotor units' reverse thrust, the drive wheels' driving torque, and the adhesion force offered by the coating material to power the robot through a coupled control strategy. Based on the Newton–Euler equations, the robot's kinematic characteristics during the ground-wall internal transition motion were analyzed, the safe adhesion conditions were obtained, and a dynamics model of the robot's ground-wall transition was established. This provided the basis for the coupling control between different power units. Finally, an internal transition PID control strategy based on DP-Climb was proposed. Through mechanical and aerodynamic characteristic experiments, it is verified that the robot's actual output pulling force can meet the transition motion demand. The experimental results show that the proposed strategy can enable the DP-Climb to complete the ground-wall mutual transition motion smoothly with a speed of 0.12 m/s. The robot's maximum wall motion speed can reach 0.45 m/s, which verifies that the hybrid adhesion system can flexibly and quickly reach the specified position in a target area flexibly and quickly. The robustness and adaptability of WCR to complex application environments are improved.

Keywords: hybrid adhesion system; wall-climbing robot; transition control strategy

Citation: Zhang, Q.; Gao, X.; Li, M.; Wei, Y.; Liang, P. DP-Climb: A Hybrid Adhesion Climbing Robot Design and Analysis for Internal Transition. *Machines* **2022**, *10*, 678. https://doi.org/10.3390/machines10080678

Academic Editor: Dan Zhang

Received: 8 July 2022
Accepted: 9 August 2022
Published: 10 August 2022

1. Introduction

As an important branch of special robots, wall-climbing robots (WCRs) can replace humans in high-risk or hard-to-reach environments to perform tasks such as environmental detection and location reconnaissance and assist in disaster rescue, which can ensure the safety of operators and reduce risks and costs. It has important application value in military and civilian fields [1,2]. For emergencies such as fires and earthquakes that occur in a dense urban high-rise building environment, it is particularly necessary to develop a WCR with strong environmental adaptability, the ability to move flexibly on walls of different roughness and damage, and to move freely within a designated target working area and to quickly reach the designated location to obtain the on-site information [3].

WCR has two basic abilities: adhesion and climbing. The stable adhesion of the robot to the wall mainly relies on its adhesion system. The adhesion method of WCR can be mainly divided into magnetic adhesion [4], vacuum suction cups [5], bionic materials and bionic hook and claw spine adhesion [6,7] and reverse thrust adhesion [8]. Magnetic WCR [9] has a strong adhesion force on the wall surface, but it can only be applied to metal contact surfaces, such as oil tanks. The vacuum suction cup-type WCR [10] can form a closed chamber to generate a vacuum with a strong adhesion force and can adapt to different material walls, but it can only be applied to a flat wall surface. The dry-adhesive bionic wall-climbing robot [11,12] can adapt to uneven vertical surfaces but requires frequent cleaning and maintenance, and its moving speed is limited. The hook-claw-type bionic WCR [13] can be used repeatedly, but it can damage the wall, and its adaptability to smooth walls such as glass is weak. Negative pressure-type WCR [14] can adapt to contact walls with different roughness and flatness, has good obstacle-crossing performance, can be reused, does not damage the building surface, and is more suitable for urban high-rise buildings and narrow alleyways and other environments. We compared the wall motion speed and weight properties of negative pressure adhesion WCRs [15–26] developed recently, as shown in Figure 1.

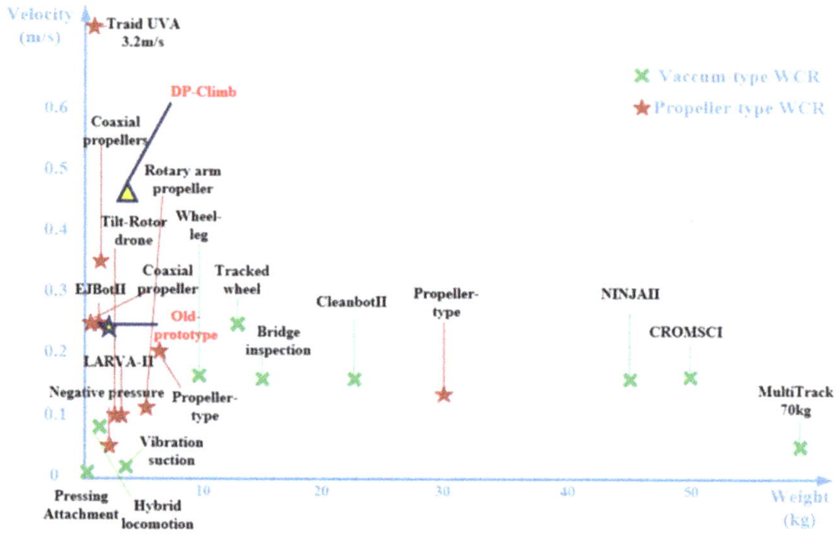

Figure 1. Property comparison of negative pressure WCRs.

According to Figure 1, propeller-type WCRs move faster on the wall and have a relatively smaller mass. Therefore, this paper designs a double propeller wall climbing robot based on the functional principle of bionic mechanical design to solve the problems of the current WCR. These include attitude adjustment that takes a long time during the transition of the ground-wall internal surface, the weak adaptability to the contact surface with different roughness and flatness, and the easy overturning of the body, which imitates the indirect flight mechanism of lepidopteran insects. When the rotor starts to rotate, the air will form a pressure difference at the upper and lower interfaces of the propeller to make the robot press against the wall. By changing the pitch angle of the rotor, the robot's motion direction can be changed to achieve wall forward and backward or static attachment action [27]. However, in the old version of WCR, which mainly uses the reverse thrust of the propeller and front drive wheels as the power, the robot appears to have a tail-flicking situation during the movement process, and it adopts an open-loop system to control the robot [28]. In practical applications, achieving accurate control of the robot on the wall surface is difficult. When the WCR needs to perform transition motion from the ground surface to the vertical wall surface, the robot is prone to appear the phenomenon of robot overturning due to stuck or excessive inclination. The lack of rolling joints of the rotor makes it difficult to meet the upward force necessary for the robot to turn and walk on the wall, making it easy to slip off. To solve this problem, Alkalla Mohamed G developed Erbo [29,30]. The two coaxial propellers provide sufficient adhesion force, but the robot motion is limited when the ground-wall transition is carried out by lifting the robot with an external joint. Myeong [31] used a multi-axis rotorcraft to hover on the wall after taking off from the ground. However, due to unpredictable wind gusts, the WCRs such as UVA will be difficult to land, and the internal drone attitude will be significantly affected in the event of a collision. With long lead times for precise positioning, the camera-mounted picture is difficult to capture, and it may even collide with the wall and damage the robot.

The flexible and fast movement of the WCR on the wall is also related to the movement mechanism of the robot. The commonly used movement methods can be classified as wheeled [32,33], crawler [34] and legged [35]. The crawler-type WCR moves faster, but the weight is generally heavier, and the crawler can cause malfunctions when encountering small obstacles. At the same time, legged structures have better obstacle-crossing performance, but the flexibility is weak, and the control is relatively complicated. Therefore, the WCR designed in this paper simplifies the limbs of the gecko into a wheeled mechanism to realize the functions of rapid movement and positioning of the robot.

This paper will design a double propeller wall climbing robot(DP-Climb) with a hybrid adhesion system that can be internally transitioned based on the principle of biomimetic design from the perspective of robot dynamics. It aims to enhance the mobility and adaptability of WCR to complex application environments, make up for the shortcomings of a single adhesion-powered WCR [36], reduce the time consumed for an attitude adjustment, and increase the endurance of the robot. The hybrid adhesion system of DP-Climb combines the clinging characteristics of different creatures. It is mainly composed of two symmetrical rotor units that imitate flapping wings and two rear drive wheels that imitate animal hind limbs. At the same time, a coating material that can enhance the friction coefficient between the wheel body and the wall surface is used on the wheel body to imitate the adhesion characteristics of the gecko feet. The three coupling controls jointly provide the adhesion and traction force required by the robot movement so that the robot can have a certain obstacle crossing ability, and the robot can move flexibly and quickly on contact surfaces of different materials and flatness without being disturbed by factors such as wall cracks, surface protrusions or holes. In addition, the DP-Climb hybrid adhesion system can help the robot adjust the ground-wall transition attitude quickly, reduce energy consumption, improve the endurance of the robot, shorten the positioning time, and enhance its stability. The main contributions of this study are as follows:

(1) A WCR with a hybrid adhesion system is designed based on the principle of biomimetic design. To solve the above problems, this manuscript designs DP-Climb, a double propeller wall climbing robot with a hybrid adhesion system. The robot uses symmetrical rotor bionic joints to adjust the pitch and roll angle of the rotor joints in real-time according to the robot's attitude. It also adopts the skeleton design to optimize the rotor fixed disk to reduce the airflow loss, utilizes the hind limb drive bionic joints to reduce the tailing of the robot and applies the coating material to improve the adhesion force of the robot. It obtains the ground-wall dynamic model of the robot based on the Newton–Euler equation. The dynamic model provides a theoretical basis for the transition motion control of the robot.

(2) A PID motion control strategy for the ground-wall internal transition based on DP-Climb is proposed. The paper designs the ground-wall PID control strategy of the robot based on the dynamic model obtained. It also realizes the autonomy of the robot through PID closed-loop control, which effectively improves the robot's autonomous motion capability and control accuracy and stability. To obtain the dynamic characteristics of the key power unit of the robot, an experimental platform for the mechanical characteristics and aerodynamic characteristics is designed, and the appropriate wheel surface coating material is selected through the tensile force test experiment and the repeatability test experiment of the attachment materials. The functional relationship between the robot's rotor power unit's current output and the actual tensile force output is obtained to compensate for the adhesion dynamic error caused by the airflow loss. The feasibility of the transition control strategy of DP-Climb with a hybrid adhesion system is verified by simulations and experiments.

The structure of this paper is as follows. Section 2 introduces the design scheme of the robot adhesion system. Section 3 establishes the dynamic model of the robot's transition motion process and discusses the control strategy of the robot's ground-wall internal transition in detail. Section 4 presents the simulation and experimental results to verify that DP-Climb can fast transition between the ground and the wall. Finally, the summary conclusions and an outlook for future work are provided in Section 5.

2. Materials and Methods

The DP-Climb adhesion system is designed based on the principle of biomimetic design using a hybrid dynamic adhesion system. The hybrid system consists of the reverse thrust generated by the rotor units, the driving force provided by the drive wheels, and the adhesion force provided by the coating material, which together powers the robot through a coupled control strategy. The hybrid adhesion mechanism of DP-Climb is shown in Figure 2.

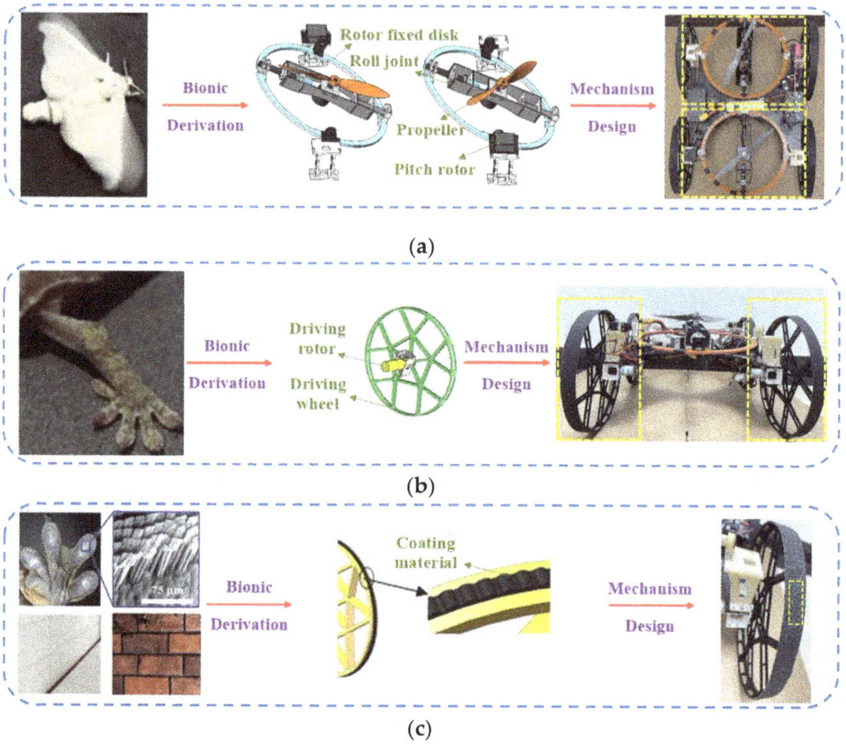

Figure 2. Design of DP-Climb hybrid adhesion system: (**a**) symmetrical rotor bionic joints; (**b**) hind limb drive bionic joints (**c**) a coating material to enhance the friction.

Lepidopteran insects such as moths fly indirectly in the air and change the direction of airflow by changing their own pitch angle. Therefore, to further improve the effective adhesion force of the rotor system, after selecting the 10-inch propellers and 1250 KV type brushless motors, two symmetrical rotor joints were designed to be mounted on the rotor fixed disk with reference to the flutter wing configuration. The rotor joints can change the reverse thrust of the robot through roll joints and pitch rotors according to the different wall postures of the robot to provide the corresponding adhesion force and traction force, as shown in Figure 2a.

When animals such as geckos transition from the ground to the wall, their limbs, especially the hind limbs, need to provide sufficient upward force. Through the technical means of bionic derivation, the limbs of the gecko can be simplified into a wheel body that can move flexibly, and the rear-wheel drive is used to provide the driving torque for the robot to climb upward. The structure is shown in Figure 2b.

In addition, the gecko has an extraordinary climbing ability through many multi-scale hierarchical structures. The Van der Waals force and other types of noncovalent forces (such as capillary forces) enable it to adhere to different types of contact surfaces, and many dry adhesive materials have been developed for climbing [37,38]. A special coating material was added to the wheel's surface to enhance the friction coefficient between the robot and the wall surface and generate a larger friction force to assist the robot in moving on the wall surface, as indicated in Figure 2c.

3. Results

3.1. Analysis of Safe Adhesive Conditions for DP-Climb Transition

WCR works mainly on different building surfaces in cities. Therefore, the robot needs to achieve stable adhesion on walls with different materials and flatness while having sufficient mobility to walk between the ground and the vertical wall surfaces.

To achieve this purpose, it is first necessary to analyze the safe adhesion conditions of the WCR ground-wall transition process. A three-dimensional global coordinate system $\{XYZO\}$ and a three-dimensional robot coordinate system $\{xyzc\}$ are established. c is the centroid of the robot, α_{10} and α_{20} represent the angle between the rotor fixed disk and the ground, respectively, α_1 and α_2 represent the angle between the reverse thrust of the rotor and the robot body, respectively, β is the pitch angle of the robot body, αs_1 and αs_2 indicate the pitch angle of the front and rear rotor fixed disks relative to the robot body, respectively, and v is the movement direction of the robot. The kinematic characteristics of DP-Climb in the ground-wall transition are shown in Figure 3.

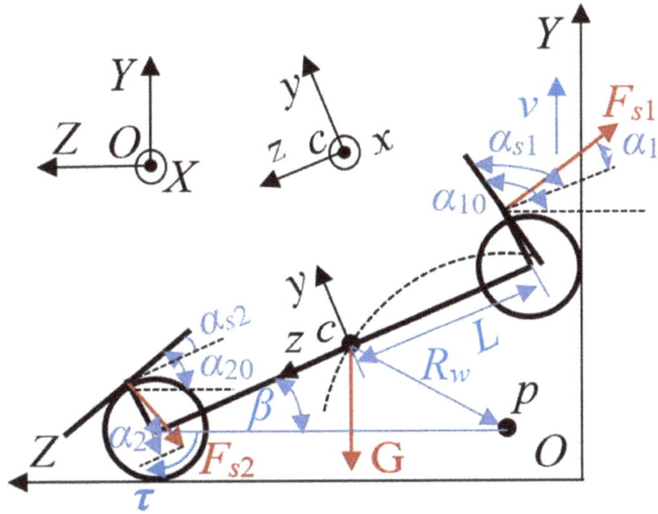

Figure 3. Kinematics model of DP-Climb internal transition.

Assume that the longitudinal and lateral slips of the robot are not considered under pure rolling conditions. It is assumed that (1) the body of DP-Climb is rigid and wheel deformation is neglected; (2) the contact wall of the robot is solid and wall deformation is neglected, and the four wheels are always in vertical contact with the contact surface, and there is only one contact point between each wheel and the surface; (3) the gravity distribution of the contact points between wheels and the surface is uniform; and (4) the robot is in a stable motion state.

The problem of adhesion conditions for the transition motion of DP-Climb can be transformed into the sliding problem of the ladder. In the process of robot ground-wall transition, DP-Climb is symmetrically arranged, c is the center of the robot's mass, p is the center of the robot's transition motion circle, and R_w is the radius of the robot's transition motion. It can be concluded that the slope of the motion trajectory at c is the time-varying situation of the robot's pitch angle β during the ground-wall transition process. The motion trajectory of point c is:

$$(y_c - R_w)^2 + (z_c - R_w)^2 = L^2, \tag{1}$$

where, (z_c, y_c) is the coordinate of the centroid c in the coordinate plane YZO, and the pitch angle β of DP-Climb during the whole ground-wall transition can be obtained as follows:

$$tan\beta = \frac{R_w - z_c}{y_c - R_w}, \tag{2}$$

where, in the process of robot ground-wall transition, i.e., $\beta \in (0, 90)^\circ$, $0^\circ \leq |\alpha_1, \alpha_2| \leq 90^\circ$, the robot pitch angle $\tan\beta$ is a continuous derivable function with respect to time t. When the positive pressure of the reverse thrust on the wall surface completely provides adhesion force for the robot, $\alpha_1, \alpha_2 = 0^\circ$, when the reverse thrust completely provides upward lift force for the robot, $\alpha_1, \alpha_2 = 90^\circ$. By analyzing the motion characteristics of DP-Climb, the safe adhesion conditions for the robot to cross the dead point in the ground-wall transition motion without front wheel overturning can be obtained:

$$\begin{cases} \theta_1 = \alpha_1 + \beta = \alpha_{10} - 90 = \alpha_{S1} - 90 + \beta \\ \theta_2 = \alpha_2 - \beta = 90 - \alpha_{20} = 90 - \alpha_{S2} - \beta \\ F_{S1}\cos\theta_1 > F_{S1}\sin\theta_1 > 0, (\theta_1 > 0) \\ F_{S1}\cos\theta_1 \geq F_{S1}\sin\theta_1 > 0, (\theta_1 = 0) \\ |F_{S2}\cos\theta_1| \geq |F_{S2}\sin\theta_1| > 0 \end{cases}. \tag{3}$$

In the robot ground-wall transition process, the front rotor platform plays a critical role. The front rotor platform of the robot needs to provide sufficient upward traction while ensuring a necessary adhesion force. Therefore, to ensure DP-Climb safety, the pitch angle of the robot needs to comply with $100^\circ \leq \alpha_{10} \leq 130^\circ$ during the transition process. If α_{10} is too small, the robot will not receive enough upward component force, and the robot will be stuck and cannot continue the ground-wall transition. If α_{10} is too large, the robot will overturn during the transition movement according to the torque at the origin point O around the X-axis.

3.2. Dynamic Modeling of DP-Climb Internal Transition

Due to the particularity of DP-Climb working on perpendicular intersecting planes, the adhesion force of the robot is easily affected by the wall environment, and the wall motion resistance also changes accordingly, which changes the wall adhesion conditions and kinematic characteristics of the robot. To obtain a more accurate dynamic model of the robot's internal transition motion, it is necessary to analyze the adhesion force, the driving torque, and the friction state between the driving wheel and the wall surface of the robot to provide a theoretical basis for the robot's motion control.

The support force of each wheel on the wall surface is not only affected by the robot's reverse thrust, but the robot's gravity and inertial force will also generate a moment on the center of mass c when the robot moves on the wall surface. Therefore, the effects of gravity, inertial force and reverse thrust should be considered separately. Then, the three solved supporting force components are superimposed to obtain the resultant force of the supporting force of each wheel to solve the dynamic equation of the motion of the centroid c of the robot on the surface. Figure 4 shows the force state analysis of DP-Climb during the ground-wall transition movement.

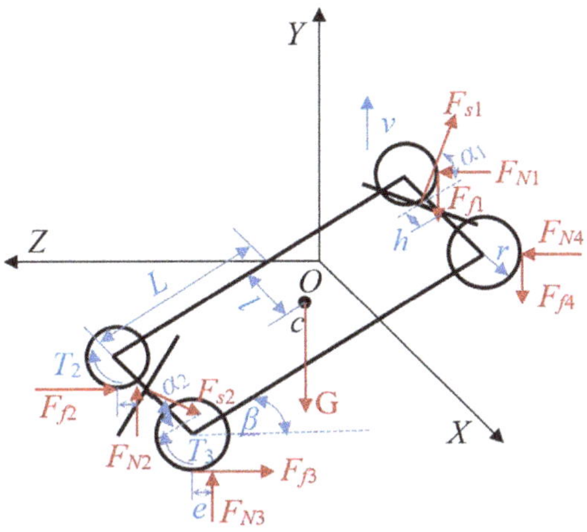

Figure 4. Dynamic analysis of DP-Climb internal transition.

2 L represents the length of the robot's body, which is also the distance between the front and rear wheels, 2 l represents the left and right wheel distances, h represents the vertical distance between the rotor fixed platform and the body, r represents the wheel radius, and e/r represents the wheel rolling resistance coefficient. F_{s1} and F_{s2} represent the reverse thrust provided by the front and rear rotor units of the robot, respectively, F_{Ni} (i = 1, 2, 3, 4) represents the positive pressures of each wheel, F_{fi} represents the sliding friction force on each wheel, G represents the total gravity of the robot, $\boldsymbol{\tau}$ is the driving torque of the driving wheels, which is $[T_2\ T_3]$, and T_2 and T_3 are the driving torques of the left and right rear wheels, respectively. Simplifying the wheel model of the robot, the longitudinal adhesion coefficient of each wheel on the wall is μ_i, and M_{fi} represents the rolling resistance moment of each wheel. Since the value of e is very small, it can be neglected in the positive pressure analysis of each wheel. The resultant forces of DP-Climb along each coordinate axis of $\{XYZO\}$ are:

$$\begin{cases} m\ddot{y} = -\mathrm{G} + F_{s1}\sin(\alpha_1+\beta) - F_{s2}\sin(\alpha_2-\beta) + (F_{N2}+F_{N3}) - (\mu_1 F_{N1} + \mu_4 F_{N4}) \\ m\ddot{z} = -F_{s1}\cos(\alpha_1+\beta) - F_{s2}\cos(\alpha_2-\beta) + (F_{N1}+F_{N4}) - (\mu_2 F_{N2} + \mu_3 F_{N3}) \\ (F_{N1}-F_{N4})l + (\mu_3 F_{N3} - \mu_2 F_{N2})l = 0 \\ (F_{N3}-F_{N2})l + (\mu_1 F_{N1} - \mu_4 F_{N4})l = 0 \end{cases}. \quad (4)$$

If the robot does not slip sideways and flip sideways, then $\mu_a = \mu_1 = \mu_4$ and $\mu_b = \mu_2 = \mu_3$. The dynamic equation of DP-Climb in the robot coordinate system $\{xyzc\}$ can be obtained:

$$\begin{cases} ma_{yc} = & F_{sl}\sin\alpha_l - F_{s2}\sin\alpha_2 - \mathrm{G}\cos\beta - 2(\mu_a\cos\beta - \sin\beta)F_{Nl} - 2(\mu_b\sin\beta - \cos\beta)F_{N2} \\ ma_{zc} = & -F_{sl}\cos\alpha_l - F_{s2}\cos\alpha_2 + \mathrm{G}\sin\beta + 2(\mu_a\sin\beta + \cos\beta)F_{Nl} - 2(\mu_b\cos\beta + \sin\beta)F_{N2} \\ m\ddot{y} = & ma_{zc}\sin\beta \\ m\ddot{z} = & ma_{zc}\cos\beta \\ J\ddot{\beta} = & (F_{N1}+F_{N4})L\sin\beta - (F_{N2}+F_{N3})L\cos\beta \\ & +\mu_b(F_{N2}+F_{N3})(L\sin\beta + r) - \mu_a(F_{N1}+F_{N4})(L\cos + r) \\ & -F_{s1}\cos(\alpha_1+\beta)(L\sin\beta + h\cos\beta) + F_{s2}\cos(\alpha_2-\beta)(L\sin\beta - h\cos\beta) \\ & +F_{s1}\sin(\alpha_1+\beta)(L\cos\beta + h\sin\beta) + F_{s2}\cos(\alpha_2-\beta)(L\cos\beta + h\sin\beta) \\ & +\left(M_{f2}+M_{f3}\right) + \left(M_{f1}+M_{f4}\right) \end{cases}, \quad (5)$$

where, I represents the total moment of inertia of the robot. $a_{yc} = 0$. Combining Equations (4) and (5), the positive pressure of robot wheels can be obtained as:

$$\begin{cases} a_1 = \mu_a \cos\beta + \sin\beta,\ a_2 = \mu_b \sin\beta + \cos\beta \\ a_3 = \sin\theta_1 \cos\beta + \cos\theta_1 \sin\beta,\ a_4 = \sin\theta_2 \cos\beta - \cos\theta_2 \sin\beta \\ b_1 = \mu_a \cos\beta - \sin\beta,\ b_2 = \mu_b \sin\beta - \cos\beta,\ b_3 = \sin 2\beta(\mu_b\mu_a - 1) \\ F_{N4} = F_{N1} = \frac{1}{2}\begin{bmatrix} \left(\frac{a_3}{a_1} + \frac{a_2 \sin\alpha_1}{b_3} - \frac{a_2 a_3 b_1}{a_1 b_3}\right)F_{s1} - \left(\frac{a_4}{a_1} + \frac{a_2 \sin\alpha_2}{b_3} - \frac{a_2 a_4 b_1}{a_1 b_3}\right)F_{s2} \\ -\left(\frac{1}{a_1} + \frac{a_2}{b_3} - \frac{a_2 b_1}{a_1 b_3}\right)\mathrm{G}\cos\beta \end{bmatrix} \\ F_{N3} = F_{N2} = \frac{1}{2}\left[\left(\frac{a_1 \sin\alpha_1}{b_3} - \frac{a_3 b_1}{b_3}\right)F_{s1} - \left(\frac{a_2 \sin\alpha_2}{b_3} - \frac{a_4 b_1}{b_3}\right)F_{s2} - \left(\frac{a_1}{b_3} - \frac{b_1}{b_3}\mathrm{G}\cos\beta\right)\right] \end{cases}. \quad (6)$$

Since the robot performs the ground-wall transition motion at an almost uniform speed, the angular acceleration of each wheel can be approximated to 0. According to the Newton–Euler equations, the dynamic model of DP-Climb can be obtained as:

$$\begin{cases} c_1 & = \sin\theta_1 - \mu_a \frac{a_3}{a_1} - \frac{\mu_a a_2 \sin\alpha_1 - a_1 \sin\alpha_1 + a_3 b_1}{b_3} + \frac{\mu_a a_2 a_3 b_1}{a_1 b_3} \\ c_2 & = \sin\theta_2 - \mu_a \frac{a_4}{a_1} - \frac{\mu_a a_2 \sin\alpha_2 - a_2 \sin\alpha_2 - a_4 b_1}{b_3} + \frac{\mu_a a_2 a_4 b_1}{a_1 b_3} \\ c_3 & = \left[1 - \frac{\mu_a \cos\beta}{a_1} - \frac{\mu_a a_2 - a_1 - b_1}{b_3}\cos\beta + \frac{\mu_a a_2 b_1}{a_1 b_3}\cos\beta\right] \\ c_4 & = -\cos\theta_1 + \frac{a_3}{a_1} + \frac{a_2 \sin\alpha_1}{b_3} - \frac{a_2 a_3 b_1}{a_1 b_3} \\ c_5 & = \cos\theta_2 + \frac{a_4}{a_1} + \frac{a_2 \sin\alpha_2}{b_3} - \frac{a_2 a_4 b_1}{a_1 b_3} \\ c_6 & = \frac{1}{a_1} + \frac{a_2}{b_3} - \frac{a_2 b_1}{a_1 b_4} \\ d_1 & = \sin\theta_1(L\cos\beta - h\sin\beta) - \cos\theta_1(L\sin\beta + h\cos\beta) + \left(\frac{a_3}{a_1} + \frac{a_2 \sin\alpha_1}{b_3} - \frac{a_2 a_3 b_1}{a_1 b_3}\right)L\sin\beta \\ & -\left(\frac{a_1 \sin\alpha_1}{b_3} - \frac{a_3 b_1}{b_3}\right)L\cos\beta - \left(\frac{a_3}{a_1} + \frac{a_2 \sin\alpha_1}{b_3} - \frac{a_2 a_3 b_1}{a_1 b_3}\right)(\mu_a L\cos\beta + \mu_a r) \\ & +\left(\frac{a_3}{a_1} + \frac{a_2 \sin\alpha_1}{b_3} - \frac{a_2 a_3 b_1}{a_1 b_3}\right)\frac{e}{r} + \left(\frac{a_1 \sin\alpha_1}{b_3} - \frac{a_3 b_1}{b_3}\right)\frac{e}{r} \\ d_2 & = \sin\theta_2(L\cos\beta + h\sin\beta) - \cos\theta_2(L\sin\beta - h\cos\beta) - \left(\frac{a_4}{a_1} + \frac{a_2 \sin\alpha_2}{b_3} - \frac{a_2 a_4 b_1}{a_1 b_3}\right)L\sin\beta \\ & +\left(\frac{a_2 \sin\alpha_2}{b_3} - \frac{a_4 b_1}{b_3}\right)L\cos\beta + \left(\frac{a_4}{a_1} + \frac{a_2 \sin\alpha_2}{b_3} - \frac{a_2 a_4 b_1}{a_1 b_3}\right)(\mu_a L\cos\beta + \mu_a r) \\ & -\left(\frac{a_4}{a_1} + \frac{a_2 \sin\alpha_2}{b_3} - \frac{a_2 a_4 b_1}{a_1 b_3}\right)\frac{e}{r} - \left(\frac{a_1 \sin\alpha_2}{b_3} - \frac{a_4 b_1}{b_3}\right)\frac{e}{r} \\ d_3 & = -\left(\frac{1}{a_1} + \frac{a_2}{b_3} - \frac{a_2 b_1}{a_1 b_3}\right)\left[L\sin\beta - \mu_a(L\cos\beta + r) + \frac{e}{r}\right] + \left(\frac{a_1}{b_3} - \frac{b_1}{b_3}\right)\left(L\cos\beta - \frac{e}{r}\right) \\ m\ddot{y} & = c_1 F_{s1} - c_2 F_{s2} - c_3 \mathrm{G} \\ m\ddot{z} & = c_4 F_{s1} - c_5 F_{s2} - c_6 \cos\beta \mathrm{G} - \frac{T_2 + T_3}{r + e} \\ I\ddot{\beta} & = d_1 F_{s1} - d_2 F_{s2} - d_3 \cos\beta \mathrm{G} + \frac{T_2 + T_3}{r + e}(L\sin\beta + r) \end{cases} \quad (7)$$

The expression of the DP-Climb dynamic model can be obtained as follows:

$$\boldsymbol{M}(q)\ddot{\boldsymbol{q}} + \boldsymbol{E}(q)\boldsymbol{F}_s + \boldsymbol{D}\mathrm{G} = \boldsymbol{B}(q)\boldsymbol{\tau}, \quad (8)$$

where, $\boldsymbol{q} = [y\ z\ \beta]^{\mathrm{T}}$, $\boldsymbol{Fs} = [F_{s1}\ F_{s2}]^{\mathrm{T}}$. Each coefficient matrix is:

$$\boldsymbol{M}(q) = \begin{bmatrix} m & 0 & 0 \\ 0 & m & 0 \\ 0 & 0 & I \end{bmatrix},\ \boldsymbol{E}(q) = \begin{bmatrix} -c_1 & c_2 \\ -c_4 & c_5 \\ -d_1 & -d_2 \end{bmatrix},\ D = \begin{bmatrix} c_3 \\ c_6\cos\beta \\ d_3\cos\beta \end{bmatrix},\ \boldsymbol{B}(q) = \begin{bmatrix} 0 & 0 \\ -\frac{1}{r+e} & -\frac{1}{r+e} \\ \frac{L\sin\beta + r}{r+e} & \frac{L\sin\beta + r}{r+e} \end{bmatrix}$$

3.3. Control Strategy of DP-Climb Internal Transition

3.3.1. Selection of Wheel Surface Adhesive Material

According to the analysis in the previous section, the adhesion coefficient between the wheel and contact surface is one of the important factors affecting the adhesion of the robot to the wall surface. Therefore, to increase the adhesion between the robot and the contact surface, it is necessary to choose a material with good repeatability and a high friction coefficient as the robot wheel body surface adhesion coating.

Through the collection of various coating materials suitable for the surface of the wheel body and the analysis of the material characteristics, this paper selects several adhesion materials that can meet the task requirements of the robot with better performance for comparison. In a windless indoor environment, an experimental platform for mechanical characteristics was set up to test the upward lift force of the robot and the adhesion force perpendicular to the wall surface using a lime medium wall as the contact surface. The established dynamic model was used to measure the friction coefficients of different adhesive materials on the wheel surface of the robot. To test the repetition rate of the adhesive materials, 10 tensile tests were repeated for each material. The experimental platform for mechanical properties is shown in Figure 5, and the experimental results are shown in Table 1.

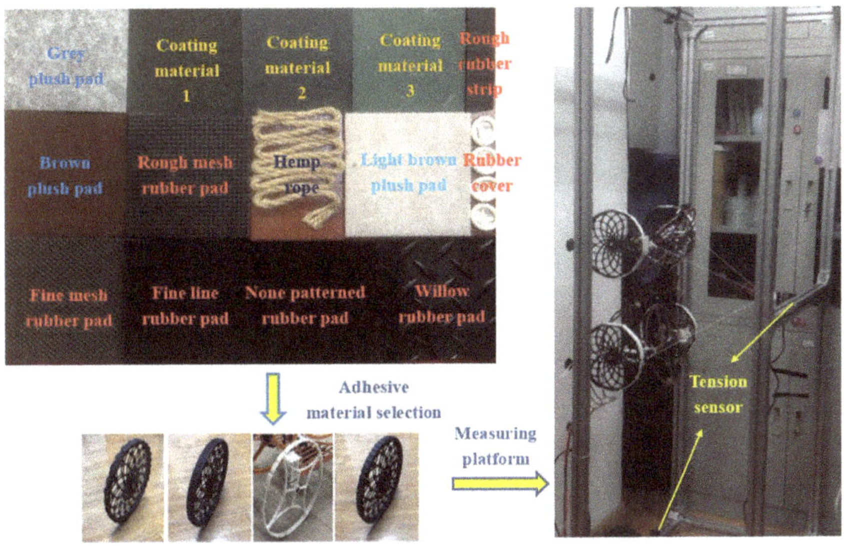

Figure 5. Selection of different adhesive materials for wheel surfaces.

Table 1. Comparison of material properties of different adhesive materials for wheel surface.

Adhesive Material	Coefficient of Friction	Repetition Rate	Adhesive Material	Coefficient of Friction	Repetition Rate
Gray plush pad	0.55	0.90	Hemp rope	0.46	1.00
Coating material 1	0.81	1.00	Light brown plush pad	0.40	0.90
Coating material 2	0.77	0.80	Rubber cover	0.75	0.40
Coating material 3	0.81	0.50	Fine mesh rubber pad	0.40	1.00
Rough rubber strip	0.56	1.00	Fine line rubber pad	0.56	1.00
Brown plush pad	0.54	1.00	Non-patterned rubber pad	0.55	1.00
Rough mesh rubber pad	0.45	1.00	Willow rubber pad	0.59	1.00

Analysis of Table 1 shows that coating material 1 has a relatively large friction coefficient and a high repetition rate, and the comprehensive performance is the best. Therefore, this material will be used as the adhesion material on the surface of the robot wheel body in subsequent research work.

3.3.2. Control Strategy of DP-Climb Ground-Wall Transition

Equation (7) shows that when the robot performs the ground-wall transition at a nearly uniform speed, $m = 3.3$ kg, $r = 0.14$ m, $L = 0.18$ m, $h = 0.08$ m, $l = 0.23$ m, and $\mu_a = 0.81$. During the DP-Climb transition process, the robot can complete the transition motion by changing the pitch joint angle of the front and rear rotor platforms of the robot according to the pitch angle of the fuselage. According to the DP-Climb dynamic model analysis, the relationship between the pitch angle of the rotor platform and the pitch angle of the fuselage is as follows:

$$\begin{cases} \alpha_{s1} = \alpha_{10} - \beta \\ \alpha_{s2} = 180 - \alpha_{10} \end{cases}. \tag{9}$$

The red vectors represent the reverse thrust provided by the front and rear rotor units, F_{s1} and F_{s2}, respectively, and the blue angle indicates the robot pitch angle β. Based on the stable adhesion conditions of the ground-wall transition motion of the robot analyzed above, the robot's ground-wall internal transition motion strategy is designed, as shown in Figure 6.

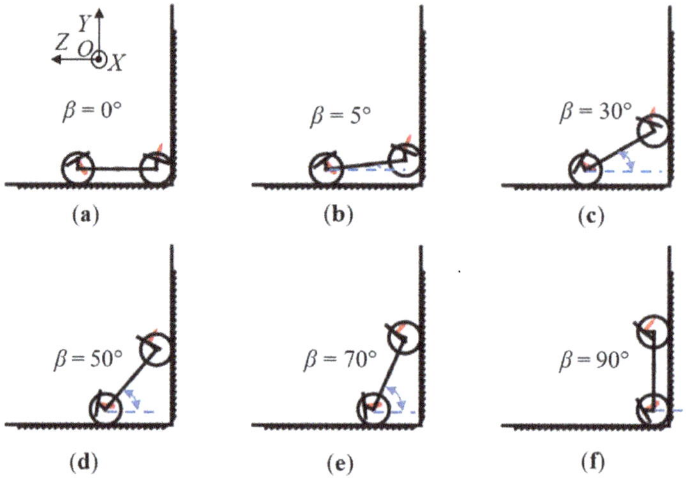

Figure 6. Control strategy of DP-Climb internal transition: (**a**) start transition movement; (**b**) the front wheels raised; (**c**) the angle of the front rotor unit corresponds to the change to ensure an upward force; (**d**) the pitch angle of the body increases, and the adhesion force of the front rotor unit increases; (**e**) the pitch angle of the body exceeds 60 degrees, and the rotor units mainly provide adhesion force; (**f**) complete transition tasks and start wall missions.

When the ranging sensor detects that DP-Climb meets the wall, the robot starts to perform a ground-wall transition motion: the front wheels of the robot start to lift, the front and rear rotor platforms change the pitch angle, and the front rotor unit mainly provides the upward lift force to the robot, the rear rotor unit mainly provides the adhesion force, and the driving wheels mainly provide the power for the robot to move forward on the ground. Then the pitch angle of the fuselage increases and the pitch angle of the front rotor platform decreases correspondingly so that the front rotor platform always provides the upward lift force and the adhesion force pointing to the wall. When the fuselage pitch angle β exceeds 30°, the front rotor platform angle decreases, and the adhesion force provided by the front rotor unit toward the wall increases. When the fuselage pitch angle β exceeds 50°, the front and rear rotor units mainly provide adhesion force pointing to the wall, and the driving wheels mainly provide the power to move forward. Finally, the robot completes the transition task and starts wall operation tasks. Figure 7 shows the pitch angle variation of the robot's front and rear rotor joints in the DP-Climb ground-wall transition control strategy.

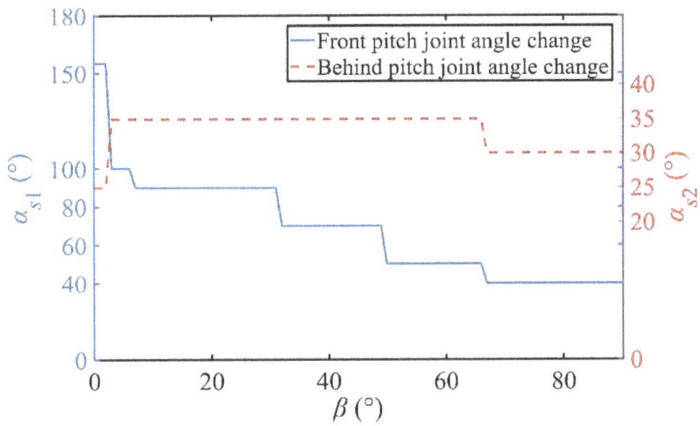

Figure 7. Variation of the pitch angle of the front and behind rotor joints of DP-Climb.

To maintain the stability of the hybrid system and reduce motion error, we introduced the control strategy into the control system. We selected the k_p, k_i and k_d coefficients of the PID controller based on the strategy to adjust the attitude of the robot. v_r and β_r are the desired speed of the robot and the desired pitch angle of the robot, respectively, based on the DP-Climb dynamic model established. The control system of DP-Climb and the PID controller is shown in Figure 8 and Table 2, respectively.

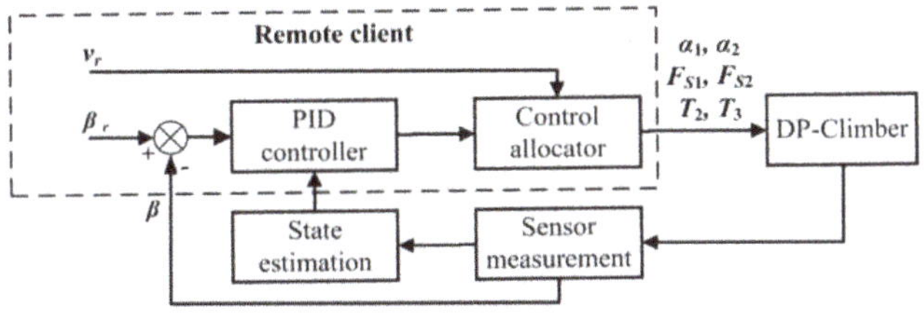

Figure 8. Control system of DP-Climb.

Table 2. k_p, k_i and k_d coefficients selected in the process of ground-wall transition.

Critical Transition Angle β (°)	k_p	k_i	k_d
0	1.80	7.00	1.50
5	1.80	7.60	1.90
30	1.80	7.20	1.90
50	1.80	7.20	1.60
70	1.80	7.00	1.60
90	1.80	7.00	1.50

The inner loop updates the attitude of the robot, the outer loop feedback control reduces the error, the sampling frequency of the inner loop is 10 Hz, and the sampling frequency of the outer loop is 5 Hz.

In summary, in the robot ground-wall transition process, the robot's head-up action mainly relies on the upward lift force provided by the front rotor platform. As the pitch angle of the fuselage increases, α_{s1} gradually decreases, and the adhesion force provided by F_{s1} gradually increases accordingly. The front rotor platform's pitch angle greatly influences the robot's motion, and the changes during the transition motion of the robot are relatively obvious. Thus, the effective output pulling force of the robot rotor unit needs to be analyzed.

3.3.3. Aerodynamic Characteristics of the DP-Climb Adhesion System

DP-Climb performs the transition task, and the wall task mainly relies on the power provided by the adhesion system, according to the analysis above. Based on the established dynamics model of DP-Climb, the output pulling force of the adhesion system required for different pitch angles in transition motion can be obtained. To compensate for the error caused by system interference and airflow disturbance, it is necessary to analyze the actual output of the pulling force of the rotor unit.

First, the robot was fixed by a fixture, and an external current sensor was used to measure the real-time current output of a single rotor unit. A host computer modulated the pulse width of the electronic speed controller of the rotor motor to control the motor speed. Finally, the measured data were collected and analyzed by wireless signal transmission to obtain the pulling force output model of the rotor unit. The aerodynamic characteristics measurement platform of the rotor unit is shown in Figure 9.

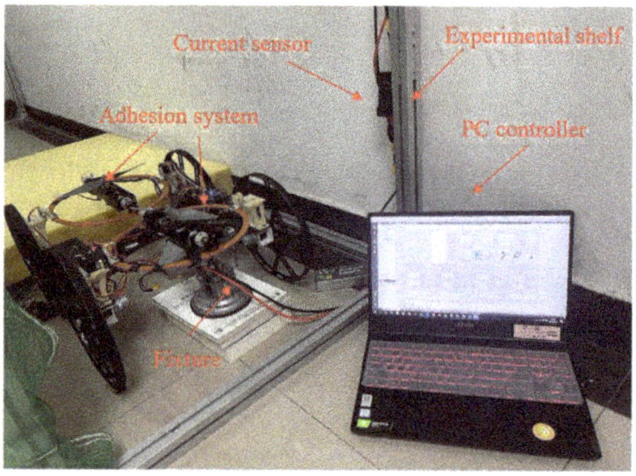

Figure 9. Aerodynamic characteristics measurement platform.

After processing the measured data by linear interpolation, the relationship between the actual current signal of a single rotor unit, the reverse thrust of the rotor and the motor speed can be obtained, as shown in Figure 10.

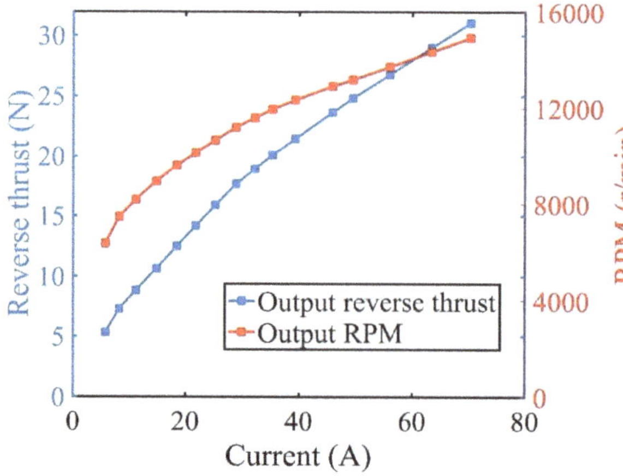

Figure 10. Aerodynamic characteristics of rotor unit.

Since there are two symmetrical rotor units in the robot's adhesion system, there is a certain interference situation when the rotor units act simultaneously, causing decreased pulling force output. Therefore, further analysis of the aerodynamic coupling characteristics of the two rotor units is needed. The RPM of the front and rear rotors were set to 10,653 r/min and 6362 r/min, respectively. Six key nodes of the robot attitude change in the ground-wall transition were selected to simulate the aerodynamic characteristics of the adhesion system at the transition nodes of 0°, 5°, 30°, 50°, 70° and 90° of the fuselage pitch angle. The simulation results of the aerodynamic characteristics of the adhesion system are shown in Figure 11.

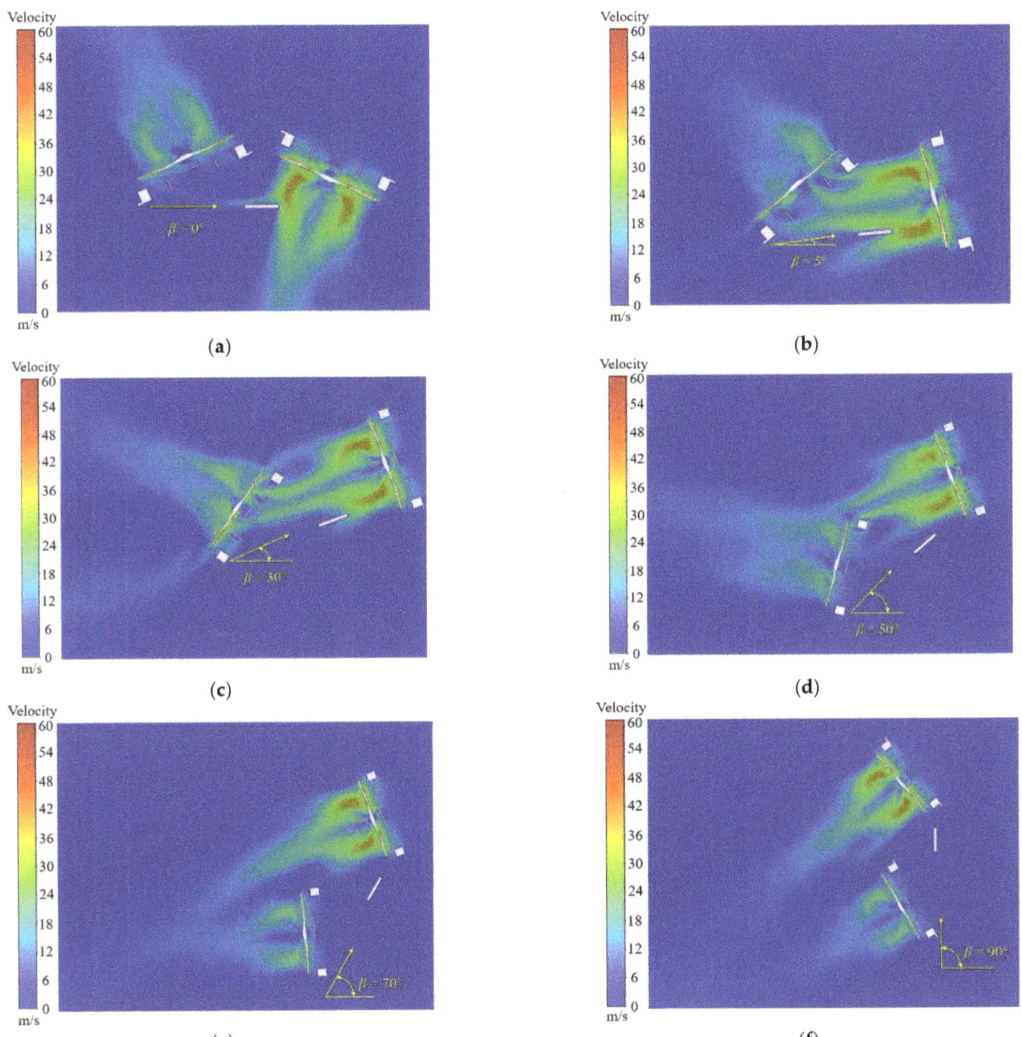

Figure 11. Velocity contours of DP-Climb critical transition nodes: (**a**) $\beta = 0°$; (**b**) $\beta = 5°$; (**c**) $\beta = 30°$; (**d**) $\beta = 50°$; (**e**) $\beta = 70°$; (**f**) $\beta = 90°$.

The relationship between the pitch angle of the robot body and the front and rear rotor platforms is shown in Table 3.

Table 3. Simulation analysis of aerodynamic coupling characteristics of adhesion system.

Rotor Unit	F (N)	Critical Transition Angle β (°)	Rotor Pitch Angle (°)	Simulation Result (N)	Error (%)
Front	15.91	0	155	15.26	4.09
		5	100	15.95	−0.25
		30	90	15.43	3.02
		50	70	15.40	3.21
		70	50	15.50	2.58
		90	40	15.63	1.76
Behind	5.32	0	25	5.53	−3.95
		5	35	5.38	−1.13
		30	35	5.77	−8.46
		50	35	5.53	−3.95
		70	35	5.52	−3.76
		90	30	5.54	−4.14

In summary, the effective pulling force error caused by the mutual coupling effect of the two rotor units is not greater than 8.46%, which can reach the reverse thrust required for the ground-wall transition calculated by the DP-Climb dynamics model. It can be seen that the proposed DP-Climb ground-wall transition strategy is feasible.

4. Discussion

To verify the feasibility of the proposed DP-Climb ground-wall transition control strategy, physical prototype experiments were carried out to validate it and to test the robot's wall operation capability. A nine-axis digital attitude sensor was selected to feed back the robot's attitude change during the ground-wall transition. At the same time, the encoder wheel was used to feed back the robot's displacement in real-time. The movement speeds of different operating postures were measured by combining the returned data information from the sensors.

An experimental safety frame was built to ensure the safety of the robot and the operator, and a safety rope was connected to the robot chassis. The safety rope was completely slack during the movement of the robot and was only used to tighten the rope to prevent accidents under special circumstances. The experiments were divided into three parts to verify the robot motion performance: (1) the transition motion of the ground wall; (2) the variable-speed motion on the wall; and (3) the transition motion of the wall-ground.

(1) Robot ground-wall transition movement. The motion process of the robot is shown in Figure 12. During this motion process, based on the proposed ground-wall transition motion strategy, the motion of the robot was controlled through the hybrid adhesion system of DP-Climb.

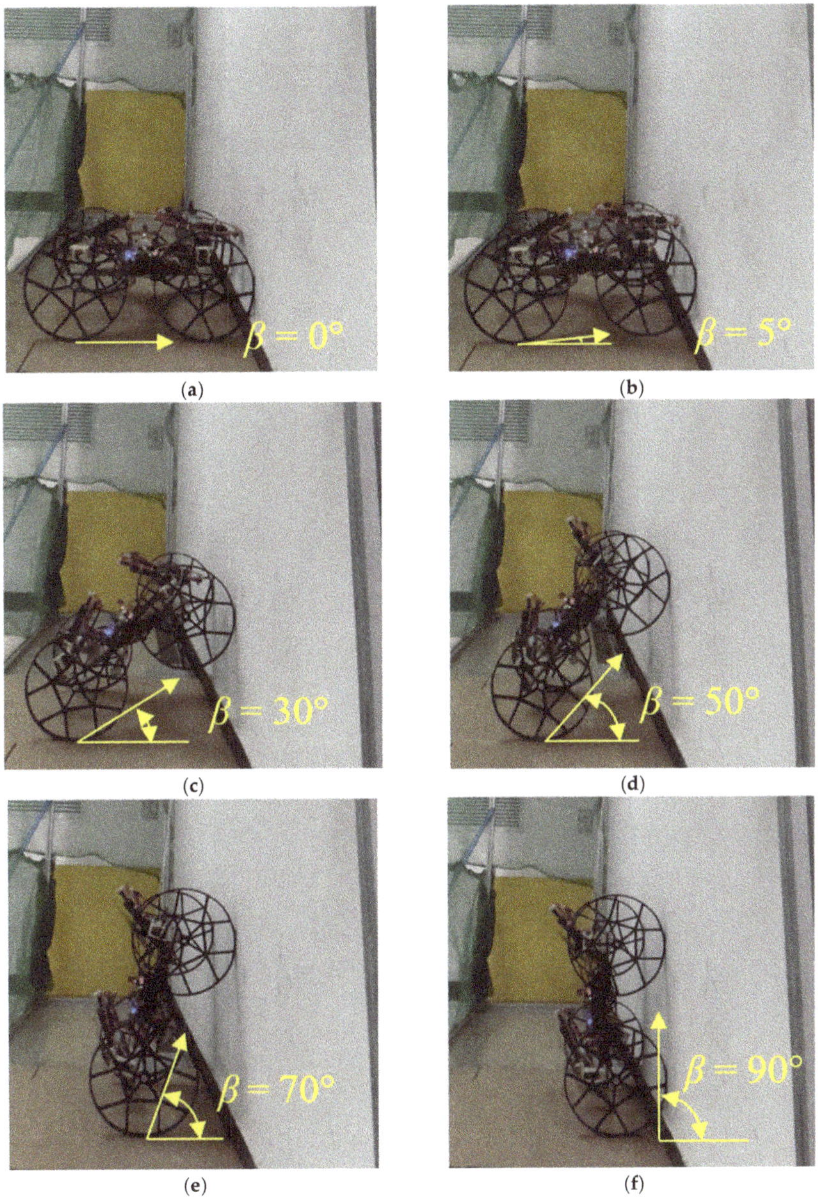

Figure 12. Ground-wall transition experiment of DP-Climb: (**a**) $\beta = 0°$; (**b**) $\beta = 5°$; (**c**) $\beta = 30°$; (**d**) $\beta = 50°$; (**e**) $\beta = 70°$; (**f**) $\beta = 90°$.

Analyzing the data information returned by the sensors, as shown in Figure 12, the robot can adjust the adhesion force and traction force of the rotor units according to its own pitch angle at different stages during the ground-wall transition motion process, which ensures that the robot's adhesion system can provide sufficient power at any angle to meet the motion demand of the robot for a smooth transition between the ground and the wall. The transition speed of the robot during the ground-wall transition is 0.12 m/s.

(2) The variable-speed motion of the robot on the wall. The movement process of the robot is shown in Figure 13. During this movement process, the robot's ability to move flexibly on the wall surface was verified by changing the movement speed and movement state of the robot.

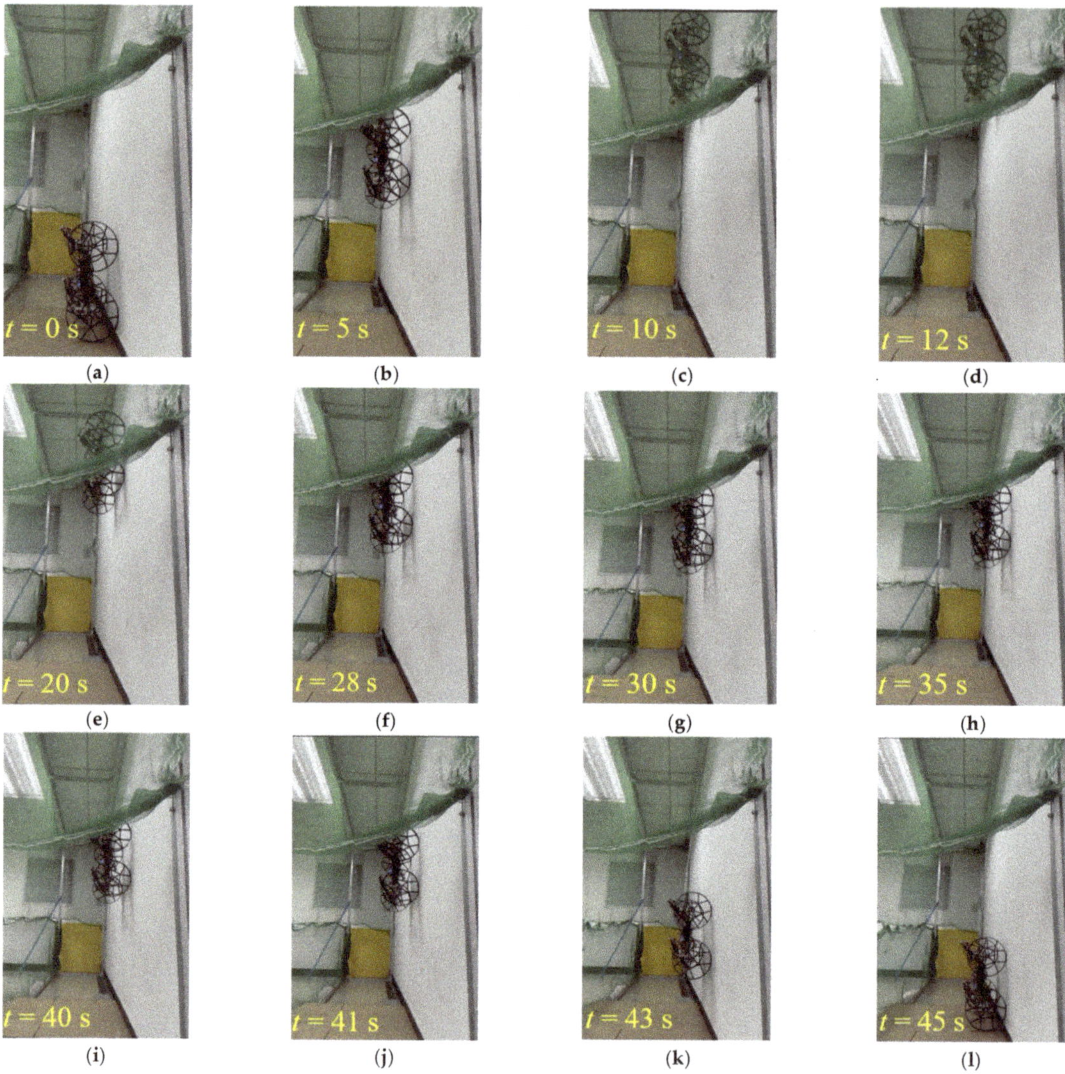

Figure 13. Wall variable-speed forward and backward experiment of DP-Climb: (**a**–**c**) wall forward movement; (**d**–**f**) wall deceleration backward movement; (**g**–**i**) wall static negative pressure adhesion; (**j**–**l**) wall acceleration backward movement.

Analyzing the data information returned by the sensors, as shown in Figure 13, during the robot's variable-speed motion on the wall, the hybrid adhesion system can ensure that the robot meets the safe adhesion conditions on the wall under different motion speeds and accelerations and realize flexible and fast movement on the wall. In Figure 13a–c), the forward speed of the robot was 0.21 m/s. In Figure 13d–f, the robot decelerated and moved backward, and the movement speed was 0.05 m/s. In Figure 13g–i, the robot decelerated to 0 and was kept on the wall with a stable negative pressure. In Figure 13j–l, the robot accelerated and moved backward, and the movement average speed reached 0.25 m/s. The DP-Climb maximum speed can reach 0.45 m/s.

(3) The wall-ground transition of the robot. The movement process of the robot is shown in Figure 14.

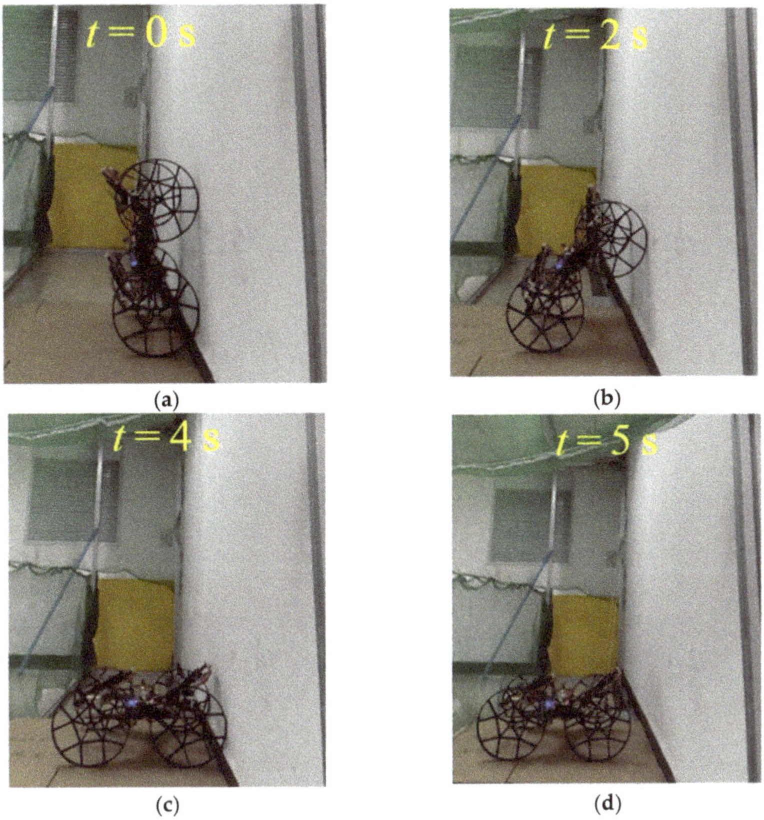

Figure 14. Wall-ground transition experiment of DP-Climb: (**a**) preparation for going down the wall; (**b**) the drive wheels reversed, and the robot went down the wall; (**c**) the robot finished going down the wall; (**d**) the robot walked on the ground.

The experiments can be seen at Supplementary Materials. Analyzing the data information returned by the sensors, as shown in Figure 14a–d, during the wall-ground transition, the robot mainly relies on the rotor units to provide sufficient adhesion force, and the driving wheels mainly provide the traction force of the robot. The transition motion speed of the robot during the wall-ground transition was 0.12 m/s.

5. Conclusions

This paper designed a double propeller wall climbing robot(DP-Climb) with a hybrid adhesion system that can be internally transitioned based on the principle of biomimetic design from the perspective of robot dynamics. The aim was to improve the mobility and adaptability of WCR to complex urban operating environments and expand the application scope of the robot according to the clinging characteristics of different creatures, this paper designed a double propeller wall-climbing robot DP-Climb with a hybrid adhesion system based on the biomimetic design principle. The motion characteristics of the robot in the transition motion inside the ground-wall surface were analyzed, and the robot's safe adhesion conditions in the robot's transition motion process were obtained. Based on the Newton–Euler equations, a DP-Climb ground-wall transition dynamic model was established, and finally, a DP-Climb ground-wall internal transition PID control strategy was introduced. Through mechanical and aerodynamic experiments, it was verified that the effective adhesion force output error of the double propeller symmetric configuration of the robot could converge within 8.46%, and the actual output adhesion force could meet the task requirements of the robot. It can flexibly and rapidly complete the ground-wall transition motion and achieve flexible switching between variable-speed forward, backward, and static adhesion motion states on the vertical wall. The transition motion speed of the robot during the ground-wall transition can reach 0.12 m/s. The wall motion maximum speed can reach 0.45 m/s. The WCR has high maneuverability and flexibility, enabling the robot to quickly complete the specified task in the target area, reducing the time consumed by the robot due to posture adjustment and extending the robot's endurance.

To verify the robot's general adaptability and system stability, the ability of DP-Climb to smoothly transition on the outer surface of the urban building with more severe damage to a slippery wall or the wall will be further studied in the future.

Supplementary Materials: The following supporting information can be downloaded at: https://www.mdpi.com/article/10.3390/machines10080678/s1, Video S1: Ground-wall transition experiment of DP-Climb; Video S2: Wall variable-speed forward and backward experiment of DP-Climb; Video S3: Wall-ground transition experiment of DP-Climb.

Author Contributions: Conceptualization, Q.Z. and X.G.; methodology, Q.Z., X.G. and M.L.; software, Q.Z. and Y.W.; validation, Q.Z., M.L. and Y.W.; formal analysis, Q.Z., M.L. and Y.W.; investigation, Q.Z. and X.G.; resources, Q.Z. and P.L.; data curation, Q.Z. and Y.W.; writing—original draft preparation, Q.Z., X.G. and M.L.; writing—review and editing, Q.Z., X.G; visualization, Q.Z.; supervision, Q.Z. and X.G; project administration, X.G.; funding acquisition, X.G. All authors have read and agreed to the published version of the manuscript.

Funding: This research was funded by the National Key Research and Department Program of China (Grant No. 2019YFB1309600).

Institutional Review Board Statement: Not applicable.

Informed Consent Statement: Not applicable.

Data Availability Statement: Not applicable.

Conflicts of Interest: The authors declare no conflict of interest.

References

1. Ge, D.; Tang, Y.; Ma, S.; Matsuno, T.; Ren, C. A Pressing attachment approach for a wall-climbing robot utilizing passive suction cups. *Robotics* **2020**, *9*, 26. [CrossRef]
2. Chu, B.; Jung, K.; Han, C.-S.; Hong, D. A survey of climbing robots: Locomotion and adhesion. *Int. J. Precis. Eng. Manuf.* **2010**, *11*, 633–647. [CrossRef]
3. Mahmood, S.K.; Bakhy, S.H.; Tawfik, M.A. Novel wall-climbing robot capable of transitioning and perching. *IOP Conf. Ser. Mater. Sci. Eng.* **2020**, *881*, 012049. [CrossRef]
4. Nagaya, K.; Yoshino, T.; Katayama, M.; Murakami, I.; Ando, Y. Wireless piping inspection vehicle using magnetic adsorption force. *IEEE/ASME Trans. Mechatron.* **2012**, *17*, 472–479. [CrossRef]
5. Nansai, S.; Mohan, R. A Survey of wall climbing robots: Recent advances and challenges. *Robotics* **2016**, *5*, 14. [CrossRef]
6. Wang, W.; Wu, S. A caterpillar climbing robot with spine claws and compliant structural modules. *Robotica* **2016**, *34*, 1553–1565. [CrossRef]
7. Chen, R.; Fu, L.; Qiu, Y.; Song, R.; Jin, Y. A gecko-inspired wall-climbing robot based on vibration suction mechanism. *Proc. Inst. Mech. Eng. Part C J. Mech. Eng. Sci.* **2019**, 233, 7132–7143. [CrossRef]
8. Chattopadhyay, P.; Ghoshal, S.K. Adhesion technologies of bio-inspired climbing robots: A survey. *Int. J. Robot. Autom.* **2018**, *33*, 654–661. [CrossRef]
9. Eto, H.; Asada, H.H. Development of a wheeled wall-climbing robot with a shape-adaptive magnetic adhesion mechanism. In Proceedings of the 2020 IEEE International Conference on Robotics and Automation (ICRA), Online Conference, 31 May–31 August 2020; pp. 9329–9335.

10. Zhu, H.; Guan, Y.; Wu, W.; Zhang, L.; Zhou, X.; Zhang, H. Autonomous pose detection and alignment of suction modules of a biped wall-climbing robot. *IEEE/ASME Trans. Mechatron.* **2015**, *20*, 653–662. [CrossRef]
11. Liu, Y.; Kim, H.; Seo, T. AnyClimb: A new wall-climbing robotic platform for various curvatures. *IEEE/ASME Trans. Mechatron.* **2016**, *21*, 1812–1821. [CrossRef]
12. Liu, Y.; Seo, T. AnyClimb-II: Dry-adhesive linkage-type climbing robot for uneven vertical surfaces. *Mech. Mach. Theory* **2018**, *124*, 197–210. [CrossRef]
13. Bian, S.; Xu, F.; Wei, Y.; Kong, D. A Novel type of wall-climbing robot with a gear transmission system arm and adhere mechanism inspired by cicada and gecko. *Appl. Sci.* **2021**, *11*, 4137. [CrossRef]
14. Liang, P.; Gao, X.; Zhang, Q.; Li, M.; Gao, R.; Xu, Y. Analysis and experimental research on motion stability of wall-climbing robot with double propellers. *Adv. Mech. Eng.* **2021**, *13*, 16878140211047726. [CrossRef]
15. Cho, K.J.; Koh, J.S.; Kim, S.; Chu, W.S.; Hong, Y.; Ahn, S.H. Climbing robots for commercial applications—A survey. *Int. J. Precis. Eng. Manuf.* **2009**, *10*, 171–181. [CrossRef]
16. Kim, H.; Kim, D.; Yang, H.; Lee, K.; Seo, K.; Chang, D.; Kim, J. Development of a wall-climbing robot using a tracked wheel mechanism. *J. Mech. Sci. Technol.* **2009**, *22*, 1490–1498. [CrossRef]
17. Fu, Y.; Li, Z.; Wang, S. A Wheel-leg hybrid wall climbing robot with multi-surface locomotion ability. In Proceedings of the 2008 IEEE International Conference on Mechatronics and Automation, Takamatsu, Japan, 5–8 August 2008; pp. 627–632.
18. Koo, I.M.; Trong, T.D.; Lee, Y.H.; Moon, H.; Koo, J.; Park, S.K.; Choi, H.R. Development of wall climbing robot system by using impeller type adhesion mechanism. *J. Intell. Robot. Syst.* **2013**, *72*, 57–72. [CrossRef]
19. Wang, Y.; Liu, S.; Xu, D.; Zhao, Y.; Shao, H.; Gao, X. Development and application of wall-climbing robots. In Proceedings of the 2013 IEEE/RSJ International Conference on Intelligent Robots and Systems, Tokyo, Japan, 3–7 November 2013; pp. 1207–1212.
20. Schmidt, D.; Berns, K. Climbing robots for maintenance and inspections of vertical structures—A survey of design aspects and technologies. *Robot. Auton. Syst.* **2013**, *61*, 1288–1305. [CrossRef]
21. Myeong, W.; Song, S.; Myung, H. Development of a wall-climbing drone with a rotary arm for climbing various-shaped surfaces. In Proceedings of the 2018 15th International Conference on Ubiquitous Robots, Honolulu, HI, USA, 26–30 June 2018; pp. 687–692.
22. Ma, B.; Liu, R.; Zhang, R.; Tian, L. Design of wall climbing robots with transition capability. In Proceedings of the 2007 IEEE International Conference on Robotics and Biomimetics (ROBIO), Sanya, China, 15–18 December 2007; pp. 1871–1875.
23. Dong, W.; Wang, H.; Li, Z.; Jiang, Y.; Xiao, J. Development of a wall-climbing robot with biped-wheel hybrid locomotion mechanism. In Proceedings of the 2013 IEEE/RSJ International Conference on Intelligent Robots and Systems, Tokyo, Japan, 3–7 November 2013; pp. 2333–2338.
24. Zhao, X.; Zhang, Q.; Wang, L.; Xie, F.; Zhang, B. A Special operation UAV in urban space. In Proceedings of the Applied Optics and Photonics China 2021, Beijing, China, 23–25 July 2021; Volume 12065, pp. 433–442.
25. Ioi, K.; Yokoi, H.; Kimura, M. Development of a compact and rapid wall-climber. In Proceedings of the 2013 18th International Conference on Methods & Models in Automation & Robotics (MMAR), Międzyzdroje, Poland, 26–29 August 2013; pp. 344–349.
26. Lee, G.; Kim, H.; Seo, K.; Kim, J.; Kim, H.S. MultiTrack: A multi-linked track robot with suction adhesion for climbing and transition. *Robot. Auton. Syst.* **2015**, *72*, 207–216. [CrossRef]
27. Liang, P.; Gao, X.; Gao, R.; Zhang, Q.; Li, M. Analysis of the aerodynamic performance of a twin-propelled wall-climbing robot based on computational fluid dynamics method. *AIP Adv.* **2022**, *12*, 015022. [CrossRef]
28. Liang, P.; Gao, X.; Zhang, Q.; Gao, R.; Li, M.; Xu, Y.; Zhu, W. Design and stability analysis of a wall-climbing robot using propulsive force of propeller. *Symmetry* **2020**, *13*, 37. [CrossRef]
29. Alkalla, M.G.; Fanni, M.A.; Mohamed, A.M.; Hashimoto, S. Tele-operated propeller-type climbing robot for inspection of petrochemical vessels. *Ind. Robot Int. J.* **2017**, *44*, 166–177. [CrossRef]
30. Alkalla, M.G.; Fanni, M.A.; Mohamed, A.F.; Hashimoto, S.; Sawada, H.; Miwa, T.; Hamed, A. EJBot-II: An optimized skid-steering propeller-type climbing robot with transition mechanism. *Adv. Robot.* **2019**, *33*, 1042–1059. [CrossRef]
31. Myeong, W.; Myung, H. Development of a wall-climbing drone capable of vertical soft landing using a tilt-rotor mechanism. *IEEE Access* **2019**, *7*, 4868–4879. [CrossRef]
32. Han, I. Development of a stair-climbing robot using springs and planetary wheels. *Proc. Inst. Mech. Eng. Part C J. Mech. Eng. Sci.* **2008**, *222*, 1289–1296. [CrossRef]
33. Oliveira, A.L.C.; Silva, M.F.; Barbosa, R.S. Architecture of an wheeled climbing robot with dynamic adjustment of the adhesion system. In Proceedings of the SISY 2010: IEEE 8th International Symposium on Intelligent Systems and Informatics, Subotica, Serbia, 10–11 September 2010; pp. 127–132.
34. Zhang, L.; Sun, J.; Yin, G.; Zhao, J.; Han, Q. A Cross Structured Light Sensor and Stripe Segmentation Method for Visual Tracking of a Wall Climbing Robot. *Sensors* **2015**, *15*, 13725–13751. [CrossRef]
35. Ding, X.; Yu, Y.; Zhu, J.J. Trajectory linearization tracking control for dynamics of a multi-propeller and multifunction aerial robot-MMAR. In Proceedings of the 2011 IEEE International Conference on Robotics and Automation (ICRA), Shanghai, China, 9–13 May 2011; pp. 757–762.
36. Nishi, A.; Miyagi, H. Mechanism and control of propeller type wall-climbing robot. In Proceedings of the IEEE/RSJ International Conference on Intelligent Robots and Systems (IROS'94), Munich, Germany, 12–16 September 1994; Volume 3, pp. 1724–1729.

37. Hawkes, E.W.; Eason, E.V.; Christensen, D.L.; Cutkosky, M.R. Human climbing with efficiently scaled gecko-inspired dry adhesives. *J. R. Soc. Interface* **2015**, *12*, 20140675. [CrossRef]
38. Yu, J.; Chary, S.; Das, S.; Tamelier, J.; Pesika, N.S.; Turner, K.L.; Israelachvili, J.N. Gecko-inspired dry adhesive for robotic applications. *Adv. Funct. Mater.* **2011**, *21*, 3010–3018. [CrossRef]

machines

MDPI

Article

Skill Acquisition and Controller Design of Desktop Robot Manipulator Based on Audio–Visual Information Fusion

Chunxu Li [1], Xiaoyu Chen [2], Xinglu Ma [2,*], Hao Sun [2] and Bin Wang [2]

[1] College of Mechanical and Electrical Engineering, Hohai University, Changzhou 213000, China
[2] Information Science and Technology, Qingdao University of Science and Technology, Qingdao 266061, China
* Correspondence: maxinglu@qust.edu.cn

Abstract: The development of AI and robotics has led to an explosion of research and the number of implementations in automated systems. However, whilst commonplace in manufacturing, these approaches have not impacted chemistry due to difficulty in developing robot systems that are dexterous enough for experimental operation. In this paper, a control system for desktop experimental manipulators based on an audio-visual information fusion algorithm was designed. The robot could replace the operator to complete some tedious and dangerous experimental work by teaching it the arm movement skills. The system is divided into two parts: skill acquisition and movement control. For the former, the visual signal was obtained through two algorithms of motion detection, which were realized by an improved two-stream convolutional network; the audio signal was extracted by Voice AI with regular expressions. Then, we combined the audio and visual information to obtain high coincidence motor skills. The accuracy of skill acquisition can reach more than 81%. The latter employed motor control and grasping pose recognition, which achieved precise controlling and grasping. The system can be used for the teaching and control work of chemical experiments with specific processes. It can replace the operator to complete the chemical experiment work while greatly reducing the programming threshold and improving the efficiency.

Keywords: desktop experimental manipulators; skill acquisition; motion control; motion detection; speech recognition; information fusion; pose recognition

Citation: Li, C.; Chen, X.; Ma, X.; Sun, H.; Wang, B. Skill Acquisition and Controller Design of Desktop Robot Manipulator Based on Audio–Visual Information Fusion. *Machines* **2022**, *10*, 772. https://doi.org/10.3390/machines10090772

Academic Editor: Raffaele Di Gregorio

Received: 14 July 2022
Accepted: 31 August 2022
Published: 6 September 2022

1. Introduction

Modern production is subject to a great deal of uncertainty since new items are being introduced at an increasingly rapid rate, particularly those with multiple varieties and a limited lifespan. Therefore, the ability to create flexible and reconfigurable production systems is highly desired. When it comes to flexible tasks, human abilities such as quick perception and the processing of different types of information or adaptability and improvisation can be crucial success factors [1–3]. The most sensitive jobs, including operating chemical experiments, still rely entirely on physical labor, despite automated robots playing a large role in modern manufacturing lines. These lines require significant labor, and a human worker may have to spend hours drilling in screws or wheels without stopping. Modern business needs collaborative robots that can effectively aid human employees because labor is becoming more expensive due to an aging population. Therefore, giving full play to the advantages of robots and using them to complete some cumbersome and high-risk experiments will become the main direction of the intelligent development of chemical laboratories. In the same process, the use of intelligent devices such as robots and robotic arms requires the mastery of complex programming and control techniques, which is still difficult for chemical experimenters. This raises the question of how to control robots, robotic arms and other intelligent devices to complete chemical experiments in a simpler and faster way.

Chemical experiment is an indispensable link in the process of research, study and production in the chemical industry [4]. At present, the intelligence of chemical engineering procedures is not very high; the experimental process that can be completed by using intelligent mechanical equipment such as manipulators to carry out chemical experiments is relatively simple, and the programming is relatively complex. In addition, various chemical reagents are used in chemical experiments. These reagents interact with each other during the experiment to produce various harmful substances, and unpredictable dangers may occur as a result of experimental errors. Therefore, it is necessary to use more intelligent manipulators and other equipment to replace the operator to complete the relevant complex tasks.

Today's technologies enable us to send a robot to land on Mars, but not to properly control a robot shaking hands with us. The performance of most advanced robot control systems at present still cannot match that of humans' adaptability, flexibility and cooperative ability, which are urgently required by the flexible manufacturing systems used to facilitate mass customization in the context of Industry 4.0 [5,6]. Modelling human skills from a robot control view is still challenging, especially for high level versatile and collaborative skills. Robots have recently been finding their way into human industrial and daily life, an example of which is by learning motor skills from human tutors through demonstration, and then generating these learned skills [7]. Obviously, a skilled robot would be more efficient in interacting with humans and industrial productions. It is increasingly expected that robots should be capable of flexible skills in order to adapt to more complex situations. Teaching by demonstration is seen as one of the most effective ways for a robot to learn motion and manipulation skills from humans [8]. In this paper, inspired by human mechanical intelligence adaptivity to variations of tasks, both position trajectory and oral interaction are achieved for robot motion control to realize a more completed skill transfer process.

With the development of machine learning and machine vision technology, the deductive programming of the manipulator provides a new solution for human-computer interaction, which is an important way to reduce the difficulty of acquiring skills for the manipulator [9–11]. The teaching programming of the manipulator is a process of automatically learning the motion trajectory by watching and learning the teaching actions of people; teaching robots by visual signals such as motion tracking-based teleoperation, or by audio inputs such as oral command interface, plays an increasingly important role. Researchers such as Haage of Lund University used an RGB-D camera-based sensing module to track human movements when implementing robot teaching. The robot is mainly used to install industrial parts and inspection equipment [12]. Li C et al. designed a LeapMotion sensor-based controller for tracking the operator's hand movements to achieve the real-time robot teaching. The end-effector of the robot is actually held for demonstration to teach the robot action. Meanwhile, a neural network (NN)-based adaptive controller has successfully been developed for the remote manipulation of the DLR-HIT II robot hand [13]. In [14], the authors developed a robot learning method by modelling the motor skills of a human operator using dynamic motor primitives (DMP) and integrating the speech recognition, wherein people could easily teach the robot by speaking. However, only a few works [15,16] take advantage of the visual signals- and audio inputs-based robot teaching by combining them, which results in a negative effect on the data transmission latency and the diversity of motor skills.

In this paper, the arm motion detection and speech recognition are combined, and at the same time, with the help of information fusion and pose recognition, a teaching control system for the manipulator is designed [17] so that the manipulator can understand the skills taught by the experimental staff. It can complete the human-robot interaction more accurately and quickly, and then replace the operator to complete the experimental work. Compared with other robot learning methods, our proposed system possesses the following features:

1. The teaching process is simple and does not require operators to have programming skills, which is suitable for direct use by chemical operators.
2. The self-developed desktop manipulator is small in size and suitable for simple desktop chemical experiments.
3. Chemical experimental actions are decomposed into a combination of multiple basic actions, and a combination of visual action recognition and speech recognition is used to improve the accuracy of teaching.
4. Compared with the manipulator used in traditional research, our proposed method employs a low-cost manipulator with lower power consumption but embeds advanced robot learning algorithms.

2. Overall System Design

The teaching control system in this design takes the desktop experimental manipulator as the physical carrier and the Raspberry Pi operating system as the platform. It is mainly aimed at teaching the experimental robot arm to imitate the experimental movements of the human arm through the teaching of the operator in the chemical analysis experimental scene. This is then combined with the grasping pose recognition algorithm of a certain experimental instrument to assist the control, teaching and motion control. The two are combined to complete a set of experimental process combinations.

2.1. System Logical Architecture

As shown in Figure 1, the overall logical architecture of this design can be divided into three levels: From the bottom up, they are the hardware composition, device driver and application software layer. The hardware consists of a desktop manipulator (with three basic degrees of freedom, similar to a human arm), a Raspberry Pi 4B core board, a motor and motor control board, a high-resolution camera, a microphone and a robotic claw. The device driver part includes an audio driver for microphone, USB_Cam camera driver, motor driver and other related programmable logic. The application software layer is the software program running in the Raspberry Pi desktop operating system and the set of motion parameters of the manipulator, which is similar to the Linux operating system. The software program part includes a visual interface program for motor control and teaching process grasp, speech recognition [18] function module, motion detection module and the fusion and matching part of motion information.

2.2. System Function Process

The overall workflow of the teaching control system is shown in Figure 2. In the teaching process of this whole set of experimental procedures, a complete set of experimental procedures is composed of many simple experimental actions, which are called action primitives. The operator needs to dictate his movements while performing the movements. The manipulator obtains information by listening and seeing, and performs fusion verification, so as to understand the movements that need to be performed. That is, the speech recognition module and the action recognition module, respectively, combine the action primitives recognized by the whole set of experimental process actions into the set in order, and then match them with the action groups in the manipulator action set stored in the local action library where the action information is displayed. The identification information in the process is obtained through the information fusion algorithm to obtain a final set of action groups, which are saved or handed over to the manipulator to run and reproduce.

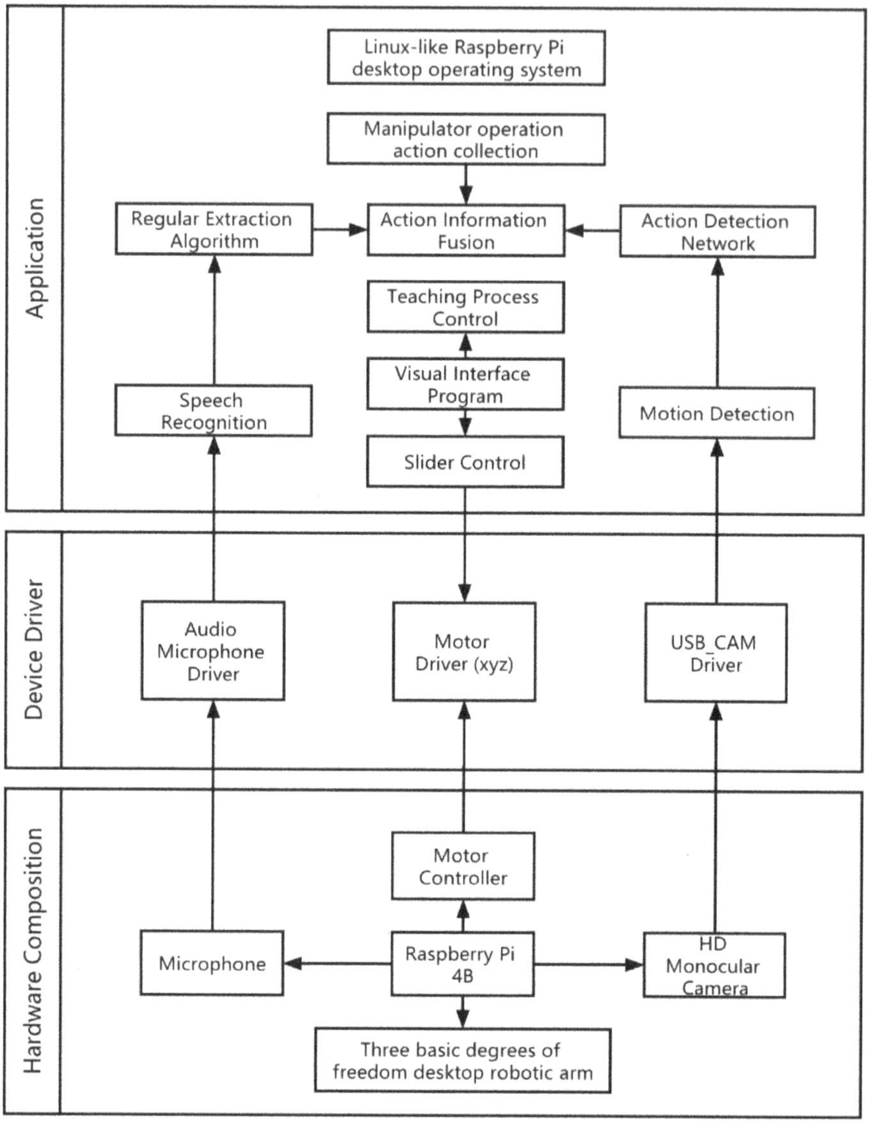

Figure 1. System logical architecture.

The system visualization interface is compatible with Windows and Linux operating systems. As shown on the right side of the figure, the action parameters are stored in an xml file, and the format is neat and simple, which is easy to read and store the action parameters. In the visual interface, three Scale and Radio buttons are set to control the steps and directions of the three motors (x, y, z, respectively), and the action group is defined and stored by sliding and adding actions. Click Start Live Teaching to turn on the camera, and then start the teaching function.

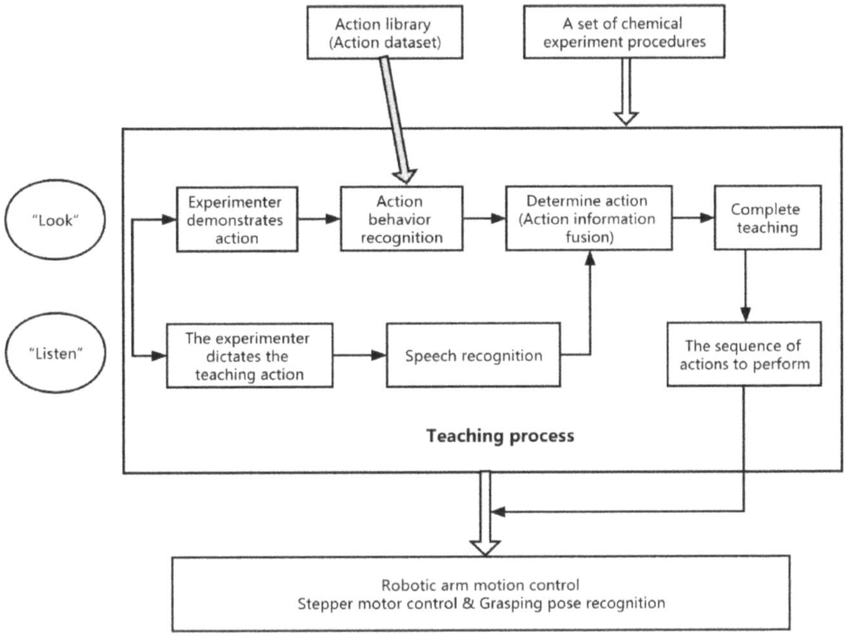

Figure 2. System function flow.

3. The Composition and Performance Parameters of the Manipulator

As shown in Figure 3, the desktop robotic arm in this design has three basic degrees of freedom and can rotate in the space above the desktop. The robotic arm is composed of a Raspberry Pi 4B as the core control board, a motor drive expansion board, a stepper motor drive module, three stepper motors, a 12 V power adapter, and a camera with a microphone. The operations can be picked and placed using robotic arm universal grippers. The maximum payload of the robotic arm is 500 g, and the max reach is 320 mm. The stepper motor adopts a high-torque 42 planetary deceleration stepper motor with a step angle of 1.8°. In the stepper motor drive module, each step (1.8°) of the motor is subdivided into 16 steps for finer control.

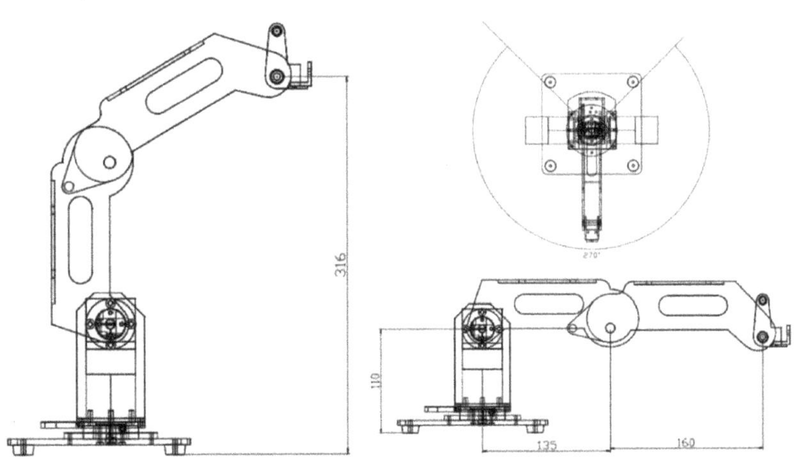

Figure 3. Structure diagram of the manipulator.

3.1. Core Control Panel

The control system runs on a non-customized version of Raspberry Pi 4B, which is essentially a tiny, embedded PC. The Raspberry Pi used in this article has a 64-bit, 1.5 GHz quad-core CPU, 1 GB of memory, and two USB 3.0 and two USB 2.0 communication ports. The programming system is a desktop Raspbian System based on the Linux operating system [19]. The control system controls the microphone and camera through the USB communication port of the Raspberry Pi and controls the self-developed stepper motor and drive module through the GPIO port.

3.2. Stepper Motor and Drive Module

The key element used to control the movement of the manipulator is the motor. There are three stepping motors used in this manipulator, which realizes the motion control of three basic degrees of freedom. The manipulator adopts a high-torque 42 planetary deceleration stepping motor. The theoretical deceleration ratio of the horizontal motion motor is 1:5.18, and the actual measurement is 11:57, which is close to the theoretical value; the deceleration ratios of the stepping telescopic motors of the two arm parts are both 1:19 (the actual measurement is 187:3591). The current of the motor is 1.7 A, the step angle is 1.8°, and its step accuracy is 5%.

As shown in Figure 4, the motor driver adopts an A4988 driver, which can drive the motor voltage of 8–35 V. This manipulator uses a 12 V power supply to power the motor. Among them, each stepper motor drive module outputs two control signals, STEP and DIR, respectively, which are connected to the Raspberry Pi pins to realize the control of the stepping pulse and direction, respectively; the MS1–3 pins are used in this design. Both are connected to a high level, and the corresponding parameters of the interface level and the number of steps is shown in Table 1. Each step (1.8°) of the motor is subdivided into 16 steps to achieve more precise control.

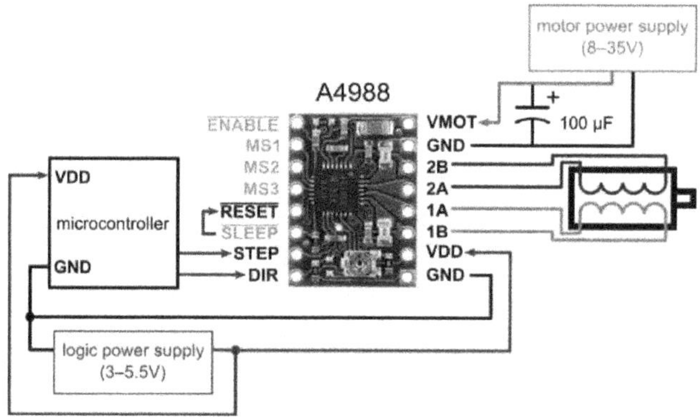

Figure 4. Stepper motor driver module.

Table 1. Corresponding parameters of MS interface.

MS1	MS2	MS3	Microstep Resolution
Low	Low	Low	Full step
High	Low	Low	Half step
Low	High	Low	Quarter step
High	High	Low	Eighth step
High	High	High	Sixteenth step

4. Skill Acquisition Module Design

4.1. Motion Detection

At the beginning, we selected the action detection of the skeleton network, but the effect was not obvious, and the earliest dual-stream network also had a very good accuracy in action detection. Based on the dual-stream network, we performed the feature extraction part and the fusion classification algorithm. The next step is to improve and expand the behavioral action recognition data set and apply it in the chemical analysis experiment business scenario.

The action detection part of this paper is mainly based on an improved two-stream convolutional network. This part inputs the preprocessed image information into the network, uses the EfficientNetv2 [20] algorithm to calculate the RGB image and optical flow image features, and then uses the extracted feature information to use linear classification. The SVM [21–24] is used to classify the behavior and obtain the identification information of the action.

As shown in Figure 5, the two-stream convolutional network divides the input video into two channels for processing, one of which is to extract the task arm and scene-related information in the RGB image by the convolutional neural network, and the other is to process the optical flow image information, which is finally normalized and fused by the Softmax function.

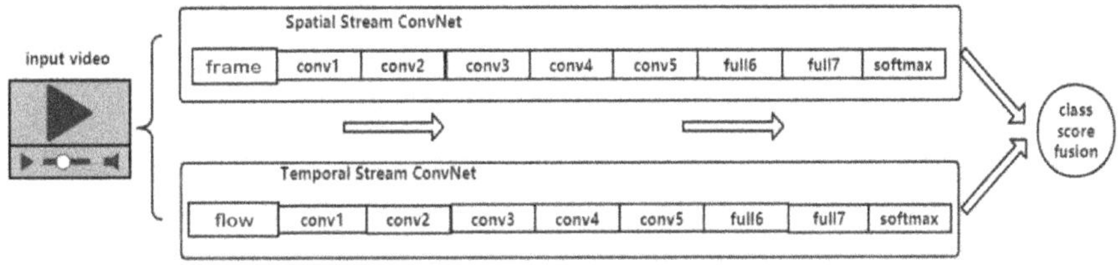

Figure 5. Two-stream convolutional network structure.

The extraction of optical flow images in the network (action video preprocessing) is obtained by gradient-based operations. The key principles of the algorithm (1) are as follows: first, set the image sequence ***I(x, y, t)***; the vector ***X*** = ***[x, y]***. The sequence is extracted from the previous and subsequent frames in a demo video, that is, when the local optical flow image of the video is basically constant. For any ***Y*** ∈ ***N(x)***, there are:

$$\frac{d}{dt}\nabla I(X,t) = \frac{\partial \nabla I}{\partial X}\frac{\partial X}{\partial t} + \frac{\partial \nabla I}{\partial t} = H(I) \cdot d + (\nabla I)_t = 0 \tag{1}$$

where ***X*** is the ***x*** vector, ***H(I)*** is the Hesse matrix of the image sequence ***I***, and the relationship between ***X*** and the offset ***d*** is introduced in (2):

$$E(X,d) = \| \ (H(I) \cdot d + (\nabla I)t) \ \|^2 \tag{2}$$

Setting the derivative equal to 0 yields (3):

$$d = -(H^T(I)H(I))^{-1}(H^T(I)(\nabla I)_t) \tag{3}$$

The above process can be summarized as analyzing the changes of the pixels in the video image on the timeline and the correlation between adjacent frame images, finding the corresponding relationship between the previous frame and the current frame, and calculating the motion information (the offset is a kind of motion information), followed by drawing the optical flow image.

In the calculation feature part, as shown in Figure 6, compared to the previous EfficientNet [25] algorithm, EfficientNetv2 uses Fused-MBConv to replace the MBConv structure, that is, the conventional 3 × 3 convolution is used to replace the 3 × 3 depth convolution in MBConv and 1 × 1 convolution to improve the calculation speed of the network.

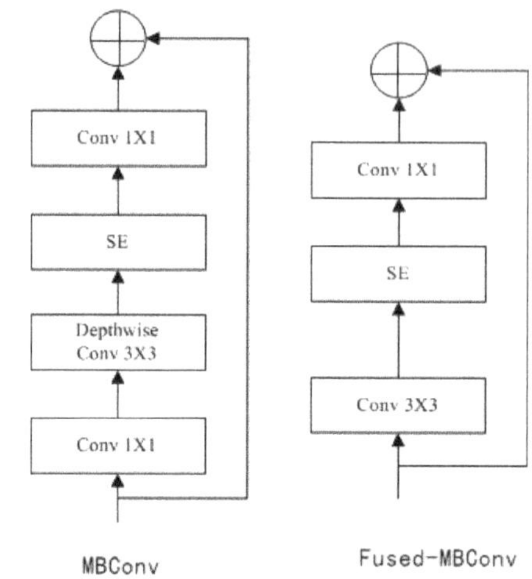

Figure 6. Schematic diagram of the improvement of EfficientNetv2 structure.

To obtain the feature information of the RGB image and optical flow image, it needs to be classified and verified. Support vector machine (SVM) is a binary classification model used to solve the separating hyperplane that can correctly divide the training data set. It also has the largest geometric interval. As shown in Figure 7, $\boldsymbol{w} \times \boldsymbol{x} + \boldsymbol{b} = 0$ is the separation hyperplane. There are generally many such hyperplanes, but the separation hyperplane with the largest interval is indeed the only one. For the optimal value among them, Formula (4) can be used to select, which is as follows:

$$\max_{w,b}(\min_{x1} \frac{y_i(w^T x_i + b_i)}{|w|}) \quad (4)$$

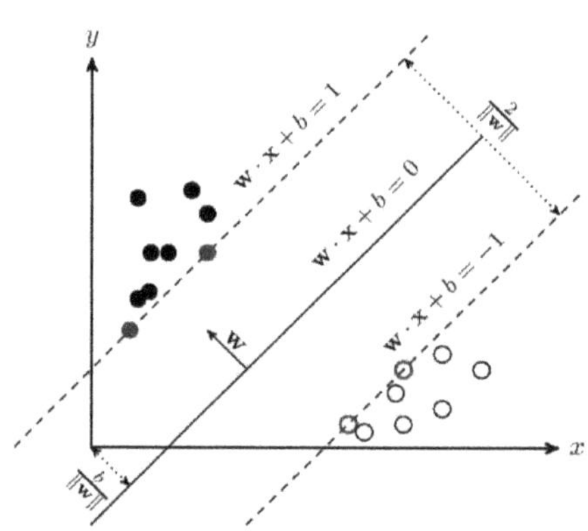

Figure 7. SVM principle.

The overall structure is shown in Figure 8. The RGB and optical flow feature extraction parts in the dual-stream convolutional network use lightweight Efficientnetv2 to perform convolution and pooling, respectively, and then combine the actions given by the SVM classifier for the two branches. The information is classified and, finally, identified action information is given. The data in the experiment are a self-made data set for chemical analysis. In the laboratory environment, a fixed camera is used to record the behaviors of stirring, picking up the experiment, putting down the experiment, taking liquid, and mixing liquid. This behavior was recorded 50 times. The video duration of the dataset is controlled within 10 s, and the recorded video changes in lighting, background, and occlusion, forming 10,000 chemical analysis action videos, with an average of 10 clip videos per video. The video format is 320 × 240, 25 fps and the audio is saved as a wav format file. The data set has a huge number of video frames, and there are data redundancy, interference, etc., which have a great impact on training and learning. Using the improved two-stream convolutional neural network, the recognition accuracy can reach 92%, which is improved in this experiment. Compared with other methods, the accuracy of the latter method is improved, but the training process still takes a significant amount of time.

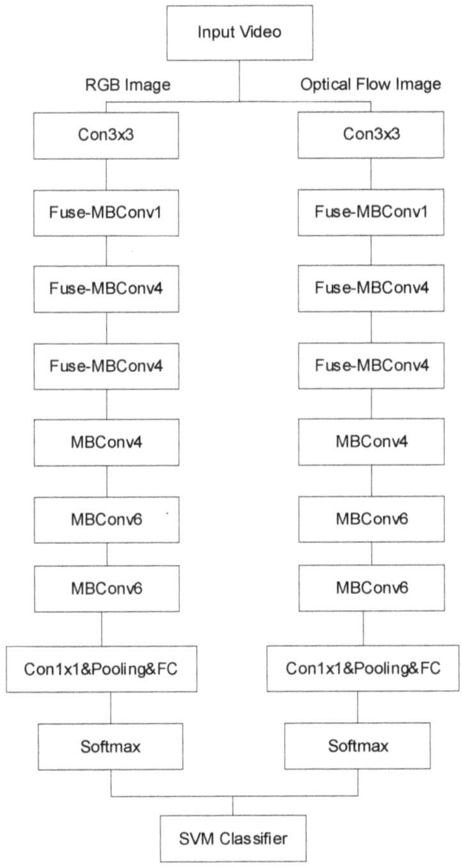

Figure 8. Action detection network structure.

After obtaining a complete set of identification information on the action, it is necessary to determine the execution sequence of the manipulator action of the module and obtain the executable action of the manipulator. First, design and name the motion primitives of the manipulator and store all the designed motion primitives in the library. The action information detected above is stored in a sequence, and the action primitives in the matching directory are used to determine the action primitive sequence in the current time. In each teaching process, the program matches the sequence combination composed of multiple action primitives, and then search for the associated actions in the manipulator action group library. The recognized action numbers are connected in series to obtain a sequence, and this sequence is used to find a complete match or the closest action group;

these are identified as the recognized experimental actions and given a coverage. After the program finds the action group with the greatest match, it continues to match with the action group in the library and stores the relevant information that the action group coverage is higher than 50% in the matching process.

4.2. Speech Recognition and Keyword Extraction

Teaching is a process of speaking while doing. The operator dictates the current action during arm movement, which requires the addition of voice technology. The key to speech technology is to process natural language, recognize speech and generate text, so that machines can listen, speak, understand and think [26]. This paper uses Baidu's real-time speech recognition, which is based on Deep Peak2 end-to-end modeling, transforms the received audio stream into text characters in real time, and then uses regular expressions to extract action keywords. The flow chart is shown in Figure 9.

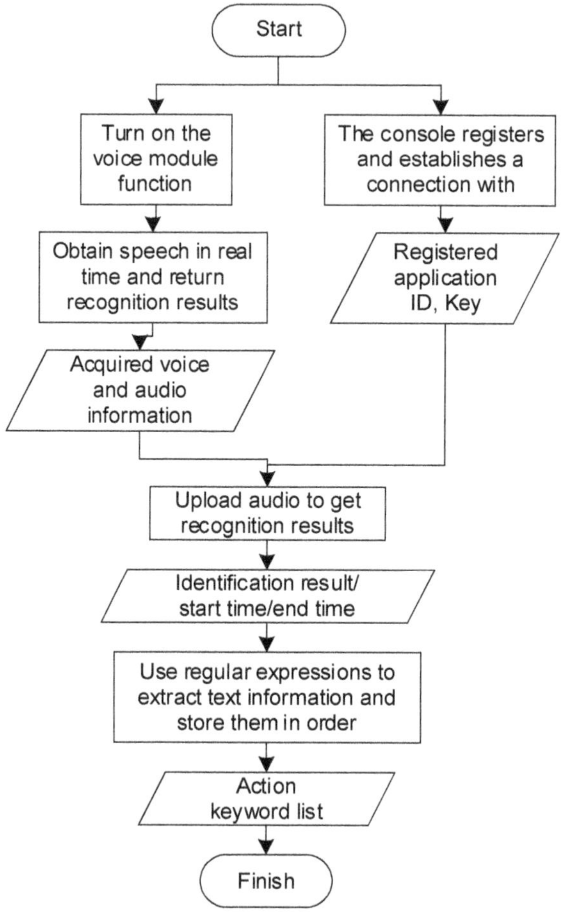

Figure 9. Flow chart of speech recognition and keyword extraction.

In this process, since it is speech recognition in chemical experiments, it is necessary to supplement the training set text corpus for many terms in chemical experiments and the speech data required by the current experimental process. The corpus data include the format of the speech files, name and text information. The corpus data set summarizes about 55 min of relevant identification content, and its audio files are converted into a direct binary sequence PCM file format after analog-to-digital conversion, which realizes the digitization of the sound and deletes the file header and end marks that differ from other file formats, in order to facilitate the concatenation of files. After supplementing the corpus data set, the accuracy of speech recognition in experimental business scenarios can be effectively improved by 7–15%.

In the above speech recognition process, the program needs to read the action file keywords manually set earlier in the text in advance and write them into memory variables. In each teaching process, the keywords recognized by the speech are matched to the text. If the match is correct, the identifier number IdentNumber is recorded immediately and stored.

Summarize all the matched speech keyword numbers in the whole teaching process in order, similar to the action detection and matching part; search for the action group of the robot arm; and search for a complete match or the closest action group. This is the speech recognition to action group, which retains the coverage information and records other related action group information with a coverage higher than 50%.

4.3. Audio and Video Information Fusion

Since the manipulator teaching program in this paper is mainly run-in embedded devices, in some scenarios, mobile terminals and embedded devices are much inferior in configuration, computing power, and device performance compared to servers or PCs [27]. Under such restrictive conditions, it is difficult to achieve high-speed and accurate recognition algorithms and teaching methods. Moreover, teaching is a dynamic process, which requires the continuous recognition of speech and actions. In this process, recognition failure will inevitably occur and affect the accuracy. Therefore, this paper draws on the information fusion of sensors [15,16,28] and writes a highly targeted algorithm to combine the key information obtained by the two separate modules of action detection and speech recognition, so as to improve the accuracy of skill acquisition and save performance.

The video recognition Information and speech recognition information obtained during the teaching process of the manipulator are an action group and its coverage. In the case of the most ideal running effect, the length of the action group given by the video and the voice is equal, and the data are shown in Table 2 as an example. The two modules in Table 2 list the highest coverage information ***ACT_G0*** generated by the current teaching process and other action coverage information greater than 50% coverage.

Table 2. Matching degree distribution table.

	ACT_G0	ACT_G1	ACT_G2	ACT_G3	ACT_G4
Video module	0.8	0.75	0.7	0.6	0.5
Audio module	0.8	0.7	0.7	0.6	0.6

If, in this teaching process, the two modules have the same maximum coverage action group—that is, the video and voice parts have selected the same action group with the highest matching degree—then the action can be regarded as a teaching action and does not need to follow the algorithmic process of fusion. The action can also be considered as being outside of this low-probability case.

After analyzing and testing the common fusion of confidence-based independent classifiers, this paper moves the fusion point to the action group category. The action group sequence given by the acquisition module is fused.

Algorithm idea: Define the ***ACT_G0*** of the video module part as ***VG0***, and then define it as ***VGn*** in turn. Similarly, the audio module part is defined as ***AG0***, ***AG1***...***AGn***. Compare ***AG0*** with 1–***n*** in the ***VG*** part and find a ***VGx*** that has a similarity of 100% with ***AG0***, according to the positive sequence—that is, if more than 50% of the ***VG*** matches ***AG0***, then record and store the coverage product of ***AG0*** and ***VGx*** ($\boldsymbol{AG0} \times \boldsymbol{VGx}, 0 < x \leq n$). The same is true for the ***AG*** part. The above two parts of the results are compared to the action group that outputs the optimal matching value (weights are equally divided), as shown in Formula (5):

$$FG = Mag[(AG0 \times VGx), (VG0 \times AGy)]x, y \in [0, n] \tag{5}$$

Among them, Mag is the function of the action group to find the maximum value, and ***FG*** is the action group of the robot arm that obtains the maximum value after the coverage rate is multiplied.

When the action group whose similarity with ***ACT_G0*** is 100% in the above process is empty, enter the following search algorithm: there is a similarity (<100%) between any two sets of actions, and the product of the original coverage is used to equalize the two. The similarity between the two is obtained, the correlation value of the two is obtained, and all the correlation values are aggregated to output the maximum action, that is, the maximum value is obtained by the fusion of two actions with different coverage, and then the two actions. In the group, select the action group that is closest to the original set of action primitive sequences (with the largest coverage), which is the final teaching action. The above process is shown in Formulas (6) and (7).

$$FQ(x,y) = AGx \times VGy \times \text{Fit}(AGx, VGy) \quad x,y \in [0,n] \tag{6}$$

$$FFG = MAX\_G\{MAX[FQ(x,y)]\} \quad x,y \in [0,n] \tag{7}$$

Among them, in Formula (6), fit is the similarity between two sets of action groups, ***FQ*** is the product of a pair of coverage and multiplied by the action-related value of the similarity between the two. In Formula (7), all the two modules are action information fusion; output a pair of actions with the optimal correlation value, and then select the action group with the highest coverage with the original action primitive sequence, that is, ***FFG***. In this way, the verification and fusion of the video module and the relevant action information of the audio module are completed, and the manipulator determines the final motor skills.

5. Manipulator Motion Control

The D-H method is usually used to build the model and analyze the motion of the mechanical arm. The D-H method is a common kinematic solution method in the field of robotics, which is beneficial to analyze and establish the kinematic model of the robotic arm and calculate the forward and inverse solutions. Through D-H modeling, the transformation matrix between each joint can be obtained, so as to obtain the transformation matrix from the base coordinate system to the claw coordinate system and the position and attitude of the end of the manipulator.

As shown in Figure 10, the specific method of establishing the link structure coordinate system for the manipulator is as follows: where $j-1$ and j represent two links, $j-1$, j and $j+1$ represent three axis joints, and the axis joint coordinates. The x-axis, y-axis, and z-axis of the system follow the right-hand rule. Among them, $\boldsymbol{a}$ is used to indicate the length of the connecting rod, $\boldsymbol{\alpha}$ is used to indicate the rotation angle of the connecting rod, $\boldsymbol{d}$ is used to indicate the offset distance of the connecting rod, and $\boldsymbol{\theta}$ is used to indicate the axis angle of the joint.

Table 3 is the attached connecting rod D-H parameters, in which the parameters of each connecting rod are d = 103 mm; a1 = 140 mm; b2 = 160 mm; a3 = 70 mm.

As shown in Figure 11, the joint coordinate system of the three-degree-of-freedom experimental manipulator is established on the basis of the D-H parameter coordinate system, including the three rotating joints of the manipulator and the position of the end effector.

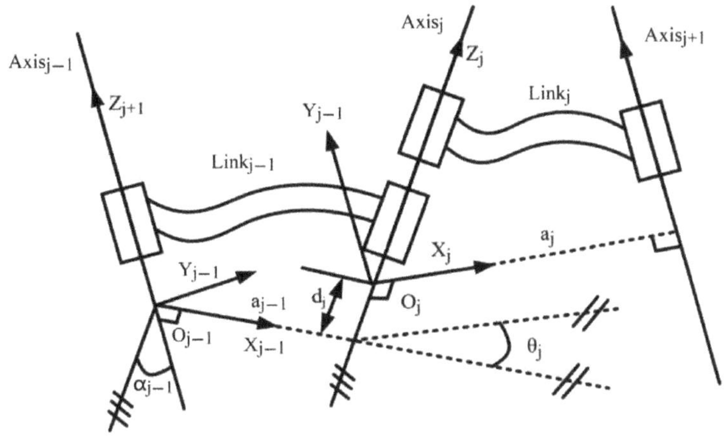

Figure 10. Robot arm link coordinate system.

Table 3. Parameters of the experimental manipulator connecting rod.

Link	Connecting Rod Angle	Connecting Rod Torsion Angle	Corner Range	Connecting Rod Distance	Connecting Rod Distance
i	$\theta_n/(\circ)$	$\alpha_n/(\circ)$	$\theta_i/(\circ)$	d_i/mm	d_i/mm
1	θ_1	0	−135~135	d	d
2	θ_2	90	−15~80	0	0
3	θ_3	0	−20~95	0	0
4	θ_4	0	−90~90	0	0

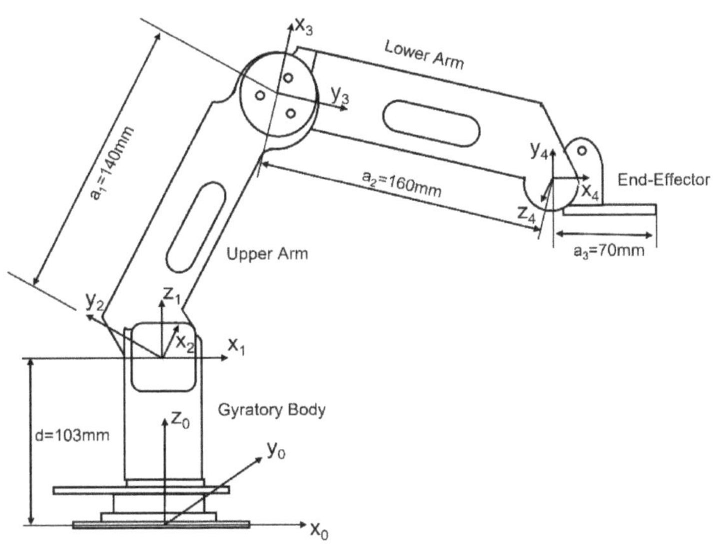

Figure 11. Robot arm model: set the coordinate system according to the D-H method.

The analysis concluded that, for the demonstration function to be implemented in this paper and the designed three-degree-of-freedom robotic arm, it is more suitable to use the geometric solution method in the inverse kinematics solution because our network can get the position where the end actuator or the clamping jaws of the robotic arm are located very simply and precisely, and the robotic arm has only three motors responsible for the operation of X, Y and Z. Figure 12 shows the simplified spatial coordinate diagram of the drawn experimental manipulator.

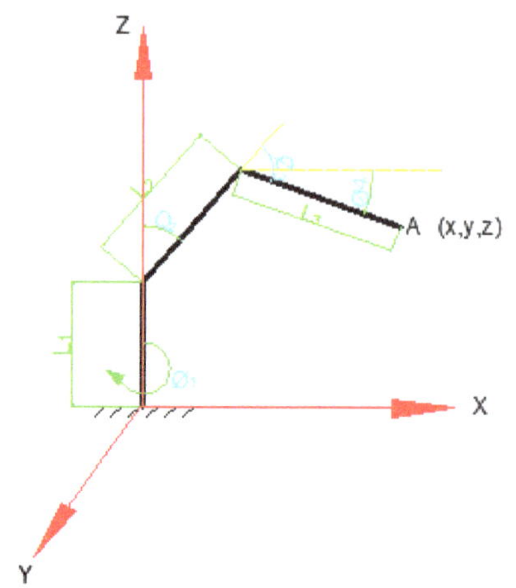

Figure 12. Space coordinate system of the three-degree-of-freedom manipulator.

There are mainly three constants in this manipulator: link length $L1$, link length $L2$ and link length $L3$; three rotation dependent variables: rotation angles $Q1$, $Q2$ and $Q3$; one variable: the general size of the gripper at the end of the manipulator; and a coordinate position (X, Y, Z).

As shown in Figure 13a, looking at it as a plane coordinate system with $\boldsymbol{x}$ as the horizontal axis and $\boldsymbol{y}$ as the vertical axis, the tangent of the connecting rod rotation angle $\boldsymbol{\theta_{cox}}$ is:

$$\tan\theta_{COX} = y/x \tag{8}$$

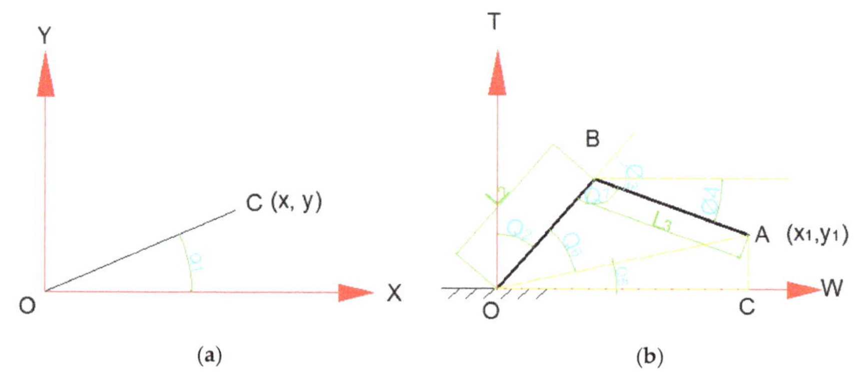

Figure 13. Coordinate system plane. (**a**) XOY plane (**b**) XOZ plane.

Reverse find:

$$\theta_1 = \theta_{COX} = \arctan y/x \tag{9}$$

From the perspective of the XOZ plane of the space coordinate system of the manipulator, as shown in Figure 13b, from the coordinates of point A ($\boldsymbol{x1}$, $\boldsymbol{z1}$), we can know that $OC = x_1$, $AC = z_1$ and obtain:

$$\theta_{AOC} = \arctan z_1/x_1 \tag{10}$$

Similarly, the cosine formula can be used to obtain:

$$\theta_{BOA} = \arctan\frac{L_2{}^2 + x_1{}^2 + z_1{}^2 - L_3{}^2}{2L_2\sqrt{x_1{}^2 + z_1{}^2}} \tag{11}$$

$$\theta_{OBA} = \arccos\frac{L_2{}^2 + L_3{}^2 - (x_1{}^2 + z_1{}^2)}{2L_2L_3} \tag{12}$$

Introduce this into formulas:

$$\theta_2 = \pi/2 - \theta_{AOC} - \theta_{BOA} \tag{13}$$

$$\theta_3 = \pi - \theta_{OBA} \tag{14}$$

$$\theta_4 = \pi/2 - \theta_2 - \theta_3 \tag{15}$$

The corresponding rotation angle of the link $L2$, the corresponding rotation angle of the link $L3$, and the horizontal angle of the link $L3$ can be obtained.

According to the above process, the inverse kinematics solution for the geometric solution of the three-degree-of-freedom experimental manipulator in this paper can be summarized as the flow chart shown in Figure 14:

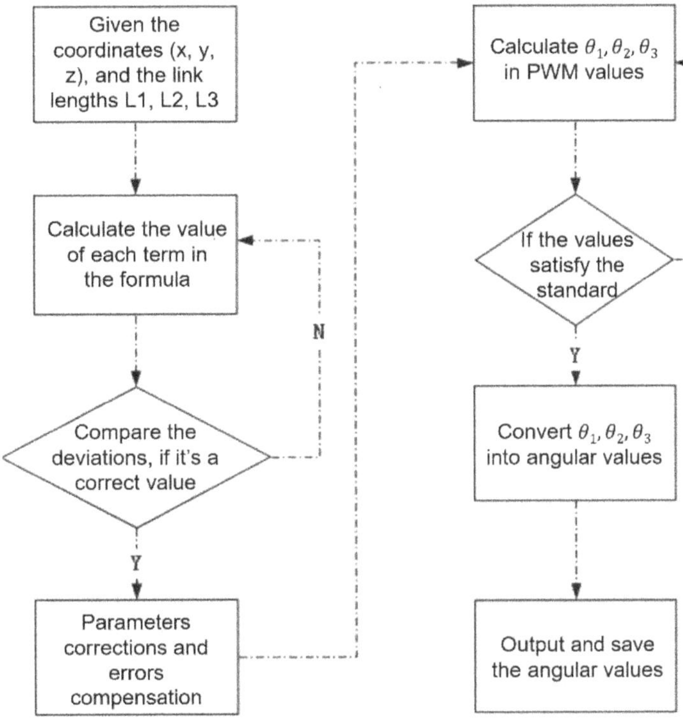

Figure 14. Inverse kinematics solution flow of mechanical arm geometry method.

After the teaching is completed, the program loads the finalized XML action parameter file. The program reads the corresponding direction and step parameters of the X, Y and Z motors in sequence, executes them in sequence, and controls the high- and low-level outputs of the related pins of the Raspberry Pi to control the operation of the motor. Among them, the rotation angle of the robot arm corresponding to the step parameter 1024 is 18°, and the maximum is 10,240.

When the manipulator is in motion, the camera installed near the gripper at the front end of the manipulator can receive real-time image data and transmit them to the Raspberry Pi through the USB port. The system combines it with the grasping pose recognition algorithm [29] for grasping. Five variables are used: (x, y, θ, h, w) to describe the gripping position and direction of the gripper when the manipulator grips the object. As shown in the rectangular box in Figure 15, (x, y) is used to represent the center position of the rectangular box; θ is used to represent the angle between the horizontal axis in the image and the current tilt position of the rectangular box; h denotes height; and w is used to represent width.

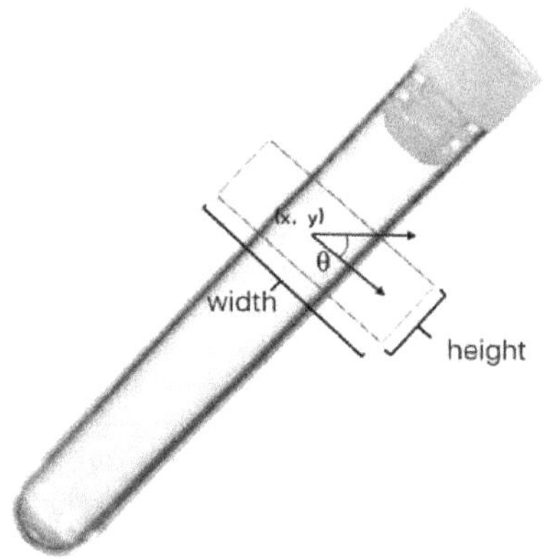

Figure 15. Grasping pose.

When the manipulator reaches the vicinity of the object, the position of the front end of the manipulator needs to be adjusted slightly, according to the position of the rectangular frame. That is, six basic adjustment actions are set: left and right, up and down, and clockwise/counterclockwise rotation. Use the above actions to adjust the mechanical arm to reach the preset position and achieve object grasping.

The grabbing pose algorithm is based on the Cornell Grasping Dataset, and on the basis of the data set, it continues to supplement and train the grabbing positions related to chemical equipment, which compensatively improves the accuracy of grabbing pose recognition.

6. System Performance Testing and Analysis

6.1. Test Experiments and Results

In the information fusion part, there is a requirement on whether the two modules generate the same number of action primitives, so an experiment is set up to record the number of action primitives in each module.

Test experiment: The experiment process was on a fixed test bench. The experimenter simulated the whole set of experimental actions and dictated the actions at the same time. Since this process does not require the movement of the robotic arm, but only observes its teaching process, the experimental program ran in the Windows 10 operating system to record the number of action primitives in the action group it generates. A total of 12 groups of different teaching tests were carried out in the process, and each group of actions was a set of coherent actions composed of 10 action primitives, as shown in Table 4, according to a certain logic and sequence. The 12 groups of action combinations are shown in Table 5. The number of action primitives of a complete set of actions detected by the speech part and the action part obtained in each teaching process is shown in Table 6.

Table 4. Action primitives.

Action Primitive Number	Experimental Action	Action Description
①	Take the test tube	Remove the test tubes from the test tube rack.
②	Shake the test tube	Hold the test tube in your hand and shake it from side to side (it is common to shake the arm repeatedly in the same direction).
③	Stir	Perform circle stirring movements with your arms (usually, the arms are kept in one position to draw a circle).
④	Liquid titration	Keeping your arms balanced and wrist vertical, perform the titration action.
⑤	Rinse the instrument	Hold the test tube in your hand for rinsing.
⑥	Take the glass rod	Retrieve the glass rod from the glass rod holder.
⑦	Gripping solids	Hold the jig to perform the action of pinching the object.
⑧	Place solids	Put the clamped solid into the test tube.
⑨	Place the test tube	Put the test tube in your hand back into the test tube rack.
⑩	Take high test tube	Hold the test tube and raise the arm to the head position.

Table 5. Composition of action primitives in action groups.

Action Group	Action Primitive
Task 1	① ④ ② ⑨
Task 2	① ⑦ ⑧ ④ ⑨
Task 3	① ④ ③ ⑨ ⑩
Task 4	① ④ ③ ⑤ ⑩ ⑨
Task 5	① ④ ⑥ ③ ⑩ ② ⑤ ⑨
Task 6	① ⑩ ⑦ ⑧ ④ ⑥ ③ ⑨
Task 7	① ④ ② ③ ⑤ ⑦ ⑧ ⑨
Task 8	① ⑦ ⑧ ⑨ ① ④ ② ⑨
Task 9	① ⑤ ⑩ ④ ②
Task 10	① ② ④ ③ ⑤ ⑦ ⑧ ⑨
Task 11	① ⑦ ⑧ ④ ③ ⑩ ② ⑤ ⑨
Task 12	① ⑤ ⑨

Table 6. Number of action primitives.

(Number/*n*)	Task 1	Task 2	Task 3	Task 4	Task 5	Task 6
Reality	4	5	5	6	8	8
Video module	4	5	5	6	6	8
Audio module	4	5	5	6	8	8
	Task 7	**Task 8**	**Task 9**	**Task 10**	**Task 11**	**Task 12**
Reality	8	8	5	8	9	3
Video module	6	7	5	8	8	3
Audio module	7	7	5	7	9	3

The data in Table 6 show that in the teaching process of some simple action groups with a small number of action primitives, the number of action primitives generated by the two modules is basically the same. According to the results, we assume that when the complexity of an action group continues to increase, with the increase in the number of action primitives that comprise the action group, the number of missed and lost ones will theoretically increase with a small confidence range. The proportional probability of the number will gradually decrease from 100% to 81.2%, and in practice, when teaching some conventional action groups, it can basically reach more than 95%—that is, the number of actions that are missed to be recognized is less than 5%, and the overall effect good.

On the basis of the above consistent situation, the coverage of the final action group is shown in Table 7 below:

Table 7. Action group coverage.

Task 1	Task 2	Task 3	Task 4
100%	80%	80%	83.3%
Task 6	**Task 8**	**Task 9**	**Task 12**
75%	71.4%	80%	100%

The results shown in the record table are supplemented with 36 teaching experiments. The teaching program is run in the Raspberry Pi Raspbian desktop operating system, four groups of basic experimental procedures are selected, and seven effective teaching experiments in each group of repeated actions are selected. Record the action group identification coverage in its process. Combining the recognition results of the speech part and the single module of the video part, the correct action primitives identified in the 28 effective teaching experiments (the experiments that are not completely executed due to the personal operation errors of the personnel) are listed in the video, speech and fusion. The comparison of numbers and accuracy rates is shown in Table 8:

Table 8. Comparison of fusion and single-module experiments.

ACT1	Reality	Video	Audio	Fusion	ACT2	Reality	Video	Audio	Fusion
Task 1	6	4	6	6	Task 1	5	5	5	5
Task 2	6	5	4	5	Task2	5	4	4	4
Task 3	6	5	4	6	Task 3	5	4	3	4
Task 4	6	6	6	6	Task 4	5	3	3	3
Task 5	6	3	4	4	Task 5	5	4	3	4
Task 6	6	4	5	5	Task 6	5	5	5	5
Task 7	6	5	5	6	Task 7	5	4	5	5
SUM	42	32	34	37	SUM	35	29	28	30
Accuracy	1	76%	82%	89%	Accuracy	1	82.8%	80%	85.7%
ACT3	**Reality**	**Video**	**Audio**	**Fusion**	**ACT4**	**Reality**	**Video**	**Audio**	**Fusion**
Task 1	4	4	4	4	Task 1	8	6	7	7
Task 2	4	4	4	4	Task 2	8	7	6	7
Task 3	4	2	3	3	Task 3	8	5	6	6
Task 4	4	4	4	4	Task 4	8	7	8	7
Task 5	4	4	4	4	Task 5	8	7	7	7
Task 6	4	4	4	4	Task 6	8	5	5	6
Task 7	4	3	3	4	Task 7	8	6	6	7
SUM	28	25	26	27	SUM	56	43	45	47
Accuracy	1	89.3%	92.8%	96.4%	Accuracy	1	76.8%	80.3%	83.9%

It can be seen from Table 4 that, compared with single-module recognition, the teaching accuracy rate after fusion is improved by about 5–7%, and the result verifies the effectiveness of audio-visual fusion. According to the number of action primitives in the test task and the coverage rate of their action groups, the overall teaching coverage rate after adding information fusion is calculated to be about 87.7%. According to the number of action primitives in the test task and its action group coverage, the overall teaching accuracy is finally calculated to be about 81.4%. This result achieves good results in the field of the non-contact skill acquisition of chemical robotic arms without using support equipment.

6.2. Test Effect and Problem Analysis

At present, the system does not have a clear solution to the inconsistency, and the follow-up research needs to improve and modify the relevant algorithms in the inconsistency.

In terms of accuracy, the stability and accuracy of the manipulator teaching system depends on the accuracy of its action behavior recognition and speech recognition parts. Since it runs in an embedded device, part of its action recognition and matching algorithm needs to be lightweight, so part of the accuracy is sacrificed in the case of improving the recognition speed. The near-field Chinese Mandarin recognition accuracy rate of the speech recognition part is 95%, and the accuracy rate of further keyword matching is higher than that of simple recognition. The two modules are continuously recognized during the teaching process, and it is inevitable that there will be lost recognition, which will reduce the accuracy. However, based on the information fusion, combining the advantages of the two, the final action coverage can be obtained. Basically, as it is stable at more than 81%, this result is satisfactory at present. After the teaching is completed, the execution effect of the robot arm on the action group basically depends on the accuracy of the teaching process. When performing the robot arm execution experiment, as shown in Figure 16, it is found that each action group can be adjusted during the teaching process. If every action primitive is identified, then the execution effect of the robotic arm can achieve the expected goal. If the action primitives of the action group are different, the results of visual action recognition can often be recognized. However, if there are two actions that are highly similar in the teaching process, misrecognition is easy to occur during visual action recognition. At this time, the accuracy can be further improved by correcting the speech recognition results, which can be seen from the above experimental data table. The different actions, the light occlusion of the test bench and the noise in the environment during the experiment have a certain impact on the teaching process. In the later experiments, a better experimental environment will be built to reduce the external influence on the experiment. Different experimenters have different proficiency in movement and different execution postures and speeds, which also makes visual action recognition more difficult. Therefore, during the teaching process, the movements should be as smooth and distinguishable as possible. The recognition accuracy should also be improved.

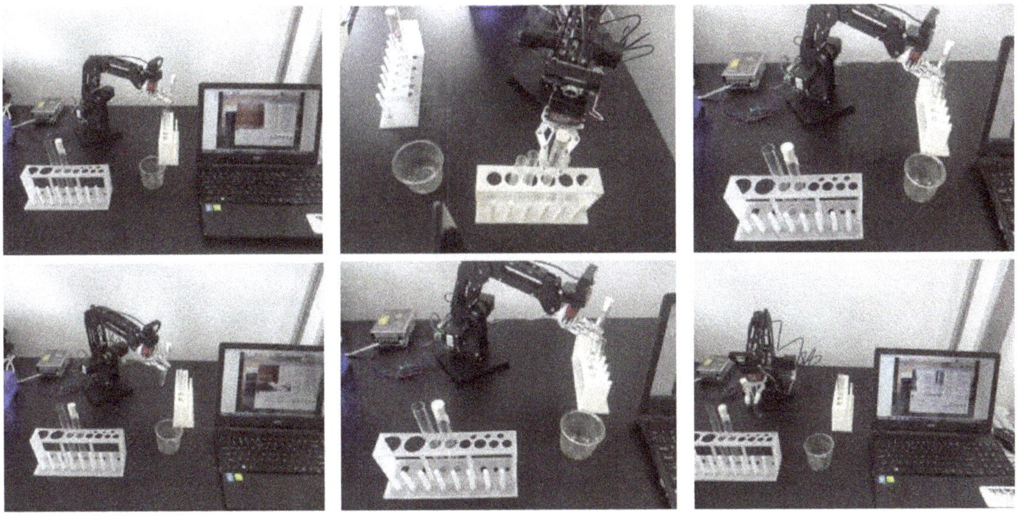

Figure 16. Experimental setup.

In terms of performance and speed, when the program is run on the Raspberry Pi system with 2 GB of RAM, the hit time of a single action primitive (the time it takes to identify the two modules) is within 1 s (the jump time of the action detection program segment is about 0.59 s), while the speech recognition is about 0.82 s. In addition, when the voice part needs to perform multiple loop verifications later, it can be changed to use the language compiled by machine code to run the loop part specifically, create a dynamic link library for this part, and use the external function library types to call, which can significantly improve the loop speed.

When performing the pose-grasping experiment, as shown in Figure 17, because the test tube and other equipment are made of transparent materials, the recognition effect is greatly affected. Try to use label recognition clamping instead (as shown in Figure 17) or use label recognition as a compensatory measure to improve the recognition efficiency.

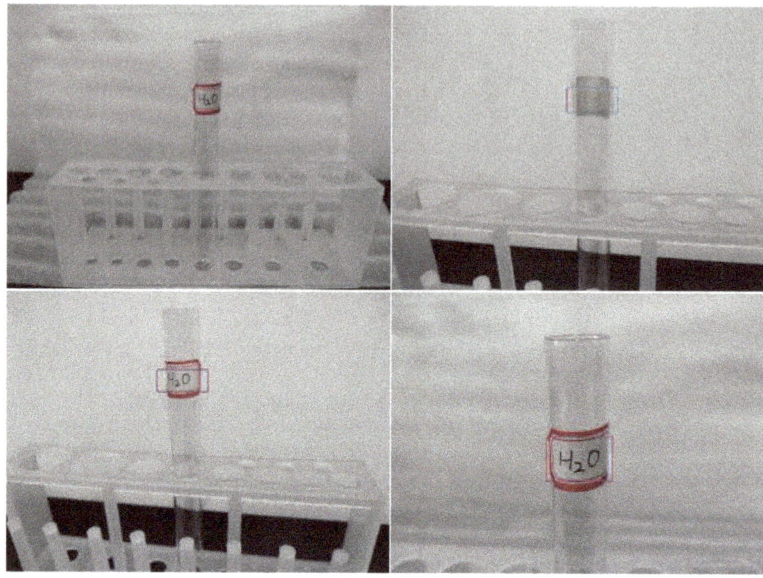

Figure 17. Label location scraping experiment.

7. Concluding Remarks

Operators need to perform various chemical experiments in chemical laboratories; these are cumbersome, and some of them are harmful to the human body. Therefore, it is necessary to use a manipulator instead of an operator to conduct experiments. However, the experimental manipulator that is currently used needs to be programmed by professional manipulator controllers, which is difficult for chemical operators. Therefore, in order to solve the above problems, this paper proposes a simple and efficient manipulator teaching system based on motion detection and speech recognition.

The operator dictates his movements during the experiment. The system uses motion detection to detect the movement of the operator's arm, matches with voice recognition, and uses algorithms related to information fusion to teach the manipulator the motor skills that should be performed. The manipulator grasps objects in combination with pose recognition during the execution process, and completes a set of experimental tasks. The accuracy rate of the system in the acquisition of motor skills can reach more than 81%.

Based on the design and experimental results of this paper, the experimental manipulator is taught and programmed to acquire and execute experimental skills, and there is a certain applicability and feasibility to use the manipulator in chemical analysis experiments. However, some problems in the system were found during the experiment: the transparent material of the test tube and light affected the recognition accuracy, the response speed of the device had a delay, similar behaviors were easily misidentified, and the recognition accuracy needed to be further improved in the actual application process, etc. Since the manipulator is a self-developed manipulator in the laboratory, no model in the corresponding simulation environment has been established, and the subsequent research and development workload is heavy. In the future, we will focus on solving the above problems, create a suitable data production environment in the laboratory, minimize the interference such as light and noise, form models in the simulation environment which are open source for everyone to use, improve the efficiency of later research, and further improve the recognition accuracy and speed. The manipulator will be replaced if necessary to achieve the purpose, but it will still be researched in the direction of low cost and low power consumption.

Author Contributions: Conceptualization, C.L. and X.M.; methodology, C.L. and X.M.; validation, X.C., H.S. and B.W.; formal analysis, X.C., H.S. and B.W.; resources, C.L. and X.M.; writing—original draft preparation, X.C. and H.S.; writing—review and editing, C.L. and X.C.; supervision, X.M.; project administration, C.L.; funding acquisition, C.L.; All authors have read and agreed to the published version of the manuscript.

Funding: This research received no external funding.

Institutional Review Board Statement: Not applicable.

Informed Consent Statement: Not applicable.

Data Availability Statement: The study did not report any data.

Conflicts of Interest: The authors declare no conflict of interest.

References

1. Lu, Y. Artificial intelligence: A survey on evolution, models, applications and future trends. *J. Manag. Anal.* **2019**, *6*, 1–29. [CrossRef]
2. Zheng, L.; Liu, S.; Wang, S. Current situation and future of Chinese industrial robot development. *Int. J. Mech. Eng. Robot. Res.* **2016**, *5*, 295–300. [CrossRef]
3. Karabegović, I. The role of industrial robots in the development of automotive industry in China. *Int. J. Eng. Work.* **2016**, *3*, 92–97.
4. Lochmüller, C.H.; Lung, K.R.; Cousins, K.R. Applications of optimization strategies in the design of intelligent laboratory robotic procedures. *Anal. Lett.* **1985**, *18*, 439–448. [CrossRef]
5. Liu, H.; Wang, L. Gesture recognition for human-robot collaboration: A review. *Int. J. Ind. Ergon.* **2018**, *68*, 355–367. [CrossRef]
6. Wang, T.M.; Tao, Y.; Liu, H. Current researches and future development trend of intelligent robot: A review. *Int. J. Autom. Comput.* **2018**, *15*, 525–546. [CrossRef]
7. Khan, A.T.; Cao, X.; Li, Z.; Li, S. Evolutionary Computation Based Real-time Robot Arm Path-planning Using Beetle Antennae Search. *EAI Endorsed Trans. AI Robot.* **2022**, *1*, 1–10. [CrossRef]
8. Li, C.; Zhu, S.; Sun, Z.; Rogers, J. BAS Optimized ELM for KUKA iiwa Robot Learning. *IEEE Trans. Circuits Syst. II Express Briefs* **2020**, *68*, 1987–1991. [CrossRef]
9. Allibert, G.; Courtial, E.; Chaumette, F. Predictive control for constrained image-based visual servoing. *IEEE Trans. Robot.* **2010**, *26*, 933–939. [CrossRef]
10. Qian, K.; Niu, J.; Yang, H. Developing a gesture based remote human-robot interaction system using kinect. *Int. J. Smart Home* **2013**, *7*, 203–208.
11. Ajit, A.; Acharya, K.; Samanta, A. A review of convolutional neural networks. In Proceedings of the 2020 International Conference on Emerging Trends in Information Technology and Engineering (ic-ETITE), Vellore, India, 24–25 February 2020; IEEE. pp. 1–5.
12. Haage, M.; Piperagkas, G.; Papadopoulos, C.; Mariolis, I.; Malec, J.; Bekiroglu, Y.; Hedelind, M.; Tzovaras, D. Teaching assembly by demonstration using advanced human robot interaction and a knowledge integration framework. *Procedia Manuf.* **2017**, *11*, 164–173. [CrossRef]
13. Li, C.; Fahmy, A.; Sienz, J. Development of a neural network-based control system for the DLR-HIT II robot hand using leap motion. *IEEE Access* **2019**, *7*, 136914–136923. [CrossRef]
14. Li, C.; Yang, C.; Annamalai, A.; Xu, Q.; Li, S. Development of writing task recombination technology based on DMP segmentation via verbal command for Baxter robot. *Syst. Sci. Control. Eng.* **2018**, *6*, 350–359. [CrossRef]
15. Esteban, J.; Starr, A.; Willetts, R.; Hannah, P.; Bryanston-Cross, P. A review of data fusion models and architectures: Towards engineering guidelines. *Neural Comput. Appl.* **2005**, *14*, 273–281. [CrossRef]
16. Liu, K.; Liu, B.; Blasch, E.; Shen, D.; Wang, Z.; Ling, H.; Chen, G. A cloud infrastructure for target detection and tracking using audio and video fusion. In Proceedings of the IEEE Conference on Computer Vision and Pattern Recognition Workshops, Boston, MA, USA, 7–12 June 2015; pp. 74–81.
17. Suay, H.B.; Toris, R.; Chernova, S. A practical comparison of three robot learning from demonstration algorithm. *Int. J. Soc. Robot.* **2012**, *4*, 319–330. [CrossRef]
18. Zhizeng, L.; Jingbing, Z. Speech recognition and its application in voice-based robot control system. In Proceedings of the 2004 International Conference on Intelligent Mechatronics and Automation, Chengdu, China, 26–31 August 2004; IEEE. pp. 960–963. [CrossRef]
19. Habil, H.J.; Al-Jarwany, Q.A.; Hawas, M.N.; Mnati, M.J. Raspberry Pi 4 and Python Based on Speed and Direction of DC Motor. In Proceedings of the 2022 4th Global Power, Energy and Communication Conference (GPECOM), Nevsehir, Turkey, 14–17 June 2022; IEEE. pp. 541–545. [CrossRef]
20. Tan, M.; Le, Q. Efficientnetv2: Smaller models and faster training. In Proceedings of the International Conference on Machine Learning (PMLR), Virtual, 18–24 July 2021; pp. 10096–10106.
21. Schuldt, C.; Laptev, I.; Caputo, B. Recognizing human actions: A local SVM approach. In Proceedings of the 17th International Conference on Pattern Recognition (ICPR), Cambridge, UK, 26 August 2004; IEEE. Volume 3, pp. 32–36. [CrossRef]

22. Duan, Y.; Zou, B.; Xu, J.; Chen, F.; Wei, J.; Tang, Y.Y. OAA-SVM-MS: A fast and efficient multi-class classification algorithm. *Neurocomputing* **2021**, *454*, 448–460. [CrossRef]
23. Kaushik, A.; Kaur, G. Review On: Gait Recognition Technique using SVM and K-means with Gait PAL and PAL Entropy. *(IJCSIT) Int. J. Comput. Sci. Inf. Technol.* **2014**, *5*, 3699–3702.
24. Abdullah, D.M.; Abdulazeez, A.M. Machine Learning Applications based on SVM Classification A Review. *Qubahan Acad. J.* **2021**, *1*, 81–90. [CrossRef]
25. Tan, M.; Le, Q. Efficientnet: Rethinking model scaling for convolutional neural networks. In Proceedings of the International Conference on Machine Learning (PMLR), California, CA, USA, 9–15 June 2019; pp. 6105–6114. [CrossRef]
26. Ermolina, A.; Tiberius, V. Voice-controlled intelligent personal assistants in health care: International Delphi Study. *J. Med. Internet Res.* **2021**, *23*, e25312. [CrossRef]
27. Yunhuan, L.I.; Jiwei, W.E.N.; Li, P.E.N.G. High frame rate Light-Weight Siamese Network target tracking. *J. Front. Comput. Sci. Technol.* **2021**, 1–13. [CrossRef]
28. Xu, R. Path planning of mobile robot based on multi-sensor information fusion. *EURASIP J. Wirel. Commun. Netw.* **2019**, *2019*, 1–8. [CrossRef]
29. Redmon, J.; Angelova, A. Real-time grasp detection using convolutional neural networks. In Proceedings of the 2015 IEEE International Conference on Robotics and Automation (ICRA), Seattle, WA, USA, 25–30 May 2015; IEEE. pp. 1316–1322. [CrossRef]

machines

Article

Lattice Structure Design Method Aimed at Energy Absorption Performance Based on Bionic Design

Gang He *, Hu Yang, Tao Chen, Yuan Ning, Huatao Zou and Feng Zhu

Department of Mechanics, College of Mechanical & Electrical Engineering, Hohai University, Changzhou 213022, China
* Correspondence: hegang@hhu.edu.cn

Abstract: To obtain the lattice structure with excellent energy absorption performance, the structure of loofah inner fiber is studied to develop bionic design of lattice structure by experiment and simulation analysis method. From the compression experiment about the four bionic multi-cell lattice structures (bio-45, bio-60, bio-75, and bio-90) and VC lattice structures, we found that all are made of PLA and fabricated by the fused deposition modeling (FDM) 3D printer. The comprehensive performance of bio-90 lattice structure is the best in the performance of the specific volume energy absorption (SEA_v), the effective energy absorption (EA), and the specific energy absorption (SEA). Based on the experimental result, the energy absorption performance of bio-90 lattice structure is then studied by the simulation analysis of influence on multiple parameters, such as the number of cells, the relative density, the impact velocity, and the material. The results can provide a reference for the design of highly efficient energy absorption structures.

Keywords: lattice structure; loofah; energy absorption performance; bionic design; relative density

Citation: He, G.; Yang, H.; Chen, T.; Ning, Y.; Zou, H.; Zhu, F. Lattice Structure Design Method Aimed at Energy Absorption Performance Based on Bionic Design. *Machines* **2022**, *10*, 965. https://doi.org/10.3390/machines10100965

Academic Editor: César M. A. Vasques

Received: 29 August 2022
Accepted: 13 October 2022
Published: 21 October 2022

1. Introduction

Lattice structure is the space truss structure, which belongs to the scope of porous structure, and is composed of nodes, rods, and panels on a certain rule. It was first proposed by Ashby [1] of Cambridge University and Evans [2] of Harvard University in 2001. Lattice structures have gained extensive attention for their comprehensive properties such as their light weight, high strength, and impact resistance. Their open internal spaces make them have the advantage in load bearing and energy absorption, and also exhibit better properties than foam structures [3]. Several typical lattice structures [4] are shown as Figure 1. Cao [5] modified the same diameter core rod of single-cell structure into a variable cross-section core rod and proposed an improved single-cell structure. Bai [6] used PA2200 as a raw material, printed three experimental samples of body-centered cubic (BCC), rod-diameter-change graded body-centered cubic (RGBCC) and size-change graded body-centered cubic (SGBCC), and studied the effect of gradient direction on the structure. Andrew [7] discussed the influence of impact energy, relative density, plate thickness, and impact angle on the dynamic impact behavior of the lattice structure by the weight drop experiment. Hammetter [8] studied the effects of slenderness ratio, inclination angle, and single-cell layer on the compression deformation mechanism of the lattice structure.

The lattice structures are considered as the most promising structure–function integrated structures at present. Recently, the development of additive manufacturing technology has resulted in the possibility of the manufacture of a complex inner structure, and that makes it is flexible to product design. To obtain special performance [9,10], the method of design about lattice structures has become a hot research topic. Squid bones, beetle elytra, glass sponges, and bamboos are found to have porous features at the microscopic level and exhibit excellent comprehensive performance, which gives researchers substantial inspiration [11,12]. After extensive studies, the biomimetic method has been proven to be an effective way to design lattice structures with better properties.

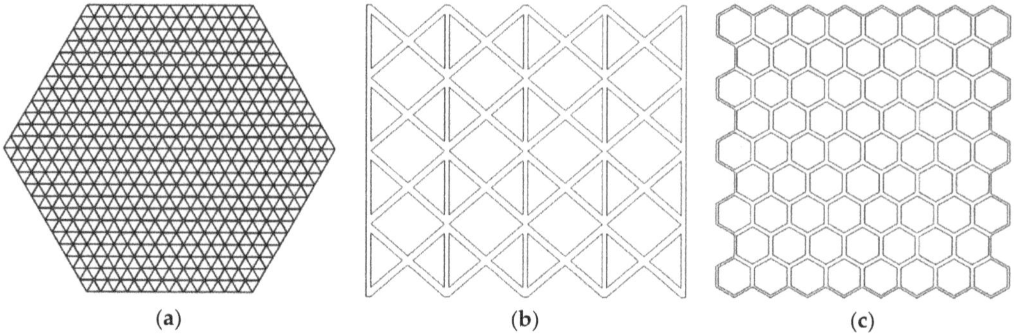

Figure 1. Several typical lattice structures: (**a**) SC lattice structure; (**b**) FC lattice structure; (**c**) Honeycomb structure.

Based on the excellent performance of the lattice structure, it is widely used in aerospace, safety protection, medical, and other fields. In the aerospace field, it has always been a major design goal that the structure is lightweight and has smaller material consumption, less fuel consumption, and higher performance. The excellent performance of lattice structure means it has great application value in the aerospace field. Helmet lining is the main component of the helmet; it can dissipate impact energy and reduce the load transmitted to the head in case of accident. SOE and HEXR [13] proposed a honeycomb-lined helmet and found that the impact energy was better distributed and absorbed by the honeycomb-lined helmet, and it had great potential in improving the safety of the helmet. The KOLLODE alliance and four innovative companies (KUPOL, TAXIX, ShapeShift3d, Numalogics) used virtual design and 3D printing technology to manufacture a helmet liner KUPOL, which can absorb energy and redirect impact force.

Loofah is highly porous material, and it has great application potential in sound absorption, shock absorption, and cushioning. Due to the unique fiber structure, loofah can bear a large load, and Zou [14] made an optimization algorithm on the displacement and stress relationship of the geometric structure. For studying the mechanical properties of loofah, Wang [15] found that the inner surface of the loofah plays a major role in supporting the axial load by conducting the tests of quasi-static compression and dynamic impact on the loofah structure. Elmadih [16] researched the ability of these lattices to provide vibration attenuation at frequencies greater than their natural frequency. Chen [17] et al. performed a multi-scale study to explore the relationship between the structure and mechanical properties of different layers of fibers (inner fibers and outer fibers) and different directions (transverse and longitudinal), and the results showed that the inner ring wall fibers contribute the most to the longitudinal properties of the loofah, while the mechanical strength of the core fiber is lower than that of the inner fiber. Qing [18] designed a new type of ultra-light bionic tube structure by combining the structure of the loofah and the pores of honeycomb hexagons, and found that its equivalent elastic modulus is from 166.9 to 180.59 MPa by the compression test.

At present, most bionic designs are aimed at the macrostructure of loofah. The fiber structure of loofah is the main factor affecting its mechanical properties. In this paper, the bionic design method is used to design the single-cell lattice structure. It is studied the influence of different parameters on the energy absorption of the lattice structure, such as the number of cells, the relative density, the impact velocity, and material. The results are useful for the design of a highly efficient energy absorption structure.

2. Modeling of Single-Cell Structure Loofah Lattice

The loofah is a natural fiber network structure [19], and it is mainly composed of the outer layer, the middle layer, the inner layer, and the core layer, as shown in Figure 2.

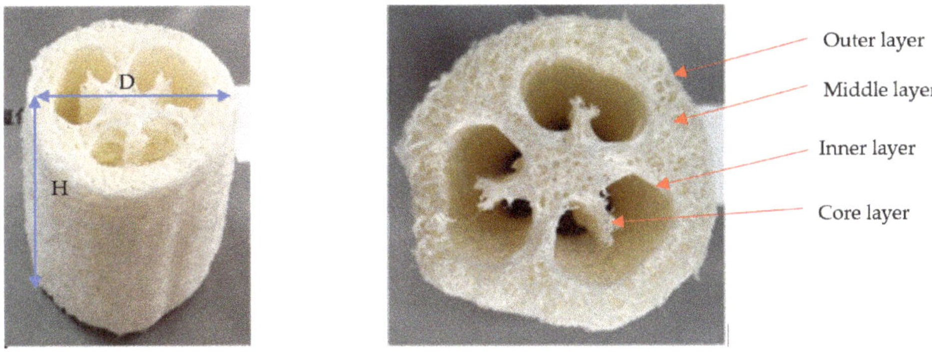

Figure 2. Overall structure of loofah.

Although the fiber structure of loofah is interlaced, its distribution is not disordered. The growth direction of fiber bundles in different layers is different. Most of the outer layer fibers of the loofah structure grow along the circumferential direction in a circular shape; the inner layer fibers mainly distribute along the longitudinal axis of the loofah structure and have the large diameter and regular texture; the fibers of the core layer are interwoven, forming a honeycomb structure, as shown in Figure 3.

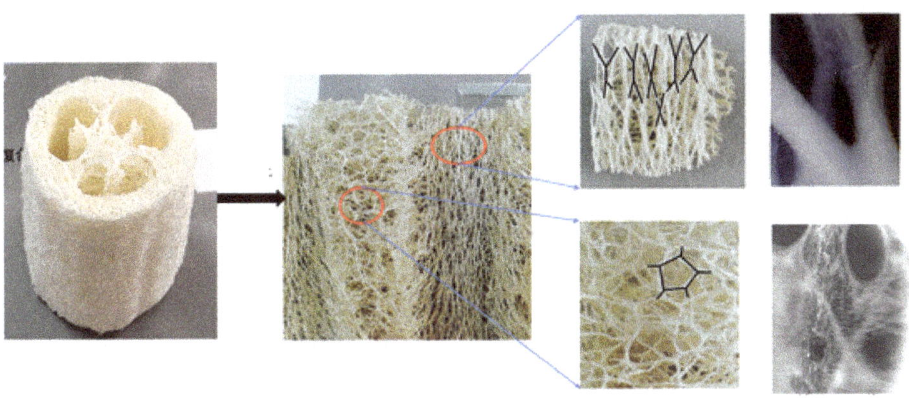

Figure 3. Regional structure of loofah sponge.

We extracted and simplified the textures of loofah fiber to obtain its structural features and summarized them into the 2D configurations of type I and type II, as Figure 4a shown. The inner layer fibers are actually interconnected in 3D space. The simplified spatial structure is shown as Figure 4b. After giving a certain rod diameter, the 3D single-cell structure is shown as Figure 4c.

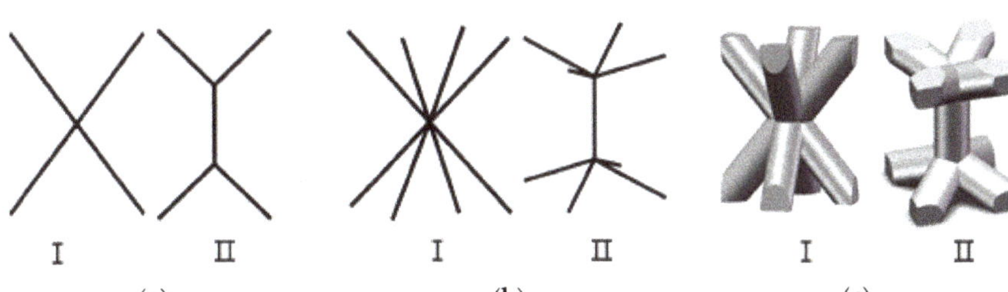

Figure 4. Evolution of bionic single-cell structure: (**a**) 2D structure; (**b**) 3D structure; (**c**) cell structure.

The difference between type I and type II structure is the values of angle. The type I represents the angle of 45° lattice structure, and given that the three lattice structures with angles of 60°, 75°, and 90° have uniform characteristics, they are represented by type II. By setting angles of 45°, 60°, 75°, and 90°, the series of the single-cell structure is built, and they are respectively named as bio-45, bio-60, bio-75, and bio-90, as shown in Table 1.

Table 1. Bionic cells with different configurations.

Number	Angle	2D Structure	3D Structure
1	45°	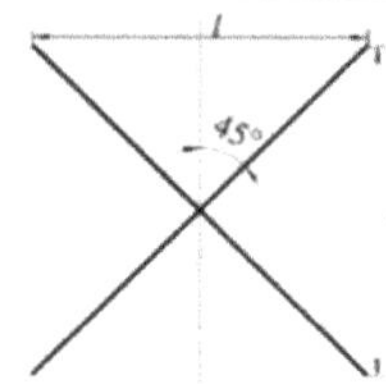	
2	60°	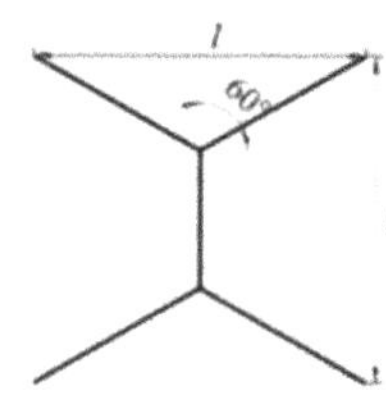	
3	75°	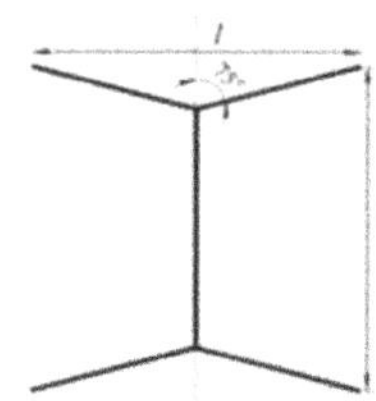	
4	90°	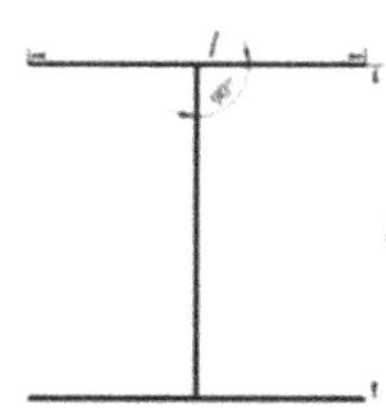	

The mechanical properties of lattice structures are mainly determined by their material and structural, and the main parameter of structural is the relative density. The relative density is the ratio of the solid volume in a single-cell structure to the volume of the cube, and its calculation formula is shown as Equation (1):

$$\overline{\rho} = \frac{V_s}{V} \times 100\% \quad (1)$$

where $\overline{\rho}$ denotes relative density of the single-cell structure, and Vs and V respectively denote the solid volume of the single-cell structure and the total volume of the cube.

The relative density is adjustable for a specific cell structure. In this section, the relative density of single-cell structure is theoretically deduced, and the specific derivation process of the bio-45 single-cell structure is given. There are two variable structural parameters in the bio-45 single-cell structure, which are the cell size and the core rod diameter. From Figure 5, the cell structure has eight of the same core rods.

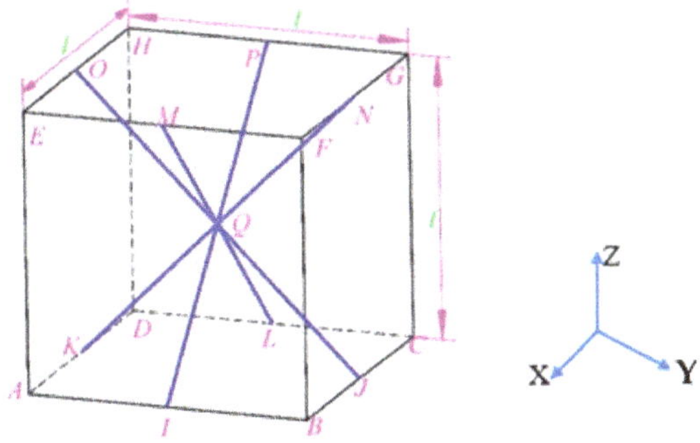

Figure 5. Lattice cell diagram.

According to the geometric relationship, the length of each rod of the bio-45 single-cell structure can be calculated as Equation (2):

$$l_0 = \frac{\sqrt{l^2 + l^2}}{2} = \frac{\sqrt{2}}{2}l \tag{2}$$

where l_0 denotes the length of each rod of the bio-45 single-cell structure, and l denotes the length, width, and height of the cell structure.

The solid volume of the single-cell structure is given as Equation (3):

$$Vs = 8\pi\frac{{d_0}^2}{4}l_0 = \sqrt{2}\pi {d_0}^2 l \tag{3}$$

where d_0 denotes the diameter of the core rod.

The spatial volume of the single-cell structure is shown as Equation (4):

$$V_{bio-45} = l^3 \tag{4}$$

Based on the Equations (3) and (4), the relative density of the single-cell structure is as follows:

$$\overline{\rho} = \frac{Vs}{V_{bio-45}} = \frac{\sqrt{2}\pi {d_0}^2}{l^2}$$

Based on theoretical derivation, the calculation formula of relative density about each of the four kinds of single-cell structure can be obtained in Table 2.

Table 2. Calculation formula of relative density of the single-cell structure.

Single-Cell Type	Relative Density
bio-45	$\overline{\rho} = \sqrt{2}\pi {d_0}^2/l^2$
bio-60	$\overline{\rho} = \left(7\sqrt{3}+3\right)\pi {d_0}^2/12l^2$
bio-75	$\overline{\rho} = 1.219\pi {d_0}^2/l^2$
bio-90	$\overline{\rho} = 3\pi {d_0}^2/4l^2$

Without considering the overlapping part of the connection between the rods and the volume at the end of rods, as shown in Figure 6, with the different edge lengths of the cube, there are differences between the calculation formula of relative density and actual relative density based on the actual relative density obtained by the volume evaluation module of the modeling software. The calculation formula of the relative density of the single-cell structures is modified by the theoretical derivation and polynomial fitting method, as shown in Table 3. The deviation between the relative density calculated by the two formulas and the actual density is shown in Figure 7. The error of the result calculated by the modified formula is within 0.08%.

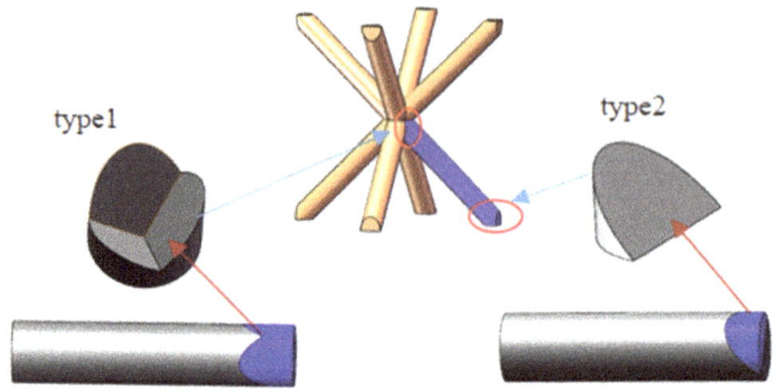

Figure 6. Volume to be subtracted in correction.

Table 3. Modified calculation formula of relative density.

Single-Cell Type	Relative Density
bio-45	$\overline{\rho}' = \overline{\rho} - \left[\begin{pmatrix} -3.564e^{-3}+1.1264e^{-3}d_0 \\ +9.72e^{-4}d_0^2+3.909{d_0}^3 \end{pmatrix}/l^3\right]$
bio-60	$\overline{\rho}' = \overline{\rho} - \left[\begin{pmatrix} 1.295-2.505d_0 \\ +1.39d_0^2+3.449{d_0}^3 \end{pmatrix}/l^3\right]$
bio-75	$\overline{\rho}' = \overline{\rho} - \left[\begin{pmatrix} 18.81-33.2d_0 \\ +18.22d_0^2+1.551{d_0}^3 \end{pmatrix}/l^3\right]$
bio-90	$\overline{\rho}' = \overline{\rho} - \left[\begin{pmatrix} -5.968e^{-4}-7.528e^{-4}d_0 \\ -1.159e^{-2}d_0^2+1.414{d_0}^3 \end{pmatrix}/l^3\right]$

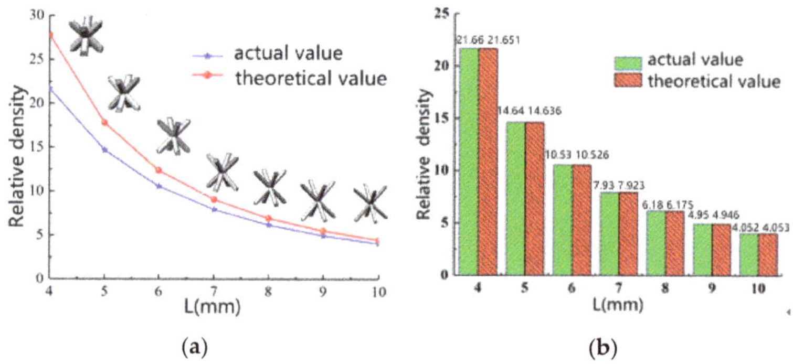

Figure 7. The differences between the theoretical and actual relative density: (**a**) primary formula; (**b**) modified formula.

3. Experiment of Static Energy Absorption Characteristics about Loofah Lattice Structure

The above four types of single-cell structure and the Vertex Cube (VC) obtained by topology optimization [20] are used to design multi-cell lattice structures, which are shown in Figure 8a, and their parameters are given in Table 4. The samples are arranged using a Solidworks software [21] and made by a fused deposition molding (FDM) 3D printer, as shown in Figure 8b. The 3D printer is the A8 equipment of the JGMaker company, as shown in Figure 9. The precision of 3D printer is 0.05~0.2mm. In order to reduce manufacturing deviation, the printing temperature, speed, printing thickness, and other parameters are all in accordance with the recommended values of the equipment. The printing direction of the printer is bottom-up, and the different printing directions of the adjacent two layers lead to structural anisotropy. The five lattice structures used in this paper all use PLA as the support materials.

Table 4. Lattice structure parameters.

Name of Lattice Structure	Length × Width × Height (mm)	Core Rod Diameter	Relative Density
bio-45	20 × 20 × 20	3.036	0.3
bio-60	20 × 20 × 20	3.305	0.3
bio-75	20 × 20 × 20	3.702	0.3
bio-90	20 × 20 × 20	4.112	0.3
VC	20 × 20 × 20	4.145	0.3

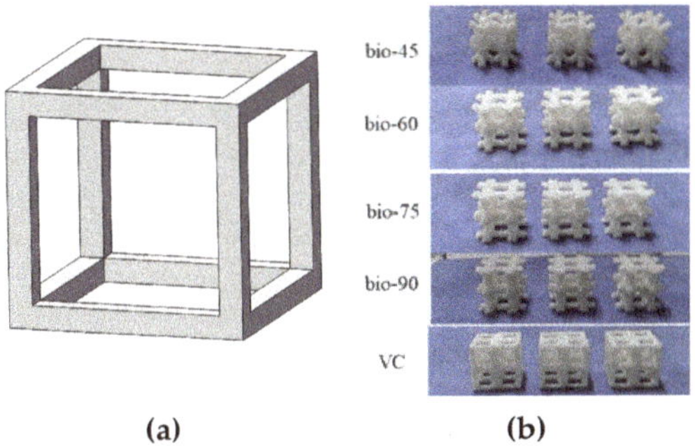

Figure 8. (**a**) 3D structure of Vertex Cube (VC); (**b**) samples of five different lattice structures.

Figure 9. Molding equipment.

By the DZ-101 machine of Dazhong Instrument Co., Ltd., this quasi-static compression experiment of the above five models (bio-45, bio-60, bio-75,bio-90, and VC) is conducted, as shown in Figure 10, and the test is divided into three steps:

(1) Put the five models on the pressure table of DZ-101 machine, and adjust the position of the compression head to make it in good contact with the model.
(2) Start the machine, and set the compression speed of the indenter as 2 mm/min; obtain the curve from the testing machine during the compression process.
(3) After the model is compacted, replace the sample and continue the above steps until all samples are tested. Stop the compression experiment and close the machine.

Figure 10. Compression test specimen clamping.

The quasi-static compression experiment can measure the mechanical response and the energy absorption characteristics of the lattice structure specimen under static load. The displacement of the compression head and the counterforce are recorded in the tests, and the stress–strain curves about five different lattice structures of bio-45, bio-60, bio-75, bio-90, and VC are shown as Figure 11. The curves are overall continuity without obvious fluctuations, which indicate that the mainly failure forms of the lattice structures are plastic deformation failures and no serious brittle fracture.

As Figure 11 shown, the compression process has gone through three stages: the elastic stage, the platform stage, and the densification stage. In the densification stage, the load changes suddenly, and the lattice structure has completely failed. During this stage, the absorbed energy has no meaning, and densification is generally regarded as the end point of energy absorption [22]. Based on the elastic phase data of the five structures, their elastic modulus and the yield strength modulus ratio can be obtained as shown in Table 5.

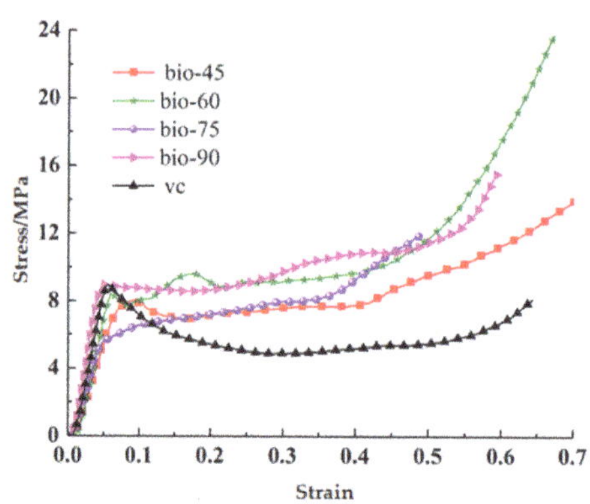

Figure 11. Testing stress–strain curves.

Table 5. Test results.

	bio-45	bio-60	bio-75	bio-90	VC
Elastic modulus (MPa)	159.19	184.11	154.73	204.92	198.73
Yield strength (MPa)	8.076	8.379	5.590	9.028	8.999

The elastic modulus and yield strength have no relationship with the design angle, as Table 5 shown. The elastic modulus of the bio-90 lattice structure is 32.44% higher than that of the bio-75, and only 3.02% higher than that of VC lattice structure. The yield strength of the bio-90 lattice structure is 61.50% higher than that of the bio-75 structure, 11.79% higher than that of the bio-45 structure, 7.75% higher than that of the bio-60 structure, and 0.32%

higher than that of the VC lattice structure. The initial stiffness and strength of the bio-90 lattice structure are the highest among the five structures.

For evaluating the energy absorption performance of lattice structure, it is necessary to define several indexes [23], such as specific volume energy absorption, effective energy absorption, and specific energy absorption. The definition of specific volume energy absorption is

$$SEAv = \int_0^{\varepsilon d} \sigma d\varepsilon \tag{5}$$

where $SEAv$ denotes specific volume energy absorption, σ denotes structural stress, and εd denotes densification strain. The definition of effective energy absorption is

$$EA = (SEAv) \times V \tag{6}$$

The definition of specific energy absorption is

$$SEA = \frac{(EA)}{m} \tag{7}$$

where m denotes mass. The energy absorption efficiency can be defined as

$$\eta(\varepsilon) = \frac{1}{\sigma(\varepsilon)} \int_0^{\varepsilon d} \sigma(\varepsilon) d\varepsilon = \frac{(SEAv)}{\sigma(\varepsilon)} \tag{8}$$

where $\sigma(\varepsilon)$ denotes load. The platforms stress can be defined as

$$\sigma_p = \frac{1}{\varepsilon_d} \int_0^{\varepsilon_d} \sigma(\varepsilon) d\varepsilon \tag{9}$$

The energy absorption efficiency–strain relationships [23,24] of the five structures are shown as Figure 12. The energy absorption efficiency of the bio-90 lattice structure is higher than that of the other lattice structures, and this difference becomes larger with the increase of strain. From Figure 13, the specific energy absorption (*SEA*), the specific volume energy absorption (SEA_v), and effective energy absorption (*EA)* values of the bio-90 lattice structure are greater than those of the other four lattice structures; the *SEA* value of the bio-90 lattice structure is 2.335 times greater than that of the VC lattice structure, the SEA_v value is 1.933 times, and the *EA* value is 1.933 times.

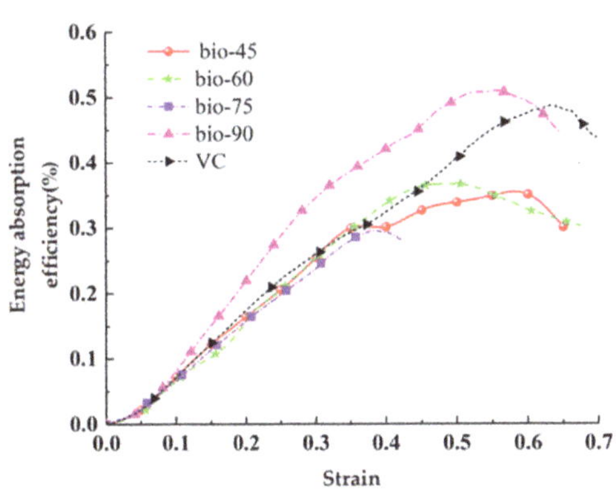

Figure 12. Energy absorption efficiency.

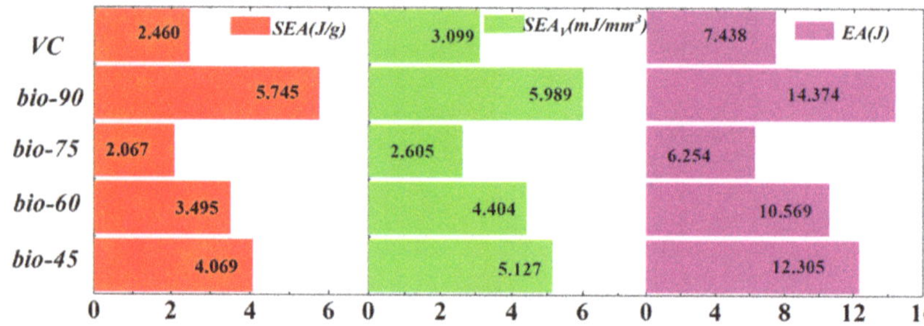

Figure 13. Comparison of *SEA*, SEA_v, and *EA*.

The *SEA*, SEA_v, and *EA* values of the bio-75 lattice structure are the smallest, which are respectively 35.98%, 43.50%, and 43.51% of those of the bio-90 lattice structure. In summary, the bio-90 lattice structure has the best energy absorption performance among the five lattice structures, not only in the stable plateau phase, but also in the specific energy absorption (*SEA*) and other indexes.

4. Analysis with the Influence of Dynamic Performance Parameters about Loofah Lattice Structure

In order to verify the accuracy of the simulation results, the quasi-static compression simulation analysis is carried out through ANSYS software, and the results are compared with the stress and strain results in Section 3. First, establish the lattice cell model in the Spaceclaim module, and obtain the overall size of 20 mm × 20 mm × 20 mm lattice structure, such as the model of bio-45 lattice structure, which is shown in Figure 14a.

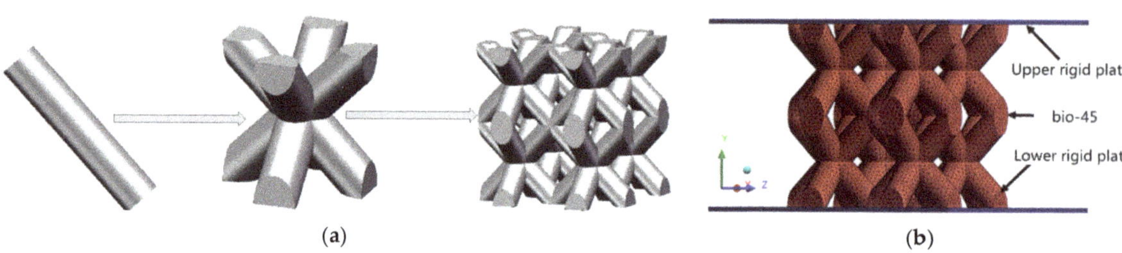

Figure 14. (**a**) The lattice structure of bio-45; (**b**) the mesh of bio-45.

The material is PLA, and the lattice structure is anisotropic, which is the same as the 3D-printed lattice structure. Two steel plates are placed on the upper and lower surfaces of the lattice structure, respectively. The steel plate and the lattice structure are in friction contact, with a friction coefficient of 0.15. The constraint condition is that the lower steel plate is fixed, and the rotation of the upper steel plate in the X, Y, and Z directions is limited. The tetrahedral element grid is adopted for grid division, with the grid size of 0.6mm and the number of elements of 106325, as shown in Figure 14b. The modeling and mesh generation processes of the bio-60, bio-75, bio-90, and VC lattice structures are the same as the bio-45, and results of the finite element analysis are shown as Figure 15.

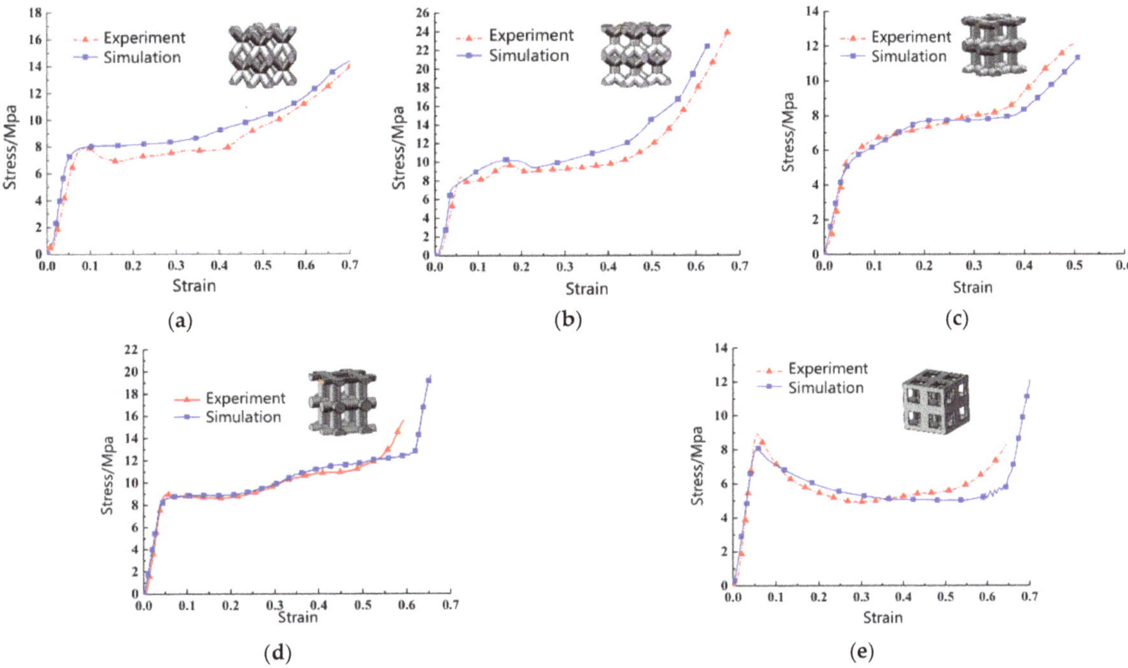

Figure 15. Stress–strain curve of experiment and simulation: (**a**) bio-45; (**b**) bio-60; (**c**) bio-75; (**d**) bio-90; (**e**) VC.

As Figure 15 shown, the relative errors of the simulation and test are within the acceptable range, and the maximum relative errors of the five structures are about 10%, indicating that the simulation results are similar to that of experiment, which verifies the reliability of the finite element model. Then, the influence of the number of cells, relative density, impact velocity, and material on the energy absorption characteristics of the lattice structure is analyzed with the finite element method.

4.1. The Influence of the Number of Cells on the Energy Absorption Characteristics

Keeping the relative density of the bio-90 bionic lattice structure at 0.3, the impact velocity is 30 m/s, and the single-cell structures are arranged by the arrays of $2 \times 2 \times 2$, $3 \times 3 \times 3$, $4 \times 4 \times 4$, $5 \times 5 \times 5$, $6 \times 6 \times 6$, and $7 \times 7 \times 7$, respectively, to get the six lattice structures of different overall sizes, the finite element model is shown as Figure 16.

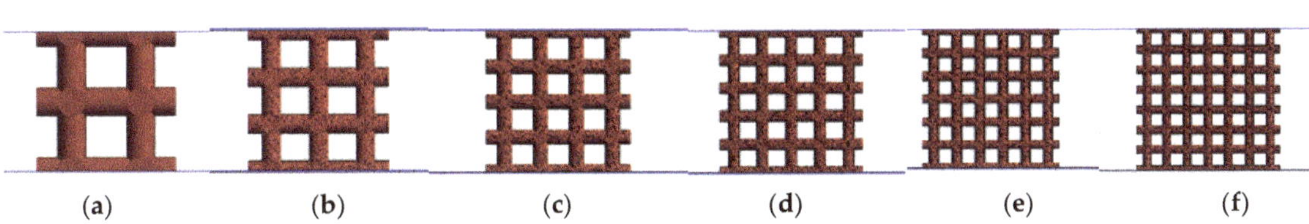

Figure 16. Finite element model of lattice structure: (**a**) 8-cell structure; (**b**) 27-cell structure; (**c**) 64-cell structure; (**d**) 125-cell structure; (**e**) 216-cell structure; (**f**) 343-cell structure.

The simulation analysis results of the bio-90 lattice structure with different cell numbers are shown in Figures 17 and 18. Figure 17 shows the stress–strain curves of the lattice structure with different cell numbers. They still have an obvious elastic phase, a plateau phase, and a densification phase at the impact velocity of 30 m/s, which are similar to the results of the quasi-static compression.

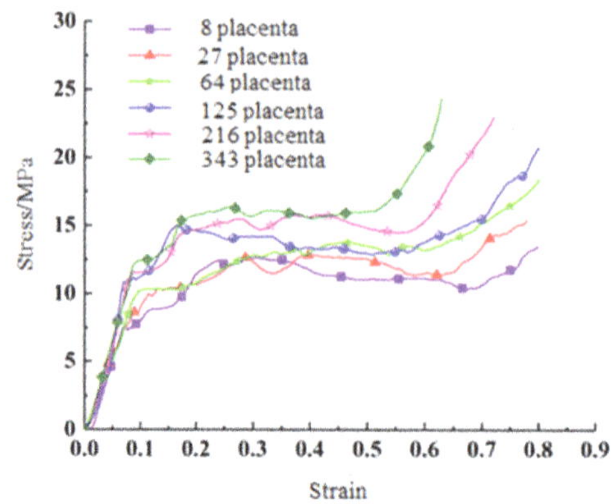

Figure 17. Stress–strain curves with different number of cells.

Crash load efficiency is defined as

$$CLE = \frac{(MCF)}{(MIF)} \times 100\% \tag{10}$$

where MCF denotes maximum current force, and MIF denotes maximum instantaneous force, which is defined as

$$MCF = \frac{(EA)}{S} \tag{11}$$

where S denotes effective displacement.

The slope of the stress–strain curves of the lattice structure with different cell numbers in the elastic phase is almost the same. This indicates that the number of single cells has little effect on the elastic phase of the structure, which is consistent with the analytical conclusion of the literature [12].

As shown in Figure 18, with the number of cell elements of the lattice structure increasing, the specific volume energy absorption (SEA_v), the specific energy absorption (SEA), the crash load efficiency (CLE), and platform stress (σ_p) increase and then decrease. However, in general, the number of single cells does not have a significant effect on the above four indexes.

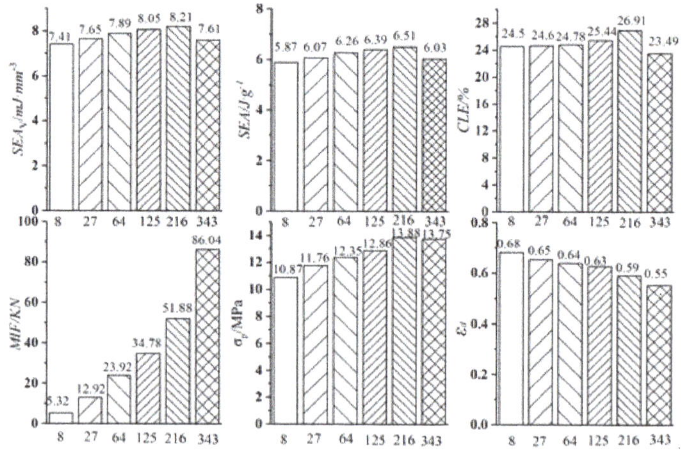

Figure 18. Energy absorption index with different number of cells.

4.2. The influence of Relative Density on Energy Absorption Characteristics

To investigate the effect of relative density on the energy absorption performance of the lattice structure, we use six relative densities of 0.09, 0.16, 0.23, 0.30, 0.37, and 0.44 with 125-grid lattice structures and the impact velocity of 30 m/s. The stress–strain curves and the indexes of SEA_v, SEA, CLE, MIF, σ_p, and ε_d at different relative densities are shown in Figures 19 and 20.

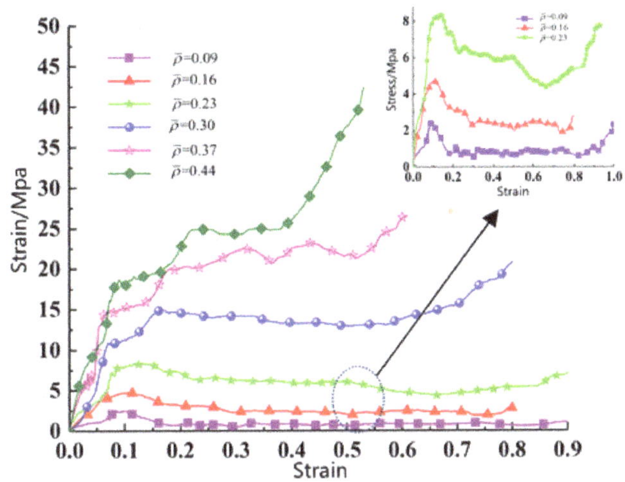

Figure 19. Stress–strain curves under different relative densities.

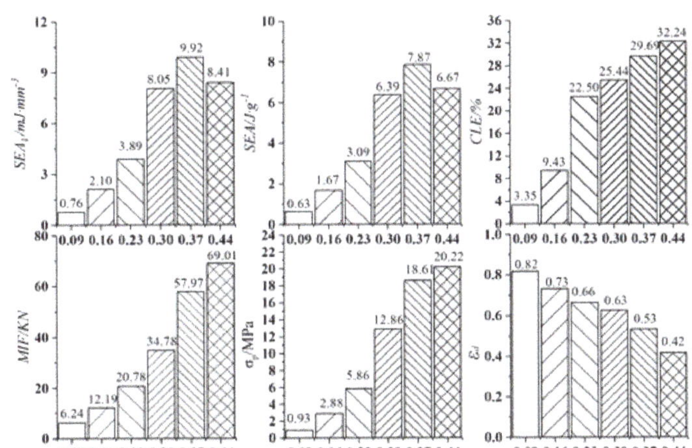

Figure 20. Energy absorption index at different relative densities.

As Figure 19 shown, at the elastic phase, the slope of the stress–strain curve becomes steeper with the relative density increasing, which indicates that the initial stiffness of the lattice structure is higher. At the platform phase, with the relative density increasing, the corresponding platform stress increases more obviously, and the overall stress–strain curve rises. At the densification phase, the overall stress–strain curve rises with the increase of the relative density, and the structure will reach densification in advance to a certain extent, which is negative to the energy absorption of the structure.

As Figure 20 shown, when the relative density is less than 0.37, with the relative density increasing, the specific volume energy absorption (SEA_v) and the specific energy absorption

(*SEA*) increase. However, when the relative density is 0.44, the specific volume energy absorption (SEA_v) and the specific energy absorption (*SEA*) are smaller than the values of those when the relative density is 0.37. This is because, although there is a relatively high-stress plateau at this point, the increase in relative density makes the strain of the structure to reach the densification stage much lower and the structure transitions to the densification stage earlier.

4.3. The influence of Impact Velocity on Energy Absorption Characteristics

In order to study the effect of impact velocity on the energy absorption performance of the lattice structure, the impact simulations of the lattice structure are performed under six different velocities of 10 m/s, 30 m/s, 50 m/s, 70 m/s, 90 m/s, and 110 m/s, with the relative density of 0.23 and cell elements of 125. The stress–strain curves of the lattice structure are shown as Figure 21. From Figure 21, at the six impact velocities, the overall trend of stress–strain curves of the lattice structure with specific relative density is relatively similar, the overall stress–strain curves under high speed are higher than those under low speed, and the height is consistent at the elastic phase of the compression process.

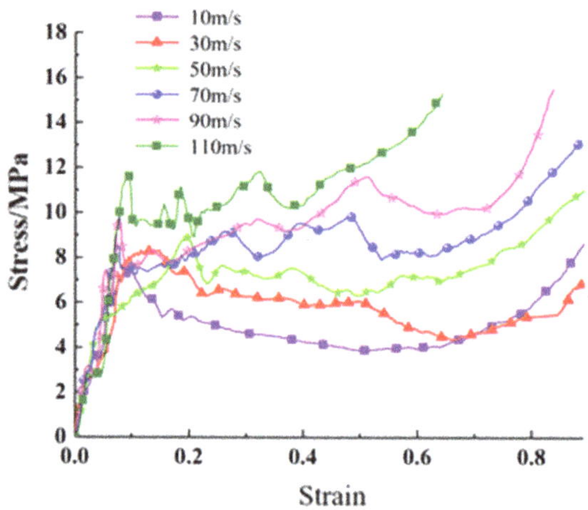

Figure 21. Stress–strain curves at different impact velocities.

With the impact velocity increasing, the strain rate sensitivity is higher. At the plastic stage, the difference between the curves gradually increases, the load becomes unstable, and there is greater fluctuation of its stress–strain curve. This phenomenon is particularly significant after v = 50 m/s.

By calculating various energy absorption and other indicators at six different impact speeds, the results about the discussion of the energy absorption characteristics of the lattice structure under dynamic impact response are shown in Figure 22.

With the impact velocities increasing from 10 m/s to 110 m/s, the densification strain of the structure increases from 0.64 to 0.74, while the specific volume energy absorption (SEA_v), specific energy absorption (*SEA*), and maximum instantaneous force (*MIF*) increase monotonically from 2.90 mJ·mm^{-3} to 6.92 mJ·mm^{-3}, 2.30J·g^{-1} to 5.49J·g^{-1}, and 18.38 kN to 29.55 kN, respectively. On the contrary, the crash load efficiency (*CLE*) always decreases. Compared with that at the velocity of 10 m/s, the crash load efficiency (*CLE*) at v = 110 m/s decreases 39.03%.The results show that the higher impact velocity, the more unstable energy absorption process of the structure, which is also consistent with the actual experiment process.

Since the different impact velocities affecting the force state of the structure at the beginning, the impact velocities will have a significant effect on the deformation mode of the dotted structure, which is the fundamental reason for changing the energy absorption performance of the structure. The deformation mode of the structure under different impact velocities is shown in Figure 23.

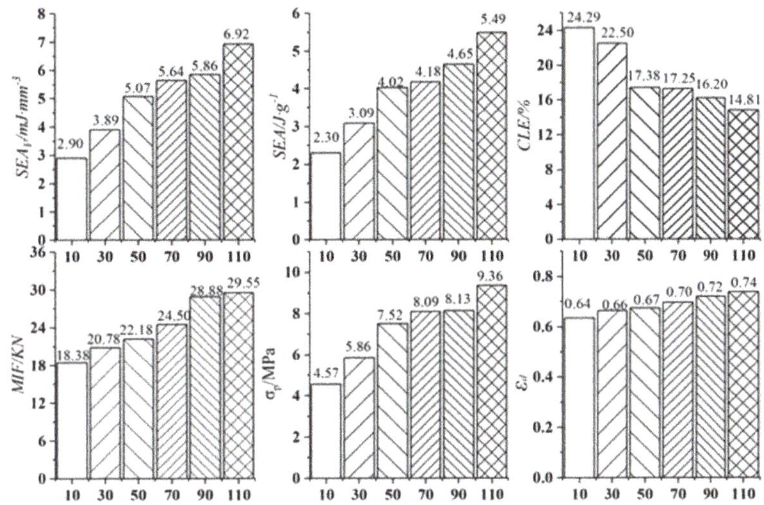

Figure 22. Energy absorption index under different impact velocities.

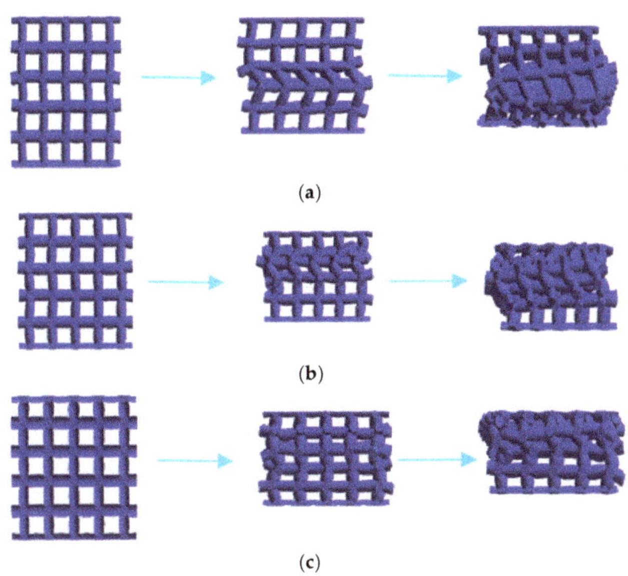

Figure 23. Deformation of structure under different impact velocities. (**a**) v = 10 m/s, (**b**) v = 50 m/s, and (**c**) v = 90 m/s.

As Figure 23a shown, when the impact velocity is v = 10 m/s, the cell element in the middle part of the structure first shows bending deformation. With the compression process continuing, the deformation is rapidly transferred to the fixed end, and the overall structural deformation presents the characteristic of tilting to one side until the structure compresses to densification. As seen in Figure 23b, when the impact velocity is 50 m/s, the initial deformation of the structure still occurs in the middle and lower part. With the compression continuing, the deformation gradually moves to the impact end, and then the impact end is first compacted. When the impact velocity is 90 m/s, the impact end is first compacted, and then the deformation gradually passes to the fixed end. The result is shown in Figure 23c.

4.4. The influence of Material on Energy Absorption Characteristics

The parameters of the base material have an important influence on the performance of the lattice structure. The influence of the base material on the dynamic impact performance of the lattice structure is studied by investigating the specific response of the lattice structure with four different base materials under dynamic impact. The parameters of four materials are listed in Table 6. The dimensions of the analyzed models are the same: all are 125-cell structures with the relative density of 0.23 and the impact velocity of 30 m/s.

Table 6. Material properties of base metal.

Materials	Elastic Modulus (GPa)	Yield Strength (MPa)	Strength Limit (MPa)	Density	Poisson's Ratio
PA2200	1.140	23.3	48.1	0.956	0.28
PLA	1.764	47.628	53.822	1.26	0.35
AlSi10Mg	25.804	170	230	2.7	0.33
20 Steel	213	245	710.67	7.80	0.3

Figure 24 shows the stress–strain curves of the lattice structure with different base materials. Two lattice structures with 20 steel and Alsi10Mg as base material deform almost simultaneously and become a drum shape on each layer. However, the deformation of the other two lattice structures occurs layer by layer, and their platform areas rise gradually without fluctuation until the structure reaches densification.

The energy absorption index results of lattice structures with different base materials are shown as Table 7, and the specific energy absorption of the structure is greatly affected by base materials. The SEA_v and SEA values of 20 steel are about 19.5 times that of PA2200, and the platform stress is about 22.1 times that of PA2200, while the peak load is much larger than PA2200.

The specific energy absorption of the four different materials of the lattice structure has a positive correlation with the strength of the materials, and it is feasible to improve the energy absorption of the lattice structure by increasing the strength of the matrices.

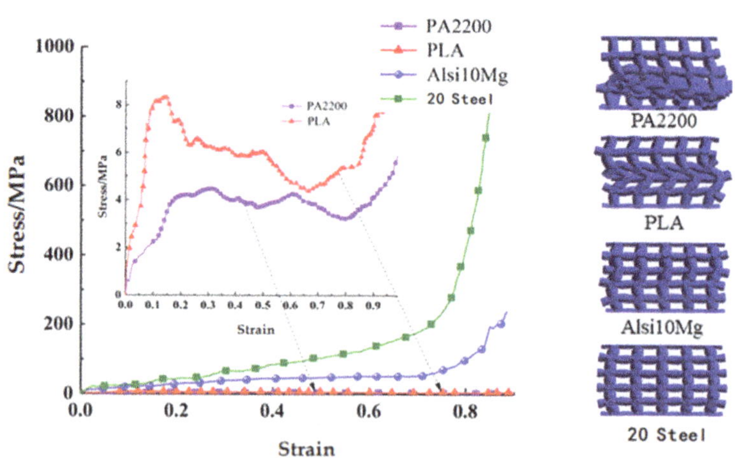

Figure 24. Stress–strain curves for different materials.

Table 7. Energy absorption index of different base metals.

Materials	SEA_v/mJ·mm^{-3}	SEA/J·g^{-1}	CLE/%	MIF/KN	σ_P/MPa	ε_d
PA2200	2.91	2.31	27.37	11.20	3.58	0.81
PLA	3.89	3.09	22.5	20.78	5.86	0.66
AlSi10Mg	29.16	23.14	35.48	166.3	38.96	0.75
20 Steel	56.86	45.12	29.73	471.75	79.76	0.71

5. Conclusions

The lattice structures have the advantages of high strength, high stiffness, and good absorption and storage of energy. It is useful to find an optimal lattice structure with better energy absorption performance by investigating the fiber structure of loofah under the numerical analysis method. By compressing the four single-cell and VC lattice structures, the bio-90 lattice structure was found to have the best energy absorption performance. We studied the influences of the number of cells, the relative density, the impact velocity, and the material on the energy absorption performance of the bio-90 lattice structure. When the number of cells exceeds a certain value, the energy absorption performance of the overall structure will be reduced. The relative density has a significantly different influence on the energy absorption performance of lattice structures, which is the maximum when the relative density is 0.37. With the impact velocity increasing, the absorption energy rises slightly, but the energy absorption process of the structure is unstable, and the crash load efficiency (*CLE*) always decreases. It is feasible to improve the energy absorption performance of the lattice structure by replacing the material. However, if the peak load is strictly limited, it can be decreased by reducing the relative density and number of cells. The research results provide useful references for the design of efficient energy absorption structure.

Author Contributions: Revision, funding acquisition, and conceptualization, G.H.; methodology, G.H. and H.Y.; writing—original draft preparation, H.Y. and T.C.; writing—review and editing, G.H. and H.Y.; software, Y.N. and H.Z.; data curation, H.Z. and F.Z.; resources, H.Y. and T.C.; validation, G.H. and H.Y.; formal analysis, T.C. and H.Z. All authors have read and agreed to the published version of the manuscript.

Funding: This work was supported by the National Natural Science Foundation of China (Grant No. 52175223 and the Key Research and Development Program of Changzhou (Grant No. CE20225044).

Institutional Review Board Statement: Not applicable.

Informed Consent Statement: Not applicable.

Data Availability Statement: Not applicable.

Conflicts of Interest: The authors declare no conflict of interest.

References

1. Deshpande, V.S.; Fleck, N.A.; Ashby, M.F. Effective properties of the octet-truss lattice material. *J. Mech. Phys. Solids* **2001**, *49*, 1747–1769. [CrossRef]
2. Evans, A.G.; Hutchinson, J.W.; Fleck, N.A.; Ashby, M.F.; Wadley, H.N.G. The topological design of multifunctional cellular metals. *Prog. Mater. Sci.* **2001**, *46*, 309–327. [CrossRef]
3. Dong, G.; Tang, Y.; Zhao, Y.F. A Survey of Modeling of Lattice Structures Fabricated by Additive Manufacturing. *J. Mech. Des.* **2017**, *2*, 1272–1279. [CrossRef]
4. Chenxi, P.; Phuong, T.; Nguyen-Xuan, H.; Ferreira, A.J.M. Mechanical performance and fatigue life prediction of lattice structures: Parametric computational approach. *Compos. Struct.* **2020**, *235*, 111821.
5. Cao, X.; Duan, S.; Liang, J.; Weibin, W.; Daining, F. Mechanical properties of an improved 3D-printed rhombic dodecahedron stainless steel lattice structure of variable cross section. *Int. J. Mech. Sci.* **2018**, *145*, 403–411. [CrossRef]
6. Bai, L.; Gong, C.; Chen, X.; Sun, Y.; Xin, L.; Pu, H.; Peng, Y.; Luo, J. Mechanical properties and energy absorption capabilities of functionally graded lattice structures: Experiments and simulations. *Int. J. Mech. Sci.* **2020**, *182*, 105–125. [CrossRef]
7. Andrew, J.J.; Schneider, J.; Ubaid, J.; Velmurugan, R.; Gupta, N.K.; Kumar, S. Energy absorption characteristics of additively manufactured plate-lattices under low- velocity impact loading. *Int. J. Impact Eng.* **2021**, *149*, 103–118. [CrossRef]

8. Hammetter, C.I.; Rinaldi, R.G.; Zok, F.W. Pyramidal Lattice Structures for High Strength and Energy Absorption. *J. Appl. Mech.* **2012**, *1*, 407–421. [CrossRef]
9. Xie, Y.; Bai, H.; Liu, Z.; Chen, N. A Novel Bionic Structure Inspired by Luffa Sponge and Its Cushion Properties. *Appl. Sci.* **2020**, *10*, 2584. [CrossRef]
10. Zhang, L.; Song, B.; Fu, J.; Wei, S.; Yang, L.; Yan, C.; Li, H.; Gao, L.; Shi, Y. Topology-optimized lattice structures with simultaneously high stiffness and light weight fabricated by selective laser melting: Design, manufacturing and characterization. *J. Manuf. Process.* **2020**, *56*, 1166–1177. [CrossRef]
11. Gibson, L.J.; Ashby, M.F.; Harley, B.A. *Cellular Materials: In Nature and Medicine*; Cambridge University Press: Cambridge, UK, 2010.
12. Du, Y.; Gu, D.; Xi, L.; Dai, D.; Gao, T.; Zhu, J.; Ma, C. Laser additive manufacturing of bio-inspired lattice structure: Forming quality, microstructure and energy absorption behavior. *Mater. Sci. Eng. A* **2020**, *773*, 138857. [CrossRef]
13. Soe, S.; Martin, P.; Jones, M.; Robinson, M.; Theobald , P. Feasibility of optimising bicycle helmet design safety through the use of additive manufactured TPE cellular structures. *Int. J. Adv. Manuf. Technol.* **2015**, *79*, 1975–1982. [CrossRef]
14. Zou, Z.; Hughes, T.; Scott, M.A.; Sauer, R.A.; Savith, E.J. Galerkin formulations of isogeometric shell analysis: Alleviating locking with Greville quadratures and higher-order elements. *Comput. Methods Appl. Mech. Eng.* **2021**, *380*, 113757. [CrossRef]
15. Wang, X.; Shen, J.; Zuo, Z.H.; Huang, X.; Zhou, S.; Xie, M.Y. Numerical investigation of compressive behaviour of luffa-filled tubes. *Compos. Part B Eng.* **2015**, *73*, 149–157. [CrossRef]
16. Elmadih, W. Additively Manufactured Lattice Structures for Vibration Attenuation. Ph.D. Thesis, University of Nottingham, Nottingham, UK, 2020.
17. Chen, Q.; Shi, Q.; Gorb, S.N.; Li, J. A multiscale study on the structural and mechanical properties of the luffa sponge from Luffa cylindrica plant. *J. Biomech.* **2014**, *47*, 1332–1339. [CrossRef] [PubMed]
18. Yong, P.L.; Wang, H.; Qing, H.Q. Bio-Inspired Design of Lightweight Metal Structure Based on Microstruc-ture of Fully Ripe Loofah. *Recent Pat. Mater. Sci.* **2015**, 69–73.
19. Boynard, C.A.; D 'Almeida, J.R.M. Morphological characterization and mechanical behavior of sponge gourd (*Luffa cylindrica*)—Polyester composite materials. *Polym. Plast. Technol. Eng.* **2010**, *39*, 489–499. [CrossRef]
20. Xiao, Z.; Yang, Y.; Xiao, R.; Bai, Y.; Song, C.; Wang, D. Evaluation of topology-optimized lattice structures manufactured via selective laser melting. *Mater. Des.* **2018**, *143*, 466–473. [CrossRef]
21. Wormser, M.; Warmuth, F.; Körner, C. Evolution of full phononic band gaps in periodic cellular structures. *Appl. Phys. A* **2017**, *123*, 661. [CrossRef]
22. Bai, L.; Zhang, J.; Xiong, Y.; Chen, X.; Sun, Y.; Gong, C.; Pu, H.; Wu, X.; Luo, J. Influence of unit cell pose on the mechanical properties of Ti6Al4V lattice structures manufactured by selective laser melting. *Addit. Manuf.* **2020**, *34*, 101–112. [CrossRef]
23. Ha, N.S.; Lu, G. A review of recent research on bio-inspired structures and materials for energy absorption applications. *Compos. Part B Eng.* **2020**, *181*, 107–122. [CrossRef]
24. Baroutaji, A.; Gilchrist, M.; Olabi, A. Quasi-static, impact and energy absorption of internally nested tubes subjected to lateral loading. *Thin-Walled Struct.* **2016**, *98 Pt B*, 337–350. [CrossRef]

machines

MDPI

Article

Bio-Inspired Artificial Receptor with Integrated Tactile Sensing and Pain Warning Perceptual Abilities

Xin Zhao [1], Gangqiang Tang [1], Chun Zhao [1], Dong Mei [1], Yujun Ji [1], Chaoqun Xiang [1], Lijie Li [2], Bo Li [3] and Yanjie Wang [1,*]

[1] Jiangsu Provincial Key Laboratory of Special Robot Technology, Changzhou Campus, Hohai University, Changzhou 213022, China
[2] Multidisciplinary Nanotechnology Centre, College of Engineering, Swansea University, Swansea SA1 8EN, UK
[3] School of Mechanical Engineering, Xi'an Jiaotong University, Xi'an 710049, China
* Correspondence: yj.wang1985@gmail.com

Abstract: Inspired by the mechanism of touch and pain in human skin, we integrated two ion-sensing films and a polydimethylsiloxane (PDMS) layer together to achieve a bionic artificial receptor with the capacity of distinguishing touch or pain perception through ion-electrical effect. The ion-sensing film provides the carrier of touch or pain perception, while the PDMS layer as a soft substrate is used to regulate the perception ability of receptor. Through a series of experiments, we investigated the effects of physical properties of the PDMS layer on the sensing ability of an artificial receptor. Further, contact area tests were performed in order to distinguish touch or pain under a sharp object. It is revealed that the pressure threshold triggering the touch and pain feedback of the artificial receptor presented an increasing trend when the elastic modulus and thickness of the PDMS substrate increase. The distinction ability of touch and pain becomes more pronounced under higher elastic modulus and larger thickness. Furthermore, the induced pain feedback becomes more intense with the decrease of the loading area under the same load, and the threshold of pain drops down from 176.68 kPa to 54.57 kPa with the decrease of the radius from 3 mm to 1 mm. This work potentially provides a new strategy for developing electronic skin with tactile sensing and pain warning. The pressure threshold and sensing range can be regulated by changing the physical properties of the middle layer, which would be advantageous to robotics and healthcare fields.

Keywords: IPMC sensor; multifunctional electronic skin; tactile sensing; damage warning; adjustable synthetic artificial receptors; PDMS

Citation: Zhao, X.; Tang, G.; Zhao, C.; Mei, D.; Ji, Y.; Xiang, C.; Li, L.; Li, B.; Wang, Y. Bio-Inspired Artificial Receptor with Integrated Tactile Sensing and Pain Warning Perceptual Abilities. *Machines* **2022**, *10*, 968. https://doi.org/10.3390/machines10110968

Academic Editor: Jan Awrejcewicz

Received: 30 August 2022
Accepted: 19 October 2022
Published: 23 October 2022

1. Introduction

The unique somatosensory system endows the human body with the ability of somatic function protection and environmental information exchange. Touch and pain are both somatic sensations, which depend on different skin receptors and nerve pathways [1]. The receptors in human skin consist of sensory cells, nerve fibers or free nerve endings that generate stress-strain and space-time electrochemical changes under mechanical load. The action potential caused by the potential difference between the inside and outside of the cell is transmitted to the brain via nerve fibers to complete the transmission of information. Figure 1 depicts the functioning mechanism of the human skin receptor. Separate neurological pathways connect nociceptors and tactile corpuscles to the brain [2]. When mechanical loads are applied to the skin, low threshold mechanoreceptors are partially deformed, which drives the movement of chemical ions (Na^+ and K^+) to produce a potential change in the receptor [3]. This electrical signal is transmitted via nerve fibers to the central nervous system to complete the final expression of tactile information [4]. The nociceptor is currently inactive and will not interfere with the functioning of the tactile receptors. Everyone has a pain sensitivity and tolerance threshold that corresponds to their health status [5]. Specifically, the upper and lower limits of the stimulus field that skin stress tissue can withstand during the perceptual process, as well as sensitivity to even the slightest sensory change within this range. Minimum stimulus and maximum stimulus are the two thresholds of absolute feeling, corresponding to the appearance and disappearance of sensation, respectively. Moreover, the chemical signals induced by external stimuli are transient, whereas the intercellular mechanical signals decay much more slowly than chemical signals, exhibiting rapid conduction and excellent directionality. The coordination of multiple receptors and the nervous system endows the human skin with a unique perception ability including touch and pain, which is crucial to the development of electronic skin. By imitating the structure and functioning mechanism of the somatosensory system [6], multifunctional electronic skins (e-skins) have been constructed. However, the e-skin cannot distinguish normal tactus and pain induced by noxious stimuli [7]. Therefore, it is necessary to investigate the working mechanism of receptors in order to achieve artificial receptors with tactile perception and warning function for injurious stimuli.

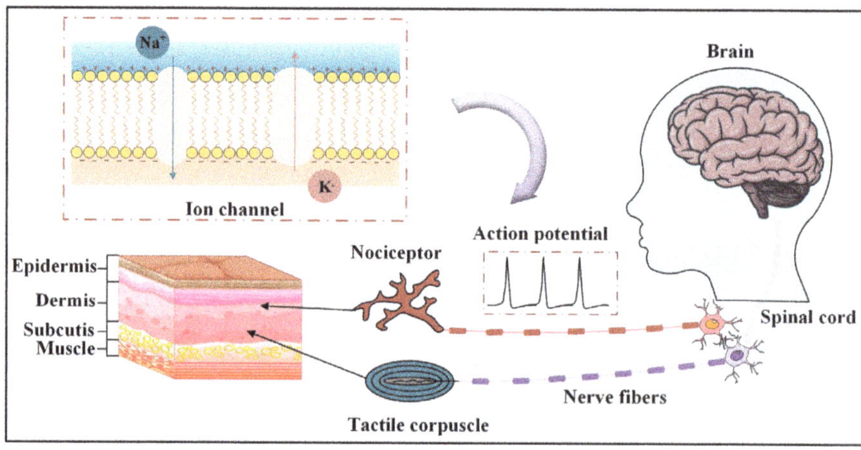

Figure 1. Schematic diagram of the mechanical-electrochemical transduction pathway of the receptor.

Recently, three strategies for the pain-perceptual emulation of e-skin have been proposed. The first strategy is to distinguish innocuous stimuli or noxious stimuli based on a specific threshold. Huang et al., for instance, designed an integrated sensing and warning multifunctional device based on the mechanical and thermal effect of porous graphene, that generate sufficient heating energy to warn the user when the detected signal reaches the threshold condition measured by the ultrasensitive strain sensor [8]. The second strategy is to integrate a tactile sensor with an electrochromic device to achieve physical force sensing and injury visualization [9]. An impressible example was a dual-mode electronic skin inspired by bioluminescent jellyfish that mimics the functions of the mechanoreceptors and nociceptors in the biological skin, respectively, by combining electrical and optical responses to quantify and map gentle tactile and injurious pressure [10]. The third strategy is based on the neural properties of memristors or transistors that mimic the properties of biological synapses [11]. The integration of artificial synaptic devices with pressure sensors permits pain perception and injury detection in the electronic skin. Utilizing the combination of stretchable pressure sensor and memristor [12], Rahman M A et al. built an integrated electronic system to replicate the feedback response of skin receptors and to detect pressure and pain stimuli. These three techniques effectively provide dangerous warnings and trigger the pain. However, unlike the human somatosensory system, current bio-inspired pain-perceptual devices are far from the sensory mechanism of the human body. In recent years, researchers have proposed using ionic polymer materials to imitate the working mechanism (ion migration—electric response) of human receptors [13], including iEAP [14], ionic gel [15] and ionic liquid polymer [16]. By combining such ionic sensing materials, we are able to integrate ionic sensors with a dielectric layer that controls the physical threshold, producing an artificial receptor with dual-mode sensing function. In comparison to sensors coupled with multiple physical signals, it is crucial for the intelligent development of sensors to detect multiple external stimuli through a single signal modality.

Herein, a bio-inspired artificial receptor with bionic-multilayer structure is constructed based on the cooperation between the ionic electrical mechanism of ionic polymer sensors (IPMC layer) and the elastic deformation property of the intermediate substrate (PDMS layer). We analyzed the signal source of the artificial receptor and illustrated its working mechanism by a finite element simulation. The prepared artificial receptor was subjected to a series of increasing force stimuli and its touch and pain sensing properties were evaluated. On the basis of the PDMS layer with varying elastic modulus and thickness, numerous artificial receptors were fabricated. The purpose is to find out the effect of PDMS layer changes on touch/pain-inducing threshold and sensing range to achieve the customized preparation of artificial receptors. To simulate the intense pain sensation caused by sharp objects on human skin, the sensing characteristics of artificial receptors under different contact areas were studied.

2. Materials and Methods

2.1. Materials

The Nafion membrane was purchased from the Dupont company, Wilmington, Delaware, USA. Auxiliary reagents such as $Pd(NH_3)_4Cl_2$, $NaBH_4$, and HCl were obtained from J&K Chemical Inc (Beijing, China). The PDMS films were provided by Hefei Keliao New Material Technology Co., Ltd., Anhui, China. The VHB^{TM} tape (thickness 0.13 mm) was purchased from Minnesota Mining and Manufacturing Corporation company (Shanghai, China).

2.2. Preparation of Receptor

Nafion membrane is widely employed owing to its unique combination of flexible, non-toxicity and ion-exchange properties. The thin metal layer deposits on both sides of the surface of Nafion as electrodes to obtain ion-polymer metal composites (IPMC) with a sandwich structure by chemical or physical methods. In this work, the ionic polymer metal composite (IPMC) films with sensing function were fabricated by immersion reduction plating and electroplating process [17]. Under externally applied pressure, the uneven stress caused by the strain in the IPMC polymer grid drives the cations to migrate to the lower stress region. Then, the subsequent formation of the space charge gradient distribution results in potential difference between the electrodes. Due to the IPMC's internal ion migration and current response characteristics [18], its sensing behavior is similar to that of bioelectrical mechanism. Therefore, IPMC is utilized to replicate the activity of deformed receptors (tactile corpuscles and nociceptors) that generate the action potential when deformed. Inspired by the multilayer structure of human skin, the bio-inspired artificial receptor is constructed of two sensory layers and one flexible PDMS layer, as shown in Figure 2. As sensory layers (thickness 0.2 mm, size 15 mm $\times$ 15 mm), IPMC lies at different depths in the artificial receptor, which is used to receive the mechanical stimuli and convert them into electrical signals. According to the position of the two sensory layers in the structure, we named them as the top IPMC sensory and the bottom IPMC sensory layer, representing the mechanoreceptor and nociceptor, respectively. For each sensory IPMC layer, there are two wires attached to both electrodes, respectively. The PDMS layer (elastic modulus, E = 2.3 MPa, thickness 0.5 mm, size 15 mm $\times$ 15 mm) has excellent chemical stability and low elastic modulus, which serves in three aspects. First, it acts as a supporting layer to position of the top and bottom sensory layers and maintain the overall shape of the artificial receptor; second, the middle layer is involved in the transport and conduction of mechanical stimuli, thereby physically regulating the pressure threshold of artificial receptors; third, the insulating properties of the PDMS layer avoids the crosstalk of two IPMC layers. The presence of VHB^{TM} flexible adhesion layer (thickness

0.13 mm) makes the adhesion between these layers very tight. In addition, the protective layers made of masking tape mirror the biological epidermis to deliver external stimuli and safeguard the interior structure. The size of the encapsulation is larger (30 mm × 30 mm) than that of the artificial receptor (15 mm × 15 mm) in order to fix the wires and eliminate interference signals caused by wire movement during testing. Figure S1a shows the front and sectional view of the artificial receptor, as well as the ion sensing film and the PDMS layer constituting the artificial receptor.

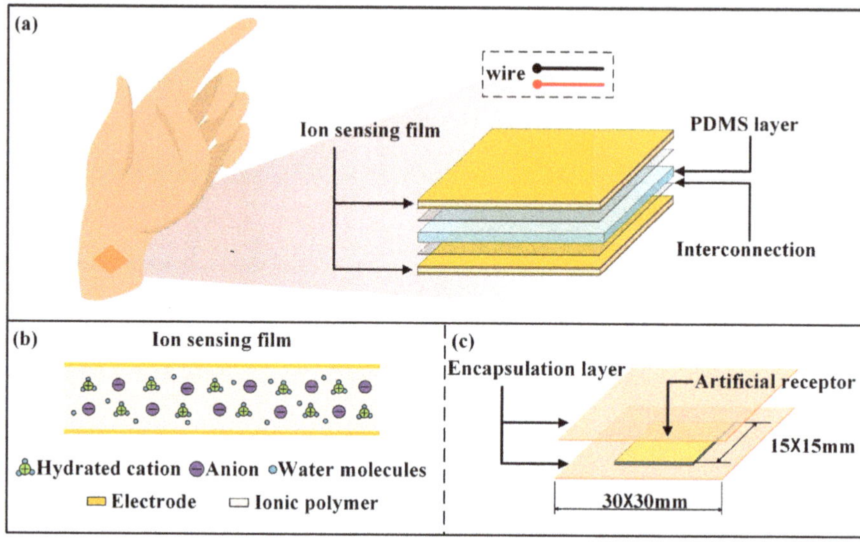

Figure 2. Preparation of bio-inspired artificial receptor. (**a**) Schematic diagram of receptor structure. (**b**) The composition of ion sensing film. (**c**) The encapsulation for receptor.

2.3. Experimental Set-Up

A self-built sensing test platform was used to characterize the sensing performance of a bio-inspired artificial receptor device, which consists of a sample fixing device, a load applying module and a signal processing module, as shown in Figure 3. First, the bio-inspired artificial receptor receives the mechanical stimuli caused by load applying module. Then, the force signal will be converted to the voltage signal via the ion-electric effect of artificial receptor. Third, the electrical signal is transferred to the computing center through signal processing module. By reading the signal, the PC device will recognize the touch or pain perception in real-time. A glass substrate (elastic modulus = 55 GPa) was used to provide the bottom fixation constraint for the artificial receptor. The pressure applied to the artificial receptor was loaded by an applying element fixed on the vibration exciter (SA-JZ002), which controlled by the signal generator (SA-SG030A). A force sensor (ZNLBM-3KG) was mounted at sharker rod to measure the actual force which was applied to the top surface of bio-inspired artificial receptor. A power generator (APS3003S-3D) was used to supply the signal amplification circuit devices with dc voltages. The Labview software on the PC records the voltage generated by the artificial receptor device and the real-time signal of the force sensor using the NI USB-6001 data acquisition.

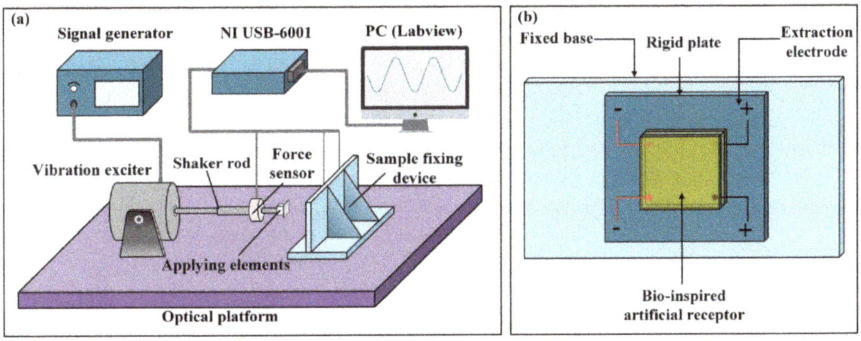

Figure 3. Schematic diagram of test platform for bio-inspired artificial receptor. (**a**) Integral test system. (**b**) The sample fixing device.

3. Results

3.1. Sensing Analysis of the Receptor

The IPMC sensory layer may generate voltage response under compression deformation or bending deformation [19,20]. In order to verify the signal source of the artificial receptor, a series of experiments were implemented on an IPMC sample (thickness 0.2 mm, size 15 mm × 15 mm). The compression load and end displacement applied to the IPMC sample are both sinusoidal signals with a frequency of 1 Hz. The voltage response of the IPMC sample under compression load with different amplitudes are shown in Figure 4a. With the increase of applied pressure, the voltage response signal of IPMC on a rigid substrate shows an increasing trend, but the growth in value is so tiny that it can be ignored. In the bending deformation experiment, a discernible response signal (0.03 mV) is generated even under the displacement load with amplitude of 0.1 mm, which is twice as much as under the compression state of 542.05 kPa (0.015 mV), as shown in Figure 4b. Therefore, it can be confirmed that the sensing signal of the artificial receptor primarily comes from the bending deformation of the IPMC sensory layer, while the voltage response caused by ion migration induced by compression deformation is negligible.

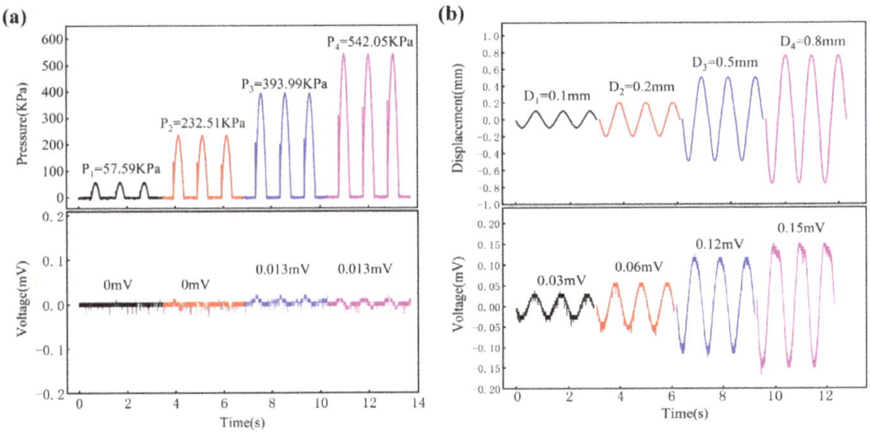

Figure 4. The experimental results of signal source analysis. (**a**) Pressure−applying signal and voltage response of IPMC sensing layer in compression mode. (**b**) Displacement−applying signal and voltage response of IPMC sensing layer in bending mode.

To theoretically analyze the working mechanism of the bio-inspired artificial receptor, we implement finite element simulations. The material characteristic parameters are shown in Table S1. Finite element mesh types use automatic meshing, as shown in Figure S2. The artificial receptor undergoes two operating modes under different load stimuli, which were applied to a circular loading region with a radius of 3 mm, as shown in Figure 5a. Figure 5b shows the deformation distributions of the top sensory layer, the PDMS layer and the bottom sensory layer. The deformation of each layer, from top to bottom, is 0.036 mm, 0.029 mm, and 0.007 mm when the pressure is 106.01 kPa. The deformation of the bottom sensory layer is only one-fifth that of the top, which is also less than the top sensory layer's deformation at 35.33 kPa (0.012 mm). Consequently, only the top IPMC layer and the PDMS layer are under compressive stress state. At this point, the top IPMC sensory layer is active while the bottom IPMC sensory layer is inactive, thus the working mode of the whole artificial receptor is tactile perception state. With the increase of applied load, mechanical signals can break through the physical threshold of the PDMS layer and transfer to the bottom IPMC layer. Figure 5c shows the deformation distributions of three layers at 353.36 kPa. The deformation of the bottom sensory layer reaches 0.024 mm, indicating that the working mode of the entire artificial receptor is nocuous perception state. In addition, the stress on each layer is relatively uniform, and the shape of the deformation area matches that of the loading area. Maximum stress is observed at the contact edge between the pressing element and the sensory layer, and it attenuates from close to far along the direction of load application. Taking into account the stress state, the IPMC electro-mechanical coupling mechanism and circuit connection, we can estimate that the output voltage of the two sensory layers will remain comparable and that the signal of the IPMC layer at the top will be stronger.

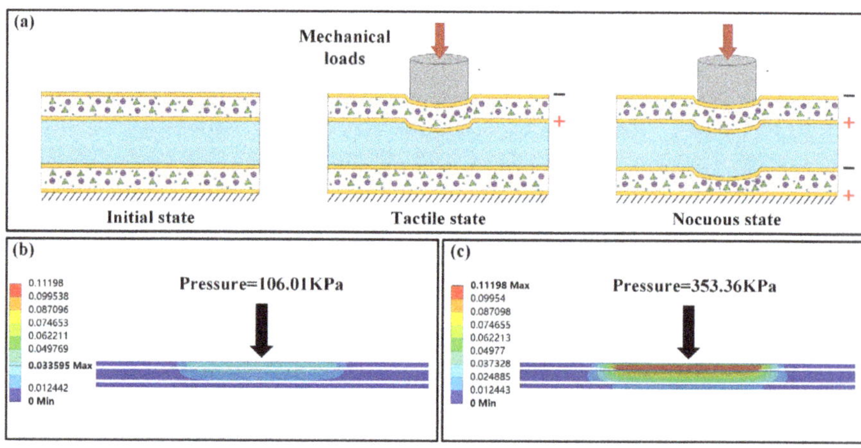

Figure 5. Working mechanism of the bio−inspired artificial receptor under mechanical stimuli with different amplitude. (**a**) Schematic of bio−inspired artificial receptor in different phases of operation. (**b**) The deformation distribution cloud diagram of the top IPMC layer, PDMS layer and bottom IPMC layer when load is 106.01 kPa. (**c**) The deformation distribution cloud diagram of the top IPMC layer, PDMS layer and bottom IPMC layer when load is 353.36 kPa.

3.2. *Characterization of Touch and Pain*

Our bionic artificial receptor can measure and discriminate touch or pain based on the voltage response induced by pressure stimuli. We evaluate the receptor's reaction to touch stimuli with different pressure amplitudes from 70.67 kPa to 176.68 kPa. As shown in Figure 6a, the top IPMC sensory layer presents increasing voltage response signals (0.0064 mV to 0.0148 mV) with the increase of the loading force amplitude. A large pressure corresponds to the large output voltage in the top sensory layer, while the bottom IPMC sensory layer has no response signals, which can be attributed to the PDMS layer absorbing all the energy. At a frequency of 1 Hz, Figure 6b shows the observed output voltage with different normal force amplitudes. With the increase of pressure from 247.35 kPa to 459.36 kPa, the bottom IPMC sensory layer begins to generate response signals and keep an upward trend to 0.0086 mV. In the real-time waveform of the output voltage, the phase of the two sensory layers is equal, validating the prediction of the prior finite element simulation. Besides, it can be observed that the signal of the top IPMC sensory layer is larger due to the static attenuation in the PDMS layer. Figure 6c presents the amplitude voltage value of top and bottom sensory IPMC layers under a series of increasing pressures. According to different pressure thresholds, the sensing range is defined as non-perceptual area, tactile sensing area and pain sensing area, which shows obvious differences among all regions. As shown in Figure 6d, the artificial receptor shows great linearity, with tactile sensitivity and pain sensitivity of 0.136 mV/Pa and 0.026 mV/Pa, respectively. The difference in the sensitivity for both sensory layers is attributed to the PDMS layer since the top sensory layer absorbs more deformation from external stimuli. Figure S1b shows the sensing signals of an artificial receptor at 176.68 kPa and 371.02 kPa. The experimental results are consistent with the experimental results in the paper.

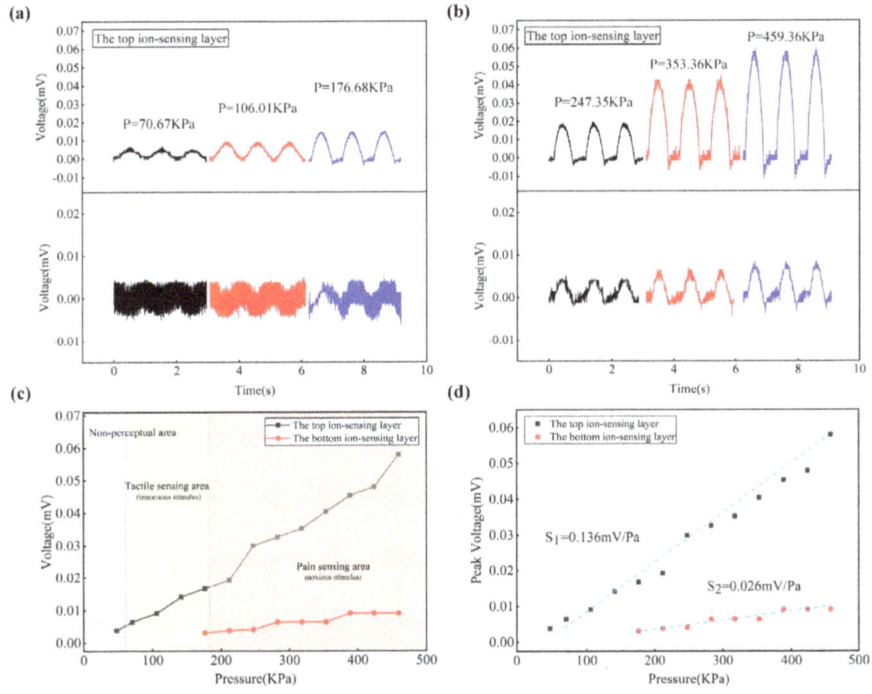

Figure 6. Experimental results of the bio−inspired artificial receptor. (**a**,**b**) Real−time waveform of the output voltage under different normal force amplitudes with a frequency of 1 Hz. (**c**) The voltage response with the increase of pressure shows different sensing modes in different pressure ranges. (**d**) Sensitivities of the bio−inspired artificial receptor during the pressure from 0 to 500 kPa.

3.3. PDMS Layer Effects

Since the permissible range of skin stimuli is governed by the threshold of pain perception, minute changes in skin structure and the distribution of receptors can affect the threshold that controls the transmission of pain signal. We investigated the effect of the elastic modulus and thickness of the PDMS layer on the sensing property of bio-inspired artificial receptor. The trend in Figure 7a represents the output voltage of artificial receptors with different elastic moduli PDMS layers under the same loading condition (353.36 kPa). With the elasticity modulus of the PDMS layer varying from 2.3 MPa to 1.2 MPa, the output voltage of bottom sensory layer changes from 0.0068 mV to 0.0579 mV, as well the voltage response of the top sensory layer increases from 0.043 mV to 0.132 mV. The increased elastic modulus of the PDMS layer will result in a stiffer structure, reducing the deformation and voltage response of the bottom sensory layers. The sensing ability of biomimetic artificial receptors assembled with PDMS layers of varying elastic modulus are shown in Figure 7b. As the elastic modulus decreases, the tactile perception range (green safety area) gradually decreases while the pain perception range (orange warning area) increases persistently. It indicates that a smaller loading force can trigger pain feedback of artificial receptors assembled by the PDMS layer with lower elastic modulus. Another discovery is that the two sensory layers have the higher voltage response when the thickness of the PDMS layer decreases. The trend in Figure 7d shows the output voltage of artificial receptors with different PDMS layers thickness under the same loading condition (353.36 kPa). The output voltage of the bottom sensory layer shows a downward trend from 0.0100 mV to 0.0035 mV with the thickness of the PDMS layer increasing from 0.1 mm to 0.7 mm, as well the output voltage of the top sensory layer changes identically as that of the bottom layer. Under the same pressure load (353.36 kPa), the deformation of bottom sensory layer behind the thicker PDMS layer is slightly lower due to the mechanical signal attenuation caused by the stiffer, thicker PDMS layer. The sensing ability of biomimetic artificial receptors assembled by PDMS layers with different thickness are shown in Figure 7e. The results indicate that the tactile sensing range reduces as the PDMS thickness decreases, while the pain sensing range has an opposite tendency. To verify the customizability of artificial receptors, we further determined the relations of the pressure threshold to physical properties of the PDMS layer using the experimental results. As the thickness increases, the pressure threshold that triggers pain increases from 70.67 kPa to 212.01 kPa, while the induced touch pressure threshold shows a small increase, as shown in Figure 7f. It can be attributed to the anabatic loss of mechanical force in the thicker PDMS layer, which makes it difficult for the bottom sensory layer to generate a voltage response at lower pressures. Similarly, with the increase of elastic modulus from 1.2 MPa to 2.3 MPa, the induced pain pressure threshold rises from 43.2 kPa to 176.68 kPa, while the tactile pressure threshold did not change significantly (Figure 7c). In addition, it can be concluded that the distinction ability of touch or pain signals becomes more obvious as the elastic modulus and thickness of the PDMS layer rise. The above results indicate that the activation of the tactile perception mode of artificial receptors is strongly correlated with the external stimulus load and the PDMS layer thickness. The pattern of pain perception is affected by both external stimuli and the PDMS layer. Therefore, the pain perception of artificial receptors can be regulated by changing the physical properties of middle layer, thus achieving the customization of the perception range.

3.4. Contract Area Effects

Typically, the human body receives pain information from acute touch with a tiny contact area. We altered the loading area of the artificial receptor and carried out a series of comparative experiments. Figure 8a shows the voltage response of artificial receptors under different loading regions at 353.36 kPa. The experimental results indicates that the output voltage of the top IPMC sensory layer falls and then increases when the radius of the applied element reduces from 3 mm to 1 mm. The sudden decrease of tactile response at R = 2 mm is attributed to the offset effect of compressive deformation in the contact area and bending deformation at the contact edge resulting from the presence of a flexible PDMS substrate [21]. As for the bottom IPMC sensory layer close to the rigid substrate, the smaller contact area causes a larger stress distribution on the surface of the artificial receptor, which increases the deformation of the whole structure along the loading direction and the output voltage of the pain response. The output peak voltage of the bottom IPMC layer representing pain perception steadily increases with the decrease of the loading area, which indicates that a smaller loading force can trigger pain perception, as shown in Figure 8b. Both the induced touch and pain threshold of artificial receptors show a descending trend gradually with the decrease of the loading area, which well simulated the intense pain of biological skin hit by a sharp object. Due to the proximity of the top sensory layer to external stimuli and the thinness of the protective masking tape layer, the metal electrode surface of the top sensory layer suffers damage after extended cycles under the minimum-area pressing element, while that of the bottom sensory layer is still intact due to the protection of thick PDMS layer. It is worth noting that this novel multiplayer dual functional sensor can be operated bi-directionally along the out-of-plane axis, i.e., it will work the same way when the device is flipped over (the bottom layer becomes the top layer).

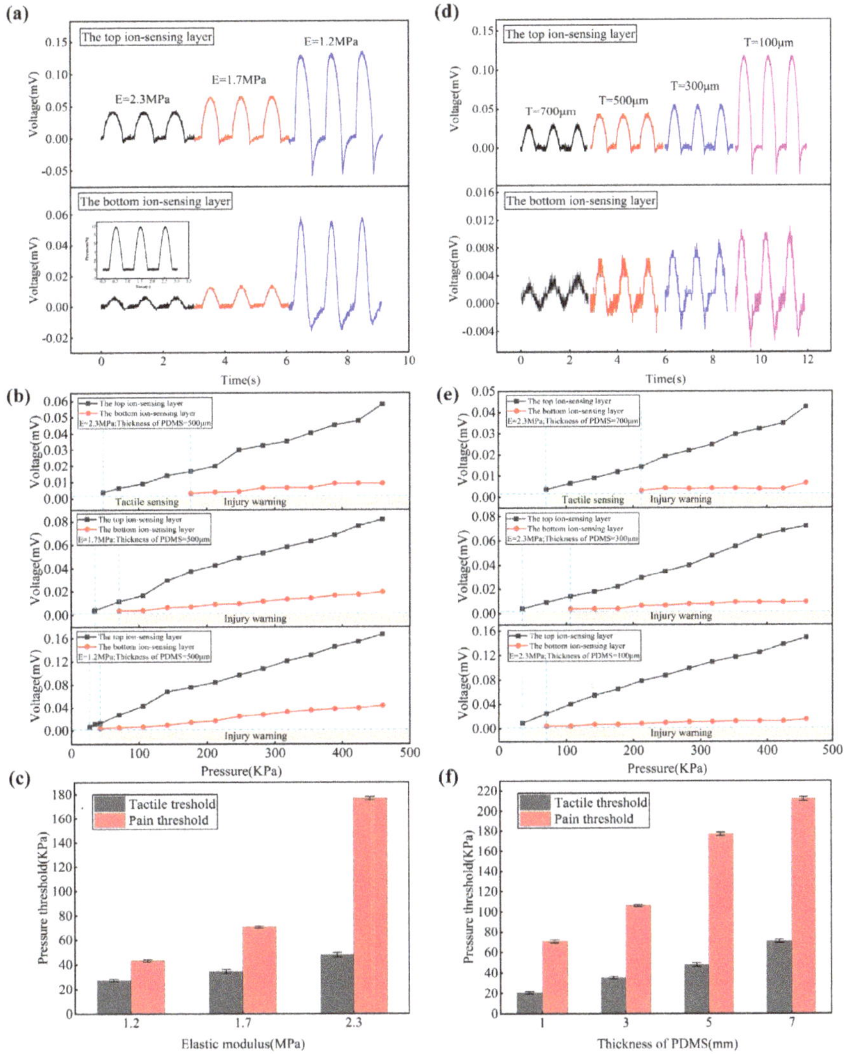

Figure 7. The experimental results. (**a**) Effect of PDMS layer with different elastic modulus on voltage response of artificial receptor. (**b**) The tactile/pain−sensing range of artificial receptors with change of PDMS elastic modulus. (**c**) The induced touch/pain threshold changes with different elastic moduli of PDMS layer. (**d**) Effect of PDMS layer with different thickness on voltage response of artificial receptor. (**e**) The tactile/pain−sensing range of artificial receptors with change of PDMS thickness. (**f**) The induced touch/pain threshold changes with different thicknesses of PDMS layer.

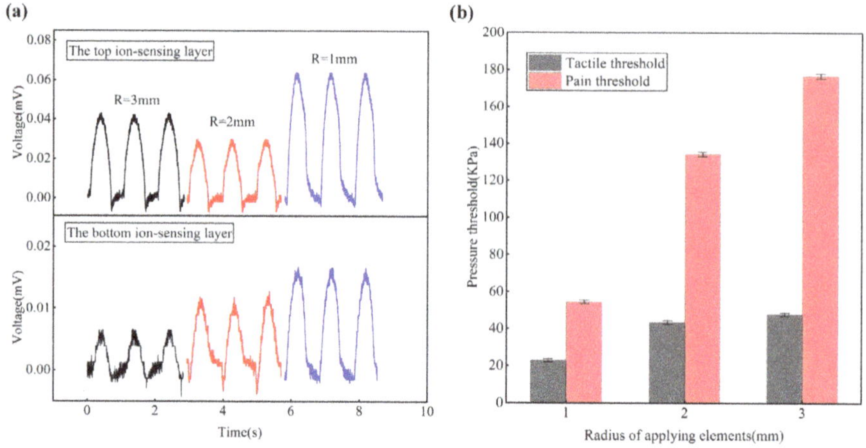

Figure 8. The experimental results. (**a**) Effect of applying elements with different radius on voltage response of artificial receptor. (**b**) The pressure threshold of artificial receptors when the radius of the contact circle increases from 1 mm to 3 mm.

4. Discussion

In this work, we report a bio-inspired artificial receptor with innocuous sensing and damage warning functions based on the coordination of ion-electric response principle and mechanical signal attenuation, which simplifies manufacturing and signal decoupling. The sensing voltage of artificial receptor primarily comes from the bending deformation of two IPMC sensory layers. According to the finite element simulation and experiments, the artificial receptor includes three operating states: non-perceptual state; tactile sensing state; and pain sensing state. The sensitivity of the touch response and pain response is 0.136 mV/Pa and 0.026 mV/Pa, respectively. Further, the pressure threshold that triggers touch and pain can be adjusted by changing the physical properties of the PDMS layer. As the PDMS thickness increases from 0.1 mm to 0.7 mm, the induced pain pressure threshold rises from 70.67 kPa to 212.01 kPa, while the induced touch pressure threshold shows a small increase from 20.46 kPa to 70.67 kPa. Similarly, the induced pain threshold increases from 43.20 kPa to 176.68 kPa with the growth of PDMS elastic modulus from 1.2 MPa to 2.3 MPa, and the induced touch threshold changes from 27.29 kPa to 47.75 kPa. The excellent performance tunability will greatly facilitate the customizable fabrication of artificial receptors. More than that, the threshold of pain drops down from 176.68 kPa to 54.57 kPa with the decrease of the radius from 3 mm to 1 mm, indicating that the artificial receptor can imitate the pain of skin in response to a sharp touch. The proposed artificial receptor with the capability of pain perception in our work will be beneficial to many fields, including electronic skins, robotics, healthcare and so on.

Supplementary Materials: The following supporting information can be downloaded at: https://www.mdpi.com/article/10.3390/machines10110968/s1, Figure S1: (a) The materials, front view and sectional view of artificial receptor (b) The pressure test experiments of artificial receptor; Figure S2: Schematic diagram of grid division; Table S1: Material characteristic parameter.

Author Contributions: Conceptualization, X.Z. and G.T.; methodology, X.Z. and Y.J.; writing—original draft preparation, X.Z.; writing—review and editing: G.T., C.X., L.L., B.L. and Y.W.; supervision, C.Z. and D.M.; project administration, Y.W. All authors have read and agreed to the published version of the manuscript.

Funding: This research was supported by the financial support from the National Natural Science Foundation of China (51975184 and 52075411), the Changzhou Sci & Tech Program (CE20215051), and the Fundamental Research Funds for the Central Universities (B210202124). The authors gratefully acknowledge the supports.

Data Availability Statement: Not applicable.

Conflicts of Interest: The authors declare no conflict of interest.

References

1. McGlone, F.; Wessberg, J.; Olausson, H. Discriminative and Affective Touch: Sensing and Feeling. *Neuron* **2014**, *82*, 737–755. [CrossRef] [PubMed]
2. Nagi, S.S.; Marshall, A.G.; Makdani, A.; Jarocka, E.; Liljencrantz, J.; Ridderström, M.; Shaikh, S.; Saade, D.; Donkervoort, S.; Olausson, H.; et al. An ultrafast system for signaling mechanical pain in human skin. *Sci. Adv.* **2019**, *5*, eaaw1297. [CrossRef]
3. Li, X.; Toyoda, H. Role of leak potassium channels in pain signaling. *Brain Res. Bull.* **2015**, *119*, 73–79. [CrossRef]
4. Wood, J.N.; Abrahamsen, B.; Baker, M.D.; Boorman, J.D.; Donier, E.; Drew, L.J.; Nassar, M.A.; Okuse, K.; Seereeram, A.; Zhao, J.; et al. Ion channel activities implicated in pathological pain. *Novartis Found. Symp.* **2004**, *261*, 32–40.
5. Deutch, A.Y. Neuroscience: Exploring the brain. *J. Clin. Psychiatry* **1999**, *60*, 59. [CrossRef]
6. De Maria, G.; Natale, C.; Pirozzi, S. Directions Toward Effective Utilization of Tactile Skin: A Review. *IEEE Sens. J.* **2014**, *14*, 4109. [CrossRef]
7. Lin, W.; Wang, B.; Peng, G.; Shan, Y.; Hu, H.; Yang, Z. Skin-Inspired Piezoelectric Tactile Sensor Array with Crosstalk-Free Row plus Column Electrodes for Spatiotemporally Distinguishing Diverse Stimuli. *Adv. Sci.* **2021**, *8*, 2002817. [CrossRef] [PubMed]
8. Huang, Y.; Tao, L.-Q.; Yu, J.; Wang, Z.; Zhu, C.; Chen, X. Integrated Sensing and Warning Multifunctional Devices Based on the Combined Mechanical and Thermal Effect of Porous Graphene. *ACS Appl. Mater. Interfaces* **2020**, *14*, 53049–53057. [CrossRef]
9. Wang, F.; Wang, M.; Liu, H.; Zhang, Y.; Lin, Q.; Chen, T.; Sun, L. Multifunctional Self-Powered E-Skin with Tactile Sensing and Visual Warning for Detecting Robot Safety. *Adv. Mater. Interfaces* **2020**, *7*, 2000536. [CrossRef]
10. Zhang, Y.; Fang, Y.; Li, J.; Zhou, Q.; Xiao, Y.; Zhang, K.; Luo, B.; Zhou, J.; Hu, B. Dual-Mode Electronic Skin with Integrated Tactile Sensing and Visualized Injury Warning. *ACS Appl. Mater. Interfaces* **2017**, *9*, 37493–37500. [CrossRef] [PubMed]
11. Li, F.; Gao, S.; Lu, Y.; Asghar, W.; Cao, J.; Hu, C.; Li, R. Bio-Inspired Multi-Mode Pain-Perceptual System (MMPPS) with Noxious Stimuli Warning, Damage Localization, and Enhanced Damage Protection. *Adv. Sci.* **2021**, *8*, 2004208. [CrossRef] [PubMed]
12. Rahman, M.A.; Walia, S.; Naznee, S.; Taha, M.; Nirantar, S.; Rahman, F.; Bhaskaran, M.; Sriram, S. Artificial Somatosensors: Feedback Receptors for Electronic Skins. *Adv. Intell. Syst.* **2020**, *2*, 2000094. [CrossRef]
13. Zhao, C.; Wang, Y.; Tang, G.; Ru, J.; Zhu, Z.; Li, B.; Guo, C.F.; Li, L.; Zhu, D. Ionic Flexible Sensors: Mechanisms, Materials, Structures, and Applications. *Adv. Funct. Mater.* **2022**, *32*, 2110417. [CrossRef]
14. Hao, M.; Wang, Y.; Zhu, Z.; He, Q.; Zhu, D.; Luo, M. A Compact Review of IPMC as Soft Actuator and Sensor: Current Trends, Challenges, and Potential Solutions from Our Recent Work. *Front. Robot. AI* **2019**, *6*, 129. [CrossRef] [PubMed]
15. Kwon, J.H.; Kim, Y.M.; Moon, H.C. Porous Ion Gel: A Versatile Ionotronic Sensory Platform for High-Performance, Wearable Ionoskins with Electrical and Optical Dual Output. *ACS Nano* **2021**, *15*, 15132–15141. [CrossRef] [PubMed]
16. Zheng, S.; Li, W.; Ren, Y.; Liu, Z.; Zou, X.; Hu, Y.; Yan, F. Moisture-Wicking, Breathable, and Intrinsically Antibacterial Electronic Skin Based on Dual-Gradient Poly (ionic liquid) Nanofiber Membranes. *Adv. Mater.* **2022**, *34*, 2106570. [CrossRef]
17. Wang, J.; Wang, Y.; Zhu, Z.; Wang, J.; He, Q.; Luo, M. The Effects of Dimensions on the Deformation Sensing Performance of Ionic Polymer-Metal Composites. *Sensors* **2019**, *19*, 2104. [CrossRef] [PubMed]
18. MohdIsa, W.; Hunt, A.; HosseinNia, S.H. Sensing and Self-Sensing Actuation Methods for Ionic Polymer–Metal Composite (IPMC): A Review. *Sensors* **2019**, *19*, 3967. [CrossRef]
19. Gudarzi, M.; Smolinski, P.; Wang, Q.-M. Bending mode ionic polymer-metal composite (IPMC) pressure sensors. *Measurement* **2017**, *103*, 250–257. [CrossRef]
20. Gudarzi, M.; Smolinski, P.; Wang, Q.-M. Compression and shear mode ionic polymer-metal composite (IPMC) pressure sensors. *Sensors Actuators A Phys.* **2017**, *260*, 99–111. [CrossRef]
21. Wang, Y.; Tang, G.; Zhao, C.; Mei, D.; Zhao, X.; Ji, Y.; Li, B. The effects of contact area on pressure sensing of ionic polymer metal composite sensor with a soft substrate. *Smart Mater. Struct.* **2022**, *31*, 8. [CrossRef]

machines

MDPI

Article

Design and Implementation of a Wire Rope Climbing Robot for Sluices

Guisheng Fang [1,*] and Jinfeng Cheng [1,2]

[1] College of Mechanical and Automotive Engineering, Zhejiang University of Water Resources and Electric Power, Hangzhou 310018, China
[2] School of Electronic Information, Hangzhou Dianzi University, Hangzhou 310018, China
* Correspondence: fanggsh@zjweu.edu.cn; Tel.: +86-13606620840

Abstract: Regular maintenance of wire rope is considered the key to ensuring the safe operation of a sluice gate. Along these lines, in this work, a six-wheeled wire rope climbing robot was proposed, which can carry cleaning and maintenance tools for online cleaning and safety inspection of the sluice wire rope, without its disassembly. The developed climbing robot is composed of separable driving and driven trolleys. It adopts the spring clamping mechanism and the wheeled movement method. Thus, it can easily adapt to the narrow working environment and different diameter ranges of the sluice wire rope. In addition, the designed six-wheeled wire rope climbing robot not only possesses a simple structure, simple control, and stable climbing speed, which are typical characteristics of wheeled climbing robots, but also a large contact area with objects and small wheel deformation, which are typical characteristics of crawler climbing robots. Structural design and mechanical analysis were also carried out, with the fabrication of a prototype robot system called WRR-II. From the acquired experimental results of the prototype's climbing speed test, load capacity test, climbing adaptability test, and obstacle-negotiation ability test, the rationality and feasibility of the designed climbing robot scheme were verified.

Keywords: climbing robot; mechanical analysis; spring clamping; sluice gate; wire rope; wheeled movement

Citation: Fang, G.; Cheng, J. Design and Implementation of a Wire Rope Climbing Robot for Sluices. *Machines* **2022**, *10*, 1000. https://doi.org/10.3390/machines10111000

Academic Editors: Yanjie Wang, Xiaofeng Liu, Aihong Ji, Shichao Niu and Bo Li

Received: 24 September 2022
Accepted: 27 October 2022
Published: 31 October 2022

1. Introduction

In water conservancy engineering facilities, sluice gates are widely used in rural and coastal river channels as the main water retaining and discharge structures. As the main load-bearing component of the hoisting sluice, the wire rope plays a vital role in the safe operation of the sluice. Due to the long-term exposure of the sluice wire rope to the outdoors, it is affected by irregular bearing, wind, rain, and sun, which will lead to various problems, such as grease hardening, local corrosion, wear, and breakage. Therefore, regular maintenance is essential for the proper operation of the wire rope. Currently, most of the daily maintenance of the sluice wire rope is done manually, leading to problems, such as high labor cost and intensity, low work efficiency, and high-risk factor. With the application of the scientific and technological developments in the field of robotics to the daily maintenance process of the wire rope, the above-mentioned problems can be easily solved. Therefore, the research and development of a wire rope climbing and maintenance robot in the water conservancy industry are anticipated to significantly improve work efficiency in this field, successfully addressing the labor issue in enterprises.

As an important branch of the mobile robot family, climbing robots have received widespread attention from the scientific community in the past two decades. As a result, a wide variety of prototype systems have been developed for specific applications, such as steel bridge climbing robots [1–4], cable-climbing robots [5–8], pole-climbing robots [9–13], tree-climbing robots [14–17], pipe-climbing robots [18,19], wall-climbing robots [20–24], among others.

In terms of rope-climbing robots, Koo et al. [25] developed a climbing robot based on the piston mechanism for a robotic competition in Malaysia. The robot consisted of a base frame and two clampers. The two clampers were used to grip the rope, and the base frame based on the piston mechanism was utilized to realize the climbing motion. The advantages of the proposed robot are low energy consumption, low cost, and ease to control and build. However, the robot is unable to climb vertically fixed ropes. When the robot is in action, it will occasionally cause vibration along the rope.

Cho et al. [26–28] designed three climbing and detection robots for hanger cables of suspension bridges, named WRC^{2}IN-I, WRC^{2}IN-I+, and WRC^{2}IN-II. WRC^{2}IN-I was composed of a wheel drive mechanism, an attachment mechanism, and a safe landing mechanism. It could climb at 50 mm/s under the condition of a 15 kg load. However, when the wheeled cable-climbing robot moved on the uneven cable surface, it would produce periodic vibration, which would affect the detection quality. Therefore, the project team improved the first-generation cable climbing robot by changing the wheeled structure into a tracked structure, which greatly reduced the vibration. In order to further simplify the installation and disassembly process of the first-generation robot and improve work efficiency, the project team developed the second-generation cable climbing robot named WRC^{2}IN-II. The robot was composed of two separable attachment modules, two driving modules, and two obstacle-surmounting sub-modules. After improvement, the robot could carry a load of 24 kg, while the installation and disassembly time only took about 5 min. Although these robots are all able to climb vertically fixed wire ropes, their dimensions are large and the applicable cable diameters are 50–90 mm.

Sun, G. [29] designed a wire rope climbing robot for the detection needs of lamps on the top of airport lighting streetlights. The robot was composed of a compression mechanism, a suspension mechanism, and a tracked type moving mechanism. The climbing robot could maneuver on a wire cable with a cross-sectional diameter varying from 10 to 14 cm with a stable and secure speed of 1 m/s. It could also lift up to 58 kg with respect to its own weight of 15.6 kg.

Ratanghayra, P. R. [30] designed a simple climbing robot for soft ropes. The robot was composed of a mounting frame and four mutually staggered wheels with motors. The wheels were pressed against the rope by the action of springs, and could adapt to the climbing tasks on ropes of different diameters. For hard wire ropes, the climbing performance of the robot will be greatly reduced.

Fang, G. [31] developed a pneumatic wire rope climbing robot, WRR-I, for the maintenance of sluice wire ropes. The robot adopted a split structure, which was composed of an upper device and a lower device. The pneumatic drive cylinder was used to realize the robot's clamping, moving, and guiding functions. Moreover, it could carry a camera and a laser cleaning device to detect and clean the sluice wire rope. The disadvantage of the robot is that its motion is discontinuous.

Under this direction, this work was mainly focused on the description of a wheeled type rope climbing robot for sluices, which was applied to carry laser cleaning and testing equipment, as well as other working tools to carry out daily maintenance of the wire rope. Therefore, the service life of the sluice wire rope can be prolonged.

The rest of this work is organized as follows. In Section 2 some considerations on robot design are analyzed, while in Section 3 the mechanical structure of the six-wheeled wire rope climbing robot is presented. In Sections 4–6, the mechanical analysis of the robot is established and verified by experiments. Finally, in Section 7 the conclusions and future work are discussed.

2. Considerations on Robot Design

Different regions and different types of hoisting sluices employ different diameters and lengths of wire ropes. For example, the sluice used in rural river channels (as the example shown in Figure 1a) uses a wire rope with a diameter of about 10–20 mm, and a length of about 5–8 m, while the sluice used in coastal rivers (as the example depicted in

Figure 1b) uses a wire rope with a diameter of about 15–30 mm, and a length about 6–15 m. By considering its versatility, the goal of the designed robot is to be able to adapt to sluice wire rope climbing tasks with diameters in the range of 10–30 mm and lengths in the range of 5–15 m.

(a) (b)

Figure 1. Sluice and wire rope working scene. (**a**) An example of the sluice used in rural river channels, and (**b**) an example of the sluice used in coastal rivers.

Differently from straight rods, wire ropes are spirally wound with multiple strands of steel wire, causing the surface of the wire rope to be uneven and flexible. Due to the long-term operation of the wire rope, its surface will have problems, such as grease hardening, broken wire, corrosion, wear, and looseness, which can effectively lead to certain changes in the diameter of the wire rope. This fact requires the designed climbing robot to have the ability to adapt to different environments, and also to overcome obstacles. According to observation, it is generally required that the robot's obstacle-crossing height should be $\geq$2 mm.

Most of the sluice wire ropes are installed vertically, and their working states are either tensioned or relaxed. While in tension, the inclination angle of the wire rope is generally 80–90°. Hence, the robot should be able to climb up and down with a load in both vertical and inclined directions, not causing damage to the wire rope. It is important to point out the fact that the sluice is hoisted by multi-strand wire ropes, through dynamic and static pulleys. The wire rope of each hanging point on the gate is arranged in four or more strands, and the distance between the two strands of wire rope is different (generally in the range of 50–300 mm). This requires that the size of the lateral structure of the designed robot should not exceed 150 mm. The wire rope is coated with grease, while the degree of hardening varies with the working time, which results in a small dynamic and static friction coefficient between the robot and the wire rope contact surface. The robot needs to be equipped with detection devices, cleaning devices, and oiling devices, which are important for maintenance operations. Thus, the weight of the robot itself should not be more than 6 kg. At the same time, to can carry work tools, the load capacity of the robot needs to be higher than 3 kg. Compared with manual maintenance, robot operation requires a certain performance improvement. Therefore, it needs to have a certain climbing operation speed, which should be $\geq$20 mm/s with a 3 kg load, and $\geq$30 mm/s without a load. In addition, to ensure the continuity of maintenance operations, the robot requires good climbing stability, with no sudden change in acceleration under normal conditions. On top of that, climbing robots work during high-altitude operations, leading to the necessity of ensuring their safety in the event of a power outage, in order that they will not slip and fall on these occasions. Another factor to be considered is that the robot needs to clean multiple wire ropes. In order to improve work efficiency and reduce non-working time, it is required that installation and disassembly are simple and convenient.

Based on the above-mentioned considerations, the designed climbing robot for sluice wire rope should meet the design requirements shown in Table 1.

Table 1. Design requirements of the climbing robot for the sluice wire rope.

Dimensions (L × W × H)	≤250 × 150 × 400 mm
Weight	≤6 kg
Load capacity	≥3 kg
Adaptable diameter	φ10–30 mm
Climbing speed	≥20 mm/s (with a 3 kg load); ≥30 mm/s (without a load)
Obstacle-crossing ability	≥2 mm
Installation time	≤5 min

3. Mechanism Design

3.1. *Choice of the Attachment and Locomotion Methods*

According to the above-mentioned analysis, the key factors to be considered in the design of the climbing robot for sluice wire rope are the choice of attachment method and the locomotion method. There are also other important considerations, which include power loss safety and flexibility of the wire rope, among others.

(1) The choice of the attachment method. Although the wire rope is a magnetically conductive material, the surface of the wire rope is both oily and uneven, which significantly attenuates its magnetic adsorption force. Hence, it can be concluded that the magnetic adsorption method is incompetent. Due to the uneven surface of the wire rope, both the clamping attachment and claw-thorn attachment methods can be applied. Compared with the claw-thorn grasping method, the clamping attachment method is simpler in structure and more adjustable in strength. For this reason, this method was chosen to be adopted for the proposed design. In terms of the selection of the specific clamping methods, two forms were considered in an earlier stage, namely, pneumatic clamping and electric clamping. After experimental verification, it was found that although the clamping force of the pneumatic clamping is large, the pneumatic control is more complex, and additional assistance, such as an air pump, is required. Besides, the electric gripping requires motors or electromagnets, resulting in excessive weight of the robot. The spring clamping method has the characteristics of adjustable clamping force, simple structure, and low cost, thus for this design, the spring clamping method was adopted.

(2) The choice of the locomotion method. For wire rope climbing, wheeled, legged, crawler, and telescopic methods can all meet the design requirements. Due to the complex control and slow speed of the legged and telescopic climbing robots, they were not considered for this design. Compared with the wheeled climbing robot, the crawler-type climbing robot needs to be specially designed in order to obtain stable vertical climbing performance on the oily wire rope, increasing the entire design cost. The wheeled robot is simple in structure, and convenient to manufacture. Thus, for the introduced design the wheeled climbing and moving method was adopted. In order to reduce the influence of the clamping mechanism on the extrusion and deformation of the wheels, a six-wheel climbing method was also adopted.

(3) Other considerations. Due to the small diameter of the sluice wire rope and the small distance between the two wire ropes, the commonly used prismatic frame structure and cylindrical frame structure equipped with three moving modules were considered as not suitable. Therefore, for the proposed design, a cuboid frame structure equipped with two moving modules was adopted. The structure has a narrow width, which is convenient for the installation and disassembly of the robot, and will also not collide and interfere with the wire rope. In terms of power loss safety considerations, the robot is driven by a DC planetary gear reduction motor. In the event of power loss, the motor has a good

locked-rotor performance, which can prevent the robot from slipping off the wire rope when it loses power.

3.2. Overall Structure Design of a Wheeled-Climbing Robot for Sluice Wire Rope

According to the aforementioned design requirements, as well as the scheme analysis, a six-wheel climbing robot for sluice wire rope named WRR-II (the second generation of wire rope climbing robot) was developed. The robot was composed of two detachable frames: a left frame composed of a driving trolley, a control box, and an upper anti-deflection guide device. The right frame is composed of a driven trolley, a trolley position adjustment mechanism, and a lower anti-deflection guide device.

The backs of the left and right frames were connected by hinges, in order that they can be easily opened and fixed onto the wire rope. The front parts of the left and right frames were clamped by pull buckles to form a closed robot, to prevent the robot from being detached from the wire rope when it is working. The control box, which was used to perform remote control operations, was equipped with several components, such as a DC power supply, a motor drive unit, a wireless control unit, etc. The upper and lower anti-deviation guide devices were composed of four mounting frames and four rollers, which play the role of anti-deviation and guidance when the robot is climbing, preventing the robot from detaching from the wire rope. Additionally, there are installation holes on the upper and lower parts of the left and right frames, which can be equipped with inspection, cleaning, oiling, and other equipment to carry out maintenance operations on the wire rope. The schematic diagram of the two- and three-dimensional structures of the designed robot is shown in Figure 2.

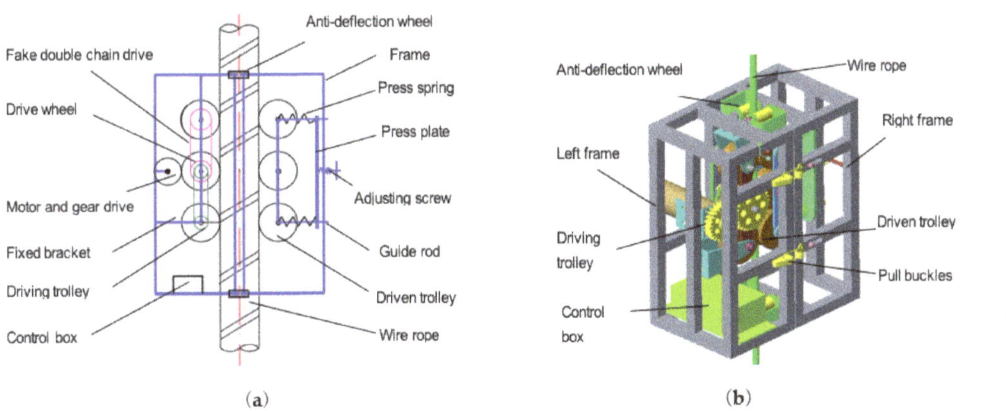

Figure 2. Wheeled climbing robot for wire rope for the sluice. (**a**) 2D schematic diagram of the robot, and (**b**) 3D CAD structure of the robot.

(1) Robot attachment device. The entire attachment device was clamped by the left-driving trolley and the right-driven trolley through the compression spring, to clamp the wire rope and its three-dimensional structure, as illustrated in Figure 3. The left-driving trolley was fixed on the left frame through the U-shaped bracket, while the position of the right driven trolley within the right frame can be adjusted through the upper and lower guide rods. When the left and right frames are enclosed, the left and right trolleys clamp the wire rope through the V-shaped rubber wheels, where the clamping force can be adjusted by manual levers, by adjusting screws, pressing plates, and springs. In order to reduce the influence of the clamping force on the deformation of the rubber wheel and the wire rope, and increase the contact area between the wheel and the wire rope, the left and right trolleys were equipped with three V-shaped rubber wheels on each wheel frame. The V-shaped rubber wheel also presents good contact and guiding effect with the wire rope.

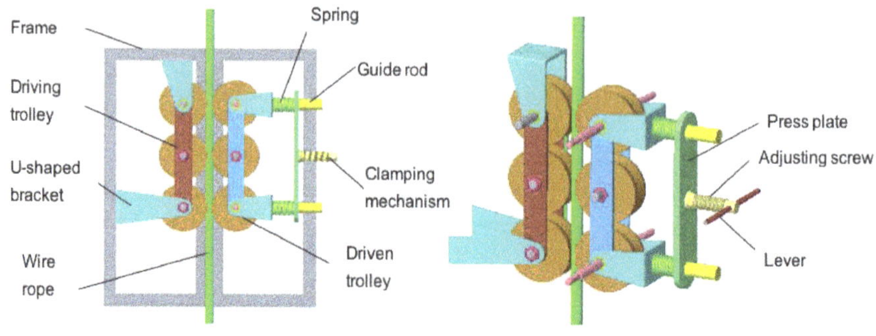

(a) (b)

Figure 3. The 3D CAD structure of the attachment device. (**a**) The external frame of the attachment device, and (**b**) the internal structure of the attachment device.

(2) Robot locomotion device. The locomotion device consists of a left-driving trolley and a right-driven trolley, as depicted in Figure 4. The left-driving trolley was composed of a DC reduction motor, a gear pair, three V-shaped rubber wheels, L-shaped and U-shaped fixed brackets, axles, bearings, wheel lateral mounting plates, and false double chain drive mechanisms. The geared motor was fixed onto the left frame by an L-shaped bracket, while the driving gear was installed coaxially with the motor. The driven gear and the intermediate driving wheel were fixed together by the axle, and the three V-shaped wheels were fixed onto the left frame through the wheel axle, the wheel frame, and the upper and lower U-shaped frames. A transmission sprocket was also installed onto the outside of each of the three axles, while each driving sprocket was fixed with the wheel axle through a locking screw. When the motor rotates, it drives the driving gear and the driven gear, thus the middle driving wheel rotates. When the middle driving wheel rotates, the upper and lower driving wheels also rotate by the action of the false double-row chain transmission. Thereby, the robot can climb up and down through the friction between the driving wheel and the wire rope. The geared motor has a reverse self-locking function, which can ensure that it does not rotate in a power-off state, thereby preventing the robot from falling. The right-driven trolley is composed of three V-shaped rubber wheels, two U-shaped fixed frames, three wheel axles, bearings, and wheel-side mounting plates. The three V-shaped wheels of the driven trolley are driven wheels, which mainly play the role of auxiliary guidance and support when the robot is running.

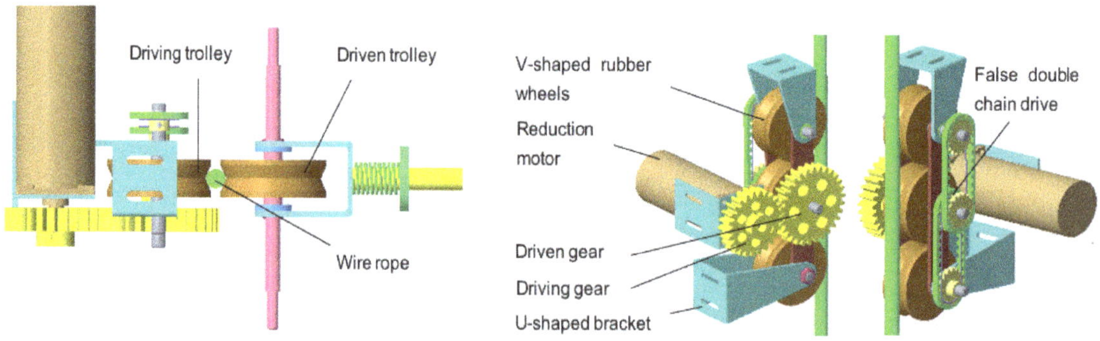

(a) (b) (c)

Figure 4. The 3D CAD structure of the locomotion device. (**a**) The top view of the mobile device, (**b**) the front structure of the driving trolley, and (**c**) the back structure of the driving trolley.

4. Mechanical Analysis

Mechanical analysis is mainly used to determine the pressing force of the robot, as well as the driving torque and several structural parameters of the motor, to provide a theoretical basis for the optimization of the robot mechanism, the selection of the motor, and the motion control. The designed robot in this work belongs to a redundant statically indeterminate structure, with a relatively complex mechanical analysis. In order to simplify its mechanical analysis process, the following assumptions were made:

(1) The wire rope is fixed at both the top and bottom, and the tensile force at both ends is large enough, and therefore, the wire rope can be regarded as an approximate rigid body. The wire rope is also inelastically elongated.

(2) During the climbing process of the robot, the wheels only roll and do not slide.

(3) The front and back sections, as well as the left and right sections, of the robot are symmetrical, with the center of mass coinciding with the origin O′ of the robot's local coordinate system, which is located at the intersection of the line that connects the centers of wheel 3 and wheel 4 and the axis of the wire rope.

(4) The stiffness coefficients of the two compression springs are the same, which are also equal to the compression lengths.

(5) In the pressed state, the wheel does not deform.

4.1. Static Analysis

4.1.1. Static Analysis of the Hovering State

In order to analyze the balance and driving conditions of the robot, a schematic diagram of the force analysis in the hovering state of the robot is established, as shown in Figure 5. In the figure, YOZ is the inertial coordinate system, Y′O′Z′ is the local coordinate system of the robot, O and O′ are their coordinate origins, F_I refers to the clamping force acting on each V-shaped wheel, N_I denotes the normal force of the wire rope to each wheel, F_{fi} represents the friction force between the wheel and the wire rope, G stands for the total weight of the robot together with the load, θ is the angle between the axle of the wire rope and the Y direction of the inertial coordinate system, F signifies the clamping force applied at the handle, k is the stiffness coefficient of the spring, r is the radius of the wire rope, R denotes the radius of the V-shaped wheel, and L is the distance between the two wheels along the axis of the wire rope.

According to the force balance equations, Equations (1) and (2) can be obtained.

$$\sum Y' = 0,\ F_1 + F_3 + F_5 + N_2 + N_4 + N_6 + G\cos\theta - F_2 - F_4 - F_6 - N_1 - N_3 - N_5 = 0 \quad (1)$$

$$\sum Z' = 0,\ F_{f1} + F_{f3} + F_{f5} + F_{f2} + F_{f4} + F_{f6} - G\sin\theta = 0 \quad (2)$$

By considering that the structure of each wheel is symmetrical and the load is balanced, then Equations (3)–(7) can be obtained.

$$N_1 = N_3 = N_5 \quad (3)$$

$$N_2 = N_4 = N_6 \quad (4)$$

$$F_1 = F_3 = F_5 \quad (5)$$

$$F_2 = F_4 = F_6 \quad (6)$$

$$F_{fi} = \mu N_I,\ i = 1 \sim 6 \quad (7)$$

In Equation (7), N_I is the normal force of the wire rope to each wheel, and μ represents the static friction coefficient between the wheel and the wire rope.

In the hover state and since $F_1 = F_2$, the values of N_1 and N_2 can be determined by Equations (8) and (9).

$$N_1 = \frac{G\sin\theta}{6\mu} + \frac{G\cos\theta}{6} \quad (8)$$

$$N_2 = \frac{G\sin\theta}{6\mu} - \frac{G\cos\theta}{6} \quad (9)$$

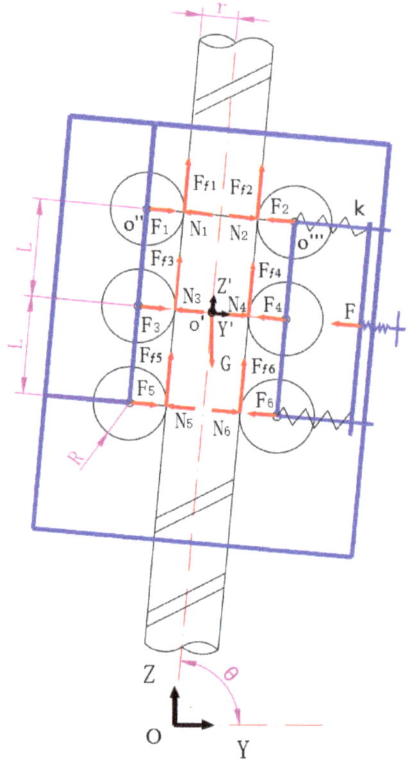

Figure 5. Force diagram of the robot in the YOZ coordinate plate.

According to the force analysis of wheel 1, Equations (10) and (11) can be obtained.

$$N_1 = F_1 + \frac{G}{3}\cos\theta \quad (10)$$

$$F_1 = N_1 - \frac{G\cos\theta}{3} = \frac{G\sin\theta}{6\mu} - \frac{G\cos\theta}{6} \quad (11)$$

The clamping force F can be expressed by Equation (12).

$$F = F_1 + F_3 + F_5 = F_2 + F_4 + F_6 = 3\,F_1 \quad (12)$$

By combining Equation (11) with Equation (12), the following is obtained:

$$F = \frac{G\sin\theta}{2\mu} - \frac{G\cos\theta}{2} \quad (13)$$

For the robot to be able to hover on the wire rope, the clamping force F applied at its handle should satisfy the following condition:

$$F \geq \frac{G\sin\theta}{2\mu} - \frac{G\cos\theta}{2} \quad (14)$$

By substituting the basic parameters of the robot as follows: $G = 40\ N$, $\mu = 0.1$, $\theta = 90°$, $k = 50\ N/cm$, the minimum value of the clamping force F is the following:

$$F \geq 200\ N \tag{15}$$

According to Hooke's law $F = 2k\Delta x$, the minimum distance that the screw needs to move can be determined by Equation (16).

$$\Delta x = \frac{F}{2k} = \frac{200}{2 \times 50} = 2\ cm \tag{16}$$

4.1.2. Static Analysis When the Robot Has an Upward Movement Trend

When driving wheel 1 of the robot rotates clockwise, the robot tends to move upward, and its force analysis is displayed in Figure 6a. At this point, the friction of the left wheel is upward, and the friction of the right wheel is downward. By considering the symmetry of the three pairs of left and right wheels, and in order to simplify the calculation process, the force analysis of the entire robot was considered equivalent to the force analysis of the top pair of wheels, as shown in Figure 6b.

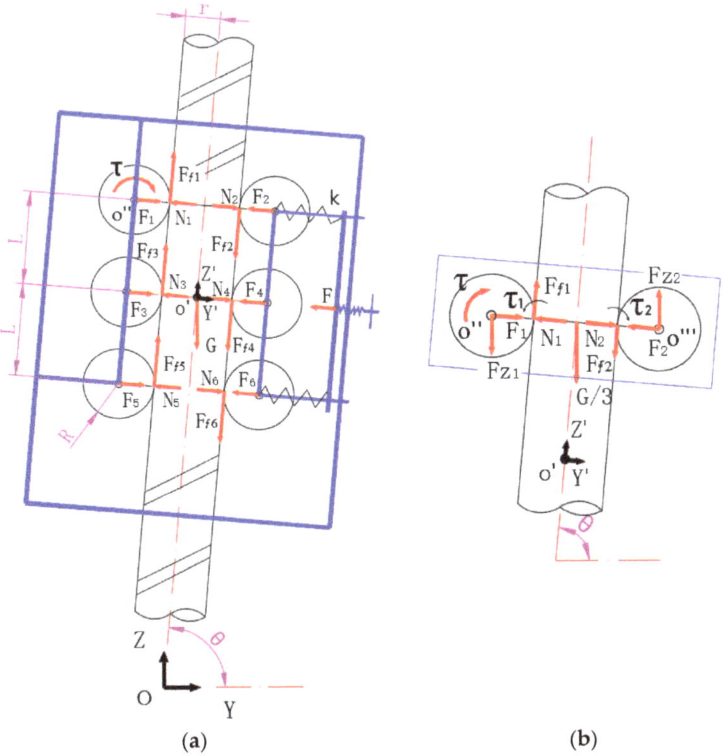

Figure 6. Force diagram of the robot when it has an upward movement trend. (**a**) Force diagram in the YOZ coordinate plane, (**b**) simplified equivalent force diagram.

According to the force and moment equilibrium conditions, and assuming that the weight of the wheel is negligible, the following balance equations can be established.

$$\sum Y' = 0,\ F_1 + N_2 + \frac{G}{3}\cos\theta - F_2 - N_1 = 0 \tag{17}$$

$$\sum Z' = 0,\ F_{f1} - F_{f2} - \frac{G}{3}\sin\theta = 0 \quad (18)$$

$$\sum M_{o''} = 0,\ F_{f1}R - F_{f2}(R + 2r) - \frac{G}{3}\sin\theta(R + r) - \tau_1 - \tau_2 = 0 \quad (19)$$

where O'' is the center of mass of wheel 1, O''' stands for the center of mass of wheel 2, R is the radius of the wheel, r is the radius of the wire rope, and τ_1,τ_2 denote the rolling friction couple moments of wheel 1 and wheel 2, respectively.

In the previous three equations, there are seven unknown values. Hence, the left and right wheels, as well as the robot framework, need to be solved separately. The force diagrams are shown in Figure 6b.

The balance Equations (20)–(22) can be obtained from wheel 1.

$$\sum M_{o''} = 0,\ F_{f1}R = \tau_1 \quad (20)$$

$$\sum Y' = 0,\ N_1 = F_1 + F_{z1}\cos\theta \quad (21)$$

$$\sum Z' = 0,\ F_{f1} = F_{z1}\sin\theta \quad (22)$$

The balance Equations (23)–(25) can be obtained from wheel 2.

$$\sum M_{o'''} = 0,\ F_{f2}R = \tau_2 \quad (23)$$

$$\sum Y' = 0,\ N_2 = F_2 \quad (24)$$

$$\sum Z' = 0,\ F_{f2} = F_{z2} \quad (25)$$

The balance Equations (26)–(27) can be obtained from the robot framework.

$$\sum Y' = 0,\ F_1 + F_{z1}\cos\theta = F_2 \quad (26)$$

$$\sum Z' = 0,\ F_{z1}\sin\theta = F_{z2} + \frac{G}{3}\sin\theta \quad (27)$$

When the wheel is in a critical equilibrium state, the rolling friction couple moment reaches the maximum value, which is equal to the following:

$$\tau_1 = \delta N_1 \quad (28)$$

$$\tau_2 = \delta N_2 \quad (29)$$

where N_i is the normal force of the wire rope to each wheel, and δ is the rolling friction coefficient between the wheel and the wire rope.

By combining Equations (17)–(29), the following can be obtained:

$$\tau \geq 3F_{f1}R = 3F_{f2}(R + 2r) + G\sin\theta(R + r) + 3\tau_1 + 3\tau_2 = 9\delta F_2 + 6\delta F_2\frac{r}{R} + G\sin\theta(R + r) \quad (30)$$

By also considering that $F_2 = \frac{F}{3}$:

$$\tau \geq 3\delta F + 2\delta F\frac{r}{R} + G\sin\theta(R + r) \quad (31)$$

Substituting the basic parameters of the robot as: $G = 40$ N, $\delta = 2$ mm, $\theta = 90°$, $R = 26$ mm, $r = 5$ mm, and $F = 300$ N, the minimum value of the drive torque τ is the following:

$$\tau \geq 3.27\ \text{n.m} \quad (32)$$

4.1.3. Static Analysis When the Robot Has a Downward Movement Trend

When driving, wheel 1 of the robot rotates counterclockwise and the robot tends to move downward—its force analysis is displayed in Figure 7a. At this point, the friction of the left wheel is downward, and the friction of the right wheel is upward. By considering the symmetry of the three pairs of left and right wheels, and in order to simplify the calculation process, the force analysis of the entire robot is considered equivalent to the force analysis of the top pair of wheels, as illustrated in Figure 7b.

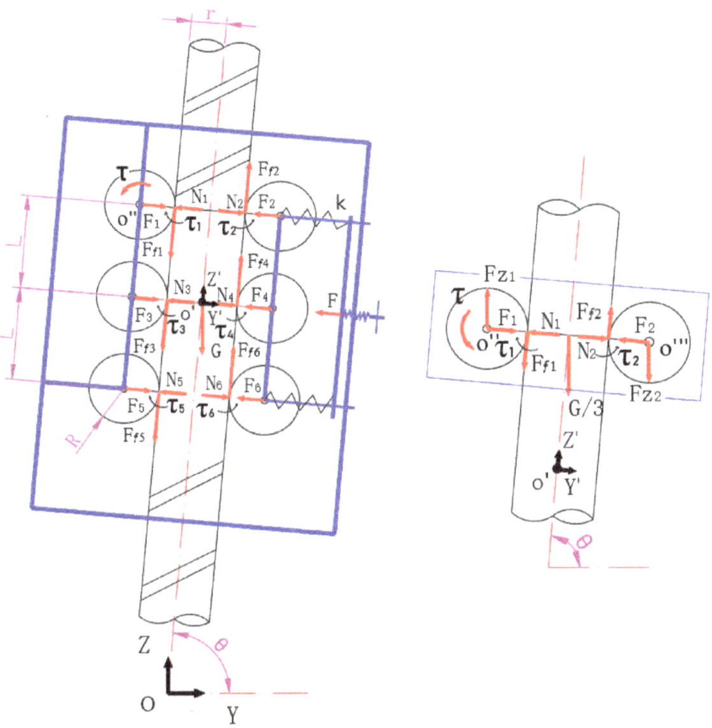

(a) (b)

Figure 7. Force diagram of the robot when it has a downward movement trend. (**a**) Force diagram in the YOZ coordinate plane, (**b**) simplified equivalent force diagram.

According to the force and moment balance conditions, and assuming that the weight of the wheel is negligible, the following equilibrium equations can be established.

$$\sum Y' = 0,\ F_1 + N_2 + \frac{G}{3}\cos\theta - F_2 - N_1 = 0 \quad (33)$$

$$\sum Z' = 0,\ F_{f2} - F_{f1} - \frac{G}{3}\sin\theta = 0 \quad (34)$$

$$\sum M_{o''} = 0,\quad -F_{f1}R + F_{f2}(R+2r) - \frac{G}{3}\sin\theta(R+r) - \tau_1 - \tau_2 = 0 \quad (35)$$

Similar to the solving method of forces for the upward motion trend of the robot, the left and right wheels and the frame were taken as research objects, and the force diagrams are shown in Figure 7b. The equilibrium equation was solved, and the following was obtained:

$$\tau \geq 3\delta F + 2\delta F\frac{r}{R} - G\sin\theta(R+r) \quad (36)$$

By substituting the basic parameters of the robot as follows: $G = 40$ N, $\delta = 2$ mm, $\theta = 90°$, $R = 26$ mm, $r = 5$ mm, and $F = 300$ N, the minimum value of the drive torque τ is the following:

$$\tau \geq 0.79 \text{ n.m} \quad (37)$$

4.2. Kinematics Analysis of the Robot

The schematic diagram of the kinematics analysis of the robot is shown in Figure 8.

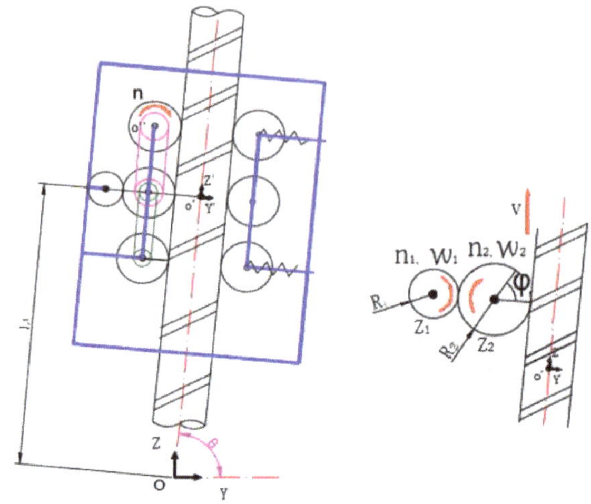

Figure 8. Schematic diagram of the kinematics analysis of the robot.

In Figure 8, L_1 is the initial position length of the robot on the wire rope, v refers to the climbing speed of the robot, n_1 is the rotation speed of the motor, n_2 is the rotation speed of the driving wheel, Z_1 denotes the number of teeth of gear 1, Z_2 is the number of teeth of gear 2, w_1 represents the rotational angular velocity of the motor, w_2 is the rotational angular velocity of the driving wheel, φ stands for the rotational angle of the driving wheel, and θ is the angle between the axle of the wire rope and the Y direction of the inertial coordinate system.

The position equation of the robot can be expressed by Equation (38).

$$\begin{cases} y = L_1\cos\theta + \varphi R_2\cos\theta \\ z = L_1\sin\theta + \varphi R_2\sin\theta \end{cases} \tag{38}$$

The speed formula of the gear transmission pairs can be expressed by the following expression:

$$\frac{n_1}{n_2} = \frac{Z_2}{Z_1} \tag{39}$$

The rotation angle φ of driving wheel 1 can be calculated as follows:

$$\varphi = w_2t = \frac{2\pi n_2}{60}t = \frac{\pi n_2}{30}t = \frac{\pi Z_1 n_1}{30Z_2}t \tag{40}$$

According to Equation (38) and Equation (40), the position equation of the robot can be obtained as the following:

$$\begin{cases} y = L_1\cos\theta + \frac{\pi Z_1 n_1}{30Z_2}R_2t\cos\theta \\ z = L_1\sin\theta + \frac{\pi Z_1 n_1}{30Z_2}R_2t\sin\theta \end{cases} \tag{41}$$

The velocity equation of the robot can be obtained as follows:

$$\begin{cases} v_y = \dot{y} = \dot{\varphi}R_2\cos\theta = w_2R_2\cos\theta = \frac{\pi Z_1 n_1}{30Z_2}R_2\cos\theta \\ v_z = \dot{z} = \dot{\varphi}R_2\sin\theta = w_2R_2\sin\theta = \frac{\pi Z_1 n_1}{30Z_2}R_2\sin\theta \end{cases} \tag{42}$$

The acceleration equation of the robot is obtained as the following:

$$\begin{cases} a_y = \ddot{y} = \ddot{\varphi} R_2 \cos\theta = \dot{w}_2 R_2 \cos\theta \\ a_z = \ddot{z} = \ddot{\varphi} R_2 \sin\theta = \dot{w}_2 R_2 \sin\theta \end{cases} \tag{43}$$

By substituting the basic parameters of the robot as follows: $Z_1 = 25$, $Z_2 = 30$, $\theta = 90°$, $R_2 = 26$ mm, and $n_1 = 23$ r/min, the theoretical value of the climbing speed of the robot is the following:

$$v_z = 52\ \text{mm/s},\ v_y = 0\ \text{mm/s} \tag{44}$$

5. Control Architecture

The current climbing robot can carry cameras and NDT equipment to detect defects in wire ropes. The electronic architecture of the WRR-II platform is presented in Figure 9. The hardware control box is depicted in Figure 10. The control system consists of two main components, namely, the user-level controller and the low-level controller. The user-level controller is on the Tablet PC platform, and it provides the user interface and data transmission from the climbing robot. The low-level controller is based on the STM32F407ZET6 main control unit (MCU), which controls the motion of the DC motors and the peripheral devices (surveillance camera, extended NDT device, etc.). The communication between the two levels was set through a Wi-Fi module. The user interface (UI) was developed based on Qt software. The robot was instructed to move upward and downward by the UI. In order to reduce the weight of the robot, the power supply was provided from the external module in our current system.

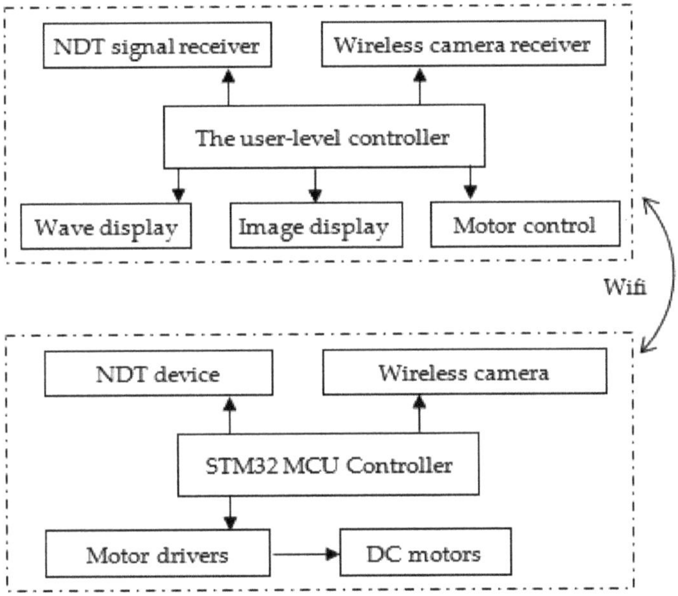

Figure 9. The control architecture of WRR-II.

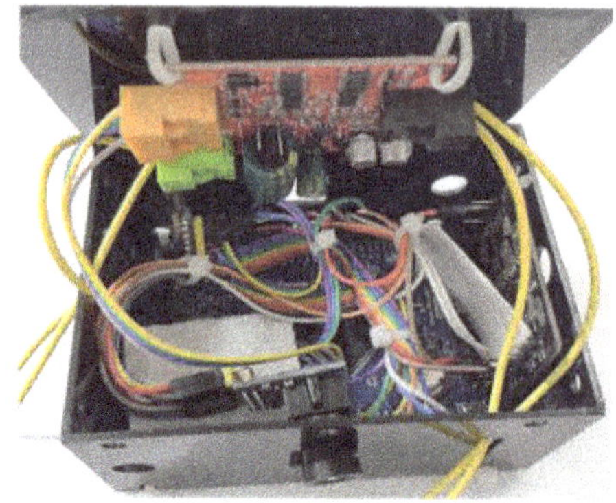

Figure 10. The control box of WRR-II.

6. Experiments

In order to verify the rationality of both the design scheme and the mechanical analysis of the robot, a prototype robot was built (WRR-II), as shown in Figure 11. In order to reduce the weight, except for the DC motor, chain drive mechanism, and V-shaped rubber wheel, the rest of the robot is made of aluminum alloy. The total mass of the robot is 3.8 kg, and the structural size of the robot is 250 × 150 × 300 mm (L × W × H).

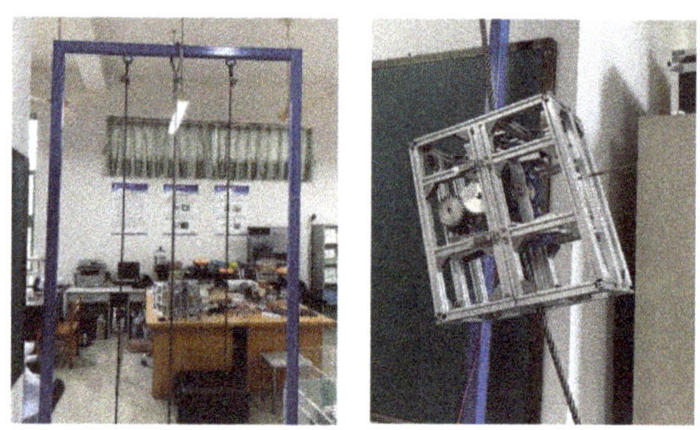

(**a**) (**b**)

Figure 11. Experimental verification of the robot. (**a**) The experimental gantry, and (**b**) the climbing robot fixed on the gantry.

In order to simulate the working scene of the outdoor sluice, an experimental gantry with a height of 2 m and a width of 1.2 m was built, as illustrated in Figure 11a. Three steel wire ropes were erected on the gantry, with the following diameters: φ10 mm, φ14 mm, and φ10 mm. A φ10 mm wire rope on the far right is adjustable for tightness, and was coated with grease. The leftmost φ10 mm and the middle φ14 mm wire ropes were fixed at both the top and bottom, and were not greased. The distance between the adjacent wire ropes was 200 mm. This arrangement can simulate and test the climbing performance of the robot under different working conditions of the wire rope.

The installation and disassembly process of the robot is very simple since it only requires opening the two pull buttons on the front of the robot to separate the left and right frames of the robot around the hinges, putting it on the wire rope, and then fastening the pull buttons. Finally, the clamping force between the wheels of the driven trolley and the wire rope is adjusted by the position adjustment mechanism. Thus, the robot can hover on the wire rope without slipping. The whole operation process can be completed within 1 min by a single person. The disassembly process is exactly the reverse of the previously described installation process.

To validate the performance of the robot, the project team carried out a series of experiments, such as a climbing speed test, a climbing adaptability test, a load capacity test, and an obstacle negotiation ability test.

6.1. Climbing Speed Test

In the case that no load is present, the climbing speed of the robot was calculated by measuring the time required for the robot to climb 1 m on φ14 mm grease-free wire rope several times. Through these experiments, it was found that when the output speed of the motor was 23 r/min, the robot can climb up at a speed of 40 mm/s, and move downward at a speed slightly higher (45 mm/s). Compared with the climbing speed of the first-generation climbing robot WRR-I developed in the early stages (26 mm/s) [31], the speed performance was significantly improved. However, when compared with the theoretical calculation speed of 52 mm/s the actual climbing speed of the robot was reduced, due to a certain slippage between the rubber wheel and the wire rope during the climbing process.

6.2. Climbing Adaptability Test

In order to test the climbing adaptability of the robot, it was installed on the wire ropes under five different working conditions, as shown in Figure 12. The climbing stroke was equal to 1 m up and down, and the test results are shown in Table 2.

(a) (b) (c)

Figure 12. The sluice wire rope under three different working conditions: (**a**) φ10 mm, without grease; (**b**) φ10 mm, with grease; (**c**) φ14 mm, without grease.

Table 2. The results of the climbing adaptability test of the robot.

	Experimental Conditions of the Wire Rope	Velocity of Upward Climbing (mm/s)	Velocity of Downward Climbing (mm/s)
1	φ10 mm, with grease, fixed at both ends	34.5	45.5
2	φ10 mm, with grease, fixed top, free bottom	32.3	47.6
3	φ10 mm, with grease, fixed at both ends, tilt angle is 60°	31.3	47.6
4	φ10 mm, without grease, fixed at both ends	38.5	45.5
5	φ14 mm, without grease, fixed at both ends	40.0	45.5

As can be seen from Table 2, the robot can stably climb on the wire rope, under various conditions. By analyzing upward climbing speeds, the robot on the non-greased wire rope was obviously faster than the one on the greased wire rope. For the same greased wire rope, the robot climbing speed was slightly faster when the two ends were fixed, when compared to the one with a single end. Regarding the angle, it is possible to conclude that the robot's climbing speed was slightly faster in the vertical case than one in the inclined case. During the downward process, the downward speed under various working conditions is generally consistent, and no obvious differences can be found. In addition, the change in the wire rope diameter has some impact on the climbing speed of the robot. As the diameter of the wire rope increased, the climbing speed increased slightly.

6.3. Load Capacity Test

In order to verify the load capacity of the robot, the robot was installed on a φ14 mm non-greased wire rope that was fixed at both ends, where loads of different weights are added on the robot, as depicted in Figure 13, with the climbing speed results shown in Figure 14. Looking at the results, it is clear that as the load increases, the upward speed of the robot slows down. When the load exceeds 10 kg, the motor is overloaded and does not move during the upward process of the robot. Regarding the downward process, the speed of the robot was relatively stable, independently of the load's weight.

Figure 13. Load capacity test setup.

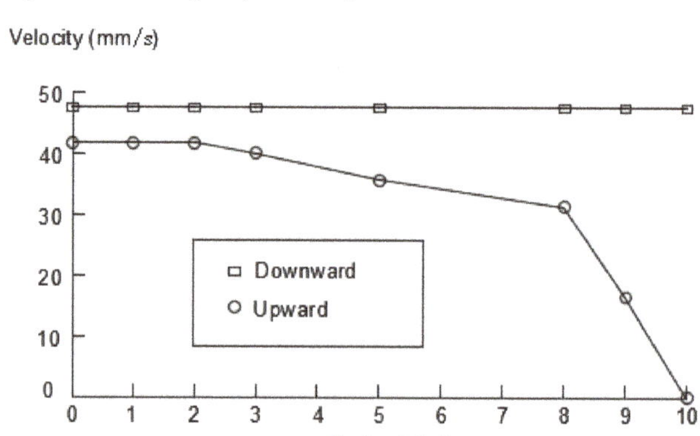

Figure 14. The results of climbing speed.

The climbing speed of the robot varies for different wire rope inclination angles and different load weights, and its relationship is depicted in Figure 15. The robot was installed on a φ10 mm non-greased wire rope that was fixed at both ends, where loads of different weights ranging from 0 to 5 kg were added onto the robot. The range of the wire rope inclination angles was from 50° to 90°. As shown in Figure 15, the climbing speed of the robot decreased significantly with the increase in load weights. For the same load, with the decrease of the inclination angle, the climbing speed of the robot decreases first and then grows.

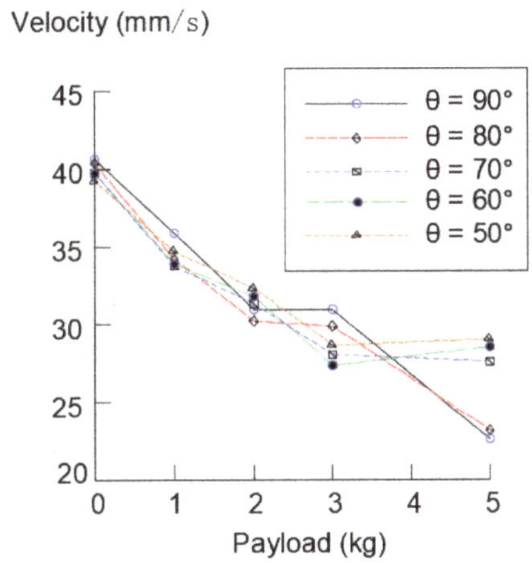

Figure 15. Depiction of the results of the climbing speed.

6.4. Obstacle Negotiation Ability Test

In order to test the robot's ability to cross obstacles, the φ14 mm wire rope was wrapped with tape to form three steps of different diameters, as shown in Figure 16. The diameters of the three steps were φ15 mm, φ16 mm, and φ17 mm, with their spacing equal to 30 mm. During the ascending process (as shown in Figure 17), the robot successfully passed three steps of different diameters but during the descending process, the robot briefly slipped at the φ17 mm step, barely passing at the end. This also reflects that wheeled climbing robots have certain deficiencies when trying to overcome obstacles.

Figure 16. Obstacle negotiation ability test setup.

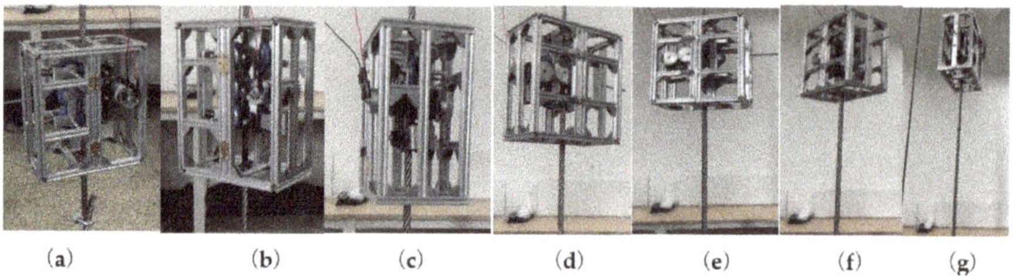

Figure 17. Process of obstacle negotiation. (a) t = 0 s, (b) t = 5 s, (c) t = 10 s, (d) t = 15 s, (e) t = 20 s, (f) t = 25 s, (g) t = 30 s.

6.5. Performance Comparison of the Robots

Table 3 lists some performance parameters of various rope-climbing robots. Compared with these climbing robots (WRR-I, WRC^{2}IN-I, WRC^{2}IN-II, as presented in Table 3), WRR-II is more competent for climbing small diameter wire ropes, and has better climbing performance.

Table 3. The performance comparison of the robots.

	WRR-I [31]	WRR-II	WRC^{2}IN-I [5]	WRC^{2}IN-II [28]
Locomotion method	Inchworm-style	Wheeled-style	Wheeled-style	Crawler-style
Attachment method	Clamp	Clamp	Clamp	Clamp
Dimensions	220 × 110 × 80 mm	250 × 150 × 300 mm	φ593 × 563 mm	328 × 507 × 701 mm
Mass	1.5 kg	3.8 kg	30 kg	26.2 kg
Payload	3 kg	8 kg	9 kg	34 kg
Diameter	10–16 mm	10–30 mm	50–90 mm	40–90 mm
Obstacle height	5 mm	3 mm	9 mm	5 mm
Climbing speed	20–26 mm/s	40–45 mm/s	35–80 mm/s	60–80 mm/s

7. Conclusions and Future Work

Both cleaning and maintenance of wire ropes have always been a major problem in the industry since there are problems regarding high labor intensity and high safety risks. The wire rope used for sluices has a small diameter and a narrow operating range, it is installed almost vertically, and is covered with grease of different degrees of hardening, which bring about greater cleaning and maintenance difficulty. Compared with the first-generation pneumatic peristaltic wire rope climbing robot WRR-I, this work proposed and described a new system, a six-wheel wire rope climbing robot (WRR-II). Under the condition of its own weight of 3.8 kg, the robot can carry a maximum of 8 kg of working tools for online laser cleaning and maintenance of steel wire ropes and visual safety inspection, thus it has a good application prospect.

The six-wheeled wire rope climbing robot proposed in this work, not only has a simple structure, a simple control, and a stable climbing speed, but it also has a large contact area and little influence on the wheel deformation of the crawler climbing robot. It was shown that it can adapt to climbing tasks of wire ropes with different diameters and different lubrication states. The theoretical analysis of the statics and kinematics of the robot, as well as the performance test of the prototype, verify the rationality and feasibility of the designed scheme. During the experiments performed with the prototype, it was also found that the V-shaped rubber wheel would have a certain slip when climbing on the surface of the wire rope covered with grease.

In future work, the project team will further optimize the structure of the rubber wheel and increase the claw-thorn structure. Therefore, it can be well adapted to the task of climbing wire rope with grease, as well as to improve the load capacity of the robot.

Some new methods that have the potential to make soft and slight robots are considered to be used to improve the robot's climbing performance, such as a fluidic rolling robot using voltage-driven oscillating liquid [32], and an active sorting of droplets by using an electro-conjugate fluid micropump [33]. What is more, it is necessary to select actual rural river sluices and coastal river sluices for outdoor field experiments, to further verify the climbing ability of the designed robot. In addition, the influence of laser cleaning devices and non-destructive testing devices on the climbing performance of the robot will also be studied, as well as the impact of wire rope maintenance.

Author Contributions: Conceptualization, G.F. and J.C.; methodology, G.F.; investigation, G.F. and J.C.; writing—original draft preparation, G.F.; writing—review and editing, G.F. and J.C.; software, J.C.; validation, G.F. and J.C.; formal analysis, G.F.; project administration, G.F.; funding acquisition, G.F. All authors have read and agreed to the published version of the manuscript.

Funding: This research was funded by the Basic Public Welfare Research Project of Zhejiang Province (grant No. LGG21F030005), and the Key R&D Program of Zhejiang Province (grant No. 2022C02035).

Institutional Review Board Statement: Not applicable.

Informed Consent Statement: Not applicable.

Data Availability Statement: Not applicable.

Conflicts of Interest: The authors declare no conflict of interest.

References

1. Nguyen, S.T.; La, H.M. A Climbing Robot for Steel Bridge Inspection. *J. Intell. Robot. Syst.* **2021**, *102*, 75. [CrossRef]
2. Nguyen, S.T.; La, H.M. Development of a Steel Bridge Climbing Robot. In Proceedings of the 2019 IEEE/RSJ International Conference on Intelligent Robots and Systems, Macau, China, 3–8 November 2019; pp. 1912–1917.
3. Nguyen, S.T.; Pham, A.Q.; Motley, C.; La, H.M. A Practical Climbing Robot for Steel Bridge Inspection. In Proceedings of the 2020 IEEE International Conference on Robotics and Automation, Paris, France, 31 May–31 August 2020; pp. 9322–9328.
4. La, H.M.; Dinh, T.H.; Pham, N.H.; Ha, Q.P.; Pham, A.Q. Automated Robotic Monitoring and Inspection of Steel Structures and Bridges. *Robotica* **2018**, *37*, 947–967. [CrossRef]
5. Cho, K.H.; Kim, H.M.; Jin, Y.H.; Liu, F.; Moon, H.; Koo, J.C.; Choi, H.R. Inspection Robot for Hanger Cable of Suspension Bridge: Mechanism Design and Analysis. *IEEE/ASME Trans. Mechatron.* **2013**, *18*, 1665–1674. [CrossRef]
6. Ding, N.; Zheng, Z.; Song, J.; Sun, Z.; Lam, T.L.; Qian, H. CCRobot-III: A Split-type Wire-driven Cable Climbing Robot for Cable-stayed Bridge Inspection. In Proceedings of the 2020 IEEE International Conference on Robotics and Automation, Paris, France, 31 May–31 August 2020; pp. 9308–9314. [CrossRef]
7. Zheng, M.; Yang, M.; Yuan, X.; Ding, N. A Light-Weight Wheel-Based Cable Inspection Climbing Robot: From Simulation to Reality. In Proceedings of the 2018 IEEE International Conference on Robotics and Biomimetics (ROBIO), Kuala Lumpur, Malaysia, 12–15 December 2018; pp. 1365–1370. [CrossRef]
8. Zheng, Z.; Hu, S.; Ding, N. A Biologically Inspired Cable Climbing Robot: CCRobot-Design and Implementation. In Proceedings of the 2018 IEEE International Conference on Robotics and Biomimetics, Kuala Lumpur, Malaysia, 12–15 December 2018; pp. 2354–2359.
9. Noohi, E.; Mahdavi, S.S.; Baghani, A.; Ahmadabadi, M.N. Wheel-Based Climbing Robot: Modeling and Control. *Adv. Robot.* **2010**, *24*, 1313–1343. [CrossRef]
10. Guan, Y.; Jiang, L.; Zhu, H.; Wu, W.; Zhou, X.; Zhang, H.; Zhang, X. Climbot: A Bio-Inspired Modular Biped Climbing Robot-System Development, Climbing Gaits, and Experiments. *J. Mech. Robot.* **2016**, *8*, 021026. [CrossRef]
11. Haynes, G.C.; Khripin, A.; Lynch, G.; Amory, J.; Saunders, A.; Rizzi, A.A.; Koditschek, D.E. Rapid Pole Climbing with a Quadrupedal Robot. In Proceedings of the 2009 IEEE International Conference on Robotics and Automation, Kobe, Japan, 12–17 May 2009; pp. 2767–2772.
12. Chen, G.; Yang, H.; Cao, H.; Ji, S.; Zeng, X.; Wang, Q. Design of an Embracing-type Climbing Robot for Variation Diameter Rod. *Ind. Robot Int. J. Robot. Res. Appl.* **2019**, *46*, 56–72. [CrossRef]
13. Xu, F.; Hu, J.L.; Jiang, G. The Obstacle-Negotiation Capability of Rod-climbing Robots and the Improved Mechanism Design. *J. Mech. Sci. Technol.* **2015**, *29*, 2975–2986. [CrossRef]
14. Lam, T.L.; Xu, Y. Biologically Inspired Tree-climbing Robot with Continuum Maneuvering Mechanism. *J. Field Robot.* **2012**, *29*, 843–860. [CrossRef]
15. Lam, T.L.; Xu, Y. Motion Planning for Tree Climbing with Inchworm-like Robots. *J. Field Robot.* **2012**, *30*, 87–101. [CrossRef]
16. Fu, G.-H.; Liu, X.-M.; Chen, Y.-F.; Yuan, J. Fast-growing Forest Pruning Robot Structure Design and Climbing Control. *Adv. Manuf.* **2015**, *3*, 166–172. [CrossRef]

17. Wibowo, T.S.; Sulistijono, I.A.; Risnumawan, A. End-to-end Coconut Harvesting Robot. In Proceedings of the 2017 Interna-tional Electronics Symposium, Denpasar, Indonesia, 29–30 September 2016; pp. 444–449.
18. Lu, X.; Guo, D.; Chen, Y. Design and Optimization of the Magnetic Adsorption Mechanism of a Pipeline-Climbing Robot. *J. Mech. Sci. Technol.* **2021**, *35*, 5161–5171.
19. Kakogawa, A.; Ma, S. Design of a Multilink-articulated Wheeled Pipeline Inspection Robot Using Only Passive Elastic Joints. *Adv. Robot.* **2017**, *32*, 37–50. [CrossRef]
20. Ahmed, M.; Eich, M.; Bernhard, F. Design and Control of MIRA: A Lightweight Climbing Robot for Ship Inspection. *Int. Lett. Chem. Phys. Astron.* **2015**, *55*, 128–135. [CrossRef]
21. Yan, C.; Sun, Z.; Zhang, W.; Chen, Q. Design of Novel Multidirectional Magnetized Permanent Magnetic Adsorption Device for Wall-climbing Robots. *Int. J. Precis. Eng. Manuf.* **2016**, *17*, 871–878. [CrossRef]
22. Gu, J.; Wang, C.; Wu, X. Self-adjusted Adsorption Strategy for an Aircraft Skin Inspection Robot. *J. Mech. Sci. Technol.* **2018**, *32*, 2867–2875. [CrossRef]
23. Zhao, Z.; Tao, Y.; Wang, J.; Hu, J. The Multi-objective Optimization Design for the Magnetic Adsorption Unit of Wall-climbing robot. *J. Mech. Sci. Technol.* **2022**, *36*, 305–316. [CrossRef]
24. Xiao, J.; Li, B.; Ushiroda, K.; Song, Q. Rise-Rover: A Wall-Climbing Robot with High Reliability and Load-Carrying Capacity. In Proceedings of the 2015 IEEE Conference on Robotics and Biomimetics, Zhuhai, China, 6–9 December 2015; pp. 2072–2077.
25. Koo, Y.C.; Elmi, A.B.; Wajdi, W.A.F. Piston Mechanism Based Rope Climbing Robot. In Proceedings of the International Symposium on Robotics and Intelligent Sensors 2012, Kuching, Malaysia, 4–6 September 2012; pp. 547–553.
26. Kim, H.M.; Cho, K.H.; Jin, Y.H.; Liu, F.; Koo, J.C.; Choi, H.R. Development of Cable Climbing Robot for Maintenance of Suspension Bridges. In Proceedings of the 2012 IEEE International Conference on Automation Science and Engineering, Seoul, Korea, 20–24 August 2012; pp. 602–607.
27. Cho, K.H.; Jin, Y.H.; Kim, H.M.; Moon, H.; Koo, J.C.; Choi, H.R. Caterpillar-based Cable Climbing Robot for Inspection of Suspension Bridge Hanger Rope. In Proceedings of the 2013 IEEE International Conference on Automation Science and Engineering, Madison, WI, USA, 17–20 August 2013; pp. 1071–1074.
28. Cho, K.H.; Jin, Y.H.; Kim, H.M.; Choi, H.R. Development of Novel Multifunctional Robotic Crawler for Inspection of Hanger Cables in Suspension Bridges. In Proceedings of the 2014 IEEE International Conference on Robotics & Automation, Hong Kong, China, 31 May–7 June 2014; pp. 2673–2678. [CrossRef]
29. Sun, G.; Li, P.; Meng, Y.; Xu, E.; Zhou, Y.; Liu, Y. A Climbing Robot for Inspection of Lamppost in the Airport: Design and Preliminary Experiments. In Proceedings of the 2017 IEEE International Conference on Robotics and Biomimetics, Macau, China, 5–8 December 2017; pp. 436–441. [CrossRef]
30. Ratanghayra, P.R.; Hayat, A.A.; Saha, S.K. *Design and Analysis of Spring-Based Rope Climbing Robot. Machines, Mechanism and robotics, Lecture Notes in Mechanical Engineering*; Springer: Singapore, 2018; pp. 453–462. [CrossRef]
31. Fang, G.; Zhang, G.; Zheng, G.; Yao, L. Development of Pneumatic Robot for Climbing Sluice Wire Rope. *Chin. Hydraul. Pneum.* **2021**, *2*, 170–176.
32. Mao, Z.; Asai, Y.; Yamanoi, A.; Seki, Y.; Wiranata, A.; Minaminosono, A. Fluidic Rolling Robot Using Voltage-driven Oscillating Liquid. *Smart Mater. Struct.* **2022**, *31*, 105006. [CrossRef]
33. Mao, Z.; Yoshida, K.; Kim, J.-W. Active Sorting of Droplets by Using an ECF (Electro-Conjugate Fluid) Micropump. *Sens. Actuators A Phys.* **2019**, *303*, 111702. [CrossRef]

machines

MDPI

Article

A Single-Joint Worm-like Robot Inspired by Geomagnetic Navigation

Dong Mei, Xin Zhao, Gangqiang Tang, Jianfeng Wang, Chun Zhao, Chunxu Li and Yanjie Wang *

Jiangsu Provincial Key Laboratory of Special Robot Technology, Hohai University, Changzhou Campus, Changzhou 213022, China
* Correspondence: yj.wang1985@gmail.com

Abstract: Inspired by identifying directions through the geomagnetic field for migrating birds, in this work, we proposed and fabricated a single-joint worm-like robot with a centimeter scale, the motion of which could be easily guided by a magnet. The robot consists of a pneumatic deformable bellow and a permanent magnet fixed in the bellow's head that will generate magnetic force and friction. Firstly, in order to clarify the actuating mechanism, we derived the relationship between the elongation of the bellows and the air pressure through the Yeoh constitutive model, which was utilized to optimize the structural parameters of the bellow. Then the casting method is introduced to fabricate the silicone bellow with a size of 20 mm in diameter and 28 mm in length. The manufacturing error of the bellow was evaluated by 3D laser scanning technology. Thereafter, the robot's moving posture was analyzed by considering the force and corresponding motion state, and the analysis model was established by mechanics theory. The experimental results show that the worm-like robot's maximum speed can reach 9.6 mm/s on the cardboard. Meanwhile, it exhibits excellent environmental adaptability that can move in pipelines with a diameter of 21 mm, 32 mm, 40 mm, and 50 mm, and surfaces with different roughness. Moreover, the robot's motion was successfully guided under the presence of the magnetic field, which shows great potential for pipeline detection applications.

Keywords: worm-like robot; magnetic navigation; pneumatic actuator; structural design; fabrication; characterization; piping application

Citation: Mei, D.; Zhao, X.; Tang, G.; Wang, J.; Zhao, C.; Li, C.; Wang, Y. A Single-Joint Worm-like Robot Inspired by Geomagnetic Navigation. *Machines* **2022**, *10*, 1040. https://doi.org/10.3390/machines10111040

Academic Editor: Dan Zhang

Received: 25 September 2022
Accepted: 4 November 2022
Published: 7 November 2022

1. Introduction

Nature provides a steady stream of inspiration for engineers to solve problems [1]. For a long time, the principle of bionic motion has been widely applied in the design of various worm-like robots and has attracted worldwide research attention [2]. Compared with the traditional wheel-legged bionic robotics, the worm-like robot, which generates motion by imitating simple changes in the body shape of worms (earthworms and inchworms), can be ideal for navigating long narrow spaces due to its compact structure, lightweight, and small movement space, and has been widely used in pipeline maintenance, medical endoscopy, and other fields [3]. The most previously developed peristaltic robots are built with rigid materials and structures [4–6]. Usually, these rigid robots have poor environmental adaptability due to the "hard" nature of their characteristics, which can only crawl in a specific pipeline environment and possibly damage the pipeline's surface due to improper operation. This significantly limits the peristaltic robot's capability to be used in environments with different pipeline sizes. With the development of materials, biomechanics, and manufacturing technologies, soft robotic technology has attracted the extensive attention of researchers. Compared with rigid robots, soft robots, which have continuous deformation, infinite degrees of freedom theoretically, more robust environmental adaptability, and safely interact with humans [7,8], are made of flexible or soft materials (a Young's modulus of 10^4–10^9 Pa) and can arbitrarily change their shape to pass the complex constrained environments. The emergence of soft robots provides considerable prospects for peristaltic robots.

Most of the existing soft worm-like robots are designed in multiple-segmented structures, which can realize the effective contraction-expansion motion like earthworms or the Ω motion like inchworms by actuating this segmented structure in a predesigned sequence [9]. For instance, Onal et al. [10] presented a worm-like origami robot driven by shape memory alloy (SMA), which can realize the crawling motion by controlling the elongation and contraction through different SMA. Pfeil et al. [11] proposed a bellows structured robot, which was actuated by dielectric elastomers (DE) and can stably realize the peristalsis on the horizontal surface. However, the efficiency of the worm-like robot's locomotion actuated by the SMA spring or DE did not exhibit satisfactory performance as a result of the intractability of controlling the SMA and DE in the motion process. In contrast, pneumatically is widely adopted for driving soft worm-like robots because of its low-cost design and easy control. Furthermore, pneumatic actuation has high driving efficiency and achieved many impressive achievements [12–14]. The standard structure of such worm-like robots has three parts: two chambers for anchoring and one drive actuator for locomotion. For example, Verma et al. [13] designed a peristaltic wall-climbing robot with three chambers, two of which are used as radial actuators, while the third is used as a liner actuator. The robot achieves bidirectional motion along pipelines by periodically changing the input air pressure in each chamber. A similar worm-like robot was also proposed by Liu et al. [14], and the vertical load capacity could reach 1 kg. However, such robots only have one degree of freedom (DOF), which cannot adjust the heading directions flexibly and actively. In order to achieve more flexible motion capabilities, such as two-dimensional steering motion, the design of the robot's structure and control system must be more complex. For example, in [14], Liu et al. superimposed multiple different pneumatic modules on the basis of the robot's peristaltic module, which realized the robot's steering in the pipeline. Zhang et al. [15] presented a multi-chamber actuator as a liner actuator that can realize the turning of the robot by inflating different chambers to bend the linear actuator. On the one hand, these complex structures undoubtedly increase the difficulty of manufacturing and controlling the robot. On the other hand, the aforementioned peristaltic robot can only crawl inside the pipeline, and its plane crawling ability has not been verified. Moreover, the diameter of the crawling pipelines is limited by the deformation range of its expansion segments. The expansion can cause the pipeline to be blocked, which dramatically limits the worm robot's application. Although, some researchers have conducted the creeping on different environments, such as pipelines and planes, based on the principle of friction anisotropy. An impressive example is a worm robot proposed by Ge et al. [16] that can realize the forward or backward movement in the pipeline and planes by changing the friction coefficient of the two end segments, but how to realize the steering is still a complex problem.

In view of the above problems, we designed a single-joint centimeter-scale magnetic navigation worm-like robot in this study inspired by the use of geomagnetic field navigation in the migration of migratory birds. As shown in Figure 1a, the migratory birds modify their flight routes according to the change of geomagnetic field during the migration process to ensure the correct migration direction. As shown in Figure 1b, the designed robot is composed of a bellows actuator and a permanent navigation magnet, which provide power and directional traction, respectively. The forward movement can be realized by the principle of friction anisotropy and the force-bending characteristics of the bellows based on the principle of inchworm peristalsis, while the robot's steering and motion path can be guided by adjusting the magnetic force direction.

In general, the main highlights of this work include the following:

(1) Based on the principle of worm motion, a single-joint centimeter-scale magnetic navigation worm-like robot that can crawl on different diameters of pipelines and surfaces with different roughness without being limited by the range of radial segment deformation is designed using the force-bending characteristics of bellows and the principle of friction anisotropy in this study.

(2) Inspired by utilizing the geomagnetic field for migrating birds to identify directions, the robot can be turned and guided by changing the direction of the magnetic field, which simplifies the robot's structural design and reduces the difficulty of manufacturing and control.

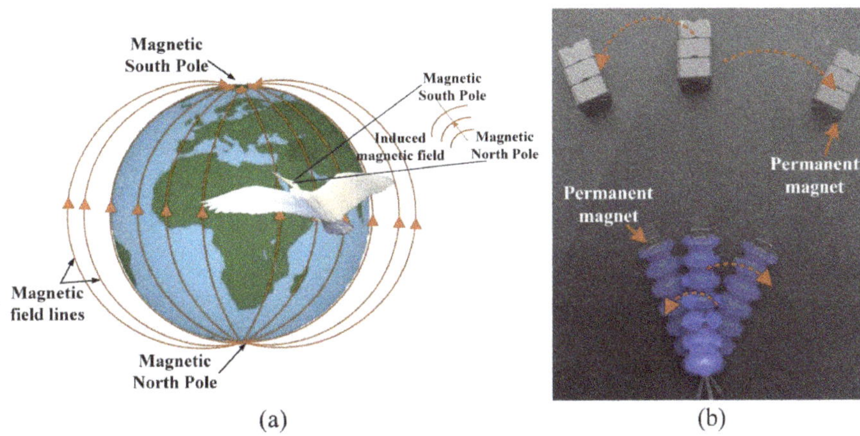

Figure 1. The schematic of the magnetic field navigation (**a**) the schematic diagram of navigation of migratory birds; (**b**) the schematic diagram of navigation of the robot.

The remainder of this work is presented as follows. The structure parameters of the pneumatic actuator and the model of air pressure and elongation are established in Section 2. Section 3 analyzes the motion principle of the robot and establishes the mechanical model. The experiments to verify the bellows characteristics, motion characteristics, and navigation characteristics are in Section 4. Finally, the conclusion and future works are concluded in Section 5.

2. Design and Fabrication

In this section, the structure and fabrication process of the robot are described, and the mathematical model between the air pressure and the elongation of the bellows actuator is established based on the Yeoh constitutive model. Then, the Abaqus is used to analyze the influence of the rectangular chamber number and wall thickness on the performance of bellows with the same conditions, and the actuator structure is optimized by its results. Finally, the robot actuator is manufactured according to the determined structural parameters.

2.1. Design Concept and Structure of the Robot

This work is inspired by migratory birds, which can adjust their flight direction according to the direction of the induced geomagnetic field. As shown in Figure 1b, the robot's structure is mainly composed of a motion module and a navigation module. The motion module is made of flexible silicone rubber that adopts a bellows actuator structure to power it. The navigation module is a rigid permanent magnet pasted on the robot head that guides the robot and furnishes a diffident friction coefficient.

As the unique power unit, the bellows actuator's performance directly determines the crawling efficiency of the robot. As shown in Figure 2, the rectangular bellows structure is adopted in this paper to realize the expansion and contraction movement of the robot by inflating and deflating. The dimensions of the actuator in the initial state are described by the height l, the single chamber height x, the chamber radius r, the thickness t, and the radius R. The dimensions of the actuator in the deformed state are described by the height l_a, the single chamber height x_a, the chamber radius r_a, the thickness t_a, and the radius R_a. The initial height l can be calculated by $l = nx$, where n is the chamber's numbers, the change of height of the actuator between the initial state and deformed state or named the actuator's elongation is termed as Δl, and the actuator's expansion is calculated as $w = 2(R_a - R)$.

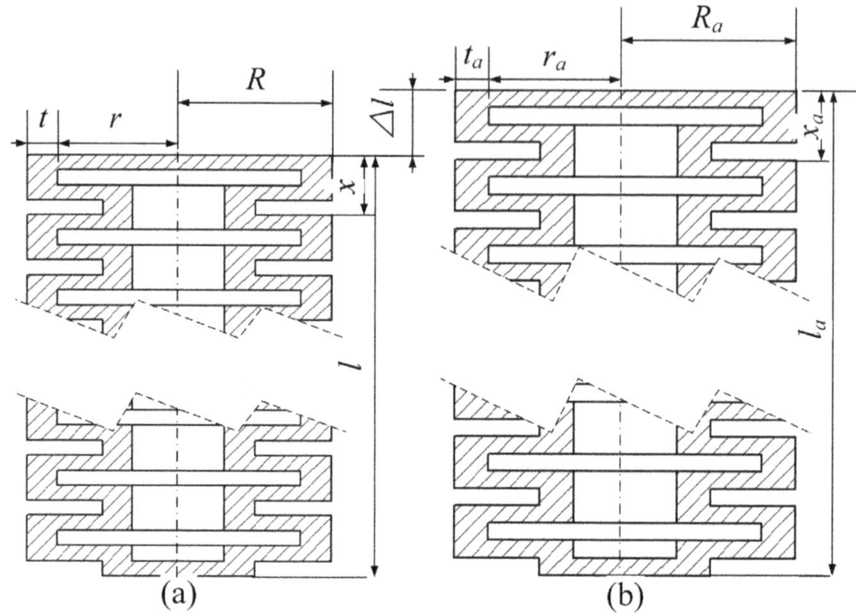

Figure 2. Cross section broken view of the bellows actuator in the initial state and deformed state (**a**) Initial state; (**b**) Deformation state.

Since silicone rubber is a kind of nonlinear material with hyperelasticity, large deformation, and incompressibility, we use the Yeoh model to analyze the relationship between pressure and elongation of the actuator. The Yeoh model strain energy function is expressed as follows:

$$W = W(I_1, I_2, I_3) \tag{1}$$

$$\begin{cases} I_1 = \lambda_1{}^2 + \lambda_2{}^2 + \lambda_3{}^2 \\ I_2 = \lambda_1{}^2\lambda_2{}^2 + \lambda_2{}^2\lambda_3{}^2 + \lambda_1{}^2\lambda_3{}^2 \\ I_3 = \lambda_1{}^2\lambda_2{}^2\lambda_3{}^2 \end{cases} \tag{2}$$

where I_1, I_2, I_3 are strain invariants, and λ_1, λ_2, λ_3 represent the main stretch ratios in three directions and it can be written as:

$$\begin{cases} \lambda_1 = \frac{l_a}{l} = \frac{nx_a}{nx} \\ \lambda_2 = \frac{r_a}{r} = \frac{R_a - t_a}{R - t} \\ \lambda_3 = t_a/t \end{cases} \tag{3}$$

Due to the incompressible nature of the silicone rubber materials, the strain invariant I_3 can be rewritten as:

$$I_3 = \lambda_1{}^2\lambda_2{}^2\lambda_3{}^2 = 1 \tag{4}$$

Assuming that the bellows base does not deform in the width direction, that is $\lambda_3 = t_a/t = 1$. Simultaneous Equations (2) and (3), it can be obtained that:

$$\begin{cases} \lambda_2 = \frac{1}{\lambda_1} \\ I_1 = I_2 = \lambda_1{}^2 + \frac{1}{\lambda_1{}^2} + 1 \end{cases} \tag{5}$$

According to Equations (1) and (5), the third-order Yeoh model can be written as:

$$W = C_1(I_1-3) + C_2(I_2-3)^2 + C_3(I_3-3)^3 = C_1\left(\frac{{l_a}^2}{l^2} + \frac{l^2}{{l_a}^2} - 2\right) + C_2\left(\frac{{l_a}^2}{l^2} + \frac{l^2}{{l_a}^2} - 2\right)^2 + 8C_3 \tag{6}$$

where C_1, C_2, C_3 are the coefficients of the material, which can be identified by least square fitting using the experiment data of uniaxial tension tests as $C_1 = 0.12$ Mpa, $C_2 = 0.023$ Mpa, and $C_3 = 0.00006$ Mpa.

According to the law of conservation of energy, the change of the strain energy is equal to the work done by the pressure [17], namely:

$$PdV_a = V_b dW_a \tag{7}$$

where P represents the pressure, $V_a = \pi(r_a - t_a)^2 l_a$ is the volume of the deformed chamber, W_a is the elastic density strain energy, dV_a and dW_a represent, respectively, the variation of V_a and W_a, and V_b is the volume of silicone after the actuator deformed. It is worth noting that V_b can be regarded as a constant value due to the incompressibility of the silicone material. It should also be noted that both V_a and W_a are unary functions of l_a, and the relationship between pressure P and the elongation l_a can be obtained with the equation above, and it can be written as:

$$P = V_b \frac{dW_a}{dV_a} = f(l_a) \tag{8}$$

The relationship between the bellows actuator elongation and air pressure can also be obtained:

$$\begin{cases} l_a = f^{-1}(p) \\ \Delta l = l_a - l \end{cases} \tag{9}$$

It can be seen from Equation (3) that when the height x of a single rectangular chamber and the radius R of the actuator are constant, the axial elongation ratio of the actuator is related to the number of rectangular chambers n, and the radial expansion ratio is related to the bellows thickness t. According to Equations (5), (6), and (9), it can be inferred that the number of rectangular chambers and the actuator thickness are transformed into a relationship with the bellows elongation. To explore more intuitively the influence of the rectangular chamber numbers and the actuator thickness on the bellows actuator performance and provides a more accurate basis for the design parameters of the bellows actuator. The bellows actuator's initial parameters are defined as $l = 28$ mm, $R = 10$ mm, $x = 3$ mm, and $t = 1$ mm. Then, the number of the rectangular chamber is made to be $n = 4$, $n = 5$, and $n = 7$, respectively, and the air pressure from 0 kpa to 50 kpa was applied to the bellows actuator for simulation calculation. The results are shown in Figure 3a,b.

As shown in Figure 3a,b, with the increase in the number of rectangular chambers, the bellows actuator elongation is positively correlated with the number of the rectangular chamber under the same conditions. Although the expansion increased with the number of rectangular chambers, there is not much difference between the expansion dimensions of the three candidates under the same thickness. Therefore, we selected the number of the rectangular chamber as $n = 7$, and simulation experiments were carried out with thickness $t = 1$ mm, $t = 2$ mm, and $t = 3$ mm, respectively. The results are shown in Figure 3c,d.

It can be seen from Figure 3c,d that the bellows actuator elongation is negatively correlated with the thickness, and its expansion also decreases with the increase of the thickness of the bellows. Nevertheless, it is worth noting that with the increase in thickness, the expansion of the bellows decreases in multiples. The maximum expansion of $t = 2$ mm is only 44.9% of the maximum expansion of $t = 1$ mm, and the maximum expansion of $t = 3$ mm is 41.8% of $t = 2$ mm. However, the numerical difference between $t = 3$ mm and $t = 2$ mm can be negligible compared with $t = 2$ mm and $t = 1$ mm. In order to avoid the influence of excessive expansion on the robot's motion, which will decrease the bending performance, the number of the rectangular chamber and thickness are selected as $n = 7$

and $t = 2$ mm, respectively, on the basis of ensuring elongation. The structural parameters of the bellows actuator are shown in Table 1.

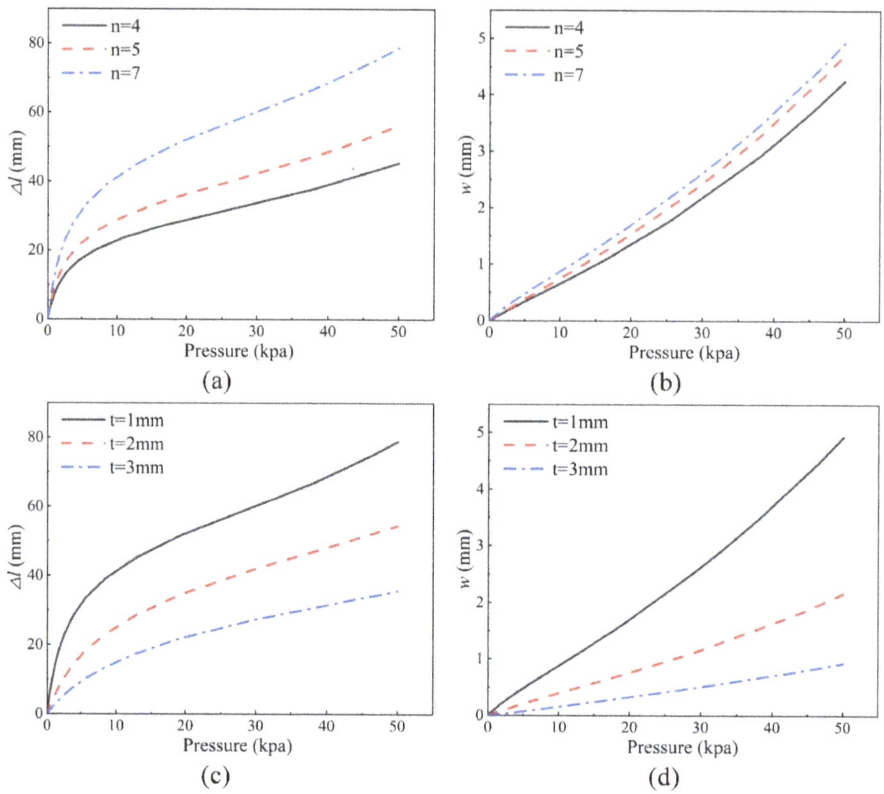

Figure 3. The influence of different structural parameters on bellows deformation. (**a**) The effect of the rectangular chamber's numbers on the elongation; (**b**) The effect of the rectangular chamber's numbers on the expansion; (**c**) The effect of the thickness of the bellows on the elongation; (**d**) the effect of the bellows thickness on the expansion.

Table 1. The bellows actuator's structure paraments.

Paraments	Value
Height (*l*. mm)	28
Radius (*r*. mm)	8
Thickness (*t*. mm)	2
Numbers (*n*)	7

2.2. Fabrication

The main existing manufacturing methods for soft robots are multi-material 3D printing techniques [18], lost wax casting methods [19], and shape deposition manufacturing [20,21]. To further simplify the manufacturing process of the bellows actuator, we adopted the traditional shape deposition method to cast the bellows, and the bodies were connected by gluing. Furthermore, the laser scanner was used to evaluate the error between the actual casting results and the ideal model. The manufacturing process is shown in Figure 4.

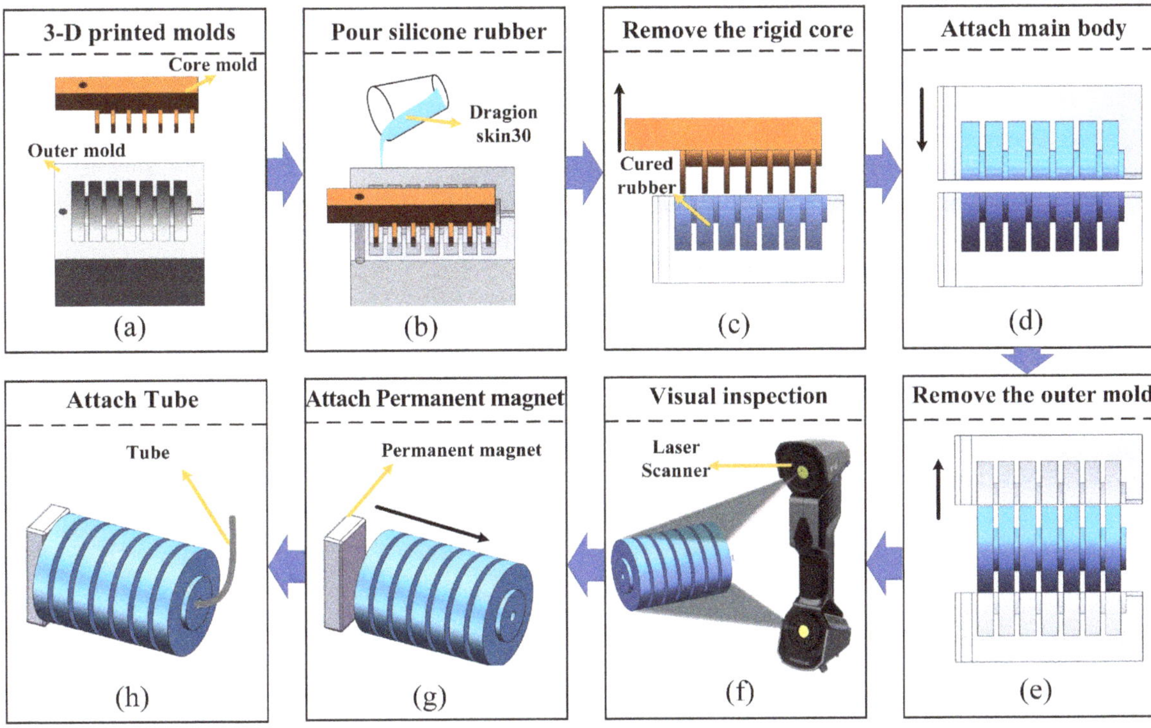

Figure 4. The manufacturing process of the robot (**a**) Prepare molds with 3-D printing; (**b**) Pour silicone rubber pre-elastomer (Dragon-skin30 from Smooth-On, Inc, Macungie, PA, USA); (**c**) Remove the rigid core; (**d**) Gluing the two parts; (**e**) Detach the outer molds; (**f**) Detect the soft body by laser scanner; (**g**) Assemble the soft body and a permanent magnet; (**h**) Insert rubber tube and sealing.

First, we 3-D printed molds with the shape of bellows and fixed the rigid core into the outer mold, and then poured silicone rubber pre-elastomer (Dragon-skin30 from Smooth-On, Inc.) into the mold (see Figure 4a,b). As the silicone rubber is cured, we removed the rigid core and fix the solidified silicone rubber into the outer mold (see Figure 4c). Then, the outer edges of the two parts of the actuator were coated with glue, and the outer mold was aligned and pressed to the fitting of the two parts (see Figure 4d). Once the glue is cured, we detach the outer molds, and then we use the laser scanner to detect the error between the cured rubber body and the ideal mold (see Figure 4e,f). Finally, we assembled the soft body and a permanent magnet (20 mm × 10 mm × 3 mm) with glue, and a rubber tube was inserted into the end of the actuator and sealed (see Figure 4g,h). Thus far, we have completed the fabrication of the bellows actuator and the worm-like robot with a diameter of 20 mm, length of 31 mm, and weight of 12.5 g, which can realize the steering by adjusting the direction of the magnetic field.

3. Force Analysis of the Worm-like Robot

In this section, the motion process of the peristaltic robot is analyzed, and the motion principle of the robot is explained theoretically according to the mechanics theory, as well as the sliding phenomenon during the motion process. Then, the mechanical models of pressure and stretching force, the elongation and stretching force, are established according to the relationship between pressure and elongation obtained in Section 2.

3.1. Motion Process Analysis

The motion process of the peristaltic robot is periodic. The robot's motion process can be divided into two stages: inflation and deflation, and can also be divided into two motion modes: extension and contraction, according to the movement state. The steering and guidance of the robot can be realized by adjusting the magnetic field position. We consider the large deformation and nonlinear characteristics of the robot's motion process and the multi-force-field coupling with the magnetic field force and the ground contact friction. It is essential to accurately describe the mechanical principles of the robot's motion process, establish its mechanical model, and control its motion. Ideally, the bellows actuator is subjected to two forces. One can divide into the trust force F_p and radial force F_e caused by pressure, and the other is the contraction force F_k generated by elastic potential energy. According to the robot's motion state, we analyzed the robot's force state and motion phenomenon during the extension, contraction, and steering processes. The robot's motion state and force direction are shown in Figure 5.

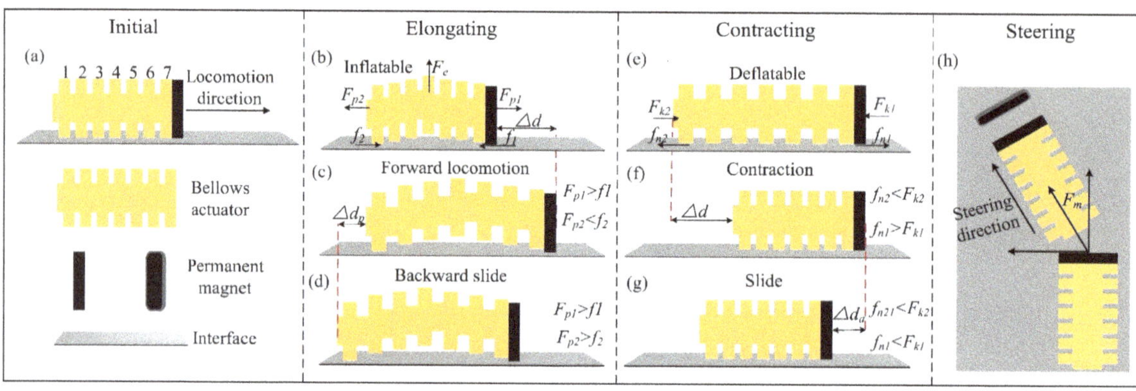

Figure 5. The mechanical theory and motion state of the worm-like robot (**a**) Initial state of the robot; (**b**) Initial inflation with without motion; (**c**) Forward movement; (**d**) Backward extension; (**e**) Deflation; (**f**) Forward contraction; (**g**) Sliding movement; (**h**) Magnetic field steering.

As shown in Figure 5a, the bellows driver is divided into seven sections, and the head is equipped with a permanent magnet. Where F_e represents the radial force, F_{p1} and F_{p2} represent the trust force, respectively, which are equal in magnitude and opposite in direction, that is $F_{p1} = F_{p2} = F_P$. f_1 and f_2 are the friction between the bellows actuator's first section and the motion surfaces, and that between the permanent magnet and the motion surfaces, respectively. F_{k1} and F_{k2} represent the contraction force, respectively, which are equal in magnitude and opposite in direction, that is $F_{k1} = F_{k2}$. f_{n1} and f_{n2} are the total friction force between the seventh section and permanent magnet with the surface, and that between the actuator's first section to the sixth section and the surface, respectively. F_m is a magnetic force, and Δd, Δd_p, and Δd_a are the theoretical forward step, backward elongation, and slip length of the robot, respectively.

During the initial inflation phase, the robot does not move due to $F_{p1} < f_1$ and $F_{p2} < f_2$. While the bellows bend upwards and deform as a result of the radial force, only the bellows first section and the permanent magnet are in contact with the crawling surface (see Figure 5b). Owing to the permanent magnet and the silicone having different friction coefficients with the crawling surface, the maximum static friction between the silicone and the crawling surface is much larger than the maximum static friction between the permanent magnet and the surface, namely $f_1 < f_2$. With the increase in pressure, the thrusts F_{p1} and F_{p2} are gradually increased. When $F_{p1} > f_1$ and $F_{p2} < f_2$, the robot moves forward, its forward extension is recorded as the theoretical step length of the robot movement Δd (see Figure 5c). If we continue to inflate and increase the pressure at this time, when $F_{p1} > f_1$ and $F_{p2} > f_2$, the bellows is stretched in both directions at the same time. The backward elongation is recorded as Δd_p (see Figure 5d), and the robot will be difficult to control at this point. Thus, this situation should be avoided in actual motion.

Nevertheless, when the actuator deflated, with the bellows thrust force and the radial force disappeared, each segment of the bellows came in contact with the crawling surface and contracted with the fourth section of the bellows as the center under the contraction force F_{k1} and F_{k2} (see Figure 5e). Ideally, when $F_{k1} < f_{n1}$ and $F_{k2} > f_{n2}$, the robot contracted along the forward direction, the contraction length is consistent with the elongation length (see Figure 5f). However, in actual motion, slippage often occurred as a result of $F_{k1} > f_{n1}$ and $F_{k2} > f_{n2}$. In this case, both ends of the robot contracted to the center simultaneously, and the shrinkage of the head towards the center is considered as the sliding length of the robot (see Figure 5g). The slippage significantly reduces the crawling efficiency of the robot, which it should avoid as much as possible in the actual movement. The steering movement of the robot mainly relies on the guidance of the magnetic field force. As shown in Figure 5h, a permanent magnetic field is applied near the robot's motion path. Since the robot head contains a permanent magnet, under the magnetic field force F_m, the head magnet generates a torque to drive the robot's motion path changed along with the direction of the magnetic field, which can achieve the navigation purpose proposed in this paper.

3.2. Mechanical Model

From the force analysis of the motion process, it can be seen that the thrust force F_p and contraction forces F_k, and the magnetic field force F_m play a decisive role in the elongation and contraction of the bellows during the robot's motion process. The thrust force can be written as:

$$F_p = PS \tag{10}$$

where P is the pressure, S represents the force-bearing area in the bellows chamber. The contraction force can be calculated by Hooke's law, and it can be written as:

$$F_k = k\Delta l = k(l_a - l) \tag{11}$$

where k represents the spring coefficient.

According to Equations (3), (5), and (9), the relationship between the force-bearing area of the bellows and the pressure can be obtained as:

$$S = \pi {r_a}^2 = \pi\left(\frac{rl}{l_a}\right)^2 = f_s(f^{-1}(p)) \tag{12}$$

Then the mathematical model between the thrust force and the pressure can be written as follows:

$$F_p = Pf_s(f^{-1}(p)) \tag{13}$$

Ignoring the friction force, the relationship between the contraction force and pressure can be obtained by the equilibrium of forces as follows:

$$F_k = F_p = Pf_s(f^{-1}(p)) \tag{14}$$

Assuming the magnets have homogenous magnetization, without eddy current, and the relative permeability is unity. Magnets placed in the magnetic field space will be subjected to magnetic force and magnetic torque [22,23]. The calculation formula of magnetic force and magnetic torque can be expressed as:

$$\vec{F} = (\vec{m} \cdot \nabla)\vec{B} \tag{15}$$

$$\vec{T} = \vec{m} \times \vec{B} \tag{16}$$

where m is the magnetic moment of the magnetic dipole, B represents the magnetic induction intensity generated by the permanent magnet, and ∇ is used to represent the magnetic induction intensity gradient.

By applying an external magnetic field to the worm-like robot, the head magnets and the external magnetic field will attract each other under the action of magnetic torque to drive the whole robot to change its motion direction, which can achieve the purpose of steering and navigation of the robot. According to Equations (15) and (16), the formula for calculating the magnetic torque can be obtained as:

$$F_m = MBV \sin \alpha \tag{17}$$

where M represents the magnetization of the magnetic material, V is the volume of the magnet, and α represents the angle between the direction of magnetization and the direction of the external magnetic field.

4. Results and Discussion

The bellows' performance is tested and compared with the simulation results in this section. Following that, the crawling performance of the robot is tested on different rough surfaces and different environments, and the force status of the robot in the motion state is verified through the experimental results. Finally, we test the navigation function of the robot in the magnetic field environment.

4.1. Characteristics of Bellows

The bellows actuator is the only driving device of the robot, which provides the main output force for the robot's movement, and its performance determines the movement performance of the robot. To perform the elongation, expansion, and thrust force with the bellows actuator, we sample multiple dates by inflating the bellows actuator to air pressure, ranging between 0 and 50 kpa, in intervals of 5 kpa, resulting in 30 pressure combinations. We repeat this procedure three times per type and take the average value as its actual value to compare with the simulation results. The results are shown in Figure 6 and Table 2.

As shown in Figure 6 and Table 2, the trend variation of the actuator's actual elongation with air pressure is almost consistent with the simulation (Figure 6a). In the low-pressure phase, the elastic potential energy overcome by the thrust force is smaller, and the rate of change of the elongation is more significant. However, the elastic potential energy overcome is more extensive as the air pressure increases. Thus, the rate of change of the expansion is reduced compared with the initial stage. Moreover, with the increase of air pressure, the error between the actual elongation and the simulation data gradually decreases, and the error rate at 50 kpa is only 1.5%. However, due to the manufacturing error, as shown in Figure 6d, the cross-sectional error is up to 0.3 mm, equivalent to the increase in its thickness t. Therefore, the actual expansion is smaller than the simulation, and the expansion error of the driver gradually increases at the low-pressure stage. But when the pressure exceeds 30 kpa, the expansion error decreases. This is because its radial force gradually increases with the pressure, but the error rate still reaches 28.7% at 50 kpa. No matter what, while ensuring the actuator elongation, the expansion is reduced to a certain extent, which positively correlates with the robot's motion performance. Assuming that the bellows do not generate radial deformation, that is, S is a constant value, it can be obtained from Equation (10) that the air pressure with a linear change in the thrust force. In this experiment, the bellows are placed in an acrylic pipeline with an internal diameter of 21 mm to limit its radial deformation. The bellows thrust force measured by the force sensor is shown in Figure 6c. With the increase of pressure, the error between the driver thrust force and the theoretical value increases gradually, but the error tends to be stable after 30 kpa and reaches 1.39N at 50 kpa. Besides the theoretical model's error, the radial deformation and the friction inside the pipeline are also responsible for the thrust reduction.

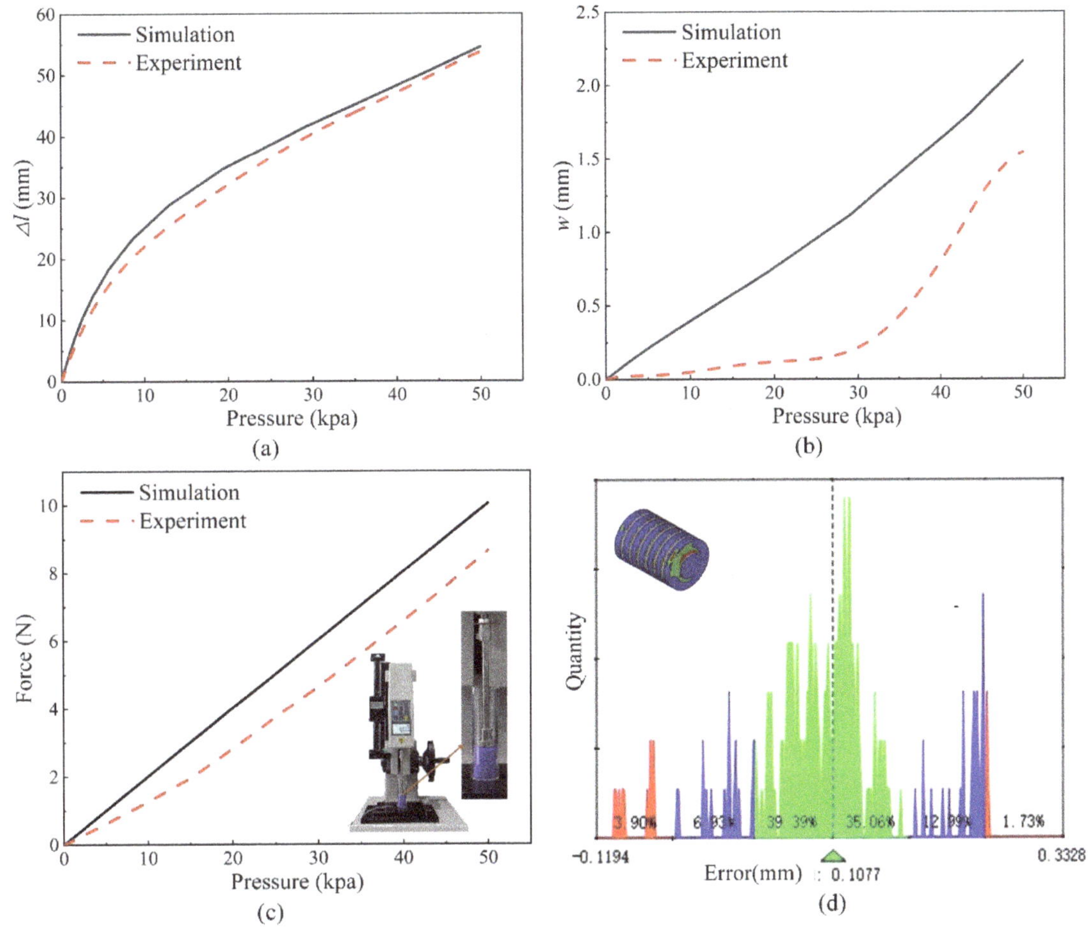

Figure 6. The bellows actuator's actual performances (**a**) The actuator's actual elongation performance; (**b**) The actuator's actual expansion performance; (**c**) The actuator's actual thrust force performance; (**d**) The histogram of the actuator's robot section manufacturing error.

Table 2. Comparison of the bellows experiment and simulation data.

Characteristics / Parameters	Max Elongation (mm)	Max Expansion (mm)	Max Force (N)
Simulation	54.5	2.16	10.05
Experiment	53.68	1.54	8.66

4.2. Characteristics of Movement

In order to demonstrate the different motion states caused by different forces during the motion process in Figure 5, we test the single-cycle motion state of the peristaltic robot on different surfaces, including the smooth cardboard and smooth acrylic pipelines, by inflating the bellows actuator to air pressure, ranging between 0 and 50 kpa, in intervals of 5 kpa, respectively. The motion efficiency under different air pressures is calculated by $\eta = \frac{(\Delta d - \Delta d_a)}{\Delta d}$ to evaluate the robot's crawling performance [24]. The results are shown in Videos S1 and S2, Figure 7, and Table 3.

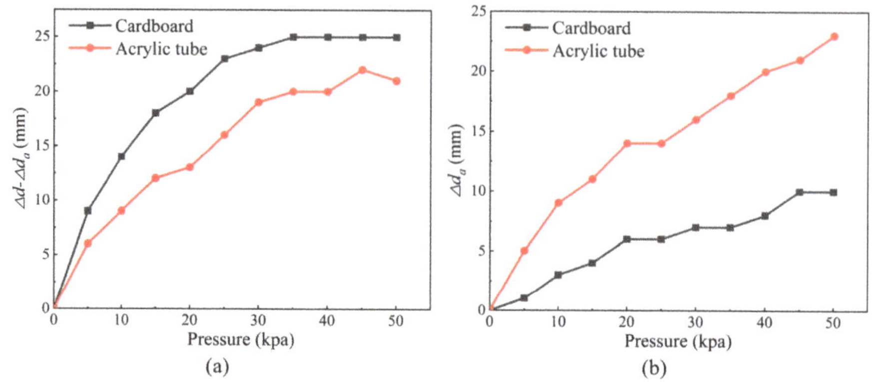

Figure 7. The robot's motion performance with different surfaces (**a**) The real step length with different surfaces; (**b**) The sliding length with different surfaces.

Table 3. The robot's motion efficiency with different surfaces.

Pressure (kpa)	5	10	15	20	25	30	35	40	45	50	Average
Cardboard	0.9	0.83	0.82	0.77	0.79	0.77	0.78	0.76	0.71	0.71	0.78
Acrylic tube	0.55	0.5	0.52	0.48	0.53	0.54	0.53	0.5	0.51	0.48	0.51

It can be seen from Video S1 that when moving on the smooth cardboard, the robot stretches forward without apparent backward sliding phenomenon as the results in $F_{p1} > f_1$ and $F_{p2} < f_2$, which is consistent with the motion state in Figure 5c. As the air pressure increased, the thrust force increased accordingly. When the air pressure exceeds 25 kpa, the robot shows an apparent backward sliding, and the sliding length increases gradually with the increase of pressure, which is consistent with the motion state in Figure 5d. Since the friction coefficient between the silicone rubber and the acrylic plate is greater than that between the silicone rubber and the smooth cardboard, that is $f_{y1} > f_{z1}$ (f_{y1} and f_{z1} are the friction force between the silicone rubber and the acrylic plate, and the friction force between the silicone rubber and the smooth cardboard, respectively), the pressure of the robot to slip backward is larger than that of the cardboard when moving on the smooth acrylic pipelines. As shown in Video S2, the robot is still without an evident backward slip phenomenon on the acrylic pipeline when the air pressure is inflated to 30 kpa, which thoroughly verifies the correctness of our force analysis during the robot movement.

According to Equations (10), (13), and (14), it can be deduced that with the growth of air pressure, the bellows contraction force will gradually increase as the thrust force increases. Thus, when f_{n1} and f_{n2} are constant, the slip length Δd_a will also increase as the air pressure increased, and the motion efficiency gradually decreases. Furthermore, the slip length of the robot in the acrylic pipeline is larger than that on the cardboard because the friction coefficient between the head permanent magnet and cardboard is greater than that between the head permanent magnet and acrylic pipeline, namely $f_{yn1} < f_{zn1}$ (f_{yn1} and f_{zn1} represent the total friction force between the seventh section and permanent magnet with acrylic pipe and smooth cardboard, respectively.). As shown in Figure 7 and Table 3, with the increase of air pressure, the actual step size of the robot in the two environments increases while the slippage gradually increased, and the slip length on the acrylic pipeline is much more considerable than that on the smooth cardboard. Moreover, with the growth of slippage, the motion efficiency of the robot gradually decreases and stabilizes in an interval, which fully proves the correctness of the mechanical model.

In order to prevent the sizeable backward movement of the robot during the movement process, which may cause instability of the movement, we select 25 kpa as the driving pressure of the robot through Figure 7 and Table 2 on the basis of ensuring the movement efficiency and adequate step size. The frequency of the driver inflating and deflating is 2 s per cycle, which was applied to evaluate the movement speed of the robot under the premise of fully considering the length of the tube and the efficiency of inflating and deflating. The results are shown in Figure 8, and the motion process is shown in Video S3. According to the equation $speed = \frac{dis\tan ce}{times}$, it can be calculated that the crawling speed of the robot in the smooth cardboard is 9.6 mm/s, and the speed in the acrylic pipeline is 5.1 mm/s.

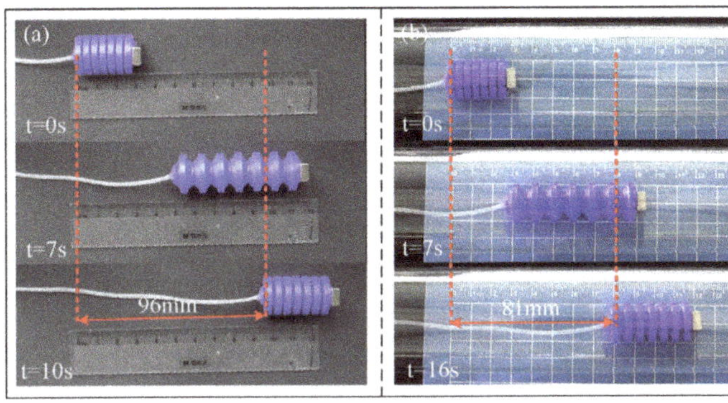

Figure 8. The robot's motion speed with different surfaces (**a**)The robot's motion speed on cardboard; (**b**) The robot's motion speed on cardboard acrylic tube.

To further demonstrate the robot's motion performance, we put the peristaltic robot into acrylic pipelines with inner diameters of 21 mm, 32 mm, 40 mm, and 50 mm for linear motion, and the slope motion was carried out in the pipes with a slope of 6°. The results are shown in Figure 9, and the motion process is shown in Video S4. Compared with the robot relying on expansion and contraction, the robot designed in this study has broader applicability. It can move on a particular slope, showing superior motion performance and environmental adaptability.

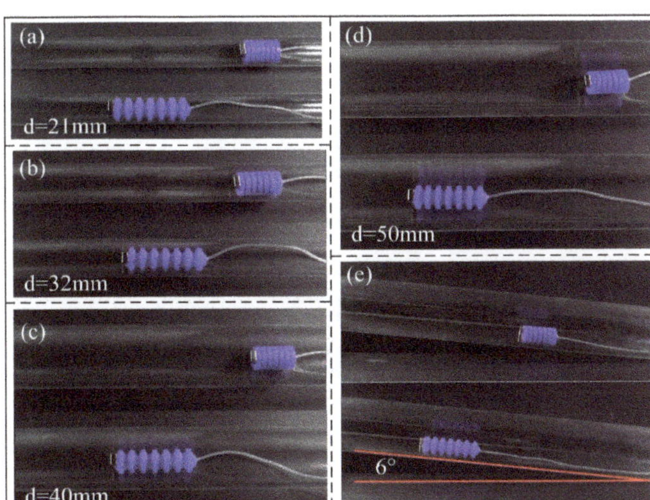

Figure 9. The crawling performance of pipelines with different diameters and slope (**a**) The inner diameter of 21 mm; (**b**) The inner diameter of 32 mm; (**c**) The inner diameter of 40 mm; (**d**) The inner diameter of 50 mm; (**e**) The slope angle with 6°.

4.3. Characteristics of Steering

Compared with conventional peristaltic robots for steering by adopting a complex multi-module and multi-chamber structure design and controller, inspired by the use of geomagnetic field navigation during the migration of migratory birds, this study utilizes an external magnetic field to guide and turn the robot and realize the navigation function. As shown in Figure 10, a permanent magnet (19 mm × 19 mm × 19 mm, a weight of 48 g) was placed on both sides and in front of the robot's motion path, and driving air pressure was applied to test the robot steering and navigation functions. The results are shown in Figure 10a–c and Video S5. When the magnetic field is the same as the initial direction, the robot moves linearly along with the initial motion direction. Then, we adjust the position of the magnetic field. The permanent magnet on the head of the robot receives the force of the magnetic field. It generates a torque along the direction of the magnetic field, which makes the robot rotate in the direction of the magnetic field and eventually along the magnetic field direction, which shows the effectiveness of magnetic field navigation.

To further verify the performance of the magnetic field navigation proposed in this paper, the robot was placed in a Y-shaped pipeline with an angle between the two straight pipelines. Its motion states with a single magnetic field and dual magnetic fields were tested, respectively. As shown in Figure 10d–f and Video S6, when a magnetic field is added to the side of the Y-shaped pipeline, the robot can realize the directional movement in the pipeline along the magnetic field direction under the traction of the magnetic field force. Furthermore, when the different intensities of the magnetic field were applied on both sides at the same distance, the robot could move in the direction of a stronger magnetic field according to the strength of the magnetic field. Compared with using multiple complex structures and controls to realize robots turning, the magnetic field navigation method proposed in this paper is more straightforward in design and stronger in environmental adaptability.

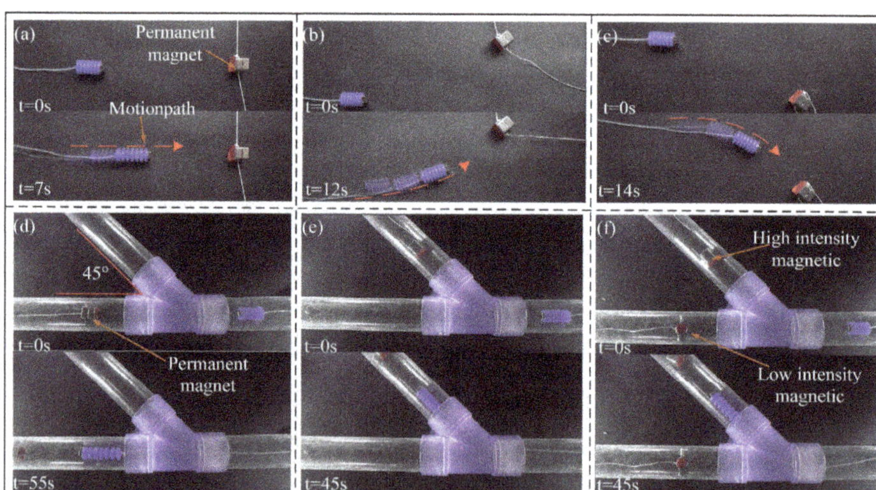

Figure 10. The magnetic navigation performance in different environments (**a**) Crawl in a straight line on the cardboard; (**b**) Turn left on the cardboard; (**c**) Turn right on the cardboard; (**d**) Crawl in a straight line on the pipeline; (**e**) Turn right on the pipeline; (**f**) Steering along the direction of the strong magnetic field under different magnetic field strengths.

5. Conclusions

In this work, inspired by the migration of migratory birds that can use the geomagnetic field for navigating, we presented a single-joint centimeter-scale peristaltic robot that can be navigated by adjusting the position of magnetic fields. The proposed robot was driven by air pressure, and the bellows were selected as the driver structure. Through the Yeoh constitutive model, we established the relationship between the elongation of the bellows and the air pressure, and the finite element software was used to analyze the influence of the chamber's number and thickness on the bellows' performance, which provides a basis for the selection of the bellows structural parameters. Moreover, the force state of the peristaltic robot under the motion states of elongation, contraction, and rotation was analyzed using the mechanics theory, and the models of air pressure and thrust force, air pressure, and contraction forces were established, providing supporting theories for experimental phenomena. Finally, through experiments, the actual bellows performance, robot motion performance, and magnetic field navigation performance have been verified, respectively, and the results are in complete agreement with the theoretical analysis. In addition, the experimental results also show the robot's superior environmental adaptability, which can move in acrylic pipelines with inner diameters of 21 mm, 32 mm, 40 mm, and 50 mm, and can make slope movements in a pipeline with a slope of 6°. As well as moving on surfaces with different roughness, its maximum speed can be 9.6 mm/s. It can also effectively realize the direction selection in planes and complex pipelines according to the magnetic field. These results effectively prove the robot's superiority in the application and structure design scope and show great potential for pipeline detection applications.

Although the robot showed excellent performance, there was still room for improvement. For example, the robot can only move on a two-dimensional plane and cannot crawl vertically because of its single-joint structure and the need to rely on friction to move. The magnetic field of the permanent magnet is difficult to control and realize direction guidance accurately, and the open loop control cannot obtain the actual motion state. In the future, we will further expand the robot's functions and use electromagnetic fields integrated with sensor feedback to achieve precise navigation and control of the robot. Some practical application experiments, such as pipeline detection and ruins search and rescue, will also be realized in future work.

Supplementary Materials: The following supporting information can be downloaded at: https://www.mdpi.com/article/10.3390/machines10111040/s1, Video S1: The single-cycle motion state of the peristaltic robot on cardboard; Video S2: The single-cycle motion state of the peristaltic robot on the acrylic pipeline; Video S3: The robot's motion speed with a different surface; Video S4: The crawling performance of pipelines with different diameters and slope; Video S5: The magnetic navigation performance in cardboard; Video S6: The magnetic navigation performance in the pipeline.

Author Contributions: Conceptualization, D.M.; methodology, D.M.; software, J.W.; validation, D.M. and X.Z.; investigation, G.T., C.Z. and D.M.; data curation, D.M., C.L. and X.Z.; writing—original draft preparation, D.M.; writing—review and editing, D.M. and Y.W.; visualization, X.Z. and G.T.; supervision, Y.W.; project administration, Y.W. All authors have read and agreed to the published version of the manuscript.

Funding: This research was supported by the National Key Research and Development Program of China (2020YFB1312900), the National Natural Science Foundation of China (51975184), the Changzhou Sci & Tech Program (CE20215051), and the Fundamental Research Funds for the Central Universities (B210202124). The authors gratefully acknowledge the support.

Data Availability Statement: Not applicable.

Acknowledgments: The authors are grateful to the editor and reviewers for their constructive comments and suggestions, which have improved this paper.

Conflicts of Interest: The authors declare no conflict of interest.

References

1. Cranford, S. Nature MADE: A Simple Guide to Biological Design Rules. *Matter* **2020**, *2*, 782–785. [CrossRef]
2. Yang, W.H.; Zhang, W.Z. A Worm-Inspired Robot Flexibly Steering on Horizontal and Vertical Surfaces. *Appl. Sci.* **2019**, *9*, 2168. [CrossRef]

3. Hemingway, E.G.; O'Reilly, O.M. Continuous models for peristaltic locomotion with application to worms and soft robots. *Biomech. Model. Mechanobiol.* **2021**, *20*, 5–30. [CrossRef] [PubMed]
4. Liu, R.; Yao, Y.A. A novel serial–parallel hybrid worm-like robot with multi-mode undulatory locomotion. *Mech. Mach. Theory* **2019**, *137*, 404–431. [CrossRef]
5. Kandhari, A.; Wang, Y.F.; Chiel, H.J.; Quinn, R.D.; Daltorio, K.A. An Analysis of Peristaltic Locomotion for Maximizing Velocity or Minimizing Cost of Transport of Earthworm-Like Robots. *Soft Robot.* **2020**, *8*, 485–505. [CrossRef]
6. Du, Z.W.; Fang, H.B.; Xu, J. Snake-worm: A Bi-modal Locomotion Robot. *J. Bionic Eng.* **2022**, *19*, 1272–1287. [CrossRef]
7. Gu, G.Y.; Zhu, J.; Zhu, L.M.; Zhu, X.Y. A survey on dielectric elastomer actuators for soft robots. *Bioinspir. Biomim.* **2017**, *12*, 011003. [CrossRef]
8. Laschi, C.; Mazzolai, B.; Cianchetti, M. Soft robotics: Technologies and systems pushing the boundaries of robot abilities. *Sci. Robot.* **2016**, *1*, eaah3690. [CrossRef]
9. Plaut, R.H. Mathematical model of inchworm locomotion. *Int. J. Non-Linear Mech.* **2015**, *76*, 56–63. [CrossRef]
10. Onal, C.D.; Wood, R.J.; Rus, D. An Origami-Inspired Approach to Worm Robots. *IEEE/ASME Trans. Mechatron.* **2013**, *18*, 430–438. [CrossRef]
11. Pfeil, S.; Henke, M.; Katzer, K.; Zimmermann, M.; Gerlach, G. A Worm-Like Biomimetic Crawling Robot Based on Cylindrical Dielectric Elastomer Actuators. *Front. Robot. AI* **2020**, *7*, 9. [CrossRef] [PubMed]
12. Rafsanjani, A.; Zhang, Y.R.; Liu, B.Y.; Rubinstein, S.M.; Bertoldi, K. Kirigami skins make a simple soft actuator crawl. *Sci. Robot.* **2018**, *3*, eaar7555. [CrossRef] [PubMed]
13. Verma, M.S.; Ainla, A.; Yang, D.; Harburg, D.; Whitesides, G.M. A Soft Tube-Climbing Robot. *Soft Robot.* **2017**, *5*, 133–137. [CrossRef] [PubMed]
14. Liu, X.M.; Song, M.Z.; Fang, Y.H.; Zhao, Y.W.; Cao, C.Y. Worm-Inspired Soft Robots Enable Adap Table Pipeline and Tunnel Inspection. *Adv. Intell. Syst.* **2022**, *4*, 2100128. [CrossRef]
15. Zhang, B.Y.; Fan, Y.W.; Yang, P.H.; Cao, T.L.; Liao, O.G. Worm-Like Soft Robot for Complicated Tubular Environments. *Soft Robot.* **2019**, *6*, 399–413. [CrossRef]
16. Ge, J.Z.; Calderon, A.A.; Chang, L.L.; Perez-Arancibia, N.O. An earthworm-inspired friction-controlled soft robot capable of bidirectional locomotion. *Bioinspir. Biomim.* **2019**, *14*, 036004. [CrossRef]
17. Wang, N.F.; Chen, B.C.; Ge, X.D.; Zhang, X.M.; Chen, W. Design, Kinematics, and Application of Axially and Radially Expandable Modular Soft Pneumatic Actuators. *J. Mech. Robot.* **2021**, *13*, 021019. [CrossRef]
18. Wallin, T.J.; Pikul, J.; Shepherd, R.F. 3D printing of soft robotic systems. *Nat. Rev. Mater.* **2018**, *3*, 84–100. [CrossRef]
19. Zhang, Z.; Yang, T.; Zhang, T.H.; Zhou, F.H.; Cen, N.; Li, T.F.; Xie, G.M. Global vision-based formation control of soft robotic fish swarm. *Soft Robot.* **2021**, *8*, 310–318. [CrossRef]
20. Gafford, J.; Ding, Y.; Harris, A.; McKenna, T.; Polygerinos, P.; Holland, D.; Moser, A.; Walsh, C. Shape deposition manufacturing of a soft, atraumatic, and deployable surgical grasper. *J. Mech. Robot.* **2015**, *7*, 021006. [CrossRef]
21. Cho, K.J.; Koh, J.S.; Kim, S.; Chu, W.S.; Hong, Y.; Ahn, S.H. Review of manufacturing processes for soft biomimetic robots. *Int. J. Precis. Eng. Manuf.* **2009**, *10*, 171–181. [CrossRef]
22. Niu, H.Q.; Feng, R.Y.; Xie, Y.W.; Jiang, B.W.; Sheng, Y.Z.; Yu, Y.; Baoyin, H.X.; Zeng, X.Y. MagWorm: A Biomimetic Magnet Embedded Worm-Like Soft Robot. *Soft Robot.* **2021**, *8*, 507–518. [CrossRef] [PubMed]
23. Tang, C.P.; Huang, H.L.; Li, B. Design and Control of a Magnetic Driven Worm-like Micro-robot. In Proceedings of the IEEE International Conference on Robotics and Biomimetics (IEEE ROBIO), Sanya, China, 27–31 December 2021; pp. 1304–1308.
24. Manwell, T.; Guo, B.J.; Back, J.; Liu, H.B. Bioinspired setae for soft worm robot locomotion. In Proceedings of the 1st IEEE-RAS International Conference on Soft Robotics (RoboSoft), Livorno, Italy, 24–28 April 2018; pp. 54–59.

machines

Article

LQR Trajectory Tracking Control of Unmanned Wheeled Tractor Based on Improved Quantum Genetic Algorithm

Xin Fan [1,2], Junyan Wang [1], Haifeng Wang [2], Lin Yang [2] and Changgao Xia [1,*]

[1] School of Automotive & Traffic Engineering, Jiangsu University, Zhenjiang 212013, China
[2] School of Automotive & Traffic Engineering, Jiangsu University of Technology, Changzhou 213001, China
[*] Correspondence: xiacg@ujs.edu.cn

Abstract: In the process of trajectory tracking using the linear quadratic regulator (LQR) for driverless wheeled tractors, a weighting matrix optimization method based on an improved quantum genetic algorithm (IQGA) is proposed to solve the problem of weight selection. Firstly, the kinematic model of the wheeled tractor is established according to the Ackermann steering model, and the established model is linearized and discretized. Then, the quantum gate rotation angle adaptive strategy is optimized to adjust the rotation angle required for individual evolution to ensure a timely jumping out of the local optimum. Secondly, the populations were perturbed by the chaotic perturbation strategy and Hadamard gate variation according to their dispersion degree in order to increase their diversity and search accuracy, respectively. Thirdly, the state weighting matrix Q and the control weighting matrix R in LQR were optimized using IQGA to obtain control increments for the trajectory tracking control of the driverless wheeled tractor with circular and double-shifted orbits. Finally, the tracking simulation of circular and double-shifted orbits based on the combination of Carsim and Matlab was carried out to compare the performance of LQR optimized by five algorithms, including traditional LQR, genetic algorithm (GA), particle swarm algorithm (PSO), quantum genetic algorithm (QGA), and IQGA. The simulation results show that the proposed IQGA speeds up the algorithm's convergence, increases the population's diversity, improves the global search ability, preserves the excellent information of the population, and has substantial advantages over other algorithms in terms of performance. When the tractor tracked the circular trajectory at 5 m/s, the root mean square error (RMSE) of four parameters, including speed, lateral displacement, longitudinal displacement, and heading angle, was reduced by about 30%, 1%, 55%, and 3%, respectively. When the tractor tracked the double-shifted trajectory at 5 m/s, the RMSE of the four parameters, such as speed, lateral displacement error, longitudinal displacement error, and heading angle, was reduced by about 32%, 25%, 37%, and 1%, respectively.

Keywords: unmanned driving; trajectory tracking; linear quadratic regulator; improved quantum genetic algorithm

Citation: Fan, X.; Wang, J.; Wang, H.; Yang, L.; Xia, C. LQR Trajectory Tracking Control of Unmanned Wheeled Tractor Based on Improved Quantum Genetic Algorithm. *Machines* **2023**, *11*, 62. https://doi.org/10.3390/machines11010062

Academic Editor: Dan Zhang

Received: 14 November 2022
Revised: 21 December 2022
Accepted: 28 December 2022
Published: 4 January 2023

1. Introduction

As a high-order and strongly coupled human–machine system, the trajectory-tracking characteristics of an unmanned wheeled tractor are highly dependent on the design of the control law. At the same time, very complex requirements are placed on the performance of the control system, which results in complicated control design and unsatisfactory control results for unmanned wheeled tractors. As the key to unmanned navigation technology, trajectory tracking has been a hot research topic. With the capability of trajectory tracking, driverless wheeled tractors can be combined with modules such as perception and decision-making to achieve even more powerful functions. In recent years, several researchers have investigated vehicle trajectory tracking control. For example, an expected trajectory is created using points collected by GPS, and the front wheel steering angle of the vehicle is controlled by a dynamics model of the vehicle [1]. The unified control of AGV trajectory tracking and energy optimization is achieved through an energy-optimized trajectory tracking control method [2]. Alternatively, the controller parameters are optimized to optimize the vehicle's path tracking at low and medium speeds while considering the path tracking accuracy and driving stability. Adding soft constraints on the side deflection angle at higher speeds ensures tracking accuracy and driving stability [3]. In addition, it also ensures using a human-like steering control approach that combines trajectory pre-scanning feedforward and state feedback to achieve optimal control of dynamic trajectory tracking and occupant comfort for intelligent vehicles changing [4].

LQR in optimal control theory has been developed over the years to provide effective guidance on the parameter configuration of the control system by solving the state feedback matrix, the closed-loop system, and the adjustment power matrix Q and R, which change the dynamic response of the system and calculate the control parameters, and are important guidelines for engineering. At present, the design of unmanned vehicle autonomous driving control methods based on the optimal control theory at home and abroad is still in the development stage, and some scholars have proposed a guiding parameter adjustment strategy after analyzing the influence of the weighting matrix parameters on the actual response. However, the adjustment method still cannot solve the problem of relying too much on experience [3]. There have also been many results on improved LQR controller design methods based on various optimization algorithms. Still, the objects are simple systems of low order and are decoupled, such as inverted pendulums and control problems with analog circuit switches. These models have in common a smaller number of input vectors and state quantities, less severe coupling problems, and simpler feedback structures. However, for higher-order coupled systems such as unmanned wheeled tractors, the rich input-output relationships result in a state feedback matrix of large dimensionality, leaving this type of design approach unproven, which is the aim of the research work undertaken in this direction. Although more weight optimization algorithms use LQR, such as GA [5–7], PSO [8–10], QGA [11–13], etc., when designing for unmanned vehicle systems, the application of the LQR design control still faces the problem that the value of the weight matrix is too dependent on engineering experience and requires a lot of time for human adjustment. At the same time, the full-state feedback of higher-order systems can significantly increase the computational load; therefore, optimizing the weight matrix parameters of LQR controllers has also been a hot issue in the field of optimization algorithms in recent years.

The contribution of this paper to the above issues is as follows.

To solve the problem of setting the rotation angle of QGA, we adopt the gradient function of the rotation angle to achieve the adaptive update of the rotation angle and solve the optimization problem of LQR weight in unmanned wheeled tractor trajectory tracking controls. At the same time, this method reduces the computing time and improves the convergence speed of the algorithm and the ability to find the optimal solution globally.

For the problem of population diversity demanded by QGA, the algorithm uses the standard deviation coefficient to analyze the population distribution law. It adopts two methods to improve the population diversity, namely chaotic perturbation strategy and Hadamard gate variation, for different population states, aiming to reduce the probability of the algorithm falling into local optimum.

To address the parameter selection problem of the power matrix, we adopt IQGA to optimize the LQR controller, jointly optimize the multiple parameters of the Q and R matrices and validate the optimized control effect using joint Matlab and CarSim simulations. The simulation results show that the optimized LQR controller has a good tracking effect and improved control accuracy compared with the four algorithms, including the traditional empirical LQR, GA, PSO, and standard QGA.

The paper is organized as follows.

Section 1 introduces the application of LQR in a variety of control systems and its advantages compared with other control algorithms. Section 2 introduces the single-track kinematic model of a wheel tractor based on the Ackermann steering model and its characteristics. Section 3 first introduces the linear quadratic (LQR) optimal control principle and the standard QGA, giving their advantages and disadvantages through theoretical and experimental analysis. Secondly, it proposes two QGA improvement strategies for their advantages and disadvantages: the quantum gate rotation angle dynamic adjustment strategy and the population diversification strategy. Finally, it constructs the fitness function and designs IQGA. Section 4 describes the experimental environment and parameter selection, gives simulation results based on a joint Matlab and CarSim simulation that compares four algorithmic controllers with the IQGA control, performs the analysis, and draws conclusions. Section 5 concludes the paper and points out the directions in which further research can be carried out. Firstly, the main data characteristics of the experimental application of IQGA are given, as well as the conclusions obtained in terms of experimental data when comparing the other control methods mentioned in this paper with IQGA. Then, the application scenarios of the tracking method are designed based on kinematics, and some problems in the process of building a mathematical model of the kinematics of the wheeled tractor are pointed out. Finally, it is pointed out that the reference trajectory also has a strong influence on the tracking effect of the wheel tractor.

2. Related Work

Recently, a GA-based LQR (GA-LQR) controller was proposed in [14] to improve the path-tracking performance of an articulated vehicle. PID was used for speed control and achieved better control accuracy and path-tracking performance by controlling the articulation angle and speed. However, only the control results are analyzed without comparing them with other good control algorithms to derive the superiority of the proposed algorithm. A steering torque control strategy is proposed in [15], which is compared with LQR and MPC for experimental scenarios with a lane change environment and two typical parking lots. It is concluded that LQR is able to stabilize the vehicle under large speed variations compared to other algorithms. Still, it is less accurate than the controller proposed in the paper. YUAN [16] et al. investigated the vehicle trajectory tracking problem based on the vehicle dynamics model and model predictive control (MPC) algorithm, conducted simulation experiments on Carsim/Simulink, and the experimental results showed that MPC had a better tracking performance under different speeds and road adhesion conditions.

In [17], LQR was applied to control the uninterruptible power supply. The weighting matrices Q and R of LQR were obtained by establishing the project characteristics, i.e., the traditional empirical method. In [18], the dynamics of the closed-loop system were shaped by penalty coefficients, and the weighting matrices Q and R of LQR were optimized by PSO. In [19], the digital LQR was applied to a voltage-source converter with an LC output filter. The weighting matrices Q and R of the LQR were then obtained by minimizing the infinite parity of the selected transfer function. It is experimentally demonstrated that the method achieves good control for different load conditions.

Liu Songyuan [20] et al. proposed an improved LQR method based on the PSO algorithm (PSO-LQR). The technique designs the index function of the PSO algorithm by constraining the system stability and obtaining the method of maximizing the influence factor of the central state feedback coefficient.

In [21], GA-LQR was used for the active vibration control of a piezoelectric beam element, and the weighting matrices Q and R of LQR were obtained by GA optimization. The effectiveness of the GA-LQR controller was verified by numerical simulations. The simulation results show that the piezoelectric beam element is very accurate in the dynamic analysis after using the GA-optimized weighting matrices Q and R when applied to the control system.

In [22], an adaptive PSO (APSO) was proposed to obtain the weighting matrices Q and R. To improve the convergence speed and accuracy of conventional PSO, an adaptive inertia weighting factor (AIWF) was introduced into the PSO speed update equation. The proposed APSO-based LQR control strategy was applied to control the pitch and yaw axis of the managed object. Experimental results show that the controller optimized with APSO reduces the tracking error and improves the tracking response, and reduces oscillations compared with PSO.

In [23], the PSO-LQR control strategy was applied to an adaptive air suspension system. The PID controller was compared with the proposed PSO-LQR controller in a simulation analysis. The simulation results show that the PSO-LQR control strategy improves the vehicle's ride comfort. The control performance of the proposed PSO-LQR control strategy is superior under the experimental conditions of random vibration.

In [24], the GA-LQR control strategy was applied to an actively supported weight compensation system. The dynamics of the system were modeled and linearized. The transient process diagrams for different initial conditions are given in the paper. Experimental simulations were performed, and the experimental results show that the proposed GA-LQR control strategy can improve mass compensation accurately and with high quality.

3. Kinematic Model for Wheeled Tractors

As shown in Figure 1, the model used in this paper is a single-track kinematic model of a wheel tractor based on Ackermann steering. The differential equations of motion for the wheel tractor are given in Equation (1).

$$\begin{bmatrix} \dot{x}_{rear} \\ \dot{y}_{rear} \\ \dot{\varphi} \end{bmatrix} = \begin{bmatrix} \cos\varphi \\ \sin\varphi \\ \tan\delta_f / l \end{bmatrix} v_r \tag{1}$$

where (x, y) are the coordinates of the rear-axle-center of the wheeled tractor, φ is the heading angle of the wheeled tractor, δ_f is the front wheel deflection angle, v_r is the rear-axle-center speed of the wheeled tractor, v_f is the front-axle-center speed of the wheeled tractor, l is the wheelbase, R is the rear wheel steering radius.

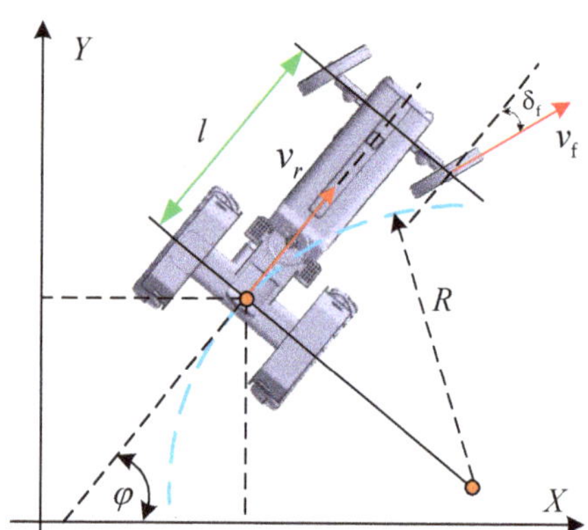

Figure 1. Single-track kinematic model of a wheeled tractor.

The system is a non-linear continuous system. The linearization of the differential Equation (1) for the system gives the Equation (2) of state:

$$\dot{\tilde{X}} = A\tilde{X} + B\tilde{u} \tag{2}$$

where, $A = \begin{bmatrix} 0 & 0 & -v_r \sin\varphi_r \\ 0 & 0 & v_r \cos\varphi_r \\ 0 & 0 & 0 \end{bmatrix}$, $B = \begin{bmatrix} \cos\varphi_r & 0 \\ \sin\varphi_r & 0 \\ \tan\delta_f / l & v_r / l\cos^2\delta_r \end{bmatrix}$, $\tilde{X} = X - X_r$, $\tilde{u} = u - u_r$, and the subscript r indicates the reference value.

The discretization of Equation (2) yields Equation (3) for the system after linearization and discretization.

$$\tilde{X}(k+1) = A_k\tilde{X}(k) + B_k\tilde{u}(k) \tag{3}$$

In the above equation, $A_k = \begin{bmatrix} 1 & 0 & -v_r \mathrm{T}\sin\varphi_r \\ 0 & 1 & v_r \mathrm{T}\cos\varphi_r \\ 0 & 0 & 1 \end{bmatrix}$, $B_k = \begin{bmatrix} \mathrm{T}\cos\varphi_r & 0 \\ \mathrm{T}\sin\varphi_r & 0 \\ \frac{T\tan\delta_f}{l} & \frac{v_r T}{l\cos^2\delta_r} \end{bmatrix}$ the wheeled tractor position and heading angle of the kinematic model in Equation (3) match very well with the CarSim output for the same speed and front wheel deflection input [25], i.e., this model is a good reflection of the kinematic characteristics of the wheeled tractor while driving.

4. LQR-Based Trajectory Tracking Control Algorithm

4.1. Linear Quadratic Optimal Control Principle

The task of the LQR in trajectory tracking is to make the actual state of the system $[x, y]$ closely follow the system's reference state under the action of the control quantity u^* and to minimize the evaluation function J of the system. This allows the unmanned wheel tractor to maintain the system state components close to the equilibrium without consuming too much energy if the system state deviates from the equilibrium state for any reason.

The quadratic evaluation function is shown in Equation (4).

$$J = \frac{1}{2}\sum_{k=0}^{N-1}\left[\tilde{X}^T(k)Q\tilde{X}(k) + u^T(k)Ru(k)\right] + \frac{1}{2}(x_N - r_N)^T Q_0 (x_N - r_N) \tag{4}$$

where Q, R, and Q_0 are the weight matrices, $Q = \begin{bmatrix} q_1 & 0 & 0 \\ 0 & q_2 & 0 \\ 0 & 0 & q_3 \end{bmatrix}$, $R = \begin{bmatrix} r_1 & 0 \\ 0 & r_2 \end{bmatrix}$, $Q_0 = \begin{bmatrix} q_4 & 0 & 0 \\ 0 & q_5 & 0 \\ 0 & 0 & q_6 \end{bmatrix}$, $q_1, q_2, q_3, q_4, q_5, q_6, r_1$ and r_2 are the coefficients to be optimized, and x_n and r_n are the terminal states and reference terminal states, respectively.

The control volume u^* at moment k of the driverless car is $u^* = u_k + \Delta u_k$, $\Delta u_k = -K\tilde{X}$, where u_k is the reference output at moment k, Δu_k is the feedback control quantity, and K is the feedback coefficient. When using Matlab for simulation, K can be obtained from Equation (5).

$$[K, S, E] = dlqr(A_k, B_k, Q, R) \tag{5}$$

The problems faced by the LQR design control law are, on the one hand, that the values of the LQR parameters require the extensive development experience of engineers and technicians, and a lot of time is needed for the human adjustment of the parameters for multiple degrees of freedom; on the other hand, because the LQR cannot directly deal with the constraint problems in the multivariate control process and when calculating the optimum for a fixed time in the future, it only calculates once and executes all the calculated control sequences; the errors generated during execution and the impact of disturbances on the system are not considered. The optimization of the power matrix parameters of the LQR controller of an unmanned wheel tractor using intelligent algorithms has therefore become a topical concern in the field of optimization algorithms in recent years, and the robustness of wheel tractor trajectory tracking systems can be improved by adding constraints to the optimization search process of the Q and R parameters.

4.2. Standard QGA

QGA is an evolutionary algorithm that combines quantum computing with GA. QGA uses state vectors to encode chromosomes and quantum logic gates to evolve and update the chromosomes, achieving better results than traditional GA.

The quantum revolving gate used in QGA is defined in Equation (6).

$$U(\theta_i) = \begin{bmatrix} \cos(\theta_i) & -\sin(\theta_i) \\ \sin(\theta_i) & \cos(\theta_i) \end{bmatrix} \tag{6}$$

The update process is as $\begin{bmatrix} \alpha'_i \\ \beta'_i \end{bmatrix} = U(\theta_i) \begin{bmatrix} \alpha_i \\ \beta_i \end{bmatrix} = \begin{bmatrix} \cos(\theta_i) & -\sin(\theta_i) \\ \sin(\theta_i) & \cos(\theta_i) \end{bmatrix} \begin{bmatrix} \alpha_i \\ \beta_i \end{bmatrix}$.
α and β are two amplitude constants that conform to equation $|\alpha|^2 + |\beta|^2 = 1$.

Quantum states are encoded using a binary, and two quantum states are encoded using one quantum bit. The chromosomes encoded using quantum bits are as follows.

$$q_i^n = \left(\begin{array}{c|c|c|c|c|c|c|c|c|c|c|c} \alpha_{11}^n & \alpha_{12}^n & \cdots & \alpha_{1k}^n & \alpha_{21}^n & \alpha_{22}^n & \cdots & \alpha_{2k}^n & \alpha_{m1}^n & \alpha_{m2}^n & \cdots & \alpha_{mk}^n \\ \beta_{11}^n & \beta_{12}^n & \cdots & \beta_{1k}^n & \beta_{21}^n & \beta_{22}^n & \cdots & \beta_{2k}^n & \beta_{m1}^n & \beta_{m2}^n & \cdots & \beta_{mk}^n \end{array} \right)$$

A quantum bit chromosome can represent multiple states simultaneously, allowing the algorithm to have better population diversity and higher computational parallelism than GA. The algorithm uses a quantum revolving gate operation for individual updates, which effectively increases the convergence speed of the algorithm. However, in the standard QGA, the rotation angle obtained from the table look-up is constant, which is not conducive to evolution in the direction favorable to the optimal determination of the solution, resulting in slow convergence and a long computation time; in addition, in the standard QGA, there is no quantum crossover, mutation, and catastrophe, the chromosomes in the population are all independent of each other, the structural information among individuals cannot be fully utilized, and the algorithm is prone to fall into local optimal solutions. Therefore, the standard QGA needs to be improved to increase the diversity of the population; as pointed out in the literature [26,27], QGA is suitable for solving combinatorial optimization problems, even only for solving backpack problems, but not for solving the optimization problems of continuous functions, especially multi-peaked functions. The application of QGA to the optimization of the weight parameters of the LQR control is a multi-peaked function optimization problem.

4.3. *Improvement Strategies*

4.3.1. Quantum Gate Rotation Angle Dynamic Adjustment Strategy

The traditional quantum revolving gate adjustment is set by table look-up, and the values of the rotation angles are fixed, lacking theoretical guidance and with obvious limitations; if the magnitude is too small, it affects the convergence speed, while too large leads to prematureness. The literature [28] proposes the random dynamic generation of rotation angles within a range. It experimentally demonstrates that this strategy is superior to fixed rotation angles but is highly random. In this paper, an adaptive dynamic rotation angle step adjustment mechanism is introduced, and the algorithm is optimally designed to adjust the rotation angle required for individual evolution promptly in the face of diverse fitness values, ensuring a timely jumping out of the local optimum and good performance in the global stage of finding the optimum.

QGA mainly occurs through the adjustment of the quantum gate to find the optimal solution. The basic idea is as follows: the contemporary optimal individual's fitness value f_{max} and the current individual i's fitness value f_i, are compared, if $f_i > f_{max}$, then the corresponding rotation angle is adjusted in favor of the i's emergence of the direction of evolution or vice versa; then, the evolution in favor of the emergence of the maximum direction is adjusted. Considering the gradient of the objective function at the search point, to make the gradient negatively correlated with the rotation angle step, the new fitness function proposed in the literature [29] considers the rate of change in the objective function. By contrast, the size of the quantum gate rotation angle step determines whether the algorithm can find the optimal solution quickly and accurately. Combining the interference and entanglement of quantum states, we used the following corner step function to realize the dynamic adjustment of the rotation angle:

$$\Delta\theta_x = sgn(M) \times \left[(\theta_{max} - \theta_{min}) \cdot exp\left(\frac{\nabla f_{max} - \nabla f_x}{\nabla f_{max} - \nabla f_{min}}\right) + \Delta\theta_0\right] \tag{7}$$

where, $M = \begin{bmatrix} \alpha_b & \alpha_x \\ \beta_b & \beta_x \end{bmatrix}$, $(\alpha_b, \beta_b)^T$ is the optimal probability magnitude within the current population and $(\alpha_x, \beta_x)^T$ is the probability magnitude of the current solution. $sgn|M|$ takes ± 1 when $|M| = 0$, i.e., $sgn|M|$ determines the direction of the quantum rotation gate. θ_{min} is the lower limit of the quantum rotation angle; θ_{max} is the upper limit of the quantum rotation angle; ∇f_x is the first order gradient of the current individual fitness value; ∇f_{max} and ∇f_{min} are the maximum and minimum values of the first order gradient of the fitness value of the current population, respectively. $\Delta\theta_0$ is the initial value of the rotation angle. By introducing the factor of gradient change in the exponential function and the fitness value, the rate of change for the objective function can be made to have an opposite trend to the change in the rotation angle step, creating the conditions for the subsequent improvement of QGA to jump out of the local optimal solution.

4.3.2. Diversification Strategies for Populations

Population diversity is critical in all evolutionary algorithms to avoid falling into local optima. In QGA, guided by the evolutionary goal, information is exchanged between individuals to improve the diversity of the next generation of populations. However, when the population size is large, the local optima increases greatly. In addition, it makes local optimum solutions the norm in the face of a multi-peaked optimization problem such as tractor LQR control. Even using single-point mutations, multi-point mutations and quantum catastrophes to regenerate individuals for the next generation of populations is not effective in reducing the impact of local optima on the global optimality-finding ability of evolutionary algorithms. Therefore, there is a need to analyze the population size and distribution trends of individual populations to use different strategies for information interactions to maximize the diversity of the population. As shown in Equation (8), the upper bound of the discrete coefficient C_D is C_{Dmax}, and the lower bound is 0, i.e., $C_D \in (0, C_{Dmax})$.

$$C_D = \frac{\sqrt{\frac{\sum_{i=1}^{n}\left(F_{ij} - \left|\frac{\sum_{i=1}^{n} F_{ij}}{n}\right|\right)}{n-1}}}{\left|\frac{\sum_{i=1}^{n} F_{ij}}{n}\right|} \times \frac{\frac{\sum_{i=1}^{n}(F_{ij} - F_\Delta)^3}{n}}{\left(\frac{\sum_{i=1}^{n}(F_{ij} - F_\Delta)^2}{n}\right)^{1.5}} \tag{8}$$

The above equation fully accounts for population dispersion and skewed distribution. C_D is the dispersion coefficient of the population fitness value; n is the number of individuals in the population; F_{ij} is the fitness value of the ith individual in the population in the jth generation and F_Δ is the mean of the fitness function. As C_{Dmax} is dynamically updated, the more it deviates towards 0, the more concentrated and closer to the normal distribution of the population fitness. Otherwise, the population is more discrete. C_{Dmax} considers the degree of dispersion and the deviation characteristics between the whole population and the normal distribution. If there are too many deviations, the population is prone to the trap of local optimality for the algorithm. In the LQR-controlled optimization problem of this paper, particular attention needs to be paid to the evolutionary process for both types of populations. Assuming that the threshold μ generally takes values in the range (0.01~0.3), during each iteration, if the value is between $(0, 0.15C_{Dmax}]$, the population falls into a local optimum and must be perturbed for this generation. If the value of C_D is between $(0.15C_{Dmax}, 0.3C_{Dmax}]$, then the population is in a discrete critical state and needs to be given a small perturbation. If the value of C_D is between $(0.3C_{Dmax}, C_{Dmax}]$, then the population is in a normal iteration.

For a critical state population, if the angle of rotation of the quantum non-gate variation is too large, it is easy to make the population, which is already close to the optimal value, further away from the optimal value, resulting in the loss of good populations. Therefore, this paper uses the Hadamard gate to perform the mutation operation. The Hadamard gate is used to apply a slight rotation to the chromosomes to prevent the generation of the local optimal solutions and increase population diversity.

If a quantum bit is denoted as $\begin{bmatrix} \cos\theta_{ij} \\ \sin\theta_{ij} \end{bmatrix}$, the variation can be carried out according to Equation (9) [30].

$$\begin{bmatrix} \frac{1}{\sqrt{2}} & \frac{1}{\sqrt{2}} \\ \frac{1}{\sqrt{2}} & -\frac{1}{\sqrt{2}} \end{bmatrix} \begin{bmatrix} \cos\theta_{ij} \\ \sin\theta_{ij} \end{bmatrix} = \begin{bmatrix} \cos\left(\theta_{ij} + \left(\frac{\pi}{4} - 2\theta_{ij}\right)\right) \\ \sin\left(\theta_{ij} + \left(\frac{\pi}{4} - 2\theta_{ij}\right)\right) \end{bmatrix} \tag{9}$$

The angle of variation for the Hadamard gate is $\left(\frac{\pi}{4} - 2\theta_{ij}\right)$, and the variables i and j are the jth position of the ith chromosome. In the case of the LQR control, the critical-state population requires particular attention because the tractor is subjected to many operating conditions. If the angle of variation is too large, it can easily lead to population oscillations and cause the optimal individuals to evolve in the opposite direction. The critical state populations, therefore, need to be stable.

For the populations that must be perturbed, the idea of adding chaotic perturbation optimization, referring to the literature [31], increases the algorithm population's diversity, thus jumping out of the local optimum. In this paper, the perturbation function is implemented using Tent mapping. The Tent expression is shown in Equation (10).

$$y_{k+1} = \begin{cases} (2y_k)\text{mod}1 & 0 \le y_k \le 0.5 \\ (2 - 2y_k)\text{mod}1 & 0.5 < y_k \le 1 \end{cases} \tag{10}$$

However, the Tent mapping iterative sequence is short and unstable, and a random perturbation needs to be applied to make the sequence x_k jump out of the minimum period to complete the chaotic state again. The conventional Tent mapping appears as a minimum period point, mainly in the case of $x_k = x_{k-a}$, $a = [1, 2, 3, 4, 5]$. In this paper, a random perturbation is performed by Equation (11).

$$y_{k+1} = \begin{cases} [2y_k + 0.2rand(0,1)]\text{mod}1 & 0 \le y_k \le 0.5 \\ [2 - 2y_k + 0.2rand(0,1)]\text{mod}1 & 0.5 < y_k \le 1 \end{cases} \tag{11}$$

The above equation allows a new $n \times m$ dimensional probability magnitude perturbation matrix to be formed, and a new population is generated by superimposing each column vector with the probability magnitude matrix of the population. Finally, the chaotic perturbations are traversed to improve the diversity of the new population and the accuracy of the subsequent search.

4.4. Construction of the Fitness Function

In GA, a chromosome corresponds to a set of coefficients in the weights Q, R, and Q_0. In order to make the tracking process smoother, a speed constraint and a front wheel deflection constraint are required, as in Equation (12).

$$\begin{cases} v_{rear,\min} \le v_{\text{rear}} \le v_{rear,\max} \\ \delta_{f,\min} \le \delta_f \le \delta_{f,\max} \\ \Delta v_{\min} \le \Delta v_{rear} \le \Delta v_{\max} \\ \Delta\delta_{f,\min} \le \Delta\delta_f \le \Delta\delta_{f,\max} \end{cases} \tag{12}$$

To simplify the form, let $U = [u^*, \Delta u_k]^T$, then Equation (12) can be expressed as $U_{\min} \leq U \leq U_{\max}$. Since LQR cannot deal with the constraint problem directly, the constraint needs to be put into the fitness function of QGA.

From this, the fitness function of IQGA can be constructed as Equation (13).

$$F_{min} = \begin{cases} \frac{1}{2}\sum_{k=0}^{N-1}\left[\tilde{X}^T(k)Q\tilde{X}(k) + \tilde{u}^T(k)R\tilde{u}(k)\right] + \frac{1}{2}(x_N - r_N)^T Q_0 (x_N - r_N), U \subseteq [U_{min}, U_{max}] \\ \frac{1}{2}\sum_{k=0}^{N-1}\left[\tilde{X}^T(k)Q\tilde{X}(k) + \tilde{u}^T(k)R\tilde{u}(k)\right] + \frac{1}{2}(x_N - r_N)^T Q_0 (x_N - r_N) + \varepsilon^2, U \subseteq (-\infty, U_{min}) \cup (U_{max}, +\infty) \end{cases} \quad (13)$$

where ε^2 is a penalty factor of a larger value. When the calculated output does not satisfy the constraint, the population is guided to evolve towards satisfying the constraint by adding a penalty factor ε^2 to the fitness function to penalize the chromosome and increase its probability of being eliminated.

4.5. Algorithms in This Paper

The principle of the control system parameter optimization based on IQGA is shown in Figure 2. According to decentralized coordinated control theory and classical inner and outer loop control system theory, the basis of unmanned wheeled tractor autopilot control is the wheeled tractor speed and heading angle control. A decoupled decentralized controller design is carried out for the wheeled tractor speed and heading angle channels, and then the control quantities are input to CarSim through the role of synergy to form a decentralized coordinated control loop to achieve control; the output signal, error signal, and control quantity signal are input to the adaptation function The output signals, error signals, and control signals are input to the fitness function. The control parameters are optimized through IQGA to form an optimization loop, which forms the optimized control system.

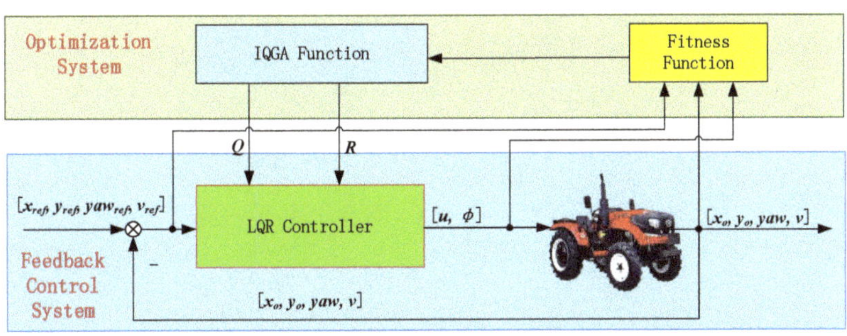

Figure 2. Schematic diagram of control parameter optimization based on IQGA.

4.6. Optimization Algorithm Process

As shown in Figure 3, the IQGA process is mainly divided into the following steps:

1. Population initialization

First, set the population size *Sizepop* and the length of the quantum chromosome *Lenchrom*, and then divide the population space into several segments of the quantum chromosome with the same probability to reduce the number of iterations for population evolution. Set the number of evolutionary generations $g = 0$ to generate a new initial population P_0.

2. Fitness measurement

According to the lateral displacement, longitudinal displacement, heading angle, speed, and other parameters uploaded by the tractor combined with the fitness function in Equation (13), each chromosomal individual in the population is measured once for its fitness. The optimal fitness and its corresponding chromosomal individual are recorded and used as the basis for the evolution of the next generation.

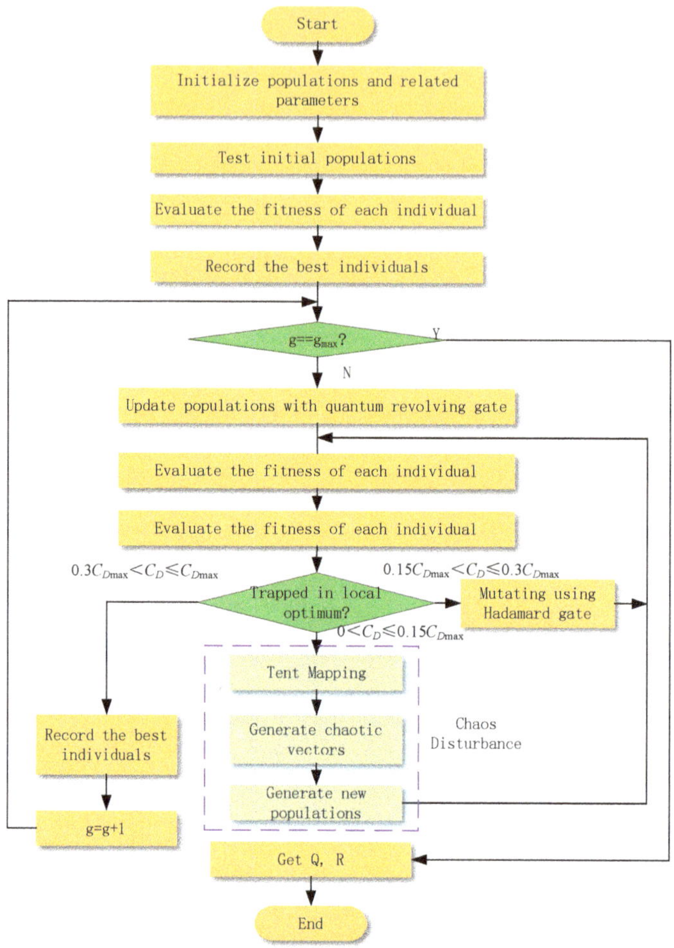

Figure 3. Flow chart of IQGA.

3. Judgment of the number of iterations

When judging whether g has reached the predetermined number of iterations, if yes, execute step 7; if not, execute step 4.

4. Dynamic updates using the Quantum Revolving Gate

The dynamic update of the quantum revolving gate for the population is according to Equation (7).

5. Fitness measurement

The fitness of each chromosome individual in the population is measured once, and the optimal fitness and its corresponding chromosomal individual are recorded and used as a basis for the evolution of the next generation.

6. Three different operations are performed according to the standard deviation coefficient

Calculate the standard deviation coefficient C_D. If $0.3C_{Dmax} < C_D \leq C_{Dmax}$, record the optimal fitness and corresponding chromosome individuals and serve as the basis for next-generation evolution. $g = g + 1$. Execute step 3. If $0.15C_{Dmax} < C_D \leq 0.3C_{Dmax}$, the Hadamard gate mutation is performed according to Formula (9). Execute step 5. If $0 < C_D \leq 0.15C_{Dmax}$, according to Formulas (10) and (11), the chaotic vector is generated and the probability amplitude perturbation matrix is formed, and then the probability amplitude perturbation matrix and the probability amplitude matrix of the population are superposed to generate a new population. Execute step 5.

7. End the program

   Obtain Q, R, and end the optimization process.

5. Experimental Results and Analysis

5.1. Experimental Environment

Intel (R) Core (TM) i7-8700K CPU @ 3.70 GHz, Memory: 16.0 GB, Windows 11 Operating System. The algorithms were run in a MATLAB 2020a, CarSim 2020.0 development environment for joint simulation.

5.2. Experimental Data and Parameter Selection

In this paper, the traditional LQR (Trad_LQR), GA, PSO, QGA, and other algorithms and IQGA are simulated and compared to test and verify the effectiveness and feasibility of the proposed IQGA for the optimization of the wheel tractor trajectory tracking control system. The relevant parameters of each algorithm are set in Table 1.

Table 1. Relevant parameters of each algorithm.

Algorithms	Parameters
Trad_LQR	Q = [10,0,0;0,10,0;0,0,100], R = [5,0;0,10]
GA	Population size = 100, Number of elites = 10, Maximum iterations = 40, Constraint termination error = 1×10^{-100}, Crossover probability $P_c = 0.4$, Mutation probability $P_m = 0.01$
PSO	Population size = 20, Maximum iterations = 40, Acceleration parameters = 2, Initial weights = 0.9, End weights = 0.4, Algorithm termination threshold = 1×10^{-25}, Iteration termination threshold = 10, PSO Algorithm Type = 0, Specify random seeds = 1
QGA	Population size = 24, Maximum iterations = 220, Binary length of the variable = 20
IQGA	Population size = 24, Maximum iterations = 220, Speed Maximum $v_{max} = 1$, Speed Minimum $v_{min} = -1$, Particle Dimension $N = 2$, Learning Factor $c_1 = 2$, $c_2 = 2$, Inertia weight maximum $\omega_{max} = 0.8$, Inertia weight minimum $\omega_{min} = 0.1$

5.3. Tracking a Circular Trajectory

To verify the performance of the designed trajectory tracking controller, we simulated and tested its tracking capability while tracking different trajectories. The kinematic model-based trajectory tracking controller is mainly used for the low-speed working condition of the tractor. According to its motion characteristics, a circular trajectory is first selected for tracking.

For a circular track, the reference speed V is 5 m/s, the reference front wheel angle is 0.106 rad, the radius is 25 m, the angular velocity is 0.2 rad/s, the center coordinate is (0, 25) m, and the reference track equation is in Equation (14).

$$\begin{cases} x_{ref} = 25 \sin 0.2t \\ y_{ref} = 25 - 25 \cos 0.2t \end{cases} \tag{14}$$

The simulation results of the tracking circular trajectory are shown in Figure 4.

Tractor transverse and longitudinal displacements determine the degree of crop neatness and are often used as visual indicators to evaluate the tractor's operational control performance. For this reason, the time domain response of the tractor displacement is compared and analyzed in this section, and the response curves are shown in Figures 5 and 6, respectively.

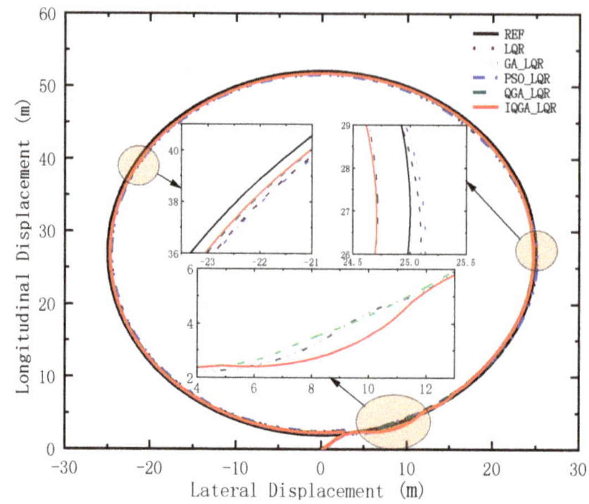

Figure 4. Reference trajectory and actual trajectory.

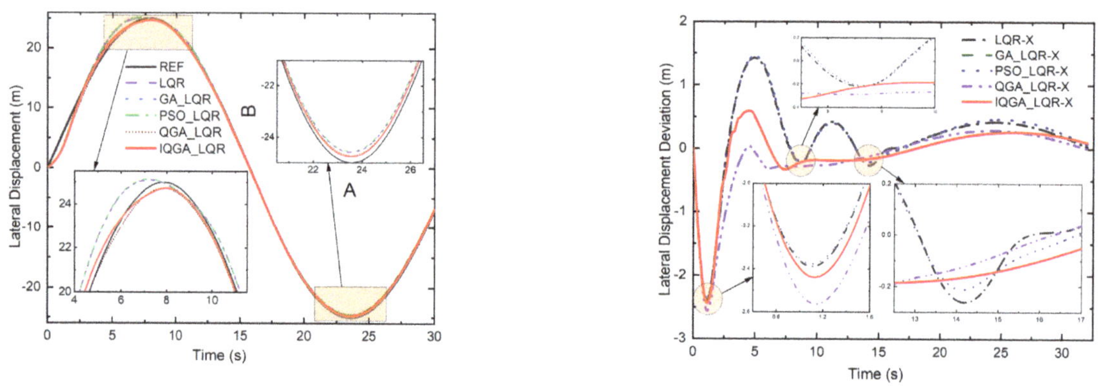

Figure 5. Lateral displacement and deviation.

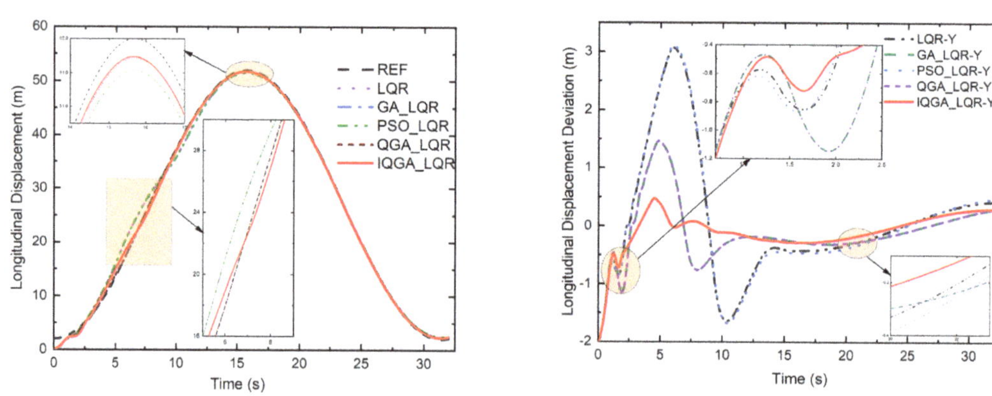

Figure 6. Longitudinal displacement and deviation.

The lateral and longitudinal displacements and their deviations using each controller are plotted in Figures 5 and 6. It can be noted that the peak transverse and longitudinal displacements of the proposed control strategy are significantly reduced, and the deviation fluctuations tend to be flat compared to the other four control methods. Figures 7 and 8 show the heading angle, tractor speed, and their deviations. In Figure 7, the heading angle of the IQGA-LQR control method enters the steady state at the earliest compared to other optimization methods. In Figure 8, the tractor speed of the IQGA-LQR control method has a slight jitter in some periods, but the overall deviation is always the smallest and the first to enter the steady state.

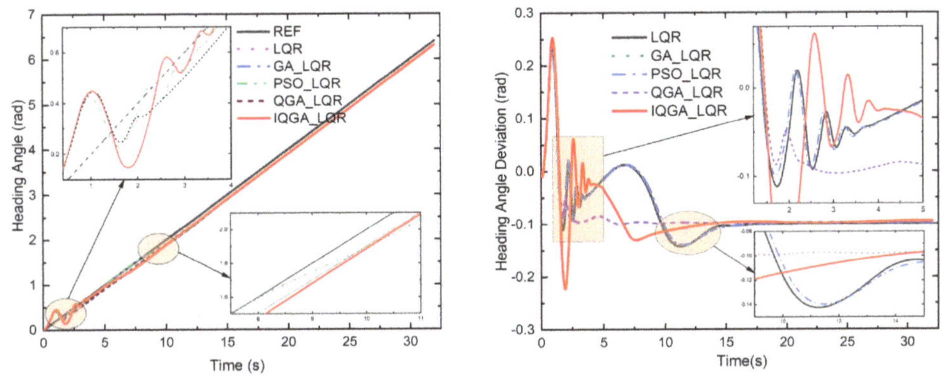

Figure 7. Heading angle and deviation.

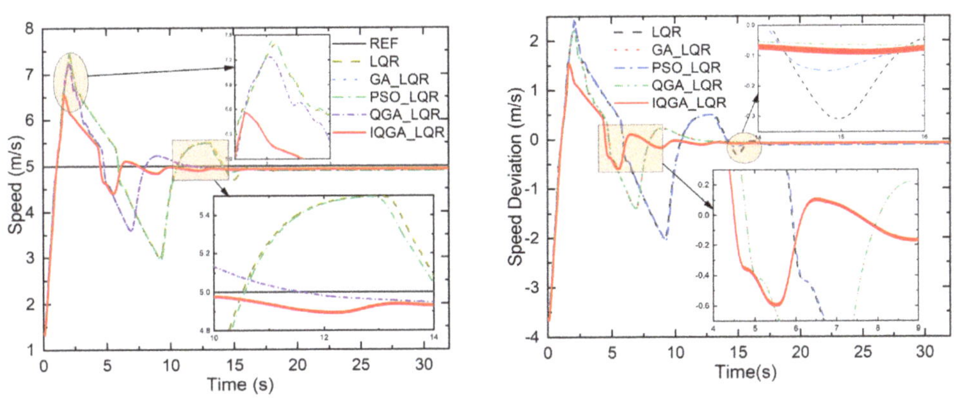

Figure 8. Tractor speed and deviation.

RMSE belongs to the L_2 norm, which is more sensitive to outliers and is widely used in engineering measurements. RMSE can measure the deviation between the actual motion trajectory of the tractor and the reference motion trajectory. The RMSE compares the advantages and disadvantages of the Trad_LQR, GA-LQR, PSO-LQR, QGA-LQR, and IQGA-LQR control methods.

To qualitatively examine the control performance of the IQGA-LQR control strategy, a comparison among the maximum, minimum, and RMSE results for each deviation is derived using the conventional LQR, GA-LQR, PSO-LQR, QGA-LQR, and IQGA-LQR methods given in Table 2.

Compared with the conventional LQR, the IQGA-LQR controller reduced the maximum value of the lateral displacement deviation and RMSE value of the tractor by 79.79% and 37.61%, respectively, and increased the minimum value of the lateral displacement deviation by 7.35%, with little increase compared to the overall result. The maximum value and RMSE value of the longitudinal displacement deviation decreased by 84.81% and 89.57%, respectively, and the minimum value of longitudinal displacement deviation was the same. The maximum value of the heading angle deviation and RMSE value increased by 3.17% and 6.45%, respectively, but the minimum value of the heading angle deviation decreased by 29.59%, which is a larger decrease; the maximum value of the speed deviation and RMSE value decreased by 34.84% and 57.65%, respectively, and the minimum value of speed deviation increased slightly by 0.7%, which is negligible.

Table 2. System deviations.

		TRAD_LQR	GA	PSO	QGA	IQGA
Lateral Deviation (m)	Maximum value	1.4238	0.5843	1.4471	0.5844	0.2878
	Minimum value	−2.3882	−2.4397	−2.3773	−2.4398	−2.56378
	RMSE	0.435	0.2724	0.4493	0.2724	0.2714
Longitudinal Deviation (m)	Maximum value	3.0731	1.4622	3.1136	1.4622	0.4668
	Minimum value	−2	−2	−2	−2	−2
	RMSE	1.2018	0.2972	1.2605	0.2972	0.1253
Heading Angle Deviation (rad)	Maximum value	0.2364	0.2526	0.2305	0.2526	0.2439
	Minimum value	−0.1426	−0.223	−0.1398	−0.223	−0.1004
	RMSE	0.0093	0.0104	0.009	0.0104	0.0099
Speed Deviation (m/s^2)	Maximum value	2.3865	2.2329	2.4391	2.2329	1.555
	Minimum value	−3.6443	−3.6444	−3.6443	−3.6444	−3.6698
	RMSE	0.6736	0.4326	0.6791	0.4326	0.2853

Compared with GA-LQR, the IQGA-LQR controller resulted in a larger decrease in the maximum value of the lateral displacement deviation of the tractor, with a reduced value of 50.74%. By contrast, the RMSE value and the minimum value of the lateral displacement deviation increased, by 0.37% and 5.09%, respectively. The maximum value of the longitudinal displacement deviation and the root-mean-square error value were greatly reduced by 68.08% and 57.84%, respectively, while the minimum value of the longitudinal displacement deviation remains unchanged; the maximum value of the heading angle deviation, root-mean-square error value and the minimum value of transverse displacement deviation was reduced by 3.44%, 54.98%, and 4.81%, respectively. The maximum value of the speed deviation and RMSE value were reduced significantly by 30.36% and 34.05%, respectively. In comparison, the minimum value of the speed deviation increased by 0.7%, which is relatively small and can also be ignored.

Compared with PSO-LQR, the IQGA-LQR controller reached the maximum value of lateral displacement deviation, and the RMSE value of the tractor had a large reduction by 80.11% and 39.59%, respectively, while the minimum value of the lateral displacement deviation did not increase much, and its upward value was 7.84%. The maximum value and RMSE of longitudinal displacement deviation decreased by 85.01% and 90.06%, respectively, while the minimum value of longitudinal displacement deviation increased by 40%; the maximum value and RMSE of the heading angle deviation increased by 5.81% and 10%, respectively, while the minimum value of the heading angle deviation decreased by 28.18%. The maximum value of the speed deviation and RMSE decreased by 36.25% and 57.99%, respectively, while the minimum value of the speed deviation increased slightly by 0.7%.

Compared with QGA-LQR, the IQGA-LQR controller reduced the maximum value of the lateral displacement deviation of the tractor by 50.75%. The RMSE value did not change much and was reduced by 0.37%. In contrast, the minimum value of the lateral displacement deviation increased, and its reduction value was 20.37%. The maximum value and RMSE of the longitudinal displacement deviation decreased by 68.08% and 57.84%, respectively, while the minimum value of the longitudinal displacement deviation did not change; the maximum value, minimum value, and RMSE of the heading angle deviation decreased by 3.44%, 54.98%, and 4.81%, respectively, with the minimum value of the heading angle deviation decreasing by a larger extent. The maximum value of speed deviation and the RMSE value decreased by 30.36% and 34.05%, respectively, and the minimum value of speed deviation showed a small increase with an upward value of 0.7%.

The RMSE values of the kinematic states under circular trajectories are given in Figure 9, from which it can be seen that among the four compared data, the IQGA-LQR RMSE data are the smallest for three and slightly larger for one. These three data are the RMSE values of lateral displacement, longitudinal displacement, the heading angle, and velocity, respectively; the slightly larger one is the lateral displacement of the RMSE value. Figure 10 gives the RMSE values of the motion states under the double-shifted trajectory, from which it can be seen that all four RMSE data of IQGA-LQR are the smallest among the four compared data; therefore, IQGA-LQR is the optimal choice from the RMSE point of view.

From the analysis of the typical numerical decrease in percentage, it can be inferred that the IQGA-LQR control strategy has a better control effect. Obviously, the tractor system controlled by this method achieved better lateral and longitudinal operation accuracy and speed tracking when performing agricultural tillage and harvesting. Thus, the effectiveness of the proposed LQR method optimized by IQGA is verified.

The controller parameters are shown in Table 3.

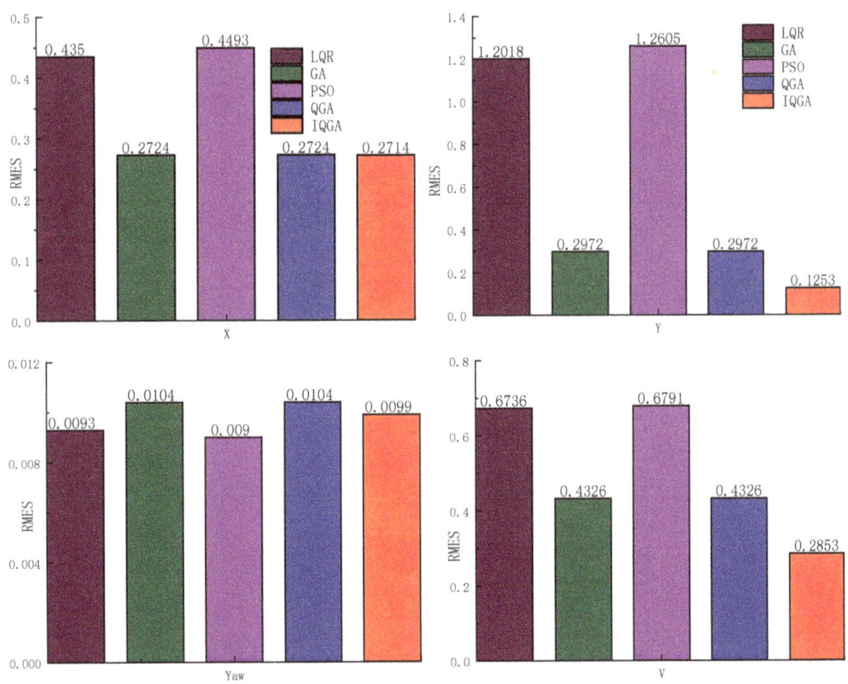

Figure 9. Motion state RMSE values when tracking circular trajectories.

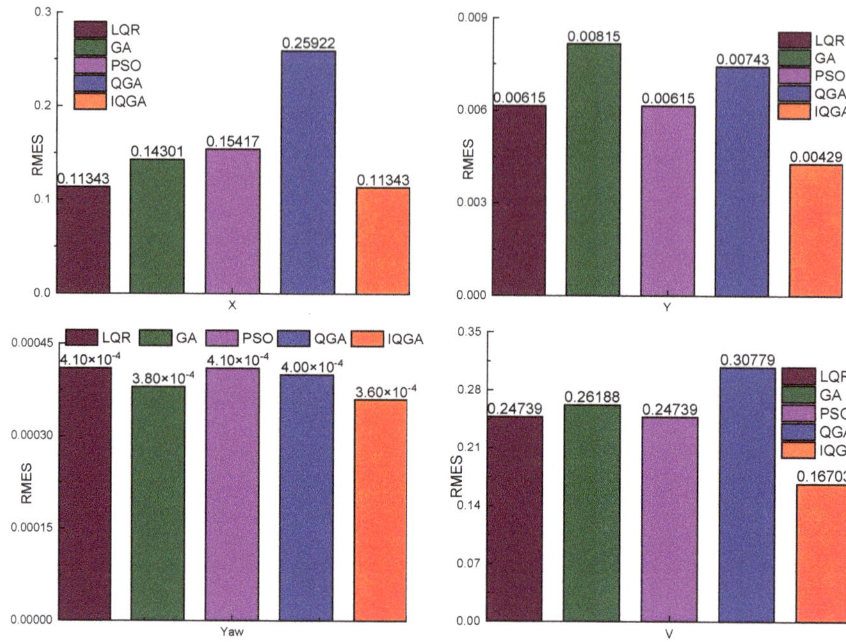

Figure 10. RMSE value of the motion State while tracking a double shift trajectory.

Table 3. Parameters of circular trajectory controller.

Parameters	Values
Q	$\begin{bmatrix} 30460.2 & 0 & 0 \\ 0 & 30095.7 & 0 \\ 0 & 0 & 27848.4 \end{bmatrix}$
R	$\begin{bmatrix} 6294.7 & 0 \\ 0 & 42094.6 \end{bmatrix}$
Q_0	$\begin{bmatrix} 41128.5 & 0 & 0 \\ 0 & 19284.6 & 0 \\ 0 & 0 & 43.8 \end{bmatrix}$
ε^2	1×10^6

5.4. Tracking Double Shift Trajectory

When the tractor is working, the evaluation of the controller is not only the tracking accuracy but also needs to focus on the stability of the tracking process. In the test of the wheeled tractor driving stability, the double-shifted line working condition is a test method used more frequently. There are also more scholars [32–34] to test the trajectory tracking ability of unmanned wheeled tractors with double-shifted trajectories. Therefore, in this paper, the designed LQR controller was simulated and tested using the double-shifted trajectory [35].

For the double-shift line trajectory, the reference tractor speed v is 5 m/s, and the reference trajectory is Equation (15).

$$\begin{cases} x_{ref} = v_t \\ y_{ref} = \frac{d_{y1}}{2}(1+\tanh(z_1)) - \frac{d_{y2}}{2}(1+\tanh(z_2)) \\ \varphi_{ref} = \arctan\left[d_{y1}\left(\frac{1}{\cosh(z_1)}\right)^2\left(\frac{1.2}{d_{x1}}\right) - d_{y2}\left(\frac{1}{\cosh(z_2)}\right)^2\left(\frac{1.2}{d_{x2}}\right)\right] \end{cases} \tag{15}$$

where, $z_1 = \frac{2.4}{25}(X - 27.19) - 1.2$, $z_1 = \frac{2.4}{21.95}(X - 56.46) - 1.2$, $d_{x1} = 25$, $d_{x2} = 21.95$, $d_{y1} = 4.05$, $d_{y2} = 5.7$.

The radius of the curvature of this double shift trajectory is calculated by Equation (16). It gives the reference front wheel rotation angle: $\delta_{ref} = \arctan(l/R)$. The simulation results of the double-shifted reference trajectory are shown in Figure 11.

$$R = \left| \frac{\left(1 + \left(\frac{dy_{ref}}{dx_{ref}}\right)^2\right)^{\frac{3}{2}}}{\frac{d^2 y_{ref}}{dx_{ref}^{\ 2}}} \right| \tag{16}$$

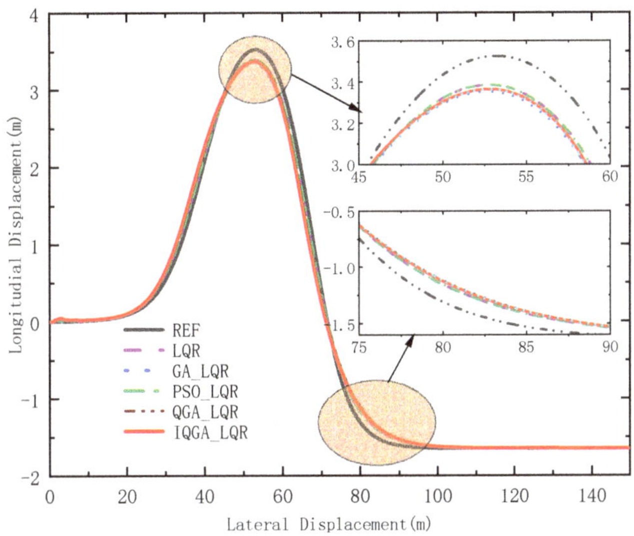

Figure 11. Reference trajectory and actual trajectory.

Similar to the analysis of circular trajectories, to qualitatively examine the control performance of the IQGA-LQR control strategy, a comparative study among the simulation results of deviation maxima, minima, and RMSEs derived with the conventional LQR, GA-LQR, PSO-LQR, QGA-LQR, and IQGA-LQR methods is given in Table 4.

Compared with the conventional LQR, the IQGA-LQR controller left the maximum value, minimum value, and RMSE value of lateral displacement deviation of the tractor unchanged; the maximum value, minimum value, and RMSE value of the longitudinal displacement deviation are reduced by 12.32%, 19.82%, and 29.51%, respectively; the maximum value, minimum value, and RMSE value of the heading angle deviation are reduced by 0.52%, 6.37%, respectively, and the RMSE value is the same. The maximum and minimum values for the deviation of the heading angle decreased by 0.52% and 6.37%, respectively, and the RMSE value was the same. The maximum value of the speed deviation was reduced by 34.44%, the minimum value did not change, and the value of RMSE was reduced by 32.5%.

Compared with GA-LQR, the IQGA-LQR controller increased the maximum value of the lateral displacement deviation of the tractor by 3.34%, reduced the minimum value of the lateral displacement deviation and RMSE value by 0.43% and 20.7%, respectively, reduced the maximum value of longitudinal displacement deviation, and RMSE value by 27.51%, 26.63%, and 47.56%, respectively. The maximum and minimum values of the heading angle deviation had a small increase of 0.35% and 1.81%, respectively, while RMSE values did not change; the maximum and RMSE values of the speed deviation decreased by 34.66% and 36.24%, respectively, and the minimum values of the speed deviation were the same.

Table 4. Systematic deviations in tracking double-shifted trajectories.

		TRAD	GA	PSO	QGA	IQGA
Lateral Deviation (m)	Maximum value	0.7250	0.7016	0.7234	0.6413	0.7250
	Minimum value	−1.6156	−1.6225	−1.6140	−1.6217	−1.6156
	RMSE	0.1134	0.1430	0.1542	0.2592	0.1134
Longitudinal Deviation (m)	Maximum value	0.1518	0.1836	0.1518	0.2161	0.1331
	Minimum value	−0.2240	−0.2448	−0.2240	−0.2269	−0.1796
	RMSE	0.0061	0.0082	0.0061	0.0074	0.0043
Heading Angle Deviation (rad)	Maximum value	0.0574	0.0569	0.0574	0.0615	0.0571
	Minimum value	−0.0361	−0.0332	−0.0361	−0.0326	−0.0338
	RMSE	0.0004	0.0004	0.0004	0.0004	0.0004
Speed Deviation (m/s^2)	Maximum value	1.49623	1.5013	1.4962	1.4825	0.9810
	Minimum value	−3.6112	−3.6112	−3.6112	−3.6112	−3.6112
	RMSE	0.2474	0.2619	0.2474	0.3078	0.1670

Compared with PSO-LQR, the IQGA-LQR controller made the maximum and minimum values of lateral displacement deviation of the tractor slightly increase by 0.22% and 0.1%, respectively, which is almost negligible, and the RMSE value decreased by 26.46%, which is larger; the maximum and minimum values of the longitudinal displacement deviation and RMSE value decreased by 12.32%, 19.82%, and 29.51%, respectively. The maximum and minimum values of longitudinal displacement deviation decreased by 12.32%, 19.82%, and 29.51%, respectively; the maximum and minimum values of the heading angle deviation decreased by 0.52% and 6.37%, respectively, and the RMSE value was the same; the maximum and RMSE values of speed deviation decreased by 34.43% and 32.5%, respectively, and the minimum value of speed deviation did not change.

Compared with QGA-LQR, the IQGA-LQR controller increased the maximum value of the lateral displacement deviation of the tractor by 13.05%, reduced the minimum value of lateral displacement deviation and the RMSE value by 0.38% and 56.25%, respectively; it reduced the maximum value of longitudinal displacement deviation, the minimum value, and RMSE value by 38.41%, 20.85%, 41.89%, respectively. It also reduced the maximum value of the heading angle deviation by 7.15%, increased the minimum value by 3.68%, and kept the RMSE value unchanged. The maximum value of the heading angle deviation decreased by 7.15%, the minimum value increased by 3.68%, and the RMSE value remained the same; the maximum and RMSE values of the speed deviation decreased by 33.83% and 45.74%, respectively, and the minimum value of speed deviation was the same.

Figures 12 and 13 are the transverse and longitudinal displacements of the wheeled tractor, and their deviations use each controller while tracking the double-shift line trajectory. The figures show that the peaks of the transverse and longitudinal displacements of the proposed control strategy are relatively smaller, and the deviations are smaller throughout the process than the other four control methods. Figures 14 and 15 show the heading angle, tractor speed, and their deviations. In Figure 14, IQGA-LQR has relatively smaller deviations at the peak compared to other optimization methods, and the advantage is not so obvious in other stages. In Figure 15, the IQGA-LQR control method of the tractor speed and their deviations are better controlled throughout the control phase, and the deviations are in a relatively small state all the time.

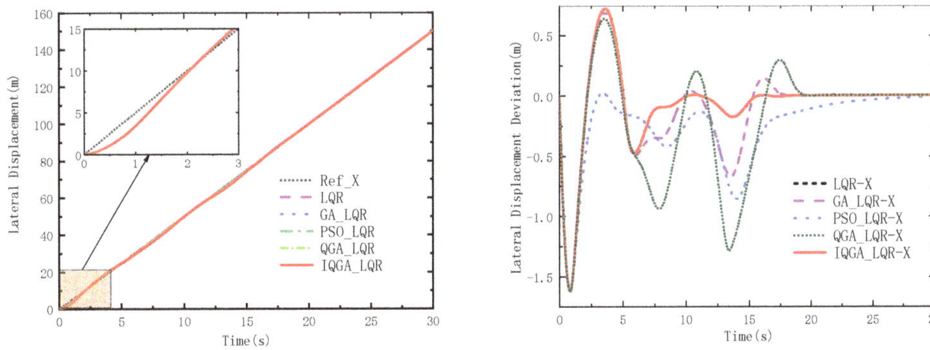

Figure 12. Lateral displacement and deviation.

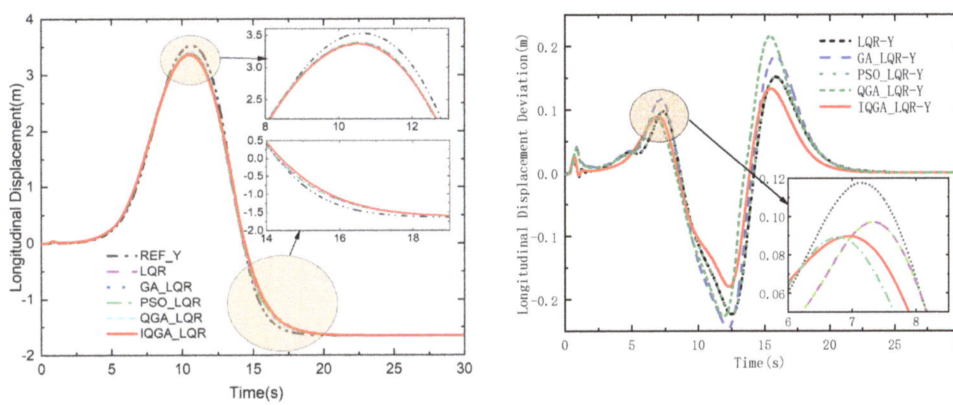

Figure 13. Longitudinal displacement and deviation.

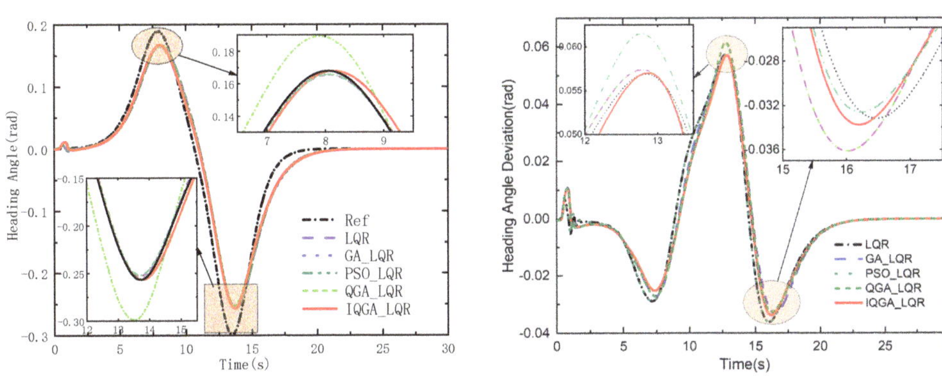

Figure 14. Heading angle and deviation.

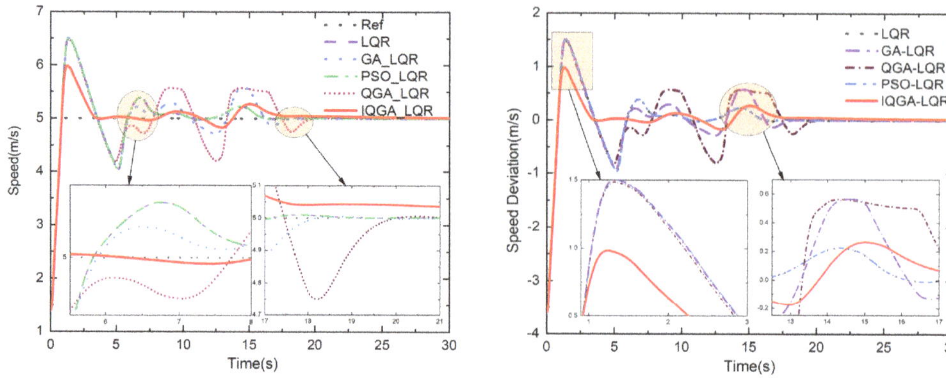

Figure 15. Tractor speed and deviation.

By comparing all system deviations in Tables 2 and 4, it can be inferred that the IQGA-LQR control strategy has better control effects in four aspects: lateral deviation, longitudinal deviation, heading angle deviation, and speed deviation, thus verifying the higher effectiveness of the proposed LQR method optimized by IQGA.

The controller parameters after IQGA optimization are shown in Table 5.

Table 5. The parameters of the dual shift trajectory controller are shown.

Parameters	Values
Q	$\begin{bmatrix} 17.6 & 0 & 0 \\ 0 & 13892.1 & 0 \\ 0 & 0 & 1996.4 \end{bmatrix}$
R	$\begin{bmatrix} 1034.1 & 0 \\ 0 & 81792.8 \end{bmatrix}$
Q_0	$\begin{bmatrix} 50752.6 & 0 & 0 \\ 0 & 23326.2 & 0 \\ 0 & 0 & 93579.0 \end{bmatrix}$
ε^2	1×10^6

6. Summary and Prospect

In this paper, a kinematic model of the wheel tractor was built based on the Ackermann steering model, the state weighting matrix in the LQR controller was optimized using IQGA, and finally, a joint simulation was performed using Carsim and MATLAB. The simulation results after comparing the other four optimization algorithms showed that:

(1) The coefficient matrix selected by IQGA had better tracking accuracy. The lateral position deviation, longitudinal position deviation, and heading angle deviation all tended to be zero, the control effect was better, and the system tended to be stable. Adding constraints to the LQR increases ride comfort. The RMSE of lateral displacement, longitudinal displacement, and the heading angle after tractor stabilization were 0.2714 m, 0.1253 m, and 0.0099 rad, respectively, when the tracking circular trajectory was at 5 m/s. The error of lateral displacement, longitudinal displacement, and the heading angle after tractor stabilization was 0.1134 m, 0.0043 m, and 0.0004 rad, respectively, when tracking a double-shift trajectory at 5 m/s.

(2) The tracking method designed based on kinematics is suitable for low-speed work scenarios. If the wheeled tractor is tracked at high speed, the situation is more complex when the wheeled tractor's kinematics cannot meet the actual demand. At the same time, the kinematic modeling of the wheel tractor is simplified, such as linearizing the nonlinear system, ignoring the disturbance term, etc., so some errors in the tracking process are difficult to eliminate. In the subsequent research, we can design the

nonlinear trajectory tracking controller based on the wheel tractor dynamics and consider the influence of the wheel tractor's lateral tilt characteristics, the tire's slip characteristics, the disturbance term, and other factors.

(3) The reference trajectory also greatly influences the tracking effect. For example, when the connection point of the trajectory is not derivable, there is an oscillation in the control data when controlling the wheel tractor. Therefore, in the subsequent study, the original reference trajectory can be sampled, and the original reference trajectory can be reprogrammed according to the kinematic constraints of the wheel tractor or wheel tractor dynamics constraints. A trajectory that meets the constraints can be reprogrammed for tracking.

Author Contributions: Methodology, X.F. and J.W.; software, H.W. and L.Y.; formal analysis, H.W. and L.Y.; data curation, X.F. and J.W.; project administration, C.X.; funding acquisition, C.X. All authors have read and agreed to the published version of the manuscript.

Funding: This research was funded by [the National Key Research and Development Program of China], grant number [2016YFD0700400] and Changzhou Science &Technology Program, grant number [CJ20220232].

Data Availability Statement: Data available on request from the authors. The data that support the findings of this study are available from the corresponding author, [Xia, C.], upon reasonable request.

Conflicts of Interest: The authors declare no conflict of interest.

References

1. Yadav, A.; Gaur, A.; Jain, S.; Chaturvedi, D.; Sharma, R. Development Navigation, Guidance & Control Program for GPS based Autonomous Ground Vehicle (AGV) using Soft Computing Techniques. *Mater. Today Proc.* **2020**, *29*, 530–535.
2. Zhang, Z.; Wu, L.; Zhang, W.; Peng, T.; Zheng, J. Energy-efficient path planning for a single-load automated guided vehicle in a manufacturing workshop. *Comput. Ind. Eng.* **2021**, *158*, 107397. [CrossRef]
3. Feilong, W.; Shiyong, G. Intelligent Vehicle Path Tracking Algorithm Based on Nonlinear Model Predictive Control. *Automob. Technol.* **2020**, 1–7.
4. Nie, Z.G.; Wang, W.Q.; Zhao, W.Q.; Huang, Z.; Zong, C.F. Dynamic trajectory planning and tracking control for lane change of intelligent vehicle based on trajectory preview. *J. Traffic Transp. Eng.* **2020**, *20*, 147–160.
5. Wongsathan, C.; Sirima, C. Application of GA to design LQR controller for an Inverted Pendulum System. In Proceedings of the IEEE International Conference on Robotics & Biomimetics, Bangkok, Thailand, 22–25 February 2009.
6. Rahimi, A.; Ghazi, R. GA-Based Optimal LQR Controller to Improve LVRT Capability of DFIG Wind Turbines. *Iran. J. Electr. Electron. Eng.* **2013**, *9*, 167–176.
7. Kafafy, M.; Rabeih, A.; Eldemerdash, S.; Elbutch, A. Active Suspension Design for Passenger Cars Using LQR and GA with PID Controller. SAE Technical Papers, St. Charles, IL, USA, 15–17 May 2007. [CrossRef]
8. Amini, F.; Hazaveh, N.; Rad, A. Wavelet PSO-Based LQR Algorithm for Optimal Structural Control Using Active Tuned Mass Dampers. *Comput.-Aided Civ. Infrastruct. Eng.* **2013**, *28*, 542–557. [CrossRef]
9. Mf, A.; Sn, A.; Ak, B. On the ability of sliding mode and LQR controllers optimized with PSO in attitude control of a flexible 4-DOF satellite with time-varying payload. *Adv. Space Res.* **2021**, *67*, 334–349.
10. Reddipogu, J.; Elumalai, V. Hardware in the Loop Testing of Adaptive Inertia Weight PSO-Tuned LQR Applied to Vehicle Suspension Control. *J. Control. Sci. Eng.* **2020**. [CrossRef]
11. Li, L.; Qiu, T.; Jia, T.; Chen, C. Stepping quantum genetic algorithm-based LQR control strategy for lateral vibration of high-speed elevator. *at-Automatisierungstechnik* **2022**, *70*, 623–634. [CrossRef]
12. Sajid, U.; Wahid, M. Topology Control of wireless sensor network using Quantum Inspired Genetic algorithm. *Int. J. Swarm Intell. Evol. Comput.* **2015**, *4*, 121. [CrossRef]
13. Marji, M.; Sumarli, E. Design and Investigation of Fuzzy Control For Independent Full Car Suspension Model in Random Road And Braking Excitation. *Int. J. Adv. Sci. Technol.* **2020**, *29*, 20–23.
14. Meng, Y.; Gan, X.; Wang, Y.; Qing, G. LQR-GA Controller for Articulated Dump Truck Path Tracking System. *J. Shanghai Jiaotong Univ. (Sci.)* **2019**, *24*, 78–85. [CrossRef]
15. Zhang, W. A robust lateral tracking control strategy for autonomous driving vehicles. *Mech. Syst. Signal Process.* **2021**, *150*, 107238. [CrossRef]
16. Yuan, S.; Zhao, P.; Zhang, Q.; Hu, X. Research on Model Predictive Control-based Trajectory Tracking for Unmanned Vehicles. In Proceedings of the 2019 4th International Conference on Control and Robotics Engineering (ICCRE), Nanjing, China, 20–23 April 2019; pp. 79–86.

17. Kanieski, J.; Carati, E.; Cardoso, R. An energy based LQR tuning approach applied for Uninterruptible Power Supplies. In Proceedings of the 2010 First IEEE Latin American Symposium on Circuits and Systems (LASCAS), Foz do Iguacu, Brazil, 24–26 February 2010; pp. 41–44.
18. Ufnalski, B.; Kaszewski, A.; Grzesiak, L. Particle Swarm Optimization of the Multioscillatory LQR for a Three-Phase Four-Wire Voltage-Source Inverter With an LC Output Filter. *IEEE Trans. Ind. Electron.* **2015**, *62*, 484–493. [CrossRef]
19. Almeida, P.; Ribeiro, A.; Souza, I.; Fernandes, M.; Fogli, G.; Ćuk, V.; Barbosa, P.; Ribeiro, P. Systematic Design of a DLQR Applied to Grid-Forming Converters. *IEEE J. Emerg. Sel. Top. Ind. Electron.* **2020**, *1*, 200–210. [CrossRef]
20. Liu, S.; Wu, W.; Jin, Q.; Zhu, Y. Design Method for Helicopter Flight Control Law Based on Particle Swarm Optimization. *J. Nanjing Univ. Aeronaut. Astronaut.* **2021**, *53*, 267–274.
21. Tian, J.; Guo, Q.; Shi, G. Laminated piezoelectric beam element for dynamic analysis of piezolaminated smart beams and GA-based LQR active vibration control. *Compos. Struct.* **2020**, *252*, 112480. [CrossRef]
22. Kumar, E.V.; Raaja, G.; Jerome, J. Adaptive PSO for optimal LQR tracking control of 2 DoF laboratory helicopter. *Appl. Soft Comput.* **2016**, *41*, 77–90. [CrossRef]
23. Gokul Prassad, S.; Malar Mohan, K. A contemporary adaptive air suspension using LQR control for passenger vehicles. *ISA Trans.* **2019**, *93*, 244–254. [CrossRef]
24. Belyaev, A.; Sumenkov, O. Hybrid control algorithm based on LQR and genetic algorithm for active support weight compensation system. *IFAC-Pap.* **2021**, *54*, 431–436. [CrossRef]
25. Jianwei, G.; Yan, J.; Wei, X. *Model Predictive Control for self-driving Vehicles*, 2nd ed.; Beijing Institute of Technology Press: Beijing, China, 2014.
26. Han, K.; Park, K.; Lee, C.; Kim, J. Parallel quantum-inspired genetic algorithm for combinatorial optimization problem. In Proceedings of the Proceedings of the 2001 Congress on Evolutionary Computation (IEEE Cat. No.01TH8546), Seoul, Korea, 27–30 May 2001.
27. Gexiang, Z.; Na, L.; Weidong, J.; Laizhao, H. A Novel Quantum Genetic Algorithm and Its Applicat. *Acta Electron. Sin.* **2004**, *32*, 476–479.
28. Zhang, X.F.; Sui, G.F.; Zheng, R.; Li, Z.N.; Yang, G.W. An Improved Quantum Genetic Algorithm of Quantum Revolving Gate. *Comput. Eng.* **2013**, *39*, 234–238.
29. He, X.; Liang, J. Genetic Algorithms Using Gradients of Object Functions. *J. Softw.* **2001**, *12*, 981–986.
30. Xu, E.; Gai, J.; Zhou, J.; Yang, F.; Liu, C. Quantum Genetic Algorithm for Hadamard Gate Mutation. *Control Eng. China* **2018**, *25*, 143–148.
31. Haiyan, G. Quantum genetic algorithm based on chaotic optimization. *Electron. Meas. Technol.* **2006**, *29*, 14–15+18.
32. Katzourakis, D.; de Winter, J.; de Groot, S.; Happee, R. Driving simulator parameterization using double-lane change steering metrics as recorded on five modern cars. *Simul. Model. Pract. Theory* **2012**, *26*, 96–112. [CrossRef]
33. Qu, X.; Feng, H.; Li, G. Cooperative Control of Vehicle Active Steering and Electronic Stability Program. *Sci. Technol. Eng.* **2021**, *21*, 13155–13162.
34. Wang, N.; Shi, J.; Zhao, C. Experimental Analysis on Handing Stability of Two Rear Drive Wheel Pure Electric Vehicle. In Proceedings of the SAECCE2020-EV025, Shanghai, China, 27 October 2020; pp. 408–413.
35. Falcone, P.; Tseng, H.E.; Borrelli, F.; Asgari, J.; Hrovat, D. MPC-based yaw and lateral stabilisation via active front steering and braking. *Veh. Syst. Dyn.* **2008**, *46*, 611–628. [CrossRef]

MDPI
St. Alban-Anlage 66
4052 Basel
Switzerland
www.mdpi.com

Machines Editorial Office
E-mail: machines@mdpi.com
www.mdpi.com/journal/machines

www.ingramcontent.com/pod-product-compliance
Lightning Source LLC
Chambersburg PA
CBHW041942260926
51850CB00018B/493
* 9 7 8 3 0 3 6 5 7 2 6 4 2 *